AF411446

New Prognostic and Predictive Markers in Cancer Progression

New Prognostic and Predictive Markers in Cancer Progression

Editors

Susan Costantini
Alfredo Budillon

MDPI • Basel • Beijing • Wuhan • Barcelona • Belgrade • Manchester • Tokyo • Cluj • Tianjin

Editors
Susan Costantini
Istituto Nazionale Tumori
Italy

Alfredo Budillon
Istituto Nazionale Tumori
Italy

Editorial Office
MDPI
St. Alban-Anlage 66
4052 Basel, Switzerland

This is a reprint of articles from the Special Issue published online in the open access journal *International Journal of Molecular Sciences* (ISSN 1422-0067) (available at: https://www.mdpi.com/journal/ijms/special_issues/Cancers_Markers).

For citation purposes, cite each article independently as indicated on the article page online and as indicated below:

LastName, A.A.; LastName, B.B.; LastName, C.C. Article Title. *Journal Name* **Year**, *Volume Number*, Page Range.

ISBN 978-3-03943-977-5 (Hbk)
ISBN 978-3-03943-978-2 (PDF)

Contents

About the Editors

Susan Costantini graduated in Chemistry and achieved the Specialty Certification in Clinical Pathology and Biochemistry at University Federico II of Naples (Italy). She obtained her PhD in Computational Biology at the Second University of Naples and Master's degree in Environment and Cancer at the University of Sannio. She is a contract professor of Biology, Biochemistry, and Bioinformatics at University Federico II of Naples and Vanvitelli Campania. Since 2008, she has been a researcher at the Mercogliano Laboratory of the National Cancer Institute of Naples "Pascale Foundation". She is the author of more than 140 publications (Scopus H-index: 31). Her research activity is currently focused on the study of the cytokinomic and metabolomic/lipidic profile by 1H-NMR in various cellular systems, biological fluids, and tissues of cancer patients. Moreover, her research activities include the application of systems biology approaches to integrate "omics" data by computational and bioinformatics methods.

Alfredo Budillon graduated in Medicine and achieved the Specialty Certification in Oncology at University Federico II, Naples, Italy. He obtained his PhD in Oncology from University G. D'Annunzio of Chieti, Italy, and Master's degree in Health Economy, Management, and Bioethics. He gained experience in biochemistry and antitumor experimental therapeutics as a post-doc during a stage fellowship at the National Cancer Institute, Bethesda, USA. Dr. Budillon is currently the chief of the Experimental Pharmacology Unit, Istituto Nazionale Tumori Fondazione G. Pascale, Naples, Italy. He is the author of more than 170 publications (Scopus H-index: 45). His research interest is focused on the molecular mechanism of novel anticancer drugs and the potential combination approaches between conventional and targeted-based drugs, particularly epigenetic drugs. His other scientific interests are related to cancer-associated extracellular vesicles as well as proteomics/metabolomics approaches in defining new molecular diagnostic/prognostic biomarkers and therapeutic targets.

International Journal of
Molecular Sciences

Editorial

New Prognostic and Predictive Markers in Cancer Progression

Susan Costantini * and Alfredo Budillon *

Unità di Farmacologia Sperimentale - Laboratori di Mercogliano, Istituto Nazionale Tumori "Fondazione G. Pascale"—IRCCS, 80131 Napoli, Italy
* Correspondence: s.costantini@istitutotumori.na.it (S.C.); a.budillon@istitutotumori.na.it (A.B.);
  Tel.: +39-0825-1911729 (S.C.); +39-081-5903292 (A.B.)

Received: 4 November 2020; Accepted: 17 November 2020; Published: 17 November 2020

Biomarkers are a critical medical need for oncologists to predict and detect disease and to determine the best course of action for cancer patient care. Prognostic markers are used to evaluate a patient's outcome and cancer recurrence probability after initial interventions such as surgery or drug treatments and hence to select follow-up and further treatment strategies. On the other hand, predictive markers are increasingly being used to evaluate the probability of benefit from clinical intervention(s), driving personalized medicine. Evolving technologies and the increasing availability of "multi-omics" data are leading to the selection of numerous potential biomarkers based on DNA, RNA, miRNA, protein, and metabolic alterations within cancer cells or tumor microenvironments, which may be combined with clinical pathological data to greatly improve the prediction of both cancer progression and therapeutic treatment responses. Indeed, the search for new prognostic and predictive cancer biomarkers is the object of many studies performed on preclinical cancer models, as well as biofluids and tissue samples from cancer patients. However, few biomarkers have progressed in the last years from discovery to become validated tools to be used in clinical practice [1].

This Special Issue comprises eight review articles and five original studies on novel potential prognostic and predictive markers for different cancer types.

Among the reviews, two are focused on miRNAs that represent a family of small non-coding RNAs recently emerging as potential predictors of prognosis and/or anticancer drug efficacy, as well as novel anticancer targets. MiRNAs have been described as onco-suppressors or onco-genes, depending on the cellular context and their biological targets [2]. In recent years, many investigations have confirmed the presence of a stable form of circulating miRNAs in human body fluids, including blood, saliva, and urine [3]. In this context, Takeuchi et al. summarized the miRNAs known to be deregulated in head and neck squamous cell carcinoma (HNSCC), with a specific focus on laryngeal cancers, describing miRNas predicting initiation, progression, and prognosis; those associated withradio-, chemo-, and thermal- resistance; and also those correlated with infection (e.g., HPV) and life habit. Notably, the authors concluded that the simultaneous evaluation of more miRNAs regulating multiple target genes might have higher diagnostic, prognostic and therapeutic performance, as well as higher sensitivity than individual miRNA assays, because they better recapitulate the multistep process leading to cancer. On the other hand, Motti et al. provided an updated overview about miRNA deregulation in melanoma, suggesting their critical role as putative diagnostic and/or prognostic biomarkers in this disease. Moreover, since some miRNAs have been found to regulate the mitogen-activated protein kinase (MAPK) signaling pathway [4] or immune checkpoint expression [5], the authors discussed in more detail which miRNAs play an important role in melanoma cell resistance to MAPK and/or immune checkpoint inhibitors, evidencing the predictive potentialities of circulating miRNAs to detect and monitor melanoma responsiveness to targeted and immune therapies.

Anyhow, immune checkpoint inhibitors (ICI), beyond melanoma are used today in clinical practice to treat a large number of tumor types. However, ICI are effective in a small subgroup of cancer patients and, hence, a great effort has been dedicated to the identification and development of predictive biomarkers for ICI response [6]. In this regard, in a review article, Sabbatino et al. focused on the putative role of the human leukocyte antigen (HLA) system as a predictive biomarker for ICI-based cancer immunotherapy. The authors described: (i) the HLA system's structure and function; (ii) HLA defects with their clinical significance in cancer patients, underlining the potential predictive role of the HLA as a biomarker of response to checkpoint-based immunotherapy in cancer patients; and (iii) the main molecules/drugs able to restore HLA function and usefulness to implement novel therapeutic strategies in cancer patients. Moreover, in a research paper, Baek et al. re-classified the urothelial cancer subtypes, focusing on ICI responsiveness. Several previous studies subdivided urothelial cancer patients in two immunotherapy-associated subgroups: the genomically unstable (GU) subtype of the Lund classification and the neuronal subtype in The Cancer Genome Atlas (TCGA) classification [7,8]; on this basis, the authors performed both hierarchical clustering and survival analysis using gene expression profiling in the IMvigor 210 cohort comprising 298 urothelial cancer patients, and evidenced that patients with upregulation of the cell cycle, DNA replication, and DNA damage and downregulation of the TGFβ and YAP/TAZ pathway were more responsive to ICI therapy. In another paper, Quagliariello et al. focused on Nod-like receptor (NLR) family pyrin domain containing 3 (NLRP3), a key player inimmune-related events involving viral and bacterial infection [9], suggesting that it could be a putative marker of cardiotoxicity induced by the ICI. In detail, with the aim of studying if hyperglycemia could affect ipilimumab-induced anticancer efficacy and toxicity, taking advantage of in vitro co-culture of human peripheral blood mononuclear cells (hPBMCs) with either cardiomyocytes or estrogen-responsive (MCF-7) and triple-negative (MDA-MB-231) breast cancer cells, the authors evaluated ipilimumab's effect under different glucose concentrations. Interestingly, high levels of glucose during ipilimumab treatment increased cardiotoxicity and NRLP3 levels, and induced decreased cancer cell mortality; on the other hand, co-treatment with ipilimumab and empagliflozin, a sodium glucose co-transporter 2 inhibitor, under high glucose or shifting from high glucose to low glucose, was able to increase ipilimumab responsiveness and to decrease cardiotoxicity and NRLP3 levels.

Acetylation represents a post-translational modification regulating protein expression and function and it is regulated by acetyltransferases catalyzing the transfer of acetyl residues on proteins and by deacetylases that remove those residues.In this regard, in a review article, Kim et al. focused on N-α-acetyltransferase 10 (NAA10), targeting the *N*-terminal α-aminogroup of nascent proteins as well as internal protein lysine residues. The authors showed that NAA10 is an interesting player to regulate cell proliferation, differentiation, migration, autophagy, and apoptosis. They also highlighted NAA10 overexpression in different cancer types and correlation with overall survival rates and disease recurrence.

Several reviews and papers in this Special Issue reported data related to biomarkers of specific cancer types.

In a review article, Majewska et al. focused on head and neck paragangliomas (HNPGL) and, on the basis of PubMed and ScienceDirect databases, highlighted known genetic changes as well as epigenetic modifications associated with HNPGL, as well the potential practical applications of such alterations and also in the light of next-generation sequencing (NGS) technology. The authors analyzed both somatic and germline mutations, evidencing that, among others, succinate dehydrogenase complex iron sulfur subunit B (SDHB) mutations lead to metastasis development in 40% or more of patients. Overall, they concluded that fumarate hydratase (FH); the proto-oncogene tyrosine-protein kinase receptor RET; succinate dehydrogenase complex, subunits A, B, C, and D (SDHA, SDHB, SDHC, SDHD); and von Hippel–Lindau tumor suppressor (VHL) should be routinely determined in HNPGL patients in order to discover genetic syndromes and for correct prognostic evaluation.

Regarding prostate cancer, in a review article, Shorning et al. focused their attention on the PI3K–AKT–mTOR pathway, which is dysregulated in a high proportion of prostate cancer patients and is associated with castrationresistance [10]. The authors, by reviewing the genetic alterations leading to the activation of this pathway, discussed the interplay with androgen receptor (AR), MAPK, and WNT, which cooperate to promote prostate tumorigenesis and therapy-resistant disease progression, concluding that a clear knowledge of thePI3K–AKT–mTOR signaling network can be useful for improvingpatient stratification and to identify more targeted therapeutic approaches. In a research paper, Polo et al. applied some integrated and bioinformatics approaches to study the network of interactions between the 25 proteins belonging to the selenoprotein family, named selenoproteins, to identify the more correlated nodes (HUB nodes) and to analyze the correlation between selenoprotein gene expression and patient outcome in 10 solid tumors. To confirm some of the correlations suggested by the bioinformatics analyses, they evaluated the gene expression level of the 25 selenoproteins and of the HUB nodes identified in two androgen receptor-positive and two androgen receptor-negative prostate cancer cell lines, compared with normal human prostate epithelial cells by RT-qPCR analysis. In this way, the authors identified some selenoproteins, such as GPX2, MSRB1, SELENOK, SELENOI, and SELENOS, which are correlated with HUB nodes and are involved in prostate cancer, concluding that their combined evaluation could improve prostate cancer patient prognosis and outcome predictions.

About breast cancer, in a review article, Cocco et al. reviewed known and emerging prognostic and predictive biomarkers in triple-negative breast cancer (TNBC). This breast cancer subtype is very heterogeneous and is characterized by high aggressivity, distant recurrence risk, and poor survival. The authors summarized the main known genetic (i.e., BRCA1/2) and protein biomarkers used for TNBC prognostic evaluation as well as for their potential to stratify patients for response to the growing number of novel targeted or immunotherapy drugs available to treat this disease. In a research article, Masuda et al. hypothesized that it is possible to predict new breast cancer metastasis (NM) and to guide the treatment of recurrent breast cancer patients by monitoring tumor mesenchymal status in real-time using liquid biopsy. In detail, the authors demonstrated that: (i) N-cadherin and vimentin expression levels were higher in the NM cases compared withpre-existing metastasis cases; (ii) N-cadherin was expressed mainly in polymorphonuclear leukocytes in peripheral blood; and (iii) the preoperative expression levels of N-cadherin were high in breast cancer blood and tissues and associated in a significant statistically way with poor recurrence-free survival. Therefore, N-cadherin mRNA levels in blood were suggested as new putative prognostic biomarkers capable ofpredicting new metastases, including recurrence, in breast cancer patients.

In a review article, Mottini and Cardone focused on the Oncogenic v-Ki-ras2 Kirsten rat sarcoma viral oncogene homolog (KRAS) dependency of pancreatic cancer beyond the mutational status of this oncogene. Indeed, KRAS mutation in pancreatic ductal adenocarcinoma (PDAC) represents a common genetic event that ismutated in almost 88% of cases; however, mutational status is not sufficient to determine whether the tumor is dependent on KRAS and thus is potentially responsive to the novel KRAS inhibitors recently entered into the clinic [11]. Therefore, the authors described the state of KRAS dependency on the basis of transcriptomic and metabolomic profiling studies, highlighting potential molecular biomarkers driven by KRAS mutations in the different PDAC subtypes for a tailored therapy.

Finally, in a research paper, Troiano et al. investigated, in oral squamous cell carcinomas (OSCC), the expression of Musashi 2 (MSI2), a RNA-binding protein that is involved in migration, invasion, epithelial–mesenchymal transition, cell proliferation, and drug resistanceby bioinformatic analysis and immunohistochemical evaluation. TCGA data analysis of 241 OSCC patients showed that the MSI2 gene was not mutated but was rather hypermethylated in all samples analyzed, with higher methylation status correlating with the age of patients and the lowexpression of MSI2 mRNA. Conversely MIS2 mRNA expression levels, being higher in males, correlated with overall survival and grading. Interestingly, although the immunohistochemical evaluation conducted on 108 tissues showeda weak expression of this protein in both OSCC samples and in healthy oral mucosae, the authors highlighted that MSI2 correlated with Cyclin-D1 expression, suggesting an indirect role of MSI2 in OSCC genesis and progression.

In conclusion, although several novel potential biomarkers have been proposed, several authors underlined that robust, well-validated biomarkers are crucial to enabling effective decision-making. Therefore, it is critical to increase the quality and the standardization of the methodology in the development pipeline to select validated and useful prognostic/predictive biomarkers [1].

Author Contributions: S.C. and A.B. contributed to writing the paper. All authors have read and agreed to the published version of the manuscript.

Funding: This research received no external funding.

Acknowledgments: We thank Danica Dong, Contact Editor of our Special Issue, for her great help, collaboration, and suggestions. Moreover, we thank the other IJMS editors and all the reviewers for their support in selecting the more scientifically interesting papers among 47 submissions received for our Special Issue.

Conflicts of Interest: The authors declare no conflict of interest.

References

1. Van Gool, A.J.; Bietrix, F.; Caldenhoven, E.; Zatloukal, K.; Scherer, A.; Litton, J.E.; Meijer, G.; Blomberg, N.; Smith, A.; Mons, B.; et al. Bridging the translational innovation gap through good biomarker practice. *Nat. Rev. Drug Discov.* **2017**, *16*, 587–588. [CrossRef] [PubMed]

2. Ali Syeda, Z.; Langden, S.S.S.; Munkhzul, C.; Lee, M.; Song, S.J. Regulatory Mechanism of MicroRNA Expression in Cancer. *Int. J. Mol. Sci.* **2020**, *3*, 21. [CrossRef] [PubMed]

3. Sohel, M.M.H. Circulating microRNAs as biomarkers in cancer diagnosis. *Life Sci.* **2020**, *248*, 117473. [CrossRef] [PubMed]

4. Fattore, L.; Ruggiero, C.F.; Pisanu, M.E.; Liguoro, D.; Cerri, A.; Costantini, S.; Capone, F.; Acunzo, M.; Romano, G.; Nigita, G.; et al. Reprogramming miRNAs global expression orchestrates development of drug resistance in BRAF mutated melanoma. *Cell Death Differ.* **2019**, *26*, 1267–1282. [CrossRef] [PubMed]

5. Huber, V.; Vallacchi, V.; Fleming, V.; Hu, X.; Cova, A.; Dugo, M.; Shahaj, E.; Sulsenti, R.; Vergani, E.; Filipazzi, P.; et al. Tumor-derived microRNAs induce myeloid suppressor cells and predict immunotherapy resistance in melanoma. *J. Clin. Investig.* **2018**, *128*, 5505–5516. [CrossRef] [PubMed]

6. Bai, R.; Lv, Z.; Xu, D.; Cui, J. Predictive biomarkers for cancer immunotherapy with immune checkpoint inhibitors. *Biomark. Res.* **2020**, *8*, 34. [CrossRef] [PubMed]

7. Kamoun, A.; de Reynies, A.; Allory, Y.; Sjodahl, G.; Robertson, A.G.; Seiler, R.; Hoadley, K.A.; Groeneveld, C.S.; Al-Ahmadie, H.; Choi, W.; et al. A Consensus Molecular Classification of Muscle-invasive Bladder Cancer. *Eur. Urol.* **2020**, *77*, 420–433. [CrossRef] [PubMed]

8. Kim, J.; Kwiatkowski, D.; McConkey, D.J.; Meeks, J.J.; Freeman, S.S.; Bellmunt, J.; Getz, G.; Lerner, S.P. The Cancer Genome Atlas Expression Subtypes Stratify Response to Checkpoint Inhibition in Advanced Urothelial Cancer and Identify a Subset of Patients with High Survival Probability. *Eur. Urol.* **2019**, *75*, 961–964. [CrossRef] [PubMed]

9. Kelley, N.; Jeltema, D.; Duan, Y.; He, Y. The NLRP3 Inflammasome: An Overview of Mechanisms of Activation and Regulation. *Int. J. Mol. Sci.* **2019**, *20*, 3328. [CrossRef] [PubMed]

10. Pearson, H.B.; Li, J.; Meniel, V.S.; Fennell, C.M.; Waring, P.; Montgomery, K.G.; Rebello, R.J.; Macpherson, A.A.; Koushyar, S.; Furic, L.; et al. Identification of Pik3ca Mutation as a Genetic Driver of Prostate Cancer That Cooperates with Pten Loss to Accelerate Progression and Castration-Resistant Growth. *Cancer Discov.* **2018**, *8*, 764–779. [CrossRef] [PubMed]

11. Kim, D.; Xue, J.Y.; Lito, P. Targeting KRAS(G12C): From Inhibitory Mechanism to Modulation of Antitumor Effects in Patients. *Cell* **2020**, *183*, 850–859. [CrossRef] [PubMed]

Publisher's Note: MDPI stays neutral with regard to jurisdictional claims in published maps and institutional affiliations.

International Journal of
Molecular Sciences

Review

Insight toward the MicroRNA Profiling of Laryngeal Cancers: Biological Role and Clinical Impact

Takashi Takeuchi [1,2], Hiromichi Kawasaki [1,3], Amalia Luce [1], Alessia Maria Cossu [1,4], Gabriella Misso [1], Marianna Scrima [4], Marco Bocchetti [1,4], Filippo Ricciardiello [5], Michele Caraglia [1,4,*] and Silvia Zappavigna [1]

[1] Department of Precision Medicine, University of Campania "Luigi Vanvitelli", 80138 Naples, Italy; takashi.takeuchi@unicampania.it (T.T.); Hiromichi.KAWASAKI@unicampania.it (H.K.); amalia.luce@unicampania.it (A.L.); alessiamaria.cossu@biogem.it (A.M.C.); gabriella.misso@unicampania.it (G.M.); marco.bocchetti@unicampania.it (M.B.); silvia.zappavigna@unicampania.it (S.Z.)
[2] Molecular Diagnostics Division, Wakunaga Pharmaceutical Co., Ltd., Hiroshima 739-1195, Japan
[3] Drug Discovery Laboratory, Wakunaga Pharmaceutical Co., Ltd., Hiroshima 739-1195, Japan
[4] Biogem Scarl, Institute of Genetic Research, Laboratory of Molecular and Precision Oncology, 83031 Ariano Irpino, Italy; marianna.scrima@biogem.it
[5] Division of Otorhinolaryngology, "A. Cardarelli" Hospital, 80131 Naples, Italy; filipporicciardiello@virgilio.it
* Correspondence: michele.caraglia@unicampania.it; Tel.: +39-081-5665874; Fax: +39-081-5665863

Received: 17 April 2020; Accepted: 22 May 2020; Published: 24 May 2020

Abstract: Head and neck squamous cell carcinoma (HNSCC), a heterogeneous disease arising from various anatomical locations including the larynx, is a leading cause of death worldwide. Despite advances in multimodality treatment, the overall survival rate of the disease is still largely dismal. Early and accurate diagnosis of HNSCC is urgently demanded in order to prevent cancer progression and to improve the quality of the patient's life. Recently, microRNAs (miRNAs), a family of small non-coding RNAs, have been widely reported as new robust tools for prediction, diagnosis, prognosis, and therapeutic approaches of human diseases. Abnormally expressed miRNAs are strongly associated with cancer development, resistance to chemo-/radiotherapy, and metastatic potential through targeting a large variety of genes. In this review, we summarize on the recent reports that emphasize the pivotal biological roles of miRNAs in regulating carcinogenesis of HNSCC, particularly laryngeal cancer. In more detail, we report the characterized miRNAs with an evident either oncogenic or tumor suppressive role in the cancers. In addition, we also focus on the correlation between miRNA deregulation and clinical relevance in cancer patients. On the basis of intriguing findings, the study of miRNAs will provide a new great opportunity to access better clinical management of the malignancies.

Keywords: microRNA; biomarkers; head and neck cancer; laryngeal cancer; prediction; prognosis; metastasis; lifestyle habit; chemo-/radio resistance; therapeutic target

---

1. Introduction

Head and neck squamous cell carcinoma (HNSCC) represents the sixth most common cancer in the world, and is a histologically and genetically heterogeneous disease. It arises from multiple anatomical sites including oral mucosa, tongue, salivary glands, nasopharyngeal, pharyngeal, and laryngeal carcinomas [1,2]. Despite substantial progress in both treatment strategies and diagnostic techniques, the overall survival rate as well as the mortality rate of the disease is still largely unimproved over the last decades [3,4]. Smoking tobacco and drinking excessive alcohol are the predominant risk factors of initiation and aggressive progression of HNSCC [2,5].

Both local and distant metastases usually result in poor prognosis of HNSCC, thus representing one of the most adverse phenomena in the patients [6].

The survival rate of HNSCC patients is still poor since the disease is often at advanced stages when diagnosed. Thus, early and precise detection in the initial and pre-clinical stages of the tumor is crucial to achieve high survival rate of HNSCC patients, but the early diagnosis is often hampered by present lack of definite symptoms. Implementation of early intervention is greatly dependent on clinical judgment by diagnostic instruments, histological results, and usable markers. However, there is still no truly reliable and identifiable tools with sufficient sensitivity and specificity at the initial stages. To overcome the issue, easily and universally available protocols have been long investigated in order to prevent initial and secondary malignant tumors. For example, many biomolecule markers, such as proteins, DNA, and RNA molecules, have been studied but there is still a big challenge to use them as a certain biomarker for the clinic purpose in HNSCC patients. Another issue, protocols for therapeutic clinical management, have not been yet established to accurately diagnose the presence of small metastases that cannot be evidenced by conventional radio-imaging analysis. Additionally, there is currently a lack of information on clinically available molecular-based biomarkers for both prediction and prognosis of aggressive malignancies in a non-invasive way. In fact, a misleading of false positive/negative metastases cannot be completely avoided, resulting in expanding tumor invasion (under treatment) or undergoing an unnecessary surgical intervention (over treatment). Hence, it is necessary to urgently address these issues by either early prediction and detection indicators or improved therapeutic treatments.

Recently, miRNAs have emerged as robust predictors and prognosticators, pharmaceutical drugs, personalized medicine, and therapeutic biomolecular targets for the effective treatment of diseases. This small group of highly conserved non-protein coding RNAs (approximately 20 nucleotides in length) is essential for maintaining physiological conditions. miRNAs negatively and post-transcriptionally regulate divergent genes involved in (signal) transduction pathways, through promotion of mRNA destabilization or prevention of protein translation machinery by either perfect or nearly perfect binding to the complementary site of mRNA at 3′ untranslated regions [7]. Biogenesis of miRNAs has been currently studied and summarized in many reviews, but not fully elucidated due to its complicated procedures. Conventionally, biogenesis pathway involves two cleavage steps, one nuclear and one cytoplasmic (Figure 1), but other alternative biogenesis pathways that require a different number of cleavage events and key enzymes are possible. The mechanism by which microRNA precursors are sorted to the different pathways is unclear but probably depends on the origin site of the microRNA, its sequence and thermodynamic stability. However, mature miRNAs usually exhibit a tissue-specific pattern of expression but an apparent tissue-specific pattern for their corresponding primary transcripts has not been found [7–9]. Consequently, miRNAs may act either as onco-promotors or tumor suppressor genes, depending on their biological targets and the cellular context [10]. Numerous studies have indicated that miRNA dysregulation patterns could be useful for prediction of clinical outcome in HNSCC patients.

In recent years, investigations of circulating extracellular miRNAs have rapidly progressed. Although miRNAs were thought to be unstable in extracellular environment due to their low stability, it has been confirmed the presence of a stable form of circulating miRNAs in human body fluids including blood, saliva, and urine [11–17]. Among patients, the amount of circulating miRNAs is also changed and closely linked to pathological events including metastases. There are several advantages of circulating miRNAs including easy and safe accessibility, ready detection and quantization, and good reproducibility. It is likely that circulating miRNAs are suitable as an informative application for the diagnostic examination and are able to monitor the real-time disease status by detecting their specific signatures. Here, we briefly summarize recent miRNA studies regarding biological functions, comprehensive expression profiles, and clinical relevance in HNSCC patients, particularly focused on laryngeal carcinoma.

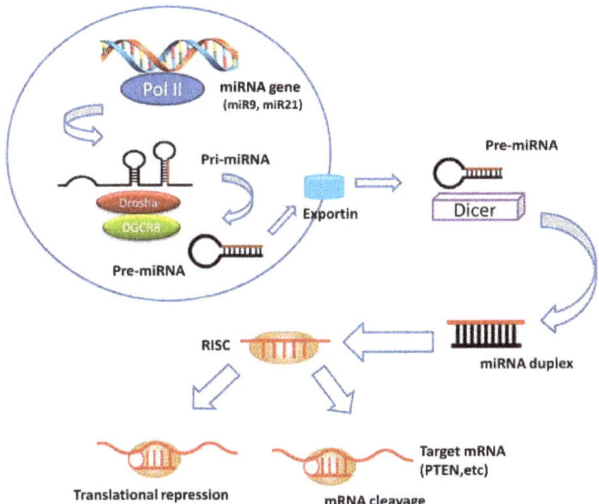

Figure 1. MiRNA biogenesis. miRNA gene is transcribed to generate a primary miRNA (pri-miRNA) precursor molecule that undergoes nuclear cleavage to form a second precursor miRNA (pre-miRNA). The pre-miRNA is exported in the cytoplasm and cleaved to generate a microRNA duplex containing the mature miRNA. The duplex unwinds and the mature miRNA assembles into RISC. The miRNA base-pairs with target mRNA to perform gene silencing trough mRNA cleavage or translation repression.

2. miRNA Deregulation

2.1. Laryngeal Cancer

Laryngeal cancer (LCa) is one of the most common types of neoplasms in the head and neck region and accounts for approximately 1–2% of all malignancies. Over 90% of LCa is histologically classified as squamous cell carcinoma. LCa is divided into glottis (60–65%), supraglottic (30–35%), or subglottic carcinoma (1–2%). Cigarette smoking, and heavy alcohol consumptions are the most detrimental risks of the disease [18–21]. In addition, it has been reported that the presence of nodal metastases is the most significant prognostic factor of LCa patients. Prediction of the overall survival rate in LCa patients is more related to node metastasis than tumor extension [22]. Metastases are often responsible for the aggressiveness and cancer-related mortality. Partial laryngectomy, chemo-radiation, and combination therapy are effective treatments of laryngeal carcinomas at an early clinical stage. However, in the advanced cases of LCa patients, total laryngectomy is often required as a therapeutic intervention, but such surgical modality strongly negatively affects patient's quality of life. Furthermore, cancer recurrence may appear at the laryngeal region or other distant sites even though all parts of the malignant larynx were correctly removed by the laryngectomy. Therefore, it is highly and urgently demanded to enhance conventional methods and to set up both more effective treatment management and earlier nodal metastasis prediction techniques in order to reduce cancer morbidity and mortality and to improve the quality of life. Regarding the diagnostic approaches, one of the possible ways is the establishment of new reliable biomarkers to predict LCa progression and to prognosticate the disease. As mentioned above, many researchers have recently focused on miRNAs as an excellent tool in the biological, clinical, and medical field. So far, a number of research groups have widely investigated that miRNA dysregulation, which is up/down expression, is observed in LCa when comparing to precancerous lesions, benign, or noncancerous counterparts, using microarray or quantitative real-time PCR (qRT-PCR) analysis.

Several miRNAs are aberrantly upregulated in LCa and they can contribute to aggressiveness of the cancer. MiR-9 was upregulated in LCa tissues and cell lines by influencing malignant development at the larynx through direct regulation of PTEN [23].

miR-10b is overexpressed in LCa, showing a role in the initiation of cancer invasion and migration in the laryngeal carcinoma cell line Hep-2. Zhang et al. unveiled that miR-10b accelerates the epithelial-mesenchymal transition (EMT) in Hep-2 cells through directly targeting E-cadherin (E-cad). The authors studied both epithelial (E-cad and ZO-1) and mesenchymal (Vim, N-cad, and FN) markers in Hep-2 cells with ectopic miR-10b expression using western blot analysis, and they showed that miR-10b-transfected cells possess reduced epithelial and increased mesenchymal markers. The regulation of CDH1 (E-Cad gene) by miR-10b was also confirmed in this study [24].

miR-21 is one of the most well-studied miRNAs and exerts oncogenic roles in different types of malignancies. It was exhibited that miR-21 is upregulated in laryngeal tumor tissues and its inhibition by antago-miRs caused growth inhibition of Hep-2 cells via the suppression of BTG2, a validated miR-21 target, leading to inhibition of cell cycle progression without influencing cell apoptosis. Other targets of miR-21, PTEN, TPM1 and PDCD4, were also downregulated [25].

Additionally, a repressor of cell cycle progression, the cyclin-dependent kinase 2-associated protein 1 (CDK2AP1), which is a G1/S transition inhibitor, was found to be a negative regulator of LCa. CDK2AP1 was reported as a target gene of miR-21 in pathological human oral keratinocytes [26] and a recent investigation showed that this gene was downregulated by miR-205, indicating that miR-205 regulates CDK2AP1 in LCa. Hence, miR-205 works as an oncogene, through the suppression of CDK2AP1, and affects cell proliferation and motility by promoting MMP2 and MMP9 activities and both c-Myc and CyclinD1 over-expression in LCa cells [27].

Overexpression of miR-93 was found in LCa tissues by global miRNA screening and was confirmed by qRT-PCR set [28].

Another manuscript reported that miR-93 was inversely correlated with cyclin G2 (CCNG2) levels in clinical samples from LCa patients. A gain of function analysis showed that miR-93 promotes cell proliferation and metastases, while suppression of miR-93 reduces these malignant processes. Further studies revealed that miR-93 accelerates cancer progression in LCa through directly inhibiting CCNG2 [29].

Plasma miR-126 was associated with clinical differentiation of LCa. The authors also showed that miR-126 in part regulated calmodulin-regulated spectrin-associated protein 1 (Camsap1), and the introduction of miR-126 mimic decreased LCa tumorigenesis in xenograft mouse models [30].

Lian and coworkers found that miR-132 was strongly overexpressed in LCa tissues and cells and directly targeted FOXO1, which is a class of human forkhead box O (FOXO) proteins and works as important effectors of PI3K/Akt signaling. This study thus demonstrated the oncogenic role of miR-132 in LCa by mediating PI3K/AKT/FOXO1 pathway [31].

miR-221 was identified as an oncogene through the modulation of several signaling pathways. Inhibition of endogenous miR-221 expression reduced cell proliferation and induced both apoptosis and cell cycle arrest of LCa cells. In addition, the study of miR-221 in LCa also showed to stimulate apoptosis resistance by affecting apoptotic protease activating factor-1 (Apaf-1). Interestingly, miR-221 inhibitor had anti-cancer effects as confirmed in xenograft mouse models inoculated with Hep-2 cells, showing that miR-221 suppression led to reduced tumor size and weights and to increased survival rate [32].

Additionally, overexpression of miR-302-3p was found in LCa tissues and cells and disclosed to be involved in EMT processes. The authors identified that tumor repressor Smad4 is a direct downstream target of miR-302-3p [33].

Guan and coworkers observed an up-regulation of miR-423-3p in primary LCa cell lines. The same group also confirmed that miR-423-3p plays an oncogenic role. In silico prediction algorithms and further validation confirmed that adiponectin receptor 2 (AdipoR2) is directly regulated by miR-423-3p [34].

Several reports exhibited lower miRNA expression in LCa, and the miRNAs could suppress tumor development and metastasis through multiple direct/indirect target oncogenes.

Many studies have recently shown that miR-1 is often downregulated in various tumors and is an anticancer gene. miR-1 downregulation and its clinical impact were also found in HNSCC specimens including LCa.

It was reported that miR-1 is significantly under-expressed in LCa tissues and ectopic miR-1 expression induces the suppression of cell growth and metastatic potential in Hep-2 cells. Target identification assay showed that miR-1 directly regulates fibronectin 1 (FN1), which has cancer metastasis related functions. Thus miR-1 exerts an anti-oncologic role in LCa [25,35].

miR-24 was significantly underexpressed in LCa tissues or cell lines compared to paired normal tissues or normal human keratinocyte cell lines. Reintroduction of miR-24 inhibited both cell growth and colony formation and promoted apoptosis. The research group revealed that miR-24 binds to the 3′-UTR of XIAP mRNA [36].

A well-known ubiquitous onco-suppressive miRNA, miR-34, is frequently downregulated in a variety of cancers [37].

miR-34a was inversely correlated to cyclin D1 (CCND1) levels, nodal metastases, and clinical stage in LCa. It was also found that miR-34a regulates CCND1 gene [38]. miR-34c is also downregulated in LCa and functions as a tumor suppressor. Replacement of miR-34c expression in LCa cells induced both inhibition of growth and invasion by directly targeting c-Met [39].

Another study in Hep-2 cells demonstrated that miR-34a/c directly regulates UDP-*N*-acetyl-α-D-galactosamine:polypeptide-*N*-acetylgalactosaminyltransferase 7 (GALNT7) a member of glycosyltransferases involved in cancer spreading and metastasization [40].

miR-144 expression was found to be downregulated in LCa, and to have anti-oncogenic functions through the suppression of insulin receptor substrate 1 (IRS1). The latter stimulates the phosphatidylinositol 3-kinase (PI3K)/Akt pathway that is then inhibited by miR-144 [41].

miR-340 is another anti-oncogene in LCa facilitating p27 expression and blocking PI3K/Akt signaling through the suppression of the histone methyltransferase EZH2, thus resulting in the inhibition of cell proliferation, metastatic abilities, and apoptosis [42].

The association between Let-7a downregulation and high-mobility group protein A2 (HMGA2) mRNA overexpression in LCa indicates that it is involved in tumor aggressiveness and metastasis. HMGA2 is able to change the DNA structure through its interaction, thus affecting DNA transcription. The expression levels of Let-7a and HMGA2 were strongly associated with clinical stage, differentiation grade, and lymph node metastases in LCa [43].

Taken together, these reports open the new opportunity to use miRNAs as either biomarkers or therapeutic approaches for patients affected by LCa.

2.2. Other HNSCC

Deregulation of miRNAs in other HNSCC, such as oral squamous cell cancer (OSCC), tongue, and salivary gland tumors [44–148] is summarized in Table 1. Unfortunately, data demonstrating the role of miRNAs in the other HNSCCs are still not sufficient to provide an accurate overview. miR-155 was found to be upregulated in OSCC tissues and its overexpression reduced the intracellular levels of cell division cycle 73 (CDC73), a target of a tumor suppressor gene [47]. Another study also reported miR-155 overexpression in OSCC cells and tissues [48]. Upregulated miR-155-5p induced cancer metastases to neck lymph nodes and was associated with poor overall survival rate in OSCC patients. Inhibition of miR-155-5p promoted the epithelial marker E-cadherin, reduced signal transducer and activator of transcription 3 (STAT3), and activated suppressor of cytokine signaling 1 (SOCS1) in an OSCC cell line [49,50].

Table 1. miRNA deregulation in head and neck cancer and validated targets.

Cancer Type	miRNA	Regulation	Role	Target Gene	Ref.
Laryngeal cancer (LCa)	Let-7a	Down	Tumor-suppressive	*HMGA2*	[43]
	miR-1	Down	Tumor-suppressive	*FN1*	[25,35]
	miR-9	Up	Oncogenic	*PTEN*	[23]
	miR-9	Up	Oncogenic	-	[109]
	miR-10b	Up	Oncogenic	*CDH1*	[24]
	miR-21	Up	Oncogenic	*BTG2, PTEN, TPM1, PDCD4, CDK2AP1*	[25]
	miR-23a	Up	Oncogenic	-	[107]
	miR-24	Down	Tumor-suppressive	*XIAP*	[36]
	miR-34a	Down	Tumor-suppressive	*CCND1, GALNT7*	[38,40]
	miR-34c	Down	Tumor-suppressive	*c-Met, GALNT7*	[39,40]
	miR-93	Up	Oncogenic	*CCNG2*	[28,29]
	miR-101	Down	Tumor-suppressive	*CDK8*	[108]
	miR-126	-	Tumor-suppressive	*Camsap1*	[30]
	miR-132	Up	Oncogenic	*FOXO1*	[31]
	miR-138-2-3p	Up	Oncogenic	-	[122]
	miR-144	Down	Tumor-suppressive	*IRS1*	[41]
	miR-203	Down	Tumor-suppressive	-	[121]
	miR-205	Up	Oncogenic	*CDK2AP1*	[27]
	miR-221	-	Oncogenic	*ER-α, p27, p57, c-kit, Apaf-1*	[32]
	miR-296-5p	Up	Oncogenic	-	[110]
	miR-302-3p	Up	Oncogenic	*Smad4*	[33]
	miR-340	Down	Tumor-suppressive	*p27, EZH2*	[42]
	miR-423-3p	Up	Oncogenic	*AdipoR2*	[34]
Head and neck squamous cell carcinoma (HNSCC)	miR-1	Down	Tumor-suppressive	*TAGLN2*	[65]
	miR-1	Down	Tumor-suppressive	*EGFR, c-Met*	[70]
	miR-10b	Down	Tumor-suppressive	-	[45]
	miR-99	Down	Tumor-suppressive	*IGF1R, mTOR, Akt1*	[76,77]
	miR-100	Down	Tumor-suppressive	*IGF1R, mTOR, Akt1*	[76,77]
	miR-101	Down	Tumor-suppressive	*EZH2, rap1GAP*	[81,82]
	miR-146a	Down	Tumor-suppressive	-	[50]
	miR-155	Down	Tumor-suppressive	-	[50]
	miR-196a	Up	Oncogenic	-	[45]
	miR-203	Up	Oncogenic	-	[54]
	miR-203	Down	Tumor-suppressive	-	[56]
	miR-204	Down	Tumor-suppressive	*Brd4*	[91]
	miR-205	Up	Oncogenic	-	[54]
	miR-206	Down	Tumor-suppressive	*EGFR, c-Met*	[70]
Oral squamous cell cancer (OSCC)	miR-1	Down	Tumor-suppressive	*Slug*	[68]
	miR-1-3p	Down	Tumor-suppressive	*DKK1*	[69]
	miR-26a/b	Down	Tumor-suppressive	*TMEM184B*	[71]
	miR-34a	Down	Tumor-suppressive	-	[74]
	miR-99-3b	Down	Tumor-suppressive	*glycogen synthase kinase-3β*	[80]
	miR-99a	Down	Tumor-suppressive	*Myotubularin-related protein 3*	[78,79]
	miR-104-5p	Down	Tumor-suppressive	*PAK4*	[86]
	miR-155	Up	Oncogenic	-	[141]
	miR-155	Up	Oncogenic	*CDC73*	[47–49]
	miR-181a	Down	Tumor-suppressive	*K-ras*	[52]
	miR-181a/b	Up	Oncogenic	-	[51]
	miR-204	Down	Tumor-suppressive	*Sox4, Slug*	[88]
	miR-204-5p	Down	Tumor-suppressive	*CXCR4*	[89]
	miR-222	Up	Oncogenic	*PUMA*	[60]
	miR-222	Down	Tumor-suppressive	*PUMA*	[127]
	miR-223	Up	Oncogenic	-	[62]
	miR-223	Up	Oncogenic	*FBXW7*	[63]
	miR-494	Down	Tumor-suppressive	*HOXA10*	[90]
Oral	miR-10b	Up	Oncogenic	-	[44]
Oral tongue squamous cell carcinoma (OTSCC)	miR-222	Down	Tumor-suppressive	*MMP1, SOD2*	[61]
Hypopharyngeal squamous cell carcinoma (HSCC)	miR-140-5p	Down	Tumor-suppressive	*ADAM10*	[85]
Maxillary sinus	miR-1	Down	Tumor-suppressive	*TAGLN2, PNP*	[66]
Nasopharyngeal carcinoma (NPC)	miR-1	Down	Tumor-suppressive	*ET-1*	[67]
	miR-10b	Down	Tumor-suppressive	-	[46]
	miR-101	Down	Tumor-suppressive	*ITGA3*	[83]
	miR-204	Down	Tumor-suppressive	*Cdc42*	[90]
	miR-494-3p	Up	Oncogenic	*Sox7*	[93]
Salivary adenoid cystic carcinoma (SACC)	miR-104-5p	Down	Tumor-suppressive	*Survivin*	[87]
	miR-181a	Down	Tumor-suppressive	*MAP2K1, MAPK1, Snai2*	[53]

Table 1. *Cont.*

Cancer Type	miRNA	Regulation	Role	Target Gene	Ref.
Salivary gland adenoid cystic carcinoma (SGACC)	miR-101-3p	Down	Tumor-suppressive	*Pim-1*	[128]
Sinonasal squamous cell carcinomas (SN-SCC)	miR-34a	Down	Tumor-suppressive	*BCL-2*	[75]
Tongue	miR-21	Up	Oncogenic	-	[148]
	miR-183	Up	Oncogenic	-	[148]
Tongue squamous cell carcinoma (TSCC)	miR-26a	Down	Tumor-suppressive	*DNMT3B*	[72]
	miR-26a	Down	Tumor-suppressive	*PAK1*	[73]
	miR-26b	Down	Tumor-suppressive	*PAK1*	[73]
	miR-140-5p	-	Tumor-suppressive	*LAMC1, HDAC7, PAX6*	[84]
	miR-181a	Down	Tumor-suppressive	*Twist1*	[126]

Liu et al. showed that miR-222 expression was downregulated in oral tongue squamous cell carcinoma (OTSCC) harboring highly metastatic potential compared to the lower invasive one. Ectopic expression of miR-222 led to a decrease in cell invasive ability and reduced both Matrix Metallopeptidase 1 (MMP1) and Superoxide Dismutase 2 (SOD2) expressions in OTSCC, through direct regulation of the mRNAs [61].

Furthermore, in tongue cancer tissues, miR-26a was also downregulated in comparison with paired non-pathological ones. Restoration of miR-26a showed inhibition of cell proliferation and apoptosis induction, suggesting that miR-26a elicits anti-cancer effects in tongue carcinogenesis. Further functional studies showed that miR-26a inhibited directly DNA methyltransferase 3b (DNMT3B) transcript [72]. Another report showed downregulation of miR-26a and miR-26b in tongue squamous cell carcinoma (TSCC) tissues. In addition, overexpression of these miRNAs suppressed TSCC cell cycle, motility and glycolysis, while enhanced cell apoptosis by direct interaction with p21 Activated Kinase 1 (PAK1) [73].

Another group showed that miR-140-5p was decreased in hypopharyngeal squamous cell carcinoma (HSCC) tissues, and the downregulation was correlated with tumor size and lymph node metastases. Moreover, functional analysis showed that miR-140-5p directly suppresses ADAM10. This study demonstrated that miR-140-5p suppressed cancer invasion and migration by inhibiting ADAM10-mediated Notch1 signaling pathway [85].

Either oncogenic or tumor suppressive miRNAs and their validated target genes in LCa and other HNSCCs are reported in Table 1.

Overall, likewise the knowledge in LCa, these recent articles also provide a rationale for developing miRNA-based therapeutic weapons or understanding molecular mechanisms underlying of HNSCC biology.

3. miRNAs as Biomarkers for the Prediction of Initiation and Progression

3.1. Laryngeal Cancer

Increasing evidence has suggested the usefulness of miRNA signatures as robust predictors for early and accurate diagnosis of LCa. A number of research groups have currently shown unique and attractive candidate miRNAs as biomolecule-based markers.

A research group showed that the combination of exosomal serum miR-21 and long non-coding RNA HOTAIR exhibits high sensitivity (94.2%) and specificity (73.5%), respectively, in different malignancies from benign laryngeal disease, thereby indicating that the combination may be suitable for diagnosis of LCa [94].

Another study also reported miR-21 as a potent prediction tool. MiR-21 upregulation and miR-375 downregulation were observed in LCa samples. Interestingly, the ratio of miR-21/miR-375 expression had a robust value with great sensitivity (94%) and specificity (94%) for the prediction of LCa [95].

A study by Saito et al. showed that miR-133b, miR-455-5p, and miR-196a were abnormally expressed in laryngeal tumors if compared to their matched non-oncological tissues. Subsequent validation tests confirmed that miR-196a was over-expressed in cancer tissues against precancerous dysplasias and laryngeal benign tissues. Moreover, in situ hybridization confirmed specific expression of miR-196a in both cancer and cancer stroma cells [96].

In early laryngeal carcinoma and normal esophageal mucosa tissues, microarray-based screening showed two highly deregulated miRNAs, hsa-miR-657 and hsa-miR-1287, with good specificity and sensitivity to classify laryngeal malignancies at early stages [97].

A report showed that both miR-148a and miR-375 levels were highly upregulated in LCa tissues. Association analysis unveiled that miR-375 expression in advanced LCa (stage III/IV) was far greater than in tumors at early stage (stage I/II). On the contrary, despite increased miR-148a level in early cancer tissues, its expression was not significantly changed when comparing to advanced ones [98].

To our knowledge, Ayaz et al. have carried out the first study on circulating miRNA profiling in plasma collected from LCa patients. In this research, several circulating miRNAs were found in the liquid biopsies and abnormally expressed across plasma specimens if compared to control plasma. Five circulating miRNAs (miR-212-3p, miR-331-3p, miR-603, miR-660-5p, and miR-1303) were detectable in the oncological plasma, but they were not detected in plasma taken from healthy individuals or patients presenting with any other diseases [99].

Serum miR-378 was more highly expressed in LCa patients than healthy controls, and the expression was significantly decreased after surgical intervention. Also, miR-378 expression was correlated with clinical stage, but did not associate to tumor size [100].

In a recent study, the expression levels of serum miRNAs in LCa patients were investigated using microarray screening and furthermore validated by qRT-PCR. The results showed eight up-regulations (miR-31, miR-33, miR-141, miR-149a, miR-182, LET-7a, miR-4853p and miR-122) and three down-regulations (miR-133a, miR-145 and miR-223) in LCa patients ($n = 66$) compared to healthy volunteers ($n = 100$). Moreover, ROC curve analysis indicated that miR-31, miR-33, and LET-7a possessed high diagnostic ability for the disease with AUC (Area under curve) = 1.0 [101]. In addition, several miRNAs reviewed on the above part could be potential diagnostic markers of LCa. To summarize, these miRNAs could contribute to the development of novel predictive and diagnostic biomarkers of LCa.

3.2. Other HNSCC

Evidence has also suggested the usefulness of miRNA signatures as biomarkers for the prediction of initiation and progression of other HNSCCs [102–106] but reports referring to the other HNSCCs are still limited. A miRNA expression ratio of miR-221:375 showed great discriminatory potential, with great sensitivity (92%) and specificity (93%) between HNSCC tissues and non-pathological ones [104].

In the investigation of circulating miRNAs in oral cancer specimens, both circulating miR-196a and miR-196b were significantly overexpressed in plasma obtained from patients with oral pre-cancer lesions and in plasma from oral cancer patients. Both miRNAs showed high discriminative values between pre-cancer patients and normal, and between cancer patients and normal samples. Furthermore, the combined determination of miR-196a and miR-196b possesses high sensitivity (91%) and specificity (85%) in predicting potential malignancy at early stages [105].

4. miRNAs as Biomarkers for the Prognosis

Despite the improved therapeutic options of the past few decades, the treatment of HNSCC and the patient's quality of life are still unsatisfying due to poor prognosis of HNSCC. The deregulated miRNA expression patterns can also help clinicians to prognosticate cancer progression and outcome. Table 2 also showed several reported miRNAs, which could be useful as prognostic biomarkers of LCa and other types of HNSCC.

4.1. Laryngeal Cancer

As reported above, LCa patients with higher miR-21 or lower miR-375 levels in cancer tissues showed poorer prognostic values than those with low miR-21 or high miR-375 levels [95].

miR-23a expression was up-regulated in LCa tissues when comparing to their corresponding normal ones; additionally, miR-23a overexpression was related to lymph node metastases and the five-year survival rate of patients. Further biological analysis showed that up-regulation of miR-23a facilitated cancer migration and invasion [107].

Strikingly, miR-101 downregulation was closely associated with nodal involvement, tumor grade (T3-T4), and LCa at advanced stage. Moreover, clinical relevance was observed between lower miR-101 level and poor outcome. Ectopic miR-101 overexpression reduced cell proliferation and migration and led to cell-cycle arrest and apoptosis. Besides, miR-101 suppressed tumor growth in a xenograft model mouse with LCa, thereby suggesting that miR-101 acts as a tumor suppressor gene [108].

Another report showed that the expression of miR-9 was higher in LCa if compared to the paired normal laryngeal tissues, and the study also reported a relation between miR-9 overexpression and poorer prognosis of LCa [109].

The article showed that miR-296-5p was correlated to tumor relapse in LCa at an early stage [110].

Cappellesso et al. found a significant downregulation of miR-200a and miR-200c levels in LCa from the group of patients who developed disease recurrence [111].

Collectively, these findings suggest that miRNAs could be useful as potential biomarkers to prognosticate the clinical outcomes.

4.2. Other HNSCC

As shown in Table 2, several recently reported miRNAs in HNSCC can be useful as predictors and/or prognosticators [112–118] even if data referring to the potential role of miRNAs in other HNSCCs are still not sufficient to provide an accurate overview.

A decrease in miR-218, miR-125b, or Let-7g expression is associated with advanced clinical stage and lymph node metastasis and is a useful prognostic factor of OSCC patients [113].

High miR-93 expression was correlated with T classification, clinical stage, lymph node involvement, and poor prognosis among HNSCC patients [28,115]. In addition, elevated serum miR-93 was found and associated with prognosis of NPC patients.

Bufalino et al. reported both miR-143 and miR-145 downregulation and high activin A levels in OSCC cell lines and tissue specimens. Downregulation of a miR-143/miR-145 cluster controlled cancer invasiveness and was clinically correlated to nodal metastasis and worse survival [116].

Serum miR-9 level was significantly lower in patients with OSCC or oral leukoplakia in comparison to that of non-pathological controls, and low serum miR-9 expression was correlated to poor overall survival and disease-free survival rate [118].

Table 2. Potential predictive and/or prognostic biomarker in HNSCC.

miRNA	Regulation	Cancer	Diagnosis/Prognosis	Ref.
Let-7a	Up	LCa	Diagnosis	[101]
Let-7g	Up	OSCC	Prognosis	[113]
miR-100	Down	HNSCC	Prognosis	[117]
miR-101	Down	LCa	Prognosis	[108]
miR-10b	Up	OSCC	Diagnosis	[44]
miR-125b	Up	OSCC	Prognosis	[113]
miR-125b	Down	HNSCC	Prognosis	[117]
miR-1287	Down	LCa	Diagnosis	[97]
miR-1303	Up	LCa	Diagnosis	[99]
miR-133b	Down	LCa	Diagnosis	[96]
miR-143	Down	OSCC	Prognosis	[116]

Table 2. *Cont.*

miRNA	Regulation	Cancer	Diagnosis/Prognosis	Ref.
miR-145	Down	OSCC	Prognosis	[116]
miR-146a	Down	HNSCC	Diagnosis	[50]
miR-155	Up	OSCC	Prognosis	[48,49]
miR-15a	Down	HNSCC	Diagnosis, Prognosis	[117]
miR-181a	Down	SACC, TSCC	Prognosis	[53,126]
miR-196a	Up	LCa, OSCC	Diagnosis	[96,105]
miR-196b	Up	OSCC	Diagnosis	[105]
miR-199b	Down	HNSCC	Diagnosis, Prognosis	[117]
miR-200a	Down	LCa	Prognosis	[111]
miR-200c	Down	LCa	Prognosis	[111]
miR-203	Up	HNSCC	Prognosis	[54]
miR-203	Down	HNSCC	Prognosis	[121]
miR-204	Down	OSCC	Prognosis	[88]
miR-205	Down	HNSCC	Prognosis	[102]
miR-205	Up	HNSCC	Prognosis	[112]
miR-205-5p	Down	OSCC	Diagnosis, Prognosis	[103]
miR-21	Up	HNSCC	Prognosis	[114,117]
miR-21	Up	LCa	Diagnosis, Prognostic	[94,95]
miR-212-3p	Up	LCa	Diagnosis	[99]
miR-218	Up	OSCC	Prognosis	[113]
miR-223	Up	HNSCC	Prognosis	[114]
miR-23a	Up	LCa	Prognosis	[107]
miR-296-5p	Up	LCa	Prognosis	[110]
miR-31	Up	LCa	Diagnosis	[101]
miR-33	Up	LCa	Diagnosis	[101]
miR-331-3p	Up	LCa	Diagnosis	[99]
miR-34c	Down	HNSCC	Diagnosis, Prognosis	[117]
miR-375	Down	LCa	Diagnosis, Prognosis	[95]
miR-375	Down	LCa	Diagnosis	[98]
miR-378	Up	LCa	Diagnosis	[100]
miR-455-5p	Up	LCa	Diagnosis	[96]
miR-603	Up	LCa	Diagnosis	[99]
miR-657	Up	LCa	Diagnosis	[97]
miR-9	Up	LCa	Prognosis	[109]
miR-9	Down	OSCC	Prognosis	[118]
miR-93	Up	HNSCC	Prognosis	[28,115]
miR-99a	Down	HNSCC	Prognosis	[114]

HNSCC: Head and neck squamous cell carcinoma; LCa: Laryngeal cancer; OSCC: Oral squamous cell cancer; SACC: Salivary adenoid cystic carcinoma; TSCC: Tongue squamous cell carcinoma.

5. Resistance Related miRNAs

The resistance is one of the most important determinants of prognosis and treatment results. Resistance to chemotherapy or irradiation often blocks the effective therapeutic modality of malignant tumors, resulting in further cancer development, invasiveness, recurrence, and consequent unfavorable outcomes. Recent findings have suggested that miRNAs are closely associated with radio/chemo-resistance in human cancers through complex processes controlling huge number of genes in cell signaling network; therefore, the use of miRNAs can become a significant breakthrough to improve cancer treatment protocols.

5.1. Radio-Resistance

An overall miRNA profiling assay showed that 4 miRNAs, miR-296-5p, miR-452, miR-183*, and miR-200c, were aberrantly altered in LCa tissues taken from radio-resistant patients. Importantly, a subsequent validation set exhibited that miR-296-5p was linked to resistance to radiation [110].

Suh et al. reported that miR-196a overexpression enhanced the radio-resistance in HNSCC cells [119].

An investigation using available HNSCC samples in the cancer genome atlas (TCGA) miRNA database identified that a 5 miRNA signature (downregulated let-7e and upregulated miR-16, miR-29b, miR-150, and miR-1254) could be useful for the prediction of radiation responsiveness in HNSCC patients. Additionally, this study also indicated that higher levels of ataxia-telangiectasia mutated expression (ATM) in HNSCC patients were correlated to increased resistance to radiotherapy [120].

De Jong et al. reported a comprehensive miRNA expression profiling correlated to radio-resistance in HNSCC cells from primary LCa. Validation studies showed that reduced expression of miR-203 was correlated to local recurrence after radiation and resistance to irradiation in LCa patients. On the other hand, overexpression of miR-203 reduced EMT processes [121].

It was reported that miR-138-2-3p and miR-218 in HNSCC were associated to radio-sensitivity as revealed by two independent studies [122,123].

Ahmad et al. demonstrated that the expression levels of miR-15b-5p in HNSCC patients, treated by definitive intensity-modulated radiotherapy, were significantly high for long time of locoregional control (LRC) compared to short time of LRC. Moreover, Kaplan-Meier and multivariable Cox regression analyses showed that miR-15b-5p could be a useful as a prediction marker of radiotherapy response for the disease [124].

5.2. Chemo-Resistance

Yin and co-workers identified multiple drug resistance (MDR)-related miRNAs in drug resistant LCa cells, suggesting that miRNAs can help clinicians to predict chemo-sensitivity and to design a therapeutic strategy to overcome drug resistance in LCa [125].

Likewise, miR-181a was found to inhibit Twist1 mediating EMT, enhancing metastatic potential, and inducing cisplatin chemo-resistance in TSCC cells. In this context, Twist1 was confirmed as a direct target gene of miR-181a [126].

miR-222 downregulation by antisense transfection enhanced the sensitivity of OSCC to cisplatin through directly blocking PUMA (p53-upregulated modulator of apoptosis) gene expression, and thereby offering a combination of miR-222 antisense and cisplatin as a new powerful therapeutic approach [127].

miR-101-3p repressed cell proliferation and metastatic functions and induced apoptosis in a salivary gland adenoid cystic carcinoma cell (ACC) line. The authors also showed that miR-101-3p directly regulated Moloney murine leukemia virus 1 (Pim-1) oncogene and promoted the sensitivity to cisplatin in ACC cell lines [128].

Qin et al. revealed that exosomal miR-196b originated from cancer-associated fibroblasts conferred cisplatin resistance in head and neck cancer (HNC) cells through targeting CDKN1B and ING5. In addition, high expression of exosomal miR-196b in plasma was clinically correlated with poor overall survival and chemoresistance, suggesting the miRNA as a prediction biomarker and a therapeutic target for cisplatin resistance [129]. In Table 3 combinatorial studies between miRNAs and radio/chemotherapy are summarized.

Table 3. Combinatorial studies between miRNAs and chemio/radiotherapy.

miRNA	Cancer	Sample Type	Combinatorial Treatment	Effect	Ref.
miR-24	LCa	LSCC cell lines (Hep-2, AMC-HN-8)	Overexpressed miR-24 (pGCMV/miR-24) + Radiation	Enhanced radiosensitivity: Suppresses cell proliferation and induces cell apoptosis	[36]
miR-196a	HNSCC	HNC cell line (HN30)	miR-196a knockdown (miR-196a sponge plasmid) + Radiation	Enhanced radiosensitivity: Decreases cell viability	[119]
miR-138-2-3p	LCa	LCa stem cells (originated from Hep-2)	Overexpressed miR-138-2-3p (100 nM miR-138-2-3p mimic) + Radiation	Enhanced radiosensitivity: Inhibits cell proliferation, viability, invasion and induces cell apoptosis, cell cycle arrest and DNA damage	[122]
miR-218	Oral cancer	Oral cancer stem cells	Overexpressed miR-218 (pLV-miR-218) + Radiation	Enhanced radiosensitivity: Decreases cell viability	[123]
miR-181a	TSCC	DDP-resistant TSCC cell line (originated from CAL27)	Overexpressed miR-181a (miR-181a mimic) + Cisplatin	Enhanced chemosensitivity: Decreases IC_{50} value to cisplatin	[126]
miR-222	OSCC	OSCC cell line (UM1)	miR-222 knockdown (anti-miR-222) + Cisplatin	Enhanced chemosensitivity: Induces cell apoptosis by up-regulation of pro-apoptotic PUMA expression and reduces cell invasiveness and IC_{50} value to cisplatin	[127]
miR-101-3p	SGACC	SGACC cell lines (SACC-LM, SACC-83)	miR-101-3p knockdown (anti-miR-101-3p) + Cisplatin	Enhanced chemosensitivity: Inhibits cell growth and induces cell apoptosis	[128]
miR-196a	HNC	HNC cell line (CAL27)	miR-196a knockdown (anti-miR-196a) + Cisplatin	Enhanced chemosensitivity: Promotes cell apoptosis and decreases colony formation	[129]

LSCC: Laryngeal squamous cell carcinoma; HNSCC: Head and neck squamous cell carcinoma; LCa: Laryngeal cancer; OSCC: Oral squamous cell cancer; SACC: Salivary adenoid cystic carcinoma; TSCC: Tongue squamous cell carcinoma SGACC: Salivary gland adenoid cystic carcinoma; HNC: head and neck cancer; DDP: cisplatin.

5.3. Thermal-Resistance

Hyperthermia is a potential therapeutic regimen for treating various cancers by damaging and killing tumor cells induced by high temperature heating and is generally employed as combination therapy with an established method, especially radiation and/or chemotherapeutic approach. A major advantage of hyperthermal treatment is tolerable for patients due to minimal damage to normal tissues and very few or no severe toxic effects.

A unique study showed an interesting association between miRNAs and thermal resistance in OSCC. Comparing miRNA levels in thermal-sensitive OSCC cells against resistant cells, 5 miRNAs (downregulated miR-23a and miR-27a, upregulated miR-30a, miR-30c, and miR-203) were differentially regulated. Notably, reintroduction of miR-27a in resistant cells significantly accelerated hyperthermia-induced cell death and inhibited HSP90 and HSP110 expressions, implying that miR-27a was positively associated with thermal sensitivity through HSPs modulation [130].

miR-218 regulation can be also associated to thermal-chemotherapy in gastric cancer [131].

These studies are informative for a further understanding to control thermal treatment combined with the conventional remedies. Furthermore, this relatively new research field, which improves the efficacy of hyperthermal therapy through simultaneously targeting or replacing miRNAs as potential adjuvant therapy, may progress rapidly and deeply as an emerging attractive therapeutic approach in the future.

As summarized in Table 4, we reviewed previous papers studying miRNAs in thermo/radio/chemo-sensitivity in HNSCC.

Table 4. Relationship between miRNA deregulation and resistance in HNSCC.

miRNA	Regulation	Cancer	Resistance	Target Gene	Ref.
let-7e	Down	HNSCC	Radiation	-	[120]
miR-101-3p	Down	SGACC	Chemotherapy	*Pim-1*	[128]
miR-1254	Up	HNSCC	Radiation	-	[120]
miR-138-2-3p	Up	LCa	Radiation	-	[122]
miR-150	Up	HNSCC	Radiation	-	[120]
miR-16	Up	HNSCC	Radiation	-	[120]
miR-181a	Down	TSCC	Chemotherapy	*Twist1*	[126]
miR-183-star	Up	LCa	Radiation	-	[110]
miR-196a	Up	HNSCC	Radiation	*Annexin A1*	[119]
miR-196b	Up	HNC	Chemotherapy	*CDKN1B, ING5*	[129]
miR-200c	Up	LCa	Radiation	-	[110]
miR-203	Down	LCa	Radiation	-	[121]
miR-203	Down	OSCC	Thermotherapy	-	[130]
miR-210	Up	LCa	Chemotherapy	-	[125]
miR-218	-	OSCC	Radiation	*Bim1*	[123]
miR-222	Down	OSCC	Chemotherapy	*PUMA*	[127]
miR-23a	Up	OSCC	Thermotherapy	-	[130]
miR-24	Down	LCa	Radiation	*XIAP*	[36]
miR-25-star	Down	LCa	Chemotherapy	-	[125]
miR-27a	Up	OSCC	Thermotherapy	-	[130]
miR-296-5p	Up	LCa	Radiation	-	[110]
miR-29b	Up	HNSCC	Radiation	-	[120]
miR-30a	Down	OSCC	Thermotherapy	-	[130]
miR-30c	Down	OSCC	Thermotherapy	-	[130]
miR-424	Down	LCa	Chemotherapy	-	[125]
miR-452	Up	LCa	Radiation	-	[110]
miR-494	Down	LCa	Chemotherapy	-	[125]
miR-923	Up	LCa	Chemotherapy	-	[125]
miR-93	Down	LCa	Chemotherapy	-	[125]
miR-93-star	Down	LCa	Chemotherapy	-	[125]

6. Infection and Life Habit Related miRNAs

6.1. HPV Infection

Human papillomavirus (HPV) has been known as the main etiological cause of cervical cancer. HPV infection has also negative impacts on HNSCC patients. Among HNSCCs, it has been known that oropharyngeal carcinoma is more commonly related to HPV status. HPV-positive head and neck cancer is classified into a distinct category in comparison to the typical HNSCC because of its different histopathological and clinical parameters. However, HNSCC with the infection is still poorly characterized despite accumulating knowledge. Moreover, the virus infection alters normal miRNA expression patterns. The identification of HPV-related miRNA deregulation gives us the informative knowledge to more clearly understand the changes of underlying biological and pathological molecular processes induced by the virus and to improve clinical management and prognostic outcome. Herein we focused on the effect of HPV status on miRNA deregulation patterns in HNSCC [132–138].

Lajer et al. described that miRNA expression patterns were associated to HPV status in HNSCC patients. Their study showed that miR-15a/miR-16/miR-195/miR-497 family, miR-143/miR-145, and miR-106-363 cluster may play crucial roles in the pathogenesis of HPV [133].

Gao et al. demonstrated that the deregulation of 6 miRNAs was associated to cancer survival in oropharyngeal SCC patients. Three miRNAs (miR-142-3p, miR-146a, and miR-26b) were upregulated in surviving patients, while the remaining miRNAs (miR-31, miR-24, and miR-193b) were upregulated in patients who died. Moreover, 5 HPV-associated miRNAs (miR-9, miR-223, miR-31, miR-18a, and miR-155) signatures were identified in this investigation [135]. More recently, a combination of 5 miRNAs (let-7g-3p, miR-6508-5p, miR-210-5p, miR-4306, and miR-7161-39) was also shown to correlate with the survival rate of HPV-negative HNSCC patients [136].

Bersani et al., showed that overexpression of miR-155 in tonsillar and base of tongue cancer (TSCC/BOTSCC) was associated with HPV positivity and improved survival, while low miR-185 expression associated with HPV negativity and decreased survival for the disease. ROC curve analysis with combination of these miRNAs exhibited prognostic ability for survival with AUC of 0.71, suggesting their usefulness as prognostic markers and therapeutic targets [138]. However, reports about the role of miRNAs in the other HNSCCs, particularly, in the complex scenario of HPV-related OPSCCs are still limited.

6.2. Smoking Tobacco and Alcoholic Beverage

As described above, smoking tobacco products and consuming excess alcoholic drinks negatively influence miRNA expression, leading to their deregulation, and subsequently promote the pathogenesis and progression of tumors through the regulation of many signaling pathways. The study of miRNA aberration by the use of tobacco or alcohol is mainly focused on and reported in oral cavity, trachea, and either lung or liver regions [139,140]. In this session, we described recent reports investigating a relationship between miRNAs and the life habit.

Manikandan et al. revealed that high miR-155 levels in OSCC patients are associated with the habit of tobacco/betel quid chewing [141].

To investigate the negative impact on tobacco constituents in oral fibroblasts, Pal et al. studied abnormal miRNA alterations using oral fibroblast models exposed to cigarette smoking. As a result, miR-145 downregulation in response to smoking exposure was observed in the model [142]. This result is supported by a manuscript which exhibited a decrease of miR-145 expression in lungs of a mouse model exposed to cigarette smoke [143].

A previous report showed that miR-133a-3p was underexpressed in oropharyngeal squamous cell carcinoma tissues originated from HPV(+) smoker patients in comparison with that ones from HPV(+) non-smokers. Moreover, the downregulation of miR-133a-3p was also confirmed both in serum and metastatic lymph nodes in the same study. Further examination revealed that reduced miR-133-3p increased both EGFR and HuR mRNA expression which may lead to HPV- associated cancer progression [144].

Gong et al. demonstrated that high expression of miR-499a was associated with lower overall survival and N stage in high tobacco exposed HNSCC patients [145].

In healthy individuals, miRNA patterns are also changed by the use of smoking [146,147]. These results suggest that smoking affects both tissue miRNAand circulating miRNA profiles, resulting in pathological events.

Normal oral keratinocytes were treated with biologically relevant doses of ethanol and acetaldehyde to examine which miRNAs were related to alcohol intake. RNA-sequencing analysis identified significant upregulation of eight miRNAs in alcohol-associated HNSCC. Among them, miR-30a and miR-934 were the most significantly overexpressed miRNAs and were also confirmed by qRT-PCR. On the basis of these results, alcohol components induced strong miR-30a and miR-934 upregulation, which may cause alcohol-associated oncological events [148].

Members of miR-34 family, miR-34a and miR-34c-5p, are also clinically identified as alcohol-associated miRNAs in HNSCC [74,149].

miR-183 overexpression was associated to high alcohol intake in tongue cancer patients [150].

A comprehensive study showed the deregulation of 4 miRNAs (miR-101, 181b, miR-486, and miR-1301) in epithelial cells (HaCaT and OKF4) exposed to cigarette treatment, indicating their involvement in smoking-related HNSCC development [151].

Collectively, these miRNAs represent a new evidence to better understand molecular biology of HNSCC and its correlation with lifestyle (Table 5).

Table 5. HPV, smoking, or alcohol related miRNA deregulation.

miRNA	Regulation	Region	Infection or Habit	Ref.
miR-101	Up	HNSCC	Alcohol	[148]
miR-107	Down	Oropharyngeal SCC	HPV	[134]
miR-1266	Up	HNSCC	Alcohol	[148]
miR-133a-3p	Down	Oropharyngeal SCC	Smoking	[144]
miR-139-5p	Down	HNSCC	HPV	[133]
miR-142-5p	Down	HNSCC	HPV	[132]
miR-143	Down	HNSCC	HPV	[133]
miR-145	Down	HNSCC	HPV	[133]
miR-145	Down	Oral Fibroblast	Smoking	[142]
miR-155	Down	HNSCC	HPV	[132]
miR-155	Up	Oropharyngeal	HPV	[135]
miR-155	Up	OSCC	Smoking	[141]
miR-155	Up	TSCC/BOTSCC	HPV	[138]
miR-15a	Up	HNSCC	HPV	[133]
miR-16	Up	HNSCC	HPV	[133]
miR-181a	Down	HNSCC	HPV	[132]
miR-181b	Down	HNSCC	HPV	[132]
miR-183	Up	Tongue Cancer	Alcohol	[150]
miR-18a	Down	Oropharyngeal SCC	HPV	[135]
miR-195	Down	HNSCC	HPV	[133]
miR-218	Down	HNSCC	HPV	[132]
miR-221	Down	HNSCC	HPV	[132]
miR-222	Down	HNSCC	HPV	[132]
miR-223	Down	Oropharyngeal SCC	HPV	[135]
miR-29a	Down	HNSCC	HPV	[132]
miR-30a	Up	HNSCC	Alcohol	[148]
miR-31	Down	Oropharyngeal SCC	HPV	[135]
miR-3164	Up	HNSCC	Alcohol	[148]
miR-3178	Up	HNSCC	Alcohol	[148]
miR-324-5p	Down	Oropharyngeal SCC	HPV	[134]
miR-33	Up	HNSCC	HPV	[132]
miR-34a	Up	OSCC	Alcohol	[74]
miR-34c-5p	Up	Laryngeal Epithelial Premalignant Lesions	Alcohol	[149]
miR-363	Up	HNSCC	HPV	[132]
miR-3690	Up	HNSCC	Alcohol	[148]
miR-381	Down	HNSCC	HPV	[133]
miR-497	Up	HNSCC	HPV	[132]
miR-497	Down	HNSCC	HPV	[133]
miR-499a	Up	HNSCC	Smoking	[145]
miR-574-3p	Down	HNSCC	HPV	[133]
miR-675	Up	HNSCC	Alcohol	[148]
miR-9	Up	Oropharyngeal SCC	HPV	[135]
miR-9	Up	OSCC, Oropharyngeal SCC	HPV	[137]
miR-934	Up	HNSCC	Alcohol	[148]

7. Conclusions and Perspectives

MicroRNAs have recently emerged as great potentials for both biomarkers and therapeutic targets of human diseases including cancers. In this review, we briefly summarized biological roles, miRNA aberration, and relationship between microRNA deregulation and clinical relevance in HNSCC. Several previous studies have found deregulation of miRNAs expression in different types of HNSCC and explored their use as potential biomarkers for cancer detection and/or prognosis. Single and combinatorial miRNA in silico analysis revealed that miRNAs dysregulated targeted genes and pathways that are involved in cancers [71,152].

Although microarray-based screening assay has a limitation, which is that the evaluable number of miRNAs is restricted to the immobilized probes for miRNA sequences on the platform and is impossible to discover a novel miRNA, the technique has been widely used so far for the assessment of deregulated miRNAs. Initial microarray and subsequent validation analysis is currently the gold standard to determine miRNA expression pattern. Several well-studied miRNAs, such as miR-1, miR-21, and miR-34, were also dramatically up/down deregulated in HNSCC, thereby clearly showing that these miRNAs are importantly responsible for cancer aggressiveness or repression through their target genes. Recent researchers have also unveiled the stable presence of circulating miRNAs in liquid biopsies taken from HNSCC patients. Findings about circulating extracellular miRNAs in systemic circulation are gradually and steady accumulated, and these miRNAs have remarkable potential as minimally invasive molecule-based markers for the prediction of node metastases and response to therapy, diagnosis, and prognosis of HNSCC. In this light, our group performed a multicentric study to identify a specific miRNA expression profile for laryngeal cancer. We found 20 miRNAs specific for laryngeal cancer and a tissue-specific miRNA signature that consists of 11 miRNAs, seven of which are upregulated and four downregulated, predictive of lymph node metastases [153]. These results are enormously innovative, at least in our opinion, and lead to the definition of a group of potential specific biomarkers for LCa that will allow to improve its early diagnosis and to identify patients with minimal residual disease or recurrence; moreover, they can be used to predict prognosis in patients that show the specific miRNA signature suggestive for nodal involvement. In this case, the miRNAs will be useful to select a tailored treatment.

This review also described a correlation between resistance to cancer therapies and change of miRNA levels. These data are useful for the prediction of the response to radio/chemotherapy and help clinicians to select the best therapeutic approaches for patients.

Lifestyle habits, such as consuming smoke products and excess alcohol, may promote HNSCC through disruption of normal miRNA regulation, resulting in poor prognosis. Accumulating evidences may provide a deeper understanding of pathological mechanisms in cancers and advise us to reconsider our lifestyle in order to maintain healthy life.

Despite advances in the research of miRNAs, there are some controversies deciphering the biological function of a miRNA, which is either tumor suppressive or oncogenic action even in the same category of cancer. One of the major reasons is tumor heterogeneity, largely contributing to the complication of interpretation mainly due to different gene expression levels in each cancer cell. The discrepant roles of miRNAs in HNSCC revealed by earlier studies could be explained by the differences of pre-analytical factors, sample collection methods, storage conditions, RNA extraction, platforms for the examination, the number of evaluated samples, selection of the control for data normalization, and the used statistical analysis. In addition, blood-derived miRNAs released by tumor cells and their deregulation patterns are not frequently consistent with those in (tumor) tissues. This could be due to the presence of the contaminated miRNAs by other components derived from circulating normal cells, lysed or apoptotic cells, or other malignant tissues, across the tested liquid samples. In fact, it is difficult to correctly clarify the origin of circulating miRNAs. Hemolysis and above-described both clinical and experimental variations also influence circulating miRNA profiles, eventually giving us conflicting findings. Therefore, establishment of standardized protocols, which eliminates variability across each sample and is commonly usable in laboratories and clinics, is essential for the study of miRNAs for diagnostic and therapeutic purposes prior to the use of clinical settings. Recent reports suggest that multiple miRNA-based profiles have higher diagnostic, prognostic and therapeutic performance as well as higher sensitivity than individual miRNA assays because the combination of different miRNAs, that regulate multiple target genes, can better explain how each of them contributes to carcinogenesis and better represent the comprehensive biological effect of miRNA regulation in the multistep process leading to cancer. Furthermore, miRNA signatures, consisting of a plurality of different miRNAs, allow to better distinguish between different pathologies while single miRNAs alone are frequently not disease-specific. On the other hand, single miRNA markers are often verified by independent studies, miRNA signatures are less frequently validated. With an increasing number of validated miRNA signatures and with the advance of matured high-throughput approaches in clinical settings, specific miRNA markers are likely to contribute to human healthcare. All of these reported miRNAs have a high possibility to successfully reach the clinical setting as either diagnostic tools or therapeutic targets of these malignancies. We believe that miRNAs will be readily available as robust applications for HNSCC patients in the next future.

Author Contributions: T.T., H.K. selected studies, collected, interpreted data and prepared the manuscript; A.L., A.M.C., M.S., M.B. and G.M. implemented the manuscript; S.Z. and F.R. revised and edited the manuscript; M.C. critically revised the manuscript and finally approved the manuscript. All authors have read and agreed to the published version of the manuscript.

Funding: This research received no external funding.

Conflicts of Interest: All the authors declared no conflict of interest.

Abbreviations

miRNA	MicroRNA
HNSCC	Head and neck squamous cell carcinoma
HNC	Head and neck cancer
LCa	Laryngeal cancer
qRT-PCR	Quantitative real-time PCR
RISC	RNA-induced silencing complex
OSCC	Oral squamous cell cancer
SACC	Salivary adenoid cystic carcinoma
MSSCC	Maxillary sinus squamous cell carcinoma
EMT	Epithelial-mesenchymal transition
OTSCC	Oral tongue squamous cell carcinoma
NPC	Nasopharyngeal carcinoma
TSCC	Tongue squamous cell carcinoma
SN-SCC	Sinonasal squamous cell carcinomas
HSCC	Hypopharyngeal squamous cell carcinoma
EBV	Epstein–Barr virus
LRC	Locoregional control
MDR	Multiple drug resistance
ACC	Adenoid cystic carcinoma
Pim-1	Moloney murine leukemia virus 1
AUC	Area under curve
PUMA	p53-upregulated modulator of apoptosis
ROC	Receiver operating characteristic
HPV	Human papillomavirus
SCC	Squamous cell carcinoma
TSCC/BOTSCC	Tonsillar and base of tongue cancer

References

1. Jemal, A.; Siegel, R.; Ward, E.; Hao, Y.; Xu, J.; Murray, T.; Thun, M.J. Cancer statistics, 2008. *CA Cancer J. Clin.* **2008**, *58*, 71–96. [CrossRef]
2. Leemans, C.R.; Braakhuis, B.J.M.; Brakenhoff, R.H. Response to correspondence on the molecular biology of head and neck cancer. *Nat. Rev. Cancer* **2011**, *11*, 382. [CrossRef]
3. Kulasinghe, A.; Perry, C.; Jovanović, L.; Nelson, C.; Punyadeera, C. Circulating tumour cells in metastatic head and neck cancers. *Int. J. Cancer* **2014**, *136*, 2515–2523. [CrossRef] [PubMed]
4. Babu, J.M.; Prathibha, R.; Jijith, V.; Hariharan, R.; Pillai, M.R. A miR-centric view of head and neck cancers. *Biochim. Biophys. Acta Rev. Cancer* **2011**, *1816*, 67–72. [CrossRef]
5. Janiszewska, J.; Szaumkessel, M.; Szyfter-Harris, J. MicroRNAs are important players in head and neck carcinoma: A review. *Crit. Rev. Oncol.* **2013**, *88*, 716–728. [CrossRef] [PubMed]

6. Chen, D.; Cabay, R.J.; Jin, Y.; Wang, A.; Lu, Y.; Shah-Khan, M.; Zhou, X. MicroRNA Deregulations in Head and Neck Squamous Cell Carcinomas. *J. Oral Maxillofac. Res.* **2013**, *4*, 2. [CrossRef] [PubMed]
7. Bartel, B. MicroRNAs: Genomics, biogenesis, mechanism, and function. *Cell* **2004**, *116*, 281–297. [CrossRef]
8. Kawasaki, H.; Zarone, M.R.; Lombardi, A.; Ricciardiello, F.; Caraglia, M.; Misso, G. Early detection of laryngeal cancer: Prominence of miRNA signature as a new tool for clinicians. *Transl. Med. Rep.* **2017**, *1*, 16502. [CrossRef]
9. Misso, G.; Zarone, M.R.; Grimaldi, A.; di Martino, M.T.; Lombardi, A.; Kawasaki, H.; Stiuso, P.; Tassone, P.; Tagliaferri, P.; Caraglia, M. Non Coding RNAs: A New Avenue for the Self-Tailoring of Blood Cancer Treatment. *Curr. Drug Targets* **2016**, *18*, 35–55. [CrossRef] [PubMed]
10. Syeda, Z.A.; Langden, S.; Munkhzul, C.; Lee, M.; Song, S. Regulatory Mechanism of MicroRNA Expression in Cancer. *Int. J. Mol. Sci.* **2020**, *21*, 1723. [CrossRef]
11. Mitchell, P.S.; Parkin, R.K.; Kroh, E.M.; Fritz, B.R.; Wyman, S.K.; Pogosova-Agadjanyan, E.L.; Peterson, A.; Noteboom, J.; O'Briant, K.C.; Allen, A.; et al. Circulating microRNAs as stable blood-based markers for cancer detection. *Proc. Natl. Acad. Sci. USA* **2008**, *105*, 10513–10518. [CrossRef] [PubMed]
12. Weber, J.A.; Baxter, D.H.; Zhang, S.; Huang, D.Y.; Huang, K.-H.; Lee, M.-J.; Galas, D.J.; Wang, K. The MicroRNA Spectrum in 12 Body Fluids. *Clin. Chem.* **2010**, *56*, 1733–1741. [CrossRef] [PubMed]
13. Valadi, H.; Ekström, K.; Bossios, A.; Sjöstrand, M.; Lee, J.J.; Lötvall, J. Exosome-mediated transfer of mRNAs and microRNAs is a novel mechanism of genetic exchange between cells. *Nature* **2007**, *9*, 654–659. [CrossRef] [PubMed]
14. Zernecke, A.; Bidzhekov, K.; Noels, H.; Shagdarsuren, E.; Gan, L.; Denecke, B.; Hristov, M.; Köppel, T.; Nazari-Jahantigh, M.; Lutgens, E.; et al. Delivery of MicroRNA-126 by Apoptotic Bodies Induces CXCL12-Dependent Vascular Protection. *Sci. Signal.* **2009**, *2*, 81. [CrossRef]
15. Arroyo, J.; Chevillet, J.; Kroh, E.M.; Ruf, I.K.; Pritchard, C.C.; Gibson, D.F.; Mitchell, P.S.; Bennett, C.F.; Pogosova-Agadjanyan, E.L.; Stirewalt, D.L.; et al. Argonaute2 complexes carry a population of circulating microRNAs independent of vesicles in human plasma. *Proc. Natl. Acad. Sci. USA* **2011**, *108*, 5003–5008. [CrossRef]
16. Vickers, K.C.; Palmisano, B.T.; Shoucri, B.M.; Shamburek, R.D.; Remaley, A.T. MicroRNAs are transported in plasma and delivered to recipient cells by high-density lipoproteins. *Nature* **2011**, *13*, 423–433. [CrossRef]
17. Turchinovich, A.; Weiz, L.; Langheinz, A.; Burwinkel, B. Characterization of extracellular circulating microRNA. *Nucleic Acids Res.* **2011**, *39*, 7223–7233. [CrossRef]
18. Guénel, P.; Chastang, J.F.; Luce, D.; Leclerc, A.; Brugere, J. A study of the interaction of alcohol drinking and tobacco smoking among French cases of laryngeal cancer. *J. Epidemiol. Community Health* **1988**, *42*, 350–354. [CrossRef]
19. Falk, R.T.; Pickle, L.W.; Brown, L.M.; Mason, T.J.; A Buffler, P.; Fraumeni, J.F. Effect of smoking and alcohol consumption on laryngeal cancer risk in coastal Texas. *Cancer Res.* **1989**, *49*, 4024–4029.
20. la Vecchia, C.; Negri, E.; D'Avanzo, B.; Franceschi, S.; de Carli, A.; Boyle, P. Dietary indicators of laryngeal cancer risk. *Cancer Res.* **1990**, *50*, 4497–4500.
21. de Miguel-Luken, M.J.; Chaves-Conde, M.; Carnero, A. A genetic view of laryngeal cancer heterogeneity. *Cell Cycle* **2016**, *15*, 1202–1212. [CrossRef] [PubMed]
22. Sessions, N.G. Surgical pathology of cancer of the larynx and hypopharynx. *Laryngoscope* **1976**, *86*, 814–839. [CrossRef] [PubMed]
23. Lu, E.; Su, J.; Zeng, W.; Zhang, C. Enhanced miR-9 promotes laryngocarcinoma cell survival via down-regulating PTEN. *Biomed. Pharmacother.* **2016**, *84*, 608–613. [CrossRef] [PubMed]
24. Zhang, L.; Sun, J.; Wang, B.; Ren, J.-C.; Su, W.; Zhang, T. MicroRNA-10b Triggers the Epithelial–Mesenchymal Transition (EMT) of Laryngeal Carcinoma Hep-2 Cells by Directly Targeting the E-cadherin. *Appl. Biochem. Biotechnol.* **2015**, *176*, 33–44. [CrossRef]
25. Liu, M.; Wu, H.; Liu, T.; Li, Y.; Wang, F.; Wan, H.-Y.; Li, X.; Tang, H. Regulation of the cell cycle gene, BTG2, by miR-21 in human laryngeal carcinoma. *Cell Res.* **2009**, *19*, 828–837. [CrossRef]
26. Zheng, J.; Xue, H.; Wang, T.; Jiang, Y.; Liu, B.; Li, J.; Liu, Y.; Wang, W.; Zhang, B.; Sun, M. miR-21 downregulates the tumor suppressor P12CDK2AP1 and Stimulates Cell Proliferation and Invasion. *J. Cell. Biochem.* **2011**, *112*, 872–880. [CrossRef]
27. Zhong, G.; Xiong, X. miR-205 promotes proliferation and invasion of laryngeal squamous cell carcinoma by suppressing CDK2AP1 expression. *Boil. Res.* **2015**, *48*, 60. [CrossRef]

28. Li, G.; Ren, S.; Su, Z.; Liu, C.; Deng, T.; Huang, N.; Tian, Y.; Qiu, Y.; Liu, Y. Increased expression of miR-93 is associated with poor prognosis in head and neck squamous cell carcinoma. *Tumor Boil.* **2015**, *36*, 3949–3956. [CrossRef]

29. Xiao, X.; Zhou, L.; Cao, P.; Gong, H.; Zhang, Y. MicroRNA-93 regulates cyclin G2 expression and plays an oncogenic role in laryngeal squamous cell carcinoma. *Int. J. Oncol.* **2014**, *46*, 161–174. [CrossRef]

30. Sun, X.; Wang, Z.; Song, Y.; Tai, X.-H.; Ji, W.-Y.; Gu, H. MicroRNA-126 modulates the tumor microenvironment by targeting calmodulin-regulated spectrin-associated protein 1 (Camsap1). *Int. J. Oncol.* **2014**, *44*, 1678–1684. [CrossRef]

31. Lian, R.; Lu, B.; Jiao, L.; Li, S.; Wang, H.; Miao, W.; Yu, W. MiR-132 plays an oncogenic role in laryngeal squamous cell carcinoma by targeting FOXO1 and activating the PI3K/AKT pathway. *Eur. J. Pharmacol.* **2016**, *792*, 1–6. [CrossRef] [PubMed]

32. Sun, X.; Liu, B.; Zhao, X.-D.; Wang, L.-Y.; Ji, W. MicroRNA-221 accelerates the proliferation of laryngeal cancer cell line Hep-2 by suppressing Apaf-1. *Oncol. Rep.* **2015**, *33*, 1221–1226. [CrossRef] [PubMed]

33. Lu, Y.; Gao, W.; Zhang, C.; Wen, S.; Huangfu, H.; Kang, J.; Wang, B. Hsa-miR-301a-3p Acts as an Oncogene in Laryngeal Squamous Cell Carcinoma via Target Regulation of Smad4. *J. Cancer* **2015**, *6*, 1260–1275. [CrossRef] [PubMed]

34. Guan, G.; Zhang, D.; Zheng, Y.; Wen, L.; Yu, D.; Lu, Y.; Zhao, Y. microRNA-423-3p promotes tumor progression via modulation of AdipoR2 in laryngeal carcinoma. *Int. J. Clin. Exp. Pathol.* **2014**, *7*, 5683–5691.

35. Wang, F.; Song, G.; Liu, M.; Li, X.; Tang, H. miRNA-1 targets fibronectin1 and suppresses the migration and invasion of the HEp2 laryngeal squamous carcinoma cell line. *FEBS Lett.* **2011**, *585*, 3263–3269. [CrossRef]

36. Xu, L.; Chen, Z.; Xue, F.; Chen, W.; Ma, R.; Cheng, S.; Cui, P. MicroRNA-24 inhibits growth, induces apoptosis, and reverses radioresistance in laryngeal squamous cell carcinoma by targeting X-linked inhibitor of apoptosis protein. *Cancer Cell Int.* **2015**, *15*, 61. [CrossRef]

37. Ye, J.; Li, L.; Feng, P.; Wan, J.; Li, J. Downregulation of miR-34a contributes to the proliferation and migration of laryngeal carcinoma cells by targeting cyclin D1. *Oncol. Rep.* **2016**, *36*, 390–398. [CrossRef]

38. Maroof, H.; Salajegheh, A.; Smith, R.A.; Lam, A.K.-Y. MicroRNA-34 family, mechanisms of action in cancer: a review. *Curr. Cancer Drug Targets* **2014**, *14*, 737–751. [CrossRef]

39. Cai, K.-M.; Bao, X.-L.; Kong, X.-H.; Jinag, W.; Mao, M.-R.; Chu, J.-S.; Huang, Y.-J.; Zhao, X.-J. Hsa-miR-34c suppresses growth and invasion of human laryngeal carcinoma cells via targeting c-Met. *Int. J. Mol. Med.* **2010**, *25*, 565–571. [CrossRef]

40. Li, W.; Ma, H.; Sun, J. microRNA-34a/c function as tumor suppressors in Hep-2 laryngeal carcinoma cells and may reduce GALNT7 expression. *Mol. Med. Rep.* **2014**, *9*, 1293–1298. [CrossRef]

41. Wu, X.; Cui, C.-L.; Chen, W.-L.; Fu, Z.-Y.; Cui, X.-Y.; Gong, X. MiR-144 suppresses the growth and metastasis of laryngeal squamous cell carcinoma by targeting IRS1. *Am. J. Transl. Res.* **2016**, *8*, 1–11. [PubMed]

42. Yu, W.; Zhang, G.; Lu, B.; Li, J.; Wu, Z.; Ma, H.; Wang, H.; Lian, R. MiR-340 impedes the progression of laryngeal squamous cell carcinoma by targeting EZH2. *Gene* **2016**, *577*, 193–201. [CrossRef] [PubMed]

43. Song, F.-C.; Yang, Y.; Liu, J.-X. Expression, and significances of MiRNA Let-7 and HMGA2 in laryngeal carcinoma. *Eur. Rev. Med. Pharmacol. Sci.* **2016**, *20*, 4452–4458. [PubMed]

44. Lu, Y.-C.; Chen, Y.-J.; Wang, H.-M.; Tsai, C.-Y.; Chen, W.-H.; Huang, Y.-C.; Fan, K.-H.; Tsai, C.-N.; Huang, S.-F.; Kang, C.-J.; et al. Oncogenic Function and Early Detection Potential of miRNA-10b in Oral Cancer as Identified by microRNA Profiling. *Cancer Prev. Res.* **2012**, *5*, 665–674. [CrossRef] [PubMed]

45. Severino, P.; Brüggemann, H.; Andreghetto, F.M.; Camps, C.; Klingbeil, M.D.F.G.; Pereira, W.D.O.; Soares, R.M.; Moyses, R.; Filho, V.W.; Mathor, M.B.; et al. MicroRNA expression profile in head and neck cancer: HOX-cluster embedded microRNA-196a and microRNA-10b dysregulation implicated in cell proliferation. *BMC Cancer* **2013**, *13*, 533. [CrossRef] [PubMed]

46. Li, G.; Wu, Z.; Peng, Y.; Liu, X.; Lu, J.; Wang, L.; Pan, Q.; He, M.-L.; Li, X.-P. MicroRNA-10b induced by Epstein-Barr virus-encoded latent membrane protein-1 promotes the metastasis of human nasopharyngeal carcinoma cells. *Cancer Lett.* **2010**, *299*, 29–36. [CrossRef] [PubMed]

47. Rather, M.I.; Nagashri, M.N.; Swamy, S.S.; Gopinath, K.S.; Kumar, A. Oncogenic microRNA-155 down-regulates tumor suppressor CDC73 and promotes oral squamous cell carcinoma cell proliferation: Implications for cancer therapeutics. *J. Biol. Chem.* **2013**, *288*, 608–618. [CrossRef]

48. Ni, Y.; Huang, X.-F.; Wang, Z.-Y.; Han, W.; Deng, R.-Z.; Mou, Y.-B.; Ding, L.; Hou, Y.; Hu, Q.-G. Upregulation of a potential prognostic biomarker, miR-155, enhances cell proliferation in patients with oral squamous cell carcinoma. *Oral Surg. Oral Med. Oral Pathol. Oral Radiol.* **2014**, *117*, 227–233. [CrossRef]

49. Baba, O.; Hasegawa, S.; Nagai, H.; Uchida, F.; Yamatoji, M.; Kanno, N.I.; Yamagata, K.; Sakai, S.; Yanagawa, T.; Bukawa, H. MicroRNA-155-5p is associated with oral squamous cell carcinoma metastasis and poor prognosis. *J. Oral Pathol. Med.* **2015**, *45*, 248–255. [CrossRef]

50. Lerner, C.; Wemmert, S.; Bochen, F.; Kulas, P.; Linxweiler, M.; Hasenfus, A.; Heinzelmann, J.; Leidinger, P.; Backes, C.; Meese, E.; et al. Characterization of miR-146a and miR-155 in blood, tissue and cell lines of head and neck squamous cell carcinoma patients and their impact on cell proliferation and migration. *J. Cancer Res. Clinic. Oncol.* **2015**, *142*, 757–766. [CrossRef]

51. Yang, C.-C.; Hung, P.-S.; Wang, P.-W.; Liu, C.-J.; Chu, T.-H.; Cheng, H.-W.; Lin, S.-C. miR-181 as a putative biomarker for lymph-node metastasis of oral squamous cell carcinoma. *J. Oral Pathol. Med.* **2011**, *40*, 397–404. [CrossRef] [PubMed]

52. Shin, K.-H.; Bae, S.D.; Hong, H.S.; Kim, R.H.; Kang, M.K.; Park, N.-H. miR-181a shows tumor suppressive effect against oral squamous cell carcinoma cells by downregulating K-ras. *Biochem. Biophys. Res. Commun.* **2011**, *404*, 896–902. [CrossRef] [PubMed]

53. He, Q.; Zhou, X.; Li, S.; Jin, Y.; Chen, Z.; Chen, D.; Cai, Y.; Liu, Z.; Zhao, T.; Wang, A. MicroRNA-181a suppresses salivary adenoid cystic carcinoma metastasis by targeting MAPK–Snai2 pathway. *Biochim. Biophys. Acta Gen. Subj.* **2013**, *1830*, 5258–5266. [CrossRef] [PubMed]

54. de Carvalho, A.C.; Scapulatempo-Neto, C.; Maia, D.C.; Evangelista, A.F.; Morini, M.A.; Carvalho, A.L.; Vettore, A.L. Accuracy of microRNAs as markers for the detection of neck lymph node metastases in patients with head and neck squamous cell carcinoma. *BMC Med.* **2015**, *13*, 108. [CrossRef]

55. Arriagada, W.G.; Olivero, P.; Rodríguez, B.; Lozano-Burgos, C.; de Oliveira, C.E.; Coletta, R.D. Clinicopathological significance of miR-26, miR-107, miR-125b, and miR-203 in head and neck carcinomas. *Oral Dis.* **2018**, *24*, 930–939. [CrossRef]

56. Obayashi, M.; Yoshida, M.; Tsunematsu, T.; Ogawa, I.; Sasahira, T.; Kuniyasu, H.; Imoto, I.; Abiko, Y.; Xu, D.; Fukunaga, S.; et al. microRNA-203 suppresses invasion and epithelial-mesenchymal transition induction via targeting NUAK1 in head and neck cancer. *Oncotarget* **2016**, *7*, 8223–8239. [CrossRef]

57. Kim, J.-S.; Choi, D.W.; Kim, C.S.; Yu, S.-K.; Kim, H.-J.; Go, D.-S.; Lee, S.; Moon, S.M.; Kim, S.-G.; Chun, H.S.; et al. MicroRNA-203 Induces Apoptosis by Targeting Bmi-1 in YD-38 Oral Cancer Cells. *Anticancer Res.* **2018**, *38*, 3477–3485. [CrossRef]

58. Yang, C.-J.; Shen, W.G.; Liu, C.-J.; Chen, Y.-W.; Lu, H.-H.; Tsai, M.-M.; Lin, S.-C. miR-221 and miR-222 expression increased the growth and tumorigenesis of oral carcinoma cells. *J. Oral Pathol. Med.* **2011**, *40*, 560–566. [CrossRef]

59. Jiang, F.; Zhao, W.; Zhou, L.; Zhang, L.; Liu, Z.; Yu, D. miR-222 regulates the cell biological behavior of oral squamous cell carcinoma by targeting PUMA. *Oncol. Rep.* **2014**, *31*, 1255–1262. [CrossRef]

60. Lopes, C.B.; Magalhães, L.; Teófilo, C.R.; Alves, A.P.N.N.; Montenegro, R.C.; Negrini, M.; Ribeiro-Dos-Santos, A. Differential expression of hsa-miR-221, hsa-miR-21, hsa-miR-135b, and hsa-miR-29c suggests a field effect in oral cancer. *BMC Cancer* **2018**, *18*, 721. [CrossRef]

61. Liu, X.; Yu, J.; Jiang, L.; Wang, A.; Shi, F.; Ye, H.; Zhou, X. MicroRNA-222 regulates cell invasion by targeting matrix metalloproteinase 1 (MMP1) and manganese superoxide dismutase 2 (SOD2) in tongue squamous cell carcinoma cell lines. *Cancer Genom. Proteom.* **2009**, *6*, 131–139.

62. Tachibana, H.; Sho, R.; Takeda, Y.; Zhang, X.; Yoshida, Y.; Narimatsu, H.; Otani, K.; Ishikawa, S.; Fukao, A.; Asao, H.; et al. Circulating miR-223 in Oral Cancer: Its Potential as a Novel Diagnostic Biomarker and Therapeutic Target. *PLoS ONE* **2016**, *11*, 159693. [CrossRef] [PubMed]

63. Jiang, L.; Lv, L.; Liu, X.; Jiang, X.; Yin, Q.; Hao, Y.; Xiao, L. MiR-223 promotes oral squamous cell carcinoma proliferation and migration by regulating FBXW7. *Cancer Biomark.* **2019**, *24*, 325–334. [CrossRef] [PubMed]

64. Yeh, C.-H.; Bellon, M.; Nicot, C. FBXW7: A critical tumor suppressor of human cancers. *Mol. Cancer* **2018**, *17*, 115. [CrossRef]

65. Nohata, N.; Sone, Y.; Hanazawa, T.; Fuse, M.; Kikkawa, N.; Yoshino, H.; Chiyomaru, T.; Kawakami, K.; Enokida, H.; Nakagawa, M.; et al. miR-1 as a tumor suppressive microRNA targeting TAGLN2 in head and neck squamous cell carcinoma. *Oncotarget* **2011**, *2*, 29–42. [CrossRef]

66. Nohata, N.; Hanazawa, T.; Kikkawa, N.; Sakurai, D.; Sasaki, K.; Chiyomaru, T.; Kawakami, K.; Yoshino, H.; Enokida, H.; Nakagawa, M.; et al. Identification of novel molecular targets regulated by tumor suppressive miR-1/miR-133a in maxillary sinus squamous cell carcinoma. *Int. J. Oncol.* **2011**, *39*, 1099–1107.

67. Lu, J.; Zhao, F.P.; Peng, Z.; Zhang, M.W.; Lin, S.X.; Liang, B.; Zhang, B.; Liu, X.; Wang, L.; Li, G.; et al. EZH2 promotes angiogenesis through inhibition of miR-1/Endothelin-1 axis in nasopharyngeal carcinoma. *Oncotarget* **2014**, *5*, 11319–11332. [CrossRef]

68. Peng, C.-Y.; Liao, Y.-W.; Lu, M.-Y.; Yu, C.-H.; Yu, C.-C.; Chou, M.-Y. Downregulation of miR-1 enhances tumorigenicity and invasiveness in oral squamous cell carcinomas. *J. Formos. Med Assoc.* **2017**, *116*, 782–789. [CrossRef]

69. Wang, Z.; Wang, J.; Chen, Z.; Wang, K.; Shi, L. MicroRNA-1-3p inhibits the proliferation and migration of oral squamous cell carcinoma cells by targeting DKK1. *Biochem. Cell Boil.* **2018**, *96*, 355–364. [CrossRef]

70. Koshizuka, K.; Hanazawa, T.; Fukumoto, I.; Kikkawa, N.; Matsushita, R.; Mataki, H.; Mizuno, K.; Okamoto, Y.; Seki, N. Dual-receptor (EGFR and c-MET) inhibition by tumor-suppressive miR-1 and miR-206 in head and neck squamous cell carcinoma. *J. Hum. Genet.* **2016**, *62*, 113–121. [CrossRef]

71. Fukumoto, I.; Hanazawa, T.; Kinoshita, T.; Kikkawa, N.; Koshizuka, K.; Goto, Y.; Nishikawa, R.; Chiyomaru, T.; Enokida, H.; Nakagawa, M.; et al. MicroRNA expression signature of oral squamous cell carcinoma: Functional role of microRNA-26a/b in the modulation of novel cancer pathways. *Br. J. Cancer* **2015**, *112*, 891–900. [CrossRef] [PubMed]

72. Jia, L.-F.; Wei, S.-B.; Gan, Y.-H.; Guo, Y.; Gong, K.; Mitchelson, K.; Cheng, J.; Yu, G.-Y. Expression, regulation and roles of miR-26a and MEG3 in tongue squamous cell carcinoma. *Int. J. Cancer* **2014**, *135*, 2282–2293. [CrossRef] [PubMed]

73. Wei, Z.; Chang, K.; Fan, C.; Zhang, Y. MiR-26a/miR-26b represses tongue squamous cell carcinoma progression by targeting PAK1. *Cancer Cell Int.* **2020**, *20*, 82–114. [CrossRef] [PubMed]

74. Manikandan, M.; Rao, A.K.D.M.; Arunkumar, G.; Rajkumar, K.S.; Rajaraman, R.; Munirajan, A.K. Down Regulation of miR-34a and miR-143 May Indirectly Inhibit p53 in Oral Squamous Cell Carcinoma: A Pilot Study. *Asian Pac. J. Cancer Prev.* **2015**, *16*, 7619–7625. [CrossRef]

75. Zhao, Y.; Wang, X. MiR-34a targets BCL-2 to suppress the migration and invasion of sinonasal squamous cell carcinoma. *Oncol. Lett.* **2018**, *16*, 6566–6572. [CrossRef]

76. Chen, Z.; Jin, Y.; Yu, N.; Wang, A.; Mahjabeen, I.; Wang, C.; Liu, X.; Zhou, X. Down-regulation of the microRNA-99 family members in head and neck squamous cell carcinoma. *Oral Oncol.* **2012**, *48*, 686–691. [CrossRef]

77. Jin, Y.; Tymen, S.D.; Chen, D.; Fang, Z.J.; Zhao, Y.; Dragas, D.; Dai, Y.; Marucha, P.T.; Zhou, X. MicroRNA-99 Family Targets AKT/mTOR Signaling Pathway in Dermal Wound Healing. *PLoS ONE* **2013**, *8*, 64434. [CrossRef]

78. Wang, Q.; Yan, B.; Fu, Q.; Lai, L.; Tao, X.; Fei, Y.; Shen, J.; Chen, Z. Downregulation of microRNA 99a in oral squamous cell carcinomas contributes to the growth and survival of oral cancer cells. *Mol. Med. Rep.* **2012**, *6*, 675–681. [CrossRef]

79. Kuo, Y.-Z.; Tai, Y.-H.; Lo, H.-I.; Chen, Y.-L.; Cheng, H.-C.; Fang, W.-Y.; Lin, S.-H.; Yang, C.-L.; Tsai, S.-T.; Wu, L.-W. MiR-99a exerts anti-metastasis through inhibiting myotubularin-related protein 3 expression in oral cancer. *Oral Dis.* **2013**, *20*, 65–75. [CrossRef]

80. He, K.; Tong, D.; Zhang, S.; Cai, D.; Wang, L.; Yang, Y.; Gao, L.; Chang, S.; Guo, B.; Song, T.; et al. miRNA-99b-3p functions as a potential tumor suppressor by targeting glycogen synthase kinase-3β in oral squamous cell carcinoma Tca-8113 cells. *Int. J. Oncol.* **2015**, *47*, 1528–1536. [CrossRef]

81. Banerjee, R.; Mani, R.-S.; Russo, N.; Scanlon, C.S.; Tsodikov, A.; Jing, X.; Cao, Q.; Palanisamy, N.; Metwally, T.; Inglehart, R.C.; et al. The tumor suppressor gene rap1GAP is silenced by miR-101-mediated EZH2 overexpression in invasive squamous cell carcinoma. *Oncogene* **2011**, *30*, 4339–4349. [CrossRef] [PubMed]

82. Zheng, M.; Jiang, Y.-P.; Chen, W.; Li, K.-D.; Liu, X.; Gao, S.-Y.; Feng, H.; Wang, S.-S.; Jiang, J.; Ma, X.-R.; et al. Snail and Slug collaborate on EMT and tumor metastasis through miR-101-mediated EZH2 axis in oral tongue squamous cell carcinoma. *Oncotarget* **2015**, *6*, 6794–6810. [CrossRef] [PubMed]

83. Tang, X.-R.; Wen, X.; He, Q.-M.; Li, Y.-Q.; Ren, X.-Y.; Yang, X.-J.; Zhang, J.; Wang, Y.-Q.; Ma, J.; Liu, N. MicroRNA-101 inhibits invasion and angiogenesis through targeting ITGA3 and its systemic delivery inhibits lung metastasis in nasopharyngeal carcinoma. *Cell Death Dis.* **2017**, *8*, 2566. [CrossRef] [PubMed]

84. Kai, Y.; Peng, W.; Ling, W.; Jiebing, H.; Zhuan, B. Reciprocal effects between microRNA-140-5p and ADAM10 suppress migration and invasion of human tongue cancer cells. *Biochem. Biophys. Res. Commun.* **2014**, *448*, 308–314. [CrossRef] [PubMed]

85. Jing, P.; Sa, N.; Liu, X.; Liu, X.; Xu, W. MicroR-140-5p suppresses tumor cell migration and invasion by targeting ADAM10-mediated Notch1 signaling pathway in hypopharyngeal squamous cell carcinoma. *Exp. Mol. Pathol.* **2016**, *100*, 132–138. [CrossRef] [PubMed]

86. Peng, M.; Pang, C. MicroRNA-140-5p inhibits the tumorigenesis of oral squamous cell carcinoma by targeting p21-activated kinase 4. *Cell Boil. Int.* **2019**, *44*, 145–154. [CrossRef] [PubMed]

87. Qiao, Z.; Zou, Y.; Zhao, H. MicroRNA-140-5p inhibits salivary adenoid cystic carcinoma progression and metastasis via targeting survivin. *Cancer Cell Int.* **2019**, *19*, 301–312. [CrossRef]

88. Yu, C.-C.; Chen, P.-N.; Peng, C.-Y.; Yu, C.-H.; Chou, M.-Y. Suppression of miR-204 enables oral squamous cell carcinomas to promote cancer stemness, EMT traits, and lymph node metastasis. *Oncotarget* **2016**, *7*, 20180–20192. [CrossRef]

89. Wang, X.; Li, F.; Zhou, X. miR-204-5p regulates cell proliferation and metastasis through inhibiting CXCR4 expression in OSCC. *Biomed. Pharmacother.* **2016**, *82*, 202–207. [CrossRef]

90. Ma, L.; Deng, X.; Wu, M.; Zhang, G.; Huang, J. Down-regulation of miRNA-204 by LMP-1 enhances CDC42 activity and facilitates invasion of EBV-associated nasopharyngeal carcinoma cells. *FEBS Lett.* **2014**, *588*, 1562–1570. [CrossRef]

91. Wang, C.; Zhang, Y.; Zhou, D.; Cao, G.; Wu, Y. miR-204 enhances p27 mRNA stability by targeting Brd4 in head and neck squamous cell carcinoma. *Oncol. Lett.* **2018**, *16*, 4179–4184. [CrossRef] [PubMed]

92. Libório-Kimura, T.N.; Jung, H.M.; Chan, E.K. miR-494 represses HOXA10 expression and inhibits cell proliferation in oral cancer. *Oral Oncol.* **2015**, *51*, 151–157. [CrossRef] [PubMed]

93. He, H.; Liao, X.; Yang, Q.; Liu, Y.; Peng, Y.; Zhong, H.; Yang, J.; Zhang, H.; Yu, Z.; Zuo, Y.; et al. MicroRNA-494-3p Promotes Cell Growth, Migration, and Invasion of Nasopharyngeal Carcinoma by Targeting Sox7. *Technol. Cancer Res. Treat.* **2018**, *17*, 1533033818809993. [CrossRef] [PubMed]

94. Wang, J.; Zhou, Y.; Lu, J.; Sun, Y.; Xiao, H.; Liu, M.; Tian, L. Combined detection of serum exosomal miR-21 and Hotair as diagnostic and prognostic biomarkers for laryngeal squamous cell carcinoma. *Med. Oncol.* **2014**, *31*, 148. [CrossRef] [PubMed]

95. Hu, A.; Huang, J.-J.; Xu, W.-H.; Jin, X.-J.; Li, J.-P.; Tang, Y.-J.; Huang, X.-F.; Cui, H.-J.; Sun, G.-B. miR-21 and miR-375 microRNAs as candidate diagnostic biomarkers in squamous cell carcinoma of the larynx: Association with patient survival. *Am. J. Transl. Res.* **2014**, *6*, 604–613. [PubMed]

96. Saito, K.; Inagaki, K.; Kamimoto, T.; Ito, Y.; Sugita, T.; Nakajo, S.; Hirasawa, A.; Iwamaru, A.; Ishikura, T.; Hanaoka, H.; et al. MicroRNA-196a Is a Putative Diagnostic Biomarker and Therapeutic Target for Laryngeal Cancer. *PLoS ONE* **2013**, *8*, e71480. [CrossRef]

97. Wang, Y.; Chen, M.; Tao, Z.; Hua, Q.; Chen, S.; Xiao, B. Identification of predictive biomarkers for early diagnosis of larynx carcinoma based on microRNA expression data. *Cancer Genet.* **2013**, *206*, 340–346. [CrossRef]

98. Wu, Y.; Yu, J.; Ma, Y.; Wang, F.; Liu, H. miR-148a and miR-375 may serve as predictive biomarkers for early diagnosis of laryngeal carcinoma. *Oncol. Lett.* **2016**, *12*, 871–878. [CrossRef]

99. Ayaz, L.; Gorur, A.; Yaroğlu, H.Y.; Ozcan, C.; Tamer, L. Differential expression of microRNAs in plasma of patients with laryngeal squamous cell carcinoma: potential early-detection markers for laryngeal squamous cell carcinoma. *J. Cancer Res. Clin. Oncol.* **2013**, *139*, 1499–1506. [CrossRef]

100. Xu, Y.; Lin, Y.P.; Yang, D.; Zhang, G.; Zhou, H.F. Expression of serum microRNA-378 and its clinical significance in laryngeal squamous cell carcinoma. *Eur. Rev. Med. Pharmacol. Sci.* **2016**, *20*, 5137–5142.

101. Grzelczyk, W.L.; Szemraj, J.; Kwiatkowska, S.; Jozefowicz-Korczynska, M. Serum expression of selected miRNAs in patients with laryngeal squamous cell carcinoma (LSCC). *Diagn. Pathol.* **2019**, *14*, 49. [CrossRef] [PubMed]

102. Childs, G.; Fazzari, M.; Kung, G.; Kawachi, N.; Brandwein-Gensler, M.; McLemore, M.; Chen, Q.; Burk, R.D.; Smith, R.V.; Prystowsky, M.B.; et al. Low-Level Expression of MicroRNAs let-7d and miR-205 Are Prognostic Markers of Head and Neck Squamous Cell Carcinoma. *Am. J. Pathol.* **2009**, *174*, 736–745. [CrossRef] [PubMed]

103. Nagai, H.; Hasegawa, S.; Uchida, F.; Terabe, T.; Kanno, N.I.; Kato, K.; Yamagata, K.; Sakai, S.; Kawashiri, S.; Sato, H.; et al. MicroRNA-205-5p suppresses the invasiveness of oral squamous cell carcinoma by inhibiting TIMP-2 expression. *Int. J. Oncol.* **2018**, *52*, 841–850. [CrossRef] [PubMed]

104. Avissar, M.; Christensen, B.C.; Kelsey, K.T.; Marsit, C.J. MicroRNA expression ratio is predictive of head and neck squamous cell carcinoma. *Clin. Cancer Res.* **2009**, *15*, 2850–2855. [CrossRef] [PubMed]

105. Lu, Y.-C.; Chang, J.T.-C.; Huang, Y.-C.; Huang, C.-C.; Chen, W.-H.; Lee, L.-Y.; Huang, B.-S.; Chen, Y.-J.; Li, H.-F.; Cheng, A.-J. Combined determination of circulating miR-196a and miR-196b levels produces high sensitivity and specificity for early detection of oral cancer. *Clin. Biochem.* **2015**, *48*, 115–121. [CrossRef]

106. Philipone, E.; Yoon, A.J.; Wang, S.; Shen, J.; Ko, Y.C.K.; Sink, J.M.; Rockafellow, A.; Shammay, N.; Santella, R.M. MicroRNAs-208b-3p, 204-5p, 129-2-3p and 3065-5p as predictive markers of oral leukoplakia that progress to cancer. *Am. J. Cancer Res.* **2016**, *6*, 1537–1546. [PubMed]

107. Zhang, X.-W.; Liu, N.; Chen, S.; Wang, Y.; Zhang, Z.-X.; Sun, Y.-Y.; Qiu, G.-B.; Fu, W.-N. High microRNA-23a expression in laryngeal squamous cell carcinoma is associated with poor patient prognosis. *Diagn. Pathol.* **2015**, *10*, 22. [CrossRef]

108. Li, M.; Tian, L.; Ren, H.; Chen, X.; Wang, Y.; Ge, J.; Wu, S.; Sun, Y.; Liu, M.; Xiao, H. MicroRNA-101 is a potential prognostic indicator of laryngeal squamous cell carcinoma and modulates CDK8. *J. Transl. Med.* **2015**, *13*, 271. [CrossRef]

109. Wu, S.; Jia, S.; Xu, P. MicroRNA-9 as a novel prognostic biomarker in human laryngeal squamous cell carcinoma. *Int. J. Clin. Exp. Med.* **2014**, *7*, 5523–5528.

110. Maia, D.C.; de Carvalho, A.C.; Horst, M.A.; Carvalho, A.L.; Scapulatempo-Neto, C.; Vettore, A.L. Expression of miR-296-5p as predictive marker for radiotherapy resistance in early-stage laryngeal carcinoma. *J. Transl. Med.* **2015**, *13*, 262. [CrossRef]

111. Cappellesso, R.; Marioni, G.; Crescenzi, M.; Giacomelli, L.; Guzzardo, V.; Mussato, A.; Staffieri, A.; Martini, A.; Blandamura, S.; Fassina, A. The prognostic role of the epithelial-mesenchymal transition markers E-cadherin and Slug in laryngeal squamous cell carcinoma. *Histopathology* **2015**, *67*, 491–500. [CrossRef] [PubMed]

112. Fletcher, A.M.; Heaford, A.C.; Trask, D.K. Detection of Metastatic Head and Neck Squamous Cell Carcinoma Using the Relative Expression of Tissue-Specific Mir-205. *Transl. Oncol.* **2008**, *1*, 202. [CrossRef] [PubMed]

113. Peng, S.-C.; Liao, C.-T.; Peng, C.-H.; Cheng, A.-J.; Chen, S.-J.; Huang, C.-G.; Hsieh, W.-P.; Yen, T.-C. MicroRNAs MiR-218, MiR-125b, and Let-7g Predict Prognosis in Patients with Oral Cavity Squamous Cell Carcinoma. *PLoS ONE* **2014**, *9*, 102403. [CrossRef] [PubMed]

114. Hou, B.; Ishinaga, H.; Midorikawa, K.; Shah, S.A.; Nakamura, S.; Hiraku, Y.; Oikawa, S.; Murata, M.; Takeuchi, K. Circulating microRNAs as novel prognosis biomarkers for head and neck squamous cell carcinoma. *Cancer Boil. Ther.* **2015**, *16*, 1042–1046. [CrossRef]

115. Sun, J.; Yong, J.; Zhang, H. MicroRNA-93, upregulated in serum of nasopharyngeal carcinoma patients, promotes tumor cell proliferation by targeting PDCD4. *Exp. Ther. Med.* **2020**, *19*, 2579–2587. [CrossRef]

116. Bufalino, A.; Cervigne, N.K.; de Oliveira, C.E.; Fonseca, F.P.; Rodrigues, P.C.; Macedo, C.C.; Sobral, L.M.; Miguel, M.C.; Lopes, M.A.; Paes Leme, A.F.; et al. Low miR-143/miR-145 Cluster Levels Induce Activin a Overexpression in Oral Squamous Cell Carcinomas, Which Contributes to Poor Prognosis. *PLoS ONE* **2015**, *10*, 136599. [CrossRef]

117. Sousa, L.O.; Sobral, L.; Matsumoto, C.S.; Saggioro, F.P.; López, R.V.; Panepucci, R.A.; Curti, C.; Silva, W.A.; Greene, L.J.; Leopoldino, A.M. Lymph node or perineural invasion is associated with low miR-15a, miR-34c and miR-199b levels in head and neck squamous cell carcinoma. *BBA Clin.* **2016**, *6*, 159–164. [CrossRef]

118. Sun, L.; Liu, L.; Fu, H.; Wang, Q.; Shi, Y. Association of Decreased Expression of Serum miR-9 with Poor Prognosis of Oral Squamous Cell Carcinoma Patients. *Med. Sci. Monit.* **2016**, *22*, 289–294. [CrossRef]

119. Suh, Y.-E.; Raulf, N.; Gäken, J.; Lawler, K.; Urbano, T.G.; Bullenkamp, J.; Gobeil, S.; Huot, J.; Odell, E.; Tavassoli, M. MicroRNA-196a promotes an oncogenic effect in head and neck cancer cells by suppressing annexin A1 and enhancing radioresistance. *Int. J. Cancer* **2015**, *137*, 1021–1034. [CrossRef]

120. Liu, N.; Boohaker, R.J.; Jiang, C.; Boohaker, J.R.; Xu, B. A radiosensitivity MiRNA signature validated by the TCGA database for head and neck squamous cell carcinomas. *Oncotarget* **2015**, *6*, 34649–34657. [CrossRef]

121. de Jong, M.C.; Hoeve, J.T.; Grenman, R.; Wessels, L.; Kerkhoven, R.; Riele, H.T.; Brekel, M.W.V.D.; Verheij, M.; Begg, A.C. Pretreatment microRNA expression impacting on epithelial to mesenchymal transition predicts intrinsic radiosensitivity in head and neck cancer celllines and patients. *Clin. Cancer Res.* **2015**, *21*, 5630–5638. [CrossRef] [PubMed]

122. Zhu, Y.; Shi, L.; Lei, Y.-M.; Bao, Y.-H.; Li, Z.-Y.; Ding, F.; Zhu, G.-T.; Wang, Q.-Q.; Huang, C. Radiosensitization effect of hsa-miR-138-2-3p on human laryngeal cancer stem cells. *PeerJ* **2017**, *5*, 3233. [CrossRef] [PubMed]
123. Yang, P.-Y.; Hsieh, P.-L.; Wang, T.H.; Yu, C.-C.; Lu, M.-Y.; Liao, Y.-W.; Lee, T.-H.; Peng, C.-Y. Andrographolide impedes cancer stemness and enhances radio-sensitivity in oral carcinomas via miR-218 activation. *Oncotarget* **2016**, *8*, 4196–4207. [CrossRef] [PubMed]
124. Ahmad, P.; Sana, J.; Slavik, M.; Gurin, D.; Radova, L.; Gablo, N.A.; Kazda, T.; Smilek, P.; Horakova, Z.; Gal, B.; et al. MicroRNA-15b-5p Predicts Locoregional Relapse in Head and Neck Carcinoma Patients Treated with Intensity-modulated Radiotherapy. *Cancer Genom. Proteom.* **2019**, *16*, 139–146. [CrossRef]
125. Yin, W.; Wang, P.; Wang, X.; Song, W.; Cui, X.; Yu, H.; Zhu, W. Identification of microRNAs and mRNAs associated with multidrug resistance of human laryngeal cancer Hep-2 cells. *Braz. J. Med. Boil. Res.* **2013**, *46*, 546–554. [CrossRef]
126. Liu, M.; Wang, J.; Huang, H.; Hou, J.; Zhang, B.; Wang, A. MiR-181a–Twist1 pathway in the chemoresistance of tongue squamous cell carcinoma. *Biochem. Biophys. Res. Commun.* **2013**, *441*, 364–370. [CrossRef]
127. Jiang, F.; Zhao, W.; Zhou, L.; Liu, Z.; Li, W.; Yu, D. MiR-222 Targeted PUMA to Improve Sensitization of UM1 Cells to Cisplatin. *Int. J. Mol. Sci.* **2014**, *15*, 22128–22141. [CrossRef]
128. Liu, X.-Y.; Liu, Z.-J.; He, H.; Zhang, C.; Wang, Y.-L. MicroRNA-101-3p suppresses cell proliferation, invasion and enhances chemotherapeutic sensitivity in salivary gland adenoid cystic carcinoma by targeting Pim-1. *Am. J. Cancer Res.* **2015**, *5*, 3015–3029.
129. Qin, X.; Guo, H.; Wang, X.; Zhu, X.; Yan, M.; Wang, X.; Xu, Q.; Shi, J.; Lu, E.; Chen, W.; et al. Exosomal miR-196a derived from cancer-associated fibroblasts confers cisplatin resistance in head and neck cancer through targeting CDKN1B and ING5. *Genome Boil.* **2019**, *20*, 12. [CrossRef]
130. Kariya, A.; Furusawa, Y.; Yunoki, T.; Kondo, T.; Tabuchi, Y. A microRNA-27a mimic sensitizes human oral squamous cell carcinoma HSC-4 cells to hyperthermia through downregulation of Hsp110 and Hsp90. *Int. J. Mol. Med.* **2014**, *34*, 334–340. [CrossRef]
131. Ruan, Q.; Fang, Z.; Cui, S.; Zhang, X.-L.; Wu, Y.-B.; Tang, H.-S.; Tu, Y.-N.; Ding, Y. Thermo-chemotherapy Induced miR-218 upregulation inhibits the invasion of gastric cancer via targeting Gli2 and E-cadherin. *Tumor Boil.* **2015**, *36*, 5807–5814. [CrossRef]
132. Wald, A.I.; Hoskins, E.E.; Wells, S.I.; Ferris, R.; Khan, S.A. Alteration of microRNA profiles in squamous cell carcinoma of the head and neck cell lines by human papillomavirus. *Head Neck* **2011**, *33*, 504–512. [CrossRef] [PubMed]
133. Lajer, C.B.; Garnæs, E.; Friis-Hansen, L.; Norrild, B.; Therkildsen, M.H.; Glud, M.; Rossing, M.; Lajer, H.; Svane, D.; Skotte, L.; et al. The role of miRNAs in human papilloma virus (HPV)-associated cancers: Bridging between HPV-related head and neck cancer and cervical cancer. *Br. J. Cancer* **2012**, *106*, 1526–1534. [CrossRef] [PubMed]
134. Mirghani, H.; Ugolin, N.; Ory, C.; Goislard, M.; Lefèvre, M.; Baulande, S.; Hofman, P.; Guily, J.L.S.; Chevillard, S.; Lacave, R. Comparative analysis of micro-RNAs in human papillomavirus-positive versus -negative oropharyngeal cancers. *Head Neck* **2016**, *38*, 1634–1642. [CrossRef] [PubMed]
135. Gao, G.; Gay, H.A.; Chernock, R.D.; Zhang, T.R.; Luo, J.; Thorstad, W.L.; Lewis, J.S.; Wang, X. A microRNA expression signature for the prognosis of oropharyngeal squamous cell carcinoma. *Cancer* **2012**, *119*, 72–80. [CrossRef]
136. Heß, J.; Unger, K.; Maihöfer, C.; Schüttrumpf, L.; Wintergerst, L.; Heider, T.; Weber, P.; Marschner, S.; Braselmann, H.; Samaga, D.; et al. A Five-MicroRNA Signature Predicts Survival and Disease Control of Patients with Head and Neck Cancer Negative for HPV Infection. *Clin. Cancer Res.* **2018**, *25*, 1505–1516. [CrossRef]
137. Božinović, K.; Sabol, I.; Dediol, E.; Gašperov, N.M.; Manojlović, S.; Vojtechova, Z.; Tachezy, R.; Grce, M. Genome-wide miRNA profiling reinforces the importance of miR-9 in human papillomavirus associated oral and oropharyngeal head and neck cancer. *Sci. Rep.* **2019**, *9*, 2306. [CrossRef]
138. Bersani, C.; Mints, M.; Tertipis, N.; Haeggblom, L.; Näsman, A.; Romanitan, M.; Dalianis, T.; Ramqvist, T. MicroRNA-155, -185 and -193b as biomarkers in human papillomavirus positive and negative tonsillar and base of tongue squamous cell carcinoma. *Oral Oncol.* **2018**, *82*, 8–16. [CrossRef]
139. Amaral, N.S.D.; Melo, N.C.; Maia, B.D.M.; Rocha, R.M. Noncoding RNA Profiles in Tobacco and Alcohol-Associated Diseases. *Genes* **2016**, *8*, 6. [CrossRef]

140. Momi, N.; Kaur, S.; Rachagani, S.; Ganti, A.K.; Batra, S.K.; Rachgani, S. Smoking and microRNA dysregulation: A cancerous combination. *Trends Mol. Med.* **2013**, *20*, 36–47. [CrossRef]

141. Manikandan, M.; Rao, A.K.D.M.; Rajkumar, K.S.; Rajaraman, R.; Munirajan, A.K. Altered levels of miR-21, miR-125b-2, miR-138, miR-155, miR-184, and miR-205 in oral squamous cell carcinoma and association with clinicopathological characteristics. *J. Oral Pathol. Med.* **2014**, *44*, 792–800. [CrossRef] [PubMed]

142. Pal, A.; Melling, G.; Hinsley, E.E.; Kabir, T.; Colley, H.E.; Murdoch, C.; Lambert, D.W. Cigarette smoke condensate promotes pro-tumourigenic stromal-epithelial interactions by suppressing miR-145. *J. Oral Pathol. Med.* **2012**, *42*, 309–314. [CrossRef] [PubMed]

143. Izzotti, A.; Calin, G.A.; Arrigo, P.; Steele, V.E.; Croce, C.M.; de Flora, S. Downregulation of microRNA expression in the lungs of rats exposed to cigarette smoke. *FASEB J.* **2008**, *23*, 806–812. [CrossRef] [PubMed]

144. House, R.; Majumder, M.; Janakiraman, H.; Ogretmen, B.; Kato, M.; Erkul, E.; Hill, E.; Atkinson, C.; Barth, J.; Day, T.A.; et al. Smoking-induced control of miR-133a-3p alters the expression of EGFR and HuR in HPV-infected oropharyngeal cancer. *PLoS ONE* **2018**, *13*, e0205077. [CrossRef]

145. Gong, S.-Q.; Xu, M.; Xiang, M.-L.; Shan, Y.-M.; Zhang, H. The Expression and Effection of MicroRNA-499a in High-Tobacco Exposed Head and Neck Squamous Cell Carcinoma: A Bioinformatic Analysis. *Front. Oncol.* **2019**, *9*, 678. [CrossRef]

146. Takahashi, K.; Yokota, S.-I.; Tatsumi, N.; Fukami, T.; Yokoi, T.; Nakajima, M. Cigarette smoking substantially alters plasma microRNA profiles in healthy subjects. *Toxicol. Appl. Pharmacol.* **2013**, *272*, 154–160. [CrossRef]

147. Suzuki, K.; Yamada, H.; Nagura, A.; Ohashi, K.; Ishikawa, K.; Yamazaki, M.; Ando, Y.; Ichino, N.; Osakabe, K.; Sugimoto, K.; et al. Association of cigarette smoking with serum microRNA expression among middle-aged Japanese adults. *Fujita Med. J.* **2016**, *2*.

148. Saad, M.A.; Kuo, S.Z.; Rahimy, E.; Zou, A.E.; Korrapati, A.; Rahimy, M.; Kim, E.; Zheng, H.; Yu, M.A.; Wang-Rodriguez, J.; et al. Alcohol-dysregulated miR-30a and miR-934 in head and neck squamous cell carcinoma. *Mol. Cancer* **2015**, *14*, 181. [CrossRef]

149. Hu, Y.; Liu, H. MicroRNA-10a-5p and microRNA-34c-5p in laryngeal epithelial premalignant lesions: Differential expression and clinicopathological correlation. *Eur. Arch. Oto-Rhino-Laryngol.* **2014**, *272*, 391–399. [CrossRef]

150. Supic, G.; Zeljic, K.; Rankov, A.D.; Kozomara, R.; Nikolic, A.; Radojkovic, D.; Magic, Z. miR-183 and miR-21 expression as biomarkers of progression and survival in tongue carcinoma patients. *Clin. Oral Investig.* **2017**, *22*. [CrossRef]

151. Krishnan, A.R.; Zheng, H.; Kwok, J.G.; Qu, Y.; Zou, A.E.; Korrapati, A.; Li, P.X.; Califano, J.; Hovell, M.F.; Wang-Rodriguez, J.; et al. A comprehensive study of smoking-specific microRNA alterations in head and neck squamous cell carcinoma. *Oral Oncol.* **2017**, *72*, 56–64. [CrossRef] [PubMed]

152. Manikandan, M.; Rao, A.K.D.M.; Arunkumar, G.; Manickavasagam, M.; Rajkumar, K.S.; Rajaraman, R.; Munirajan, A.K. Oral squamous cell carcinoma: microRNA expression profiling and integrative analyses for elucidation of tumourigenesis mechanism. *Mol. Cancer* **2016**, *15*, 28. [CrossRef] [PubMed]

153. Ricciardiello, F.; Capasso, R.; Kawasaki, H.; Abate, T.; Oliva, F.; Lombardi, A.; Misso, G.; Ingrosso, D.; Leone, C.; Iengo, M.; et al. A miRNA signature suggestive of nodal metastases from laryngeal carcinoma. *Acta Otorhinolaryngol. Ital.* **2017**, *37*, 467–474. [PubMed]

International Journal of *Molecular Sciences*

Review

MicroRNAs as Key Players in Melanoma Cell Resistance to MAPK and Immune Checkpoint Inhibitors

Maria Letizia Motti [1],*, Michele Minopoli [2], Gioconda Di Carluccio [2], Paolo Antonio Ascierto [3] and Maria Vincenza Carriero [2],*

[1] Department of Motor and Wellness Sciences, University "Parthenope", 80133 Naples, Italy
[2] Neoplastic Progression Unit, Istituto Nazionale Tumori IRCCS 'Fondazione G. Pascale', 80131 Naples, Italy; m.minopoli@istitutotumori.na.it (M.M.); g.dicarluccio@istitutotumori.na.it (G.D.C.)
[3] Melanoma, Cancer Immunotherapy and Development Therapeutics Unit, Istituto Nazionale Tumori-IRCCS Fondazione "G. Pascale", 80131 Naples, Italy; p.ascierto@istitutotumori.na.it
* Correspondence: motti@uniparthenope.it (M.L.M.); m.carriero@istitutotumori.na.it (M.V.C.); Tel.: +39-081-5474670 (M.L.M.); +39-081-5903569 (M.V.C.)

Received: 30 May 2020; Accepted: 24 June 2020; Published: 26 June 2020

Abstract: Advances in the use of targeted and immune therapies have revolutionized the clinical management of melanoma patients, prolonging significantly their overall and progression-free survival. However, both targeted and immune therapies suffer limitations due to genetic mutations and epigenetic modifications, which determine a great heterogeneity and phenotypic plasticity of melanoma cells. Acquired resistance of melanoma patients to inhibitors of BRAF (BRAFi) and MEK (MEKi), which block the mitogen-activated protein kinase (MAPK) pathway, limits their prolonged use. On the other hand, immune checkpoint inhibitors improve the outcomes of patients in only a subset of them and the molecular mechanisms underlying lack of responses are under investigation. There is growing evidence that altered expression levels of microRNAs (miRNA)s induce drug-resistance in tumor cells and that restoring normal expression of dysregulated miRNAs may re-establish drug sensitivity. However, the relationship between specific miRNA signatures and acquired resistance of melanoma to MAPK and immune checkpoint inhibitors is still limited and not fully elucidated. In this review, we provide an updated overview of how miRNAs induce resistance or restore melanoma cell sensitivity to mitogen-activated protein kinase inhibitors (MAPKi) as well as on the relationship existing between miRNAs and immune evasion by melanoma cell resistant to MAPKi.

Keywords: miRNA; melanoma; melanoma resistance to MAPK/MEK inhibitors; resistance to immune checkpoint inhibitors

---

1. Introduction

Melanoma represents one of the most aggressive skin cancers with a significantly increased incidence in the last decades [1–3]. Currently, therapeutic options include surgical excision, chemotherapy, targeted and immune therapies administered as single agents or in combination, depending on the stage of the disease, location, as well as the genetic profile of the tumor [4]. In the last years, molecular targeted therapies and immunotherapies have significantly improved the overall survival of patients with metastatic disease [5,6].

In the past years, either dabrafenib or vemurafenib BRAF inhibitors (BRAFi) showed encouraging response rates, although the duration of response appeared to be limited [7,8]. BRAF inhibitor resistance depends on oncogenic signaling through reactivation of MAPK/Erk or activation of PI3K/Akt, which may be acquired by directly affecting genes in each pathway, by upregulation of receptor tyrosine kinases, or by affecting downstream signaling [9]. Thus, the combination of dabrafenib with the MEK inhibitor (MEKi) trametinib, has become employed worldwide for the care of patients with BRAF-mutant metastatic melanoma, improving their progression-free and overall survival [10,11]. Unfortunately, patients treated with dabrafenib/trametinib combination therapy also develop alterations in the same genes that support single-agent resistance including MEK1/2 mutations, BRAF amplification, BRAF alternative splicing, and NRAS mutations [12,13]. The limiting factor for these therapeutic approaches is the heterogeneity and phenotypic plasticity of melanoma cells due to genetic mutations and epigenetic modifications that may determine the paradoxical activation of the mitogen-activated protein kinase (MAPK) and thus sustain resistance to these drugs [14]. The new immune checkpoint blockade therapies improve the outcomes of patients with advanced melanoma regardless of the mutation status and several ongoing clinical trials highlight that combinations of BRAFi and MEKi with immune checkpoint inhibitors result in more durable responses in about 50% of patients [15–17]. Based on these considerations, the identification of biomarkers that monitor and/or predict an early response during melanoma therapy still represents an unmet clinical need.

Using a variety of technical approaches such as chromosomal analysis, miRNA microarrays, miRNA qPCR arrays, and high-throughput small RNA sequencing platforms, microRNA (miRNA)s have been identified to function as oncogenes or tumor repressors genes. Oncogenic miRNAs (oncomiRs) are frequently overexpressed in cancers while tumor-suppressive miRNAs are down-regulated. It has been documented that miRNAs regulate more than 30% of human protein-coding genes [18] and control, through degradation of mRNA or a translation block, numerous cancer-relevant processes including proliferation, autophagy, migration, and apoptosis [19]. Specific miRNA signatures have been found differentially expressed in normal and tumor tissues, suggesting their potential value as molecular biomarkers useful for diagnosis, staging, progression, prognosis, and response to treatments [20–22].

miRNAs are short, single-stranded, non-coding nucleotide sequences with an average 22 nucleotides in length. They are transcribed as individual genes, from introns of coding genes (intronic miRNAs) or from regions between the clusters of genes (intergenic miRNAs) while clustered miRNAs are transcribed as polycistronic transcripts [23]. miRNA genes are transcribed by RNA polymerase II into primary miRNAs (pri-miRNA)s, processed into precursor miRNA's (pre-miRNA)s and then into mature miRNAs. After processing, mature single-stranded miRNAs associate with argonaute protein family (Argo) and glycine-tryptophan proteins of 182 kDa (GW182), which are the principal constituents of the miRNA-induced silencing complex (miRISC) [24], and usually bind to the 3′UTRs of their cytosolic mRNA targets, resulting in mRNA-reduced translation or deadenylation and degradation of the mRNA transcript [25]. The interaction of miRNAs with other regions, including the 5′UTR coding sequence, and gene promoters, has also been reported [26,27]. miRNA interaction with target genes may be influenced by several factors, including the subcellular location of miRNAs, abundancy of miRNAs and/or corresponding target mRNAs, as well as the affinity of miRNA-mRNA interactions [28]. Moreover, recent studies suggest that miRNAs may be shuttled between different subcellular compartments to control the rate of translation and transcription [29] and that an individual miRNA can act on several mRNA simultaneously, modulating multiple processes in cancer cells in a cooperative manner [30,31]. Furthermore, some microRNAs are related to the expression of transmembrane oncogenes, acting directly on their expression (e.g., EGFR) [32], or acting indirectly, by regulating the expression of soluble ligands that recognize specifically the extracellular domain of the receptors [33].

It has been shown that chromosomal rearrangements, epigenetic regulation and disorders in miRNA biogenesis, result in increased or decreased expression of miRNAs in melanoma cells as compared to melanocytes [34–36]. Furthermore, miRNA altered expression has been described in different stages of melanoma progression, so that expression levels of specific miRNAs are considered as diagnostic and/or prognostic biomarkers in melanoma [37–40]. When secreted into extracellular fluids, miRNAs are stable in human fluids since they are packaged in exosomes and microvescicles or associated with RNA-binding proteins such as Argo2 or lipoprotein complexes, which protect them from degradation [28,41,42]. In a recent study, 11 miRNAs were identified as differentially expressed between healthy controls and plasma samples from different melanoma stages [43]. Therefore, miRNAs, especially those being part of the circulating transcriptome, may be useful as biomarkers for early melanoma detection and response to treatments [44]. Numerous miRNAs have been found to regulate melanoma cell behavior and gene expression acting on the MAPK signaling pathway [45], while some miRNAs have been found to regulate the expression of immune checkpoints, acting on melanoma cells or immune cells [46].

In this review, we discuss the latest progress regarding mechanisms by which miRNAs regulate melanoma cell resistance to MAPKi and immune evasion. Furthermore, the potential predictive value of circulating miRNAs for monitoring melanoma responsiveness to targeted and immune therapies is debated.

2. miRNAs Involved in the Regulation of Melanoma MAPKi-Resistance

In recent years, by next-generation sequencing, the Cancer Genome Atlas provided the analysis on the somatic aberrations underlying melanoma genesis, identifying BRAF, RAS, and NF1 mutant genetic subtypes of cutaneous melanoma, all of them being able to deregulate the MAPK/ERK pathway, leading to uncontrolled cell growth [47]. Over 50% of melanomas harbor activating mutation in the BRAF gene, which sustains proliferation and survival of melanoma cells by activating the MAPK pathway. Over 90% BRAF mutations are at codon 600 and among these, over 90% are a single nucleotide mutation resulting in substitution of the valine with a glutamic acid residue (BRAFV600E), while less common mutations are the substitutions of valine with lysine, arginine, leucine or aspartic acid residues [48]. Vemurafenib and dabrafenib BRAF inhibitors (BRAFi) have improved the outcomes of patients with BRAF-mutant metastatic melanoma [7,8]. Unfortunately, most of them develop drug resistance early as a consequence of the activation of alternative proliferation-inducing pathways, often associated to the reactivation of the MAPK pathway [49–53]. Indeed, resistance also occurs in the majority of melanoma patients treated with BRAFi and MEKi combinations, although overall and progression-free survival are prolonged compared to single-agent therapies [54,55]. Furthermore, it has to be taken into account that BRAF-mutant melanomas may acquire BRAF inhibitor resistance via up-regulation of both MAPK and PI3K/Akt pathways in about 22% of the melanoma patients [49], whereas other drugs targeting different cellular pathways may escape development of drug resistance, probably due to the extraordinary plasticity of melanoma cells [56–58].

During the progressive development of drug resistance, several deregulated miRNAs have been shown to control both tumor cell growth and melanoma cell interactions with the tumor microenvironment. Some miRNAs provoke drug resistance while others restore drug sensitivity. In Table 1, miRNAs with a potential role in regulating melanoma sensitivity and resistance to MAPKi and the underlying mechanisms of action are listed.

First, Liu and co-workers showed that miR-200c is a potential therapeutic target to restore melanoma cell sensitivity to BRAFi. They found that miR-200c reverts drug resistance to PLX4720 BRAF and U0126 MEK inhibitors by down-regulating the p16 transcriptional repressor BMI-1, which, in turn, inhibits melanoma cell growth and metastases in nude mice. Moreover, they found that miR-200c acts on ABC transporters, a superfamily of transmembrane proteins that mediate drug resistance in melanoma cells [59]. The clinical significance of miR-200c/Bmi1 axis in inhibiting acquired resistance to BRAFi was confirmed in human melanoma tissues: loss of miR-200c expression was found to correlate with development of resistance to BRAFi and promote the development of a BRAFi-resistant phenotype in melanoma cells and in melanoma tissues with a mechanism that involves MAPK and PI3K/AKT signaling pathways [60]. Like miR-200c, miR-524-5p expression appeared down-regulated in melanoma cells with activated MAPK/ERK pathway. miR-524-5p suppresses MAPK/ERK pathway-triggered melanoma cell proliferation by directly binding to the 3'-UTR of both BRAF and ERK2 [61]. Fattore L. and colleagues found that miR-579-3p is down-regulated in vemurafenib-resistant melanoma cells and that its ectopic expression impairs the establishment of drug resistance in human melanoma cells. They also showed that down-regulation of miR-579-3p in tumor tissues from melanoma patients with acquired resistance to BRAFi well correlates with a poor prognosis [62]. Mechanistically, miR-579-3p binds to the 3'UTR of either BRAF and MDM2, an E3 ubiquitin protein ligase that promotes p53 degradation [63], so that MDM2 and p53 cause a negative-feedback loop, in which p53 induces the expression of MDM2 [62]. The miR-506-514 cluster has been shown to regulate not only melanocyte transformation but also melanoma cell proliferation [64]. Stark and coworkers demonstrated that miR-514a, which is expressed in 69% of melanoma cell lines, reverts drug resistance to BRAFi by directly binding to NF1 transcripts, leading to altered NF1 protein expression and consequent decreased cell proliferation. Accordingly, overexpression of miR-514a increases survival of vemurafenib-treated BRAF(V600E) melanoma cells [65]. A microarray profiling analysis of vemurafenib-resistant and sensible A375 melanoma cells allowed Sun X. and colleagues to identify 17 dysregulated miRNAs in BRAFi resistant A375 cells. Among these, miR-7 was found to be the most down-regulated miRNA that prevents proliferation and partially reverts drug resistance of vemurafenib-resistant melanoma cells [32]. miR-7 inhibits both MAPK and PI3K/Akt signaling pathways by targeting EGFR, IGF-1R and CRAF [32]. In this regard, miR-7 could inhibit the activation of the MAPK and PI3K/AKT pathways and reverse melanoma cell resistance to BRAFi, by decreasing the expression levels of EGFR and IGF-1R. Using real time quantitative PCR and microarray analyses, Kim JA and co-workers found that up-regulation of miR-1246 associates with acquired resistance to BRAFi by A375P melanoma cells. Although the exact mechanism of action of miR-1246 in eliciting drug resistance has been not yet completely identified, Authors provided evidence that resistance to PLX4720 in miR-1246 mimic-transfected cells is mostly due to the inhibition of autophagy [66]. By miRNA expression profiling of sensible and BRAFi resistant melanoma cells, Lisa Koetz-Ploch and colleagues found that miR-125a becomes overexpressed upon acquisition of cell resistance to BRAFi. Mechanistically, miR-125a suppresses the apoptotic program in BRAFi-treated melanoma cells by targeting two components of the intrinsic pro-apoptotic pathway: BAK1 and MLK3 [67]. The finding that miR-125a is up-regulated in tissues of BRAFi-treated melanoma patients as compared to tumor samples excised before BRAF-treatment, allowed Authors to propose the use of anti-miR-125-a for preventing or overcame BRAFi resistance [67].

Melanoma cells are documented to release into the extracellular milieu different types of extracellular vesicles (EV)s, including oncosomes, ectosomes, exosomes, and melanosomes carrying protein and small RNAs cargos [68]. Comparing RNA sequences of exosomal miRNA released by a number of melanoma cell lines with clinical miRNA datasets from human melanoma tissue samples, Lunavat TR and coworkers found that the exosomal miR-214-3p, miR-199a-3p and miR-155-5p associate with melanoma progression [69]. More recently, the same Authors found that both vemurafenib and dabrafenib BRAFi significantly increase expression of miR-211-5p in EVs from melanoma cell cultures and tissues, leading to re-activation of the survival pathway. Mechanistically, overexpression of miR-211-5p depends on BRAFi-induced up-regulation of the microphthalmia-associated transcription factor (MITF) which, in turn, induces activation of the survival pathway trough the master regulator TRPM1 gene [70]. By carrying out RNA-seq analyses, Díaz-Martínez and co-workers documented in vemurafenib-resistant A375 cells very high levels of miR-204-5p and miR-211-5p when compared to parental counterparts. They found that, when engrafted in mice, sensible A375 cells transfected with both miR-204-5p and miR-211-5p became resistant to vemurafenib and were able to grow, whereas resistant cells silenced for miR-204-5p and miR-211-5p expression lost tumor growth ability and became sensible to vemurafenib [71]. Mechanistically, co-overexpression of miR-204-5p and miR-211-5p triggers Ras and MAPK up-regulation not only in response to BRAFi but also in response to inhibitors of other downstream effectors of the MAPK pathway [71]. Examination of some potential targets for these miRNAs revealed that miR-204-5p or miR-211-5p reduce significantly at mRNA and protein levels the NUAK1/ARK5 kinase [71]. Accordingly, NUAK1/ARK5 protein was consistently reduced in vemurafenib-resistant cells [71].

Overexpression of the Yes-associated protein (YAP) has been found to associate with resistance to anticancer therapies in solid tumors, including BRAFi resistant melanomas [72,73]. miR-550a-3-5p overexpression has been proven to down-regulate YAP at mRNA and protein levels and YAP down-regulation-dependent tumor-suppressive activity induces sensitization of BRAFi-resistant melanoma cells to vemurafenib [74]. Fattore L. and coworkers demonstrated that down-modulation of miR-199b-5p in drug-resistant melanoma cells causes increased VEGF release and acquisition of a pro-angiogenic status that may be reverted by restoring miR-199b-5p levels [33]. In line with these findings, the occurrence of a miRNA-dependent regulation of VEGF production in melanoma cells resistant to BRAF inhibitors was documented by Caporali and colleagues [75]. These Authors found low levels of miR-126-3p in dabrafenib-resistant melanoma cells as compared with their parental counterparts and that proliferation and invasiveness of dabrafenib-resistant cells may be reduced by restoring the miR-126-3p expression [75]. By analyzing the global miRNAome changes in sensible and BRAFi-resistant melanoma cells, Fattore L. et colleagues identified many deregulated miRNAs involved in the acquisition of drug resistance to BRAFi. They identified specific miRNA signatures capable of distinguishing drug responding from non-responding patients as well as a subset of miRNAs capable to block or revert the development of drug resistance when down- or up-regulated. Using qRT-PCR on matched tumor biopsies and serum samples from melanoma patients, the same Authors found that miR-204-5p and miR-199b-5p are down-regulated in relapsing melanomas, whereas miR-4443 and miR-4488 are up-regulated [33]. Accordingly, they found that overexpression of down-regulated miR-204-5p and miR-199b-5p reduces cell proliferation and induces apoptosis, whereas inhibition of up-regulated miR-4443 and miR-4488 with specific antagomiRs, restores inhibitory effects exerted by BRAFi. Authors also found that a reduced proliferation of A375 melanoma cells double resistant to BRAFi and MEKi, may be achieved by down-regulating simultaneously miR-204-5p, miR-199b-5p and miR-579-3p, highlighting the notion that co-targeting multiple microRNAs may be a valid approach to prevent proliferation of melanoma cells with acquired resistance to BRAFi and MEKi [33].

Table 1. microRNAs Involved in the Acquisition of Melanoma Cell Resistance to MAPK Inhibitors.

miRNAs	Expression	Target Gene/s	Mechanism/s	Tissue/Cell Lines/Blood	Reference
miR-200c	Down	BMI1, ZEB2, TUBB3, ABCG5, MDR1	p16 Transcriptional Repressor BMI-1/up-Regulation of ABC Transporters. Activation of MAPK and PI3K/AKT Signaling Cascades	Tissues, Cell lines	[59,60]
miR-579-3p	Down	BRAF and MDM2	Reduced Proliferation (by Targeting BRAF). Increased Apoptosis (by Down-Regulating of MDM2)	Tissues, Cell lines	[62]
miR-7	Down	EGFR, IGF-1R, CRAF	Inhibition of MAPK and PI3K/Akt Signaling Pathways	Cell lines	[32]
miR-550a-3-5p	Down	YAP	Reduced Proliferation through YAP Inhibition	Cell lines	[74]
miR-199b-5p	Down	HIF-1α, VEGFA	Pro-Angiogenic Activity	Tissues, Cell Lines, Plasma	[33]
miR-126-3p	Down	VEGFA, ADAM9	Increased Proliferation through the p-ERK1/2, p-Akt//VEGF axis	Cell Lines	[75]
miR-204-5p, miR-199b-5p	Down	BCL-2, FOXM1, NOTCH, VEGF	Increased Survival/Reduced Apoptosis Bcl2, HIF-1/VEGF	Tissues, Cell Lines, Plasma	[33]
miR-514a	Up	NF1	Inhibition of NF1 Increased Survival	Cell Lines	[65]
miR-1246	Up	NS	Inhibition of Autophagy	Cell Lines	[66]
miR-125a	Up	BAK1 and MLK3	Inhibition of Apoptotic Program	Tissues, Cell Lines	[67]
miR-204-5p, miR-211-5p	Up	NUAK	Up-Regulation of the Ras/MEK/ERK Pathway through MITF/Increased Survival Pathway	Tissues, Cell Lines	[70,71]
miR-4443, miR-4488	Up	Autophagy-Related Genes	Deregulation of Autophagy	Tissues, Cell Lines, Plasma	[33]

List of miRNAs involved in the melanoma resistance to MAPK inhibitors. Up/Down expression levels are referred to resistant melanoma cells. Not shown (NS) indicates that miRNAs target genes have not been identified in the corresponding studies.

3. miRNA in Melanoma Cell Resistance to Immunotherapy

There are several attempts to investigate the potential link between miRNAs expression profile and patients' response to immune checkpoint inhibitors in order to verify at the same time their potential use for monitoring efficacy of immune checkpoint blockade and improving the outcomes of patients with advanced melanoma. Although few data regarding miRNAs and immune checkpoint inhibitors relationship are available in the literature, recent studies demonstrate that some miRNAs may regulate directly or indirectly the expression of immune checkpoints, acting on tumor cells or immune cells, respectively (Table 2).

Galore-Haskel and collaborators found higher levels of miR-222 in melanoma tissues from patients that were non-responders to ipilimumab when compared to responder patients, raising the possibility that miR-222 expression could be considered a valid biomarker for predicting responsiveness of melanoma patients to ipilimumab [76]. These Authors documented that Adenosine Deaminase Acting on RNA-1 (ADAR1) overcomes melanoma immune resistance and increase proliferation of melanoma cells by regulating the biogenesis of miR-222 at transcriptional level [76]. miR-222 directly interact with 3′UTR of the Intracellular Adhesion Molecule 1 (ICAM1) mRNA [77] which, consequently, affects melanoma immune resistance by rendering melanoma cells more resistant to TIL-mediated killing mainly due to their ability to cross endothelial vessels and infiltrate tumor tissues [78,79]. Analyzing exosomal miRNAs in sera from melanoma patients, Tengda and co-workers found higher levels of miR-532-5p and miR-106b in melanoma patients with stage III–IV disease, as compared to patients with stage I–II disease and low levels of miR-532-5p and miR-106b in melanoma patients treated with pembrolizumab compared to those untreated. The Authors concluded that measurement of exosomal miRNA-532-5p and miRNA-106b in the sera from melanoma patients could be used for monitoring and/or predicting their response to immunotherapies [80].

miRNAs are also involved in the regulation of immune cells within the tumor microenvironment, including cytotoxic, CD4 or γδ T lymphocytes, natural killer (NK), macrophages and myeloid-derived suppressor cells (MDSCs).

A direct involvement of tumor-suppressor miRNAs in the control of antitumor immune response through the regulation of immune checkpoints PD-1, PD-L1, and CTLA-4 has been ascertained in tumors of different origin [81]. By a microarray-based profiling performed in PD1+ and PD1- CD4 T cells sorted from lymph nodes and spleen of melanoma-bearing mice, Li and colleagues demonstrated that miR-28 decreases PD1 expression by directly binding to its 3′UTR, suggesting that miR-28 regulates exhaustive differentiation of Treg in melanoma cells. Moreover, exhausted T cells showed a reduced secretion of IL-2, TNF-α and IFN-γ and the use of miR-28 mimics was able to restore their secretion [82]. Martinez-Usatorre and co-workers analyzed miR-155 expression in CD8$^+$ T cells isolated from tumor-infiltrated lymph nodes and tumor tissues of melanoma patients and murine models. They found that miR-155 up-regulation within the tumors correlates with increased CD8$^+$ T-cell infiltration while low expression of miR-155 targets in melanoma tumors associates with a prolonged overall survival. These findings allowed Authors to conclude that miR-155 could be considered a marker of responsiveness of CD8 T cells, as further demonstrated by its up-regulation after PD1 blockade [83].

Up-regulation of stress-induced ligands, including ULBP2, allows tumor cell recognition by immune cells trough the NKG2D receptor expressed on lymphocytes, Natural Killer cells, as well as cytotoxic, CD4 or γδ T cells [84]. miR-34a and miR-34c have been shown to enhance NK-cell killing activity against melanoma cells by targeting the UL16 binding protein 2, while miR-34 mimics led to down-regulation of ULBP2, diminishing tumor cell recognition by NK cells [85]. By using next-generation sequencing, Cobos JV and colleagues identified a repertoire of miRNAs that have a specific expression signature in M2 polarized macrophages [86]. A panel of miRNAs have been recognized to promote the conversion of monocytes into myeloid-derived suppressor cells (MDSC)s, their baseline levels being found to correlate with the clinical efficacy of immune checkpoint inhibitors. For instance, miR-125a-5p inhibits M1 polarization and promotes the alternative M2 phenotype by targeting KLF13, a transcriptional factor that is active during T lymphocyte activation [87,88]. Moreover, both miR-146a and miR-146b promote M2 polarization in human and mouse models by down-regulating pro-inflammatory responses [88]. Finally, several circulating miRNAs (let-7e, miR-99b, miR-100, miR-125a, miR-125b, miR-146a, miR-146b, and miR-155) were found to correlate with a shorter progression free and overall survival in melanoma patients treated with ipilimumab and nivolumab, thus representing the first predictive peripheral blood biomarker of resistance to immune checkpoint inhibitors [46]. These miRNAs released in the blood by melanoma EVs act by converting monocytes into MDSC and reduce the clinical efficacy of the PD-1 and CTLA-4 inhibitors [46]. Based on these findings, it will be foreseeing that combinations of miRNAs with different immune checkpoint targets could mimic or improve the effect of immune checkpoint blockade therapies.

Table 2. microRNAs Involved in the Acquisition of Melanoma Resistance to Immune Checkpoint Inhibitors.

miRNA	Tissue/Cell Lines/Blood	Target/Function/Proposed Mechanism	Reference
miR-222	Tissues, Cell Lines	ADAR1/ICAM-Dependent -Increased Trans-Endothelial Migration of T Cells. Reduced Response to Ipilimumab	[76]
miR-532-5p, miR-106b	Serum Exosomes	Reduced Response to Pembrolizumab	[80]
miR-28	Cell Lines	Reduced PD1 Expression and Response to Pembrolizumab. Increased Differentiation of Treg. Reduced Secretion of IL-2, TNF-α and IFN-γ	[82]
miR-155	Tissues, Cell Lines, PBMC	Increased CD8+ T-Cell Infiltration	[83]
miR-34a, miR-34c	Cell Lines	Target UL16 Binding Protein 2I (ULBP2). Increased NK-cell Killing Activity	[85]
miR-125a-5p	Cell Lines	Targets KLF13. Protumoral Activity Trough Macrophages	[87]
let-7e, miR-99b, miR-100, miR-125a, miR-125b, miR-146a, miR-146b, miR-155	Tissues, Blood Monocytes, Plasma	Protumoral Activity by Converting Monocytes Into MDSC. Reduced Response to PD-1 and CTLA-4 Inhibitors. Reduced Response to Ipilimumab and Nivolumab	[46]

4. Relationship between miRNAs and Immune Evasion by Melanoma Cell Resistant to MAPKi

The activation of the MAPK pathway through BRAF mutations leads to downstream production of several cytokines that promote tumor growth and immune evasion with autocrine or paracrine mechanisms. Recent studies have documented that the MAPK signaling pathway may be considered as a potential molecular target for overcoming melanoma cell evasion of the immune surveillance (Figure 1). By activating the MAPK cascade, the BRAF(V600E) mutation stimulates melanoma cells to produce a wide spectrum of chemokines and cytokines which, in turn, are responsible for the recruitment of immune and myeloid cells. For the first time, Sumimoto H. and co-workers, using the U0126 MEK inhibitor and lentiviral BRAF(V600E) RNA interference, found that the oncogenic BRAF favors melanoma immune escape increasing production of IL-6 and IL-10 which increase T-cell stimulatory function of dendritic cells [89]. Furthermore, constitutively activated BRAF(V600E) in melanoma tumor cells has been shown to initiate and sustain IL-1α/β-dependent T-cell suppression in a murine model. Mechanistically, IL-1α and IL-1β secreted by melanoma cells increase COX-2, PD-L1, and PD-L2 expression levels in tumor associated fibroblasts which, in turn, suppress the function of tumor-infiltrating T cells [90]. Jiang X. and colleagues identified the molecular mechanism by which melanoma cells resistant to BRAFi can evade the immune system via PDL-1 up-regulation. By using a panel of melanoma cell lines harboring BRAF(V600E) mutation, the Authors showed that the BRAFi resistance leads to c-Jun and STAT3-mediated increase of PD-L1 expression [91]. Conversely, the same Authors demonstrated, in vitro, that the U0126 MEK inhibitor simultaneously counteracts MAPK reactivation and reduces PD-L1 expression [91]. Analyzing several melanoma cell lines resistant to BRAFi as well as plasma and tumor samples from vemurafenib-treated melanoma patients, Vergani and coauthors found that BRAFi-resistant melanoma cells secrete higher levels of CC-chemokine ligand 2 (CCL2) then sensible counterparts. The CCL2 increase elicits up-regulation of miR-34a, miR-100 and miR-125b, which, in turn, down-regulate the canonical genetic pathway for apoptosis. Conversely, down-regulation of CCL2 and/or miR-34a restores apoptosis and melanoma sensitivity to vemurafenib [92]. More recently, miRNAs have been directly associated with melanoma resistance to treatment with immune checkpoint inhibitors (Figure 1). Audrito V. and coworkers found that PD-L1 expression is limited to a subset of patients with metastatic melanoma and unfavorable prognosis [93]. These Authors found that resistance to BRAFi and MEKi associates with induction of PD-L1 expression in BRAF(V600E)-mutated melanoma cell lines and identified the post-transcriptional circuit responsible for PD-L1 up-regulation, consisting of a direct interaction of miR-17-5p with the 3′UTR mRNA of PD-L1 [93]. Finally, miR-17-5p levels were found to inversely correlate with PD-L1 expression and thus predict sensitivity to BRAFi in patients with metastatic melanoma [93]. In this contest, modulating miRNAs impinging both MAPK pathway and immune responses could be a useful approach for treating patients with advanced melanoma.

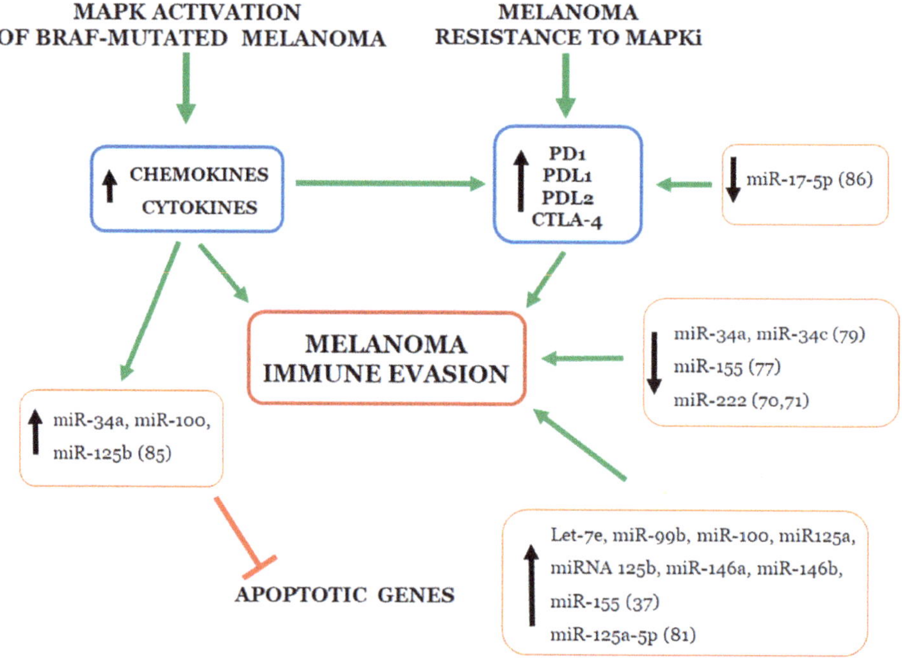

Figure 1. Schematic representation of up- (arrow pointing up) or down- (arrow pointing down) regulated miRNAs involved in evasion of immune surveillance by melanoma cells harboring BRAF mutations.

5. Predictive Value of Circulating miRNAs for Monitoring Melanoma Responsiveness to Targeted and Immune Therapies

To date, there is an urgent need to develop new non-invasive methods for monitoring disease progression or resistance to treatments of melanoma patients. In this regard, liquid biopsy may be considered a non-invasive source of biomarkers, potentially useful for monitoring responsiveness of melanoma patients to targeted and immune therapies, although the strategies for these approaches are still under investigation. In the last decade, many efforts have been made to identify diagnostic and prognostic circulating miRNA biomarkers for melanoma. Circulating miRNAs have emerged as powerful biomarkers since they are highly stable in body fluids, which are protected against enzymatic degradation thanks to their association with RNA binding proteins (Argonaute-2 and nucleophosmin-1), with high- and low-density lipoproteins, or to their embedding in membrane vesicles such as exosomes [28,41,42]. Also, they are resistant to both high or low pH, multiple freeze-thaw cycles, and long-term storage [94]. In a recent review article, Gajos-Michniewicz A summarizes studies reporting significant alterations in the miRNA expression profile in the serum and plasma of melanoma patients compared to healthy controls, suggesting circulating miRNAs as promising diagnostic melanoma biomarkers [40]. In a recent study, Solé C and colleagues found 11 miRNAs (let-7b, miR-16, miR-21, miR-92b, miR-98, miR-134, miR-320a, miR-486, miR-628, miR1180, and miR-1827) that are differentially expressed between healthy controls and plasma samples from different melanoma stages [43].

As above described, numerous miRNAs have been shown to modulate melanoma sensitivity and resistance to MAPKi (Table 1) and/or immune checkpoint inhibitors (Table 2). Among these, some miRNAs are present not only in tissue samples but also in serum or plasma of melanoma patients, thus representing soluble putative markers to monitor the therapeutic responses to MAPKi and immune treatments. miR-199b-5p expression levels were found downregulated in the plasma of melanoma patients post-MAPKi treatment as compared to plasma from the untreated ones, whereas miR-4488 levels were significantly increased in patients after MAPKi treatment, indicating that these miRNAs may represent soluble putative markers to monitor the therapeutic responses to MAPKi [68]. Svedman and co-workers identified let-7g-5p and miR-497-5p as predictive biomarkers of MAPKi treatment benefit in metastatic melanoma patients. They analyzed miRNA content in the extracellular microvesicles recovered from plasma of melanoma patients before and after the treatment with MAPKi. Both let-7g-5p and miR-497-5p levels were found to increase after the treatment with MAPKi and to correlate with a prolonged progression-free survival [95]. By performing Nanostring nCounter analysis of 48 plasma samples from individuals with or without melanoma, Van Laar R and coworkers identified a set of thirty-eight independently validated circulating miRNAs. The so-called MEL38 signature includes some miRNA (hsa-miR-34a-5p, hsa-miR-299-3p, hsa-miR-624-3p, hsa-miR-1-5p, hsa-miR-152-3p, hsa-miR-1973, hsa-miR-454-3p, hsa-miR-4532) involved in the drug/immune resistance [96]. Eight miRNAs (let-7e, miR-99b, miR-100, miR-125a, miR-125b, miR-146a, miR-146b, and miR-155) detected in patients receiving ipilimumab and nivolumab have been found to correlate with the frequency of altered myeloid cells, shorter progression-free survival as well as overall survival [46].

Finally, Tengda L. and coworkers demonstrated that miR-532-5p and miR-106b, isolated from serous exosomes as well as from total serum, were able to discriminate patients with melanoma from healthy controls, metastatic patients from those with no metastasis, patients with stage I–II disease from those with stage III–IV, and patients treated with pembrolizumab from untreated ones [80].

6. Conclusions and Future Perspectives

In melanoma, several miRNAs are deregulated because of epigenetic changes, impaired transcription, amplification, or deletion of miRNA genes as well as defects in the miRNA biogenesis machinery. It is currently accepted that distinct profiles of miRNA expression are detected at each step of melanoma development, and that an altered expression of miRNAs frequently correlates with poor prognosis and/or inadequate response to treatments. As recapitulated in this review, dysregulated miRNAs may induce and sustain or prevent melanoma cell resistance to BRAFi/MEKi and immune therapies by acting as oncogenes or tumor suppressors, respectively. A partial or complete reversion of melanoma cells resistance to BRAFi and MEKi may be achieved by restoring down-regulated miRNAs or silencing up-regulated miRNAs, suggesting that specific miRNAs or their antagonists may be considered for potential therapeutic applications to overcame melanoma cell resistance to BRAFi and MEKi. In this regard, miRNAs, especially those being part of the circulating transcriptome, may be useful as biomarkers for early melanoma response to treatments, but the strategies for these approaches are still under investigation. In melanoma, implications of microRNAs in the regulation of immune checkpoint blockade and controlling their expression for therapeutic purposes is the subject of intense ongoing research. Specific miRNA signatures associate with specific alterations of immune checkpoint pathways in the melanoma microenvironment while subsets of miRNAs directly regulate the transcription of immune checkpoints. Thus, miRNA could provide new biomarkers predicting patient response to immune checkpoint inhibition and it is reasonable to foresee that combining miRNAs with different immune checkpoint targets could mimic and possibly improve the effect of combined immune checkpoint blockade therapies. Activation of the MAPK pathway through BRAF mutations may be a potential molecular target for overcoming evasion of the immune surveillance by melanoma cells. MAPK cascade stimulates melanoma cells to secrete cytokines, chemokines and soluble growth factors that recruit immune and myeloid cells sustaining both tumor growth and immune evasion. In this contest, specific miRNAs or their antagonists may be considered for potential therapeutic use for restoring the effector function of immune cells. New approaches that look at simultaneous or sequential use of drugs targeting the MAPK pathway with immune checkpoint inhibitors are also a priority, with evidence suggesting that specific miRNAs may overcome melanoma growth and immune evasion. Thus, many questions regarding the best first- and second-line treatment and the best treatment sequence remain to be addressed.

Author Contributions: M.L.M. and M.V.C.: conception of the work. M.L.M., M.M. and G.D.C.: manuscript drafting. M.L.M., P.A.A. and M.V.C.: critical revision of the work and final version approval. All authors have read and agreed to the published version of the manuscript.

Funding: This research was supported by funds provided by the University of Naples "Parthenope" to M. L. Motti (DSMB 187, CUP I62I15000090005, 2017) and by Italian Ministry of Health to M.V.Carriero (project Ricerca Corrente 2019 M4/4 n.2611752).

Acknowledgments: The authors deeply appreciated the assistance of Sabrina Sarno for extensive literature search.

Abbreviations

MAPK	Mitogen-Activated Protein Kinase
BRAFi	BRAF Inhibitor
MEKi	MEK Inhibitor
MAPKi	MAPK Inhibitor
miRNA	microRNA
OncomiR	Oncogenic miRNA
AntagomiR	miRNA antagonist
Ar	Argonaute protein
miRISC	miRNA-induced silencing complex
NGS	Next-Generation Sequencing
EV	Extracellular Vesicle
MITF	Microphthalmia-Associated Transcription Factor
ADAR1	Adenosine Deaminase Acting on RNA-1
ICAM1	Intracellular Adhesion Molecule 1
NK	Natural Killer
MDSC	Myeloid-Derived Suppressor Cell

References

1. Siegel, R.; Ma, J.; Zou, Z.; Jemal, A. Cancer statistics, 2014. *CA Cancer J. Clin.* **2014**, *64*, 9–29. [CrossRef] [PubMed]
2. Karimkhani, C.; Green, A.C.; Nijsten, T.; Weinstock, M.A.; Dellavalle, R.P.; Naghavi, M.; Fitzmaurice, C. The global burden of melanoma: Results from the Global Burden of Disease Study 2015. *Br. J. Dermatol.* **2017**, *177*, 134–140. [CrossRef] [PubMed]

3. Matthews, N.H.; Li, W.-Q.; Qureshi, A.A.; Weinstock, M.A.; Cho, E. Epidemiology of Melanoma. In *Cutaneous Melanoma: Etiology and Therapy*; Ward, W.H., Farma, J.M., Eds.; Codon Publications: Brisbane, Australia, 2017; ISBN 978-0-9944381-4-0.

4. Domingues, B.; Lopes, J.M.; Soares, P.; Pópulo, H. Melanoma treatment in review. *Immunotargets Ther.* **2018**, *7*, 35–49. [CrossRef]

5. Ascierto, P.A.; McArthur, G.A.; Dréno, B.; Atkinson, V.; Liszkay, G.; Di Giacomo, A.M.; Mandalà, M.; Demidov, L.; Stroyakovskiy, D.; Thomas, L.; et al. Cobimetinib combined with vemurafenib in advanced BRAF(V600)-mutant melanoma (coBRIM): Updated efficacy results from a randomised, double-blind, phase 3 trial. *Lancet Oncol.* **2016**, *17*, 1248–1260. [CrossRef]

6. Wolchok, J.D.; Chiarion-Sileni, V.; Gonzalez, R.; Rutkowski, P.; Grob, J.-J.; Cowey, C.L.; Lao, C.D.; Wagstaff, J.; Schadendorf, D.; Ferrucci, P.F.; et al. Overall Survival with Combined Nivolumab and Ipilimumab in Advanced Melanoma. *N. Engl. J. Med.* **2017**, *377*, 1345–1356. [CrossRef]

7. Chapman, P.B.; Hauschild, A.; Robert, C.; Haanen, J.B.; Ascierto, P.; Larkin, J.; Dummer, R.; Garbe, C.; Testori, A.; Maio, M.; et al. Improved survival with vemurafenib in melanoma with BRAF V600E mutation. *N. Engl. J. Med.* **2011**, *364*, 2507–2516. [CrossRef]

8. Flaherty, K.T.; Puzanov, I.; Kim, K.B.; Ribas, A.; McArthur, G.A.; Sosman, J.A.; O'Dwyer, P.J.; Lee, R.J.; Grippo, J.F.; Nolop, K.; et al. Inhibition of mutated, activated BRAF in metastatic melanoma. *N. Engl. J. Med.* **2010**, *363*, 809–819. [CrossRef]

9. Luebker, S.A.; Koepsell, S.A. Diverse Mechanisms of BRAF Inhibitor Resistance in Melanoma Identified in Clinical and Preclinical Studies. *Front. Oncol.* **2019**, *9*, 268. [CrossRef]

10. Long, G.V.; Stroyakovskiy, D.; Gogas, H.; Levchenko, E.; de Braud, F.; Larkin, J.; Garbe, C.; Jouary, T.; Hauschild, A.; Grob, J.-J.; et al. Dabrafenib and trametinib versus dabrafenib and placebo for Val600 BRAF-mutant melanoma: A multicentre, double-blind, phase 3 randomised controlled trial. *Lancet* **2015**, *386*, 444–451. [CrossRef]

11. Dummer, R.; Ascierto, P.A.; Gogas, H.J.; Arance, A.; Mandala, M.; Liszkay, G.; Garbe, C.; Schadendorf, D.; Krajsova, I.; Gutzmer, R.; et al. Encorafenib plus binimetinib versus vemurafenib or encorafenib in patients with BRAF-mutant melanoma (COLUMBUS): A multicentre, open-label, randomised phase 3 trial. *Lancet Oncol.* **2018**, *19*, 603–615. [CrossRef]

12. Wagle, N.; Van Allen, E.M.; Treacy, D.J.; Frederick, D.T.; Cooper, Z.A.; Taylor-Weiner, A.; Rosenberg, M.; Goetz, E.M.; Sullivan, R.J.; Farlow, D.N.; et al. MAP kinase pathway alterations in BRAF-mutant melanoma patients with acquired resistance to combined RAF/MEK inhibition. *Cancer Discov.* **2014**, *4*, 61–68. [CrossRef] [PubMed]

13. Long, G.V.; Fung, C.; Menzies, A.M.; Pupo, G.M.; Carlino, M.S.; Hyman, J.; Shahheydari, H.; Tembe, V.; Thompson, J.F.; Saw, R.P.; et al. Increased MAPK reactivation in early resistance to dabrafenib/trametinib combination therapy of BRAF-mutant metastatic melanoma. *Nat. Commun.* **2014**, *5*, 5694. [CrossRef]

14. Arozarena, I.; Wellbrock, C. Phenotype plasticity as enabler of melanoma progression and therapy resistance. *Nat. Rev. Cancer* **2019**, *19*, 377–391. [CrossRef] [PubMed]

15. Hodi, F.S.; Chesney, J.; Pavlick, A.C.; Robert, C.; Grossmann, K.F.; McDermott, D.F.; Linette, G.P.; Meyer, N.; Giguere, J.K.; Agarwala, S.S.; et al. Combined nivolumab and ipilimumab versus ipilimumab alone in patients with advanced melanoma: 2-year overall survival outcomes in a multicentre, randomised, controlled, phase 2 trial. *Lancet Oncol.* **2016**, *17*, 1558–1568. [CrossRef]

16. Schachter, J.; Ribas, A.; Long, G.V.; Arance, A.; Grob, J.-J.; Mortier, L.; Daud, A.; Carlino, M.S.; McNeil, C.; Lotem, M.; et al. Pembrolizumab versus ipilimumab for advanced melanoma: Final overall survival results of a multicentre, randomised, open-label phase 3 study (KEYNOTE-006). *Lancet* **2017**, *390*, 1853–1862. [CrossRef]

17. Wistuba-Hamprecht, K.; Pawelec, G. KEYNOTE-006: A success in melanoma, but a long way to go. *Lancet* **2017**, *390*, 1816–1817. [CrossRef]

18. Bartel, D.P. MicroRNAs: Genomics, biogenesis, mechanism, and function. *Cell* **2004**, *116*, 281–297. [CrossRef]

19. Kim, J.; Yao, F.; Xiao, Z.; Sun, Y.; Ma, L. MicroRNAs and metastasis: Small RNAs play big roles. *Cancer Metastasis Rev.* **2018**, *37*, 5–15. [CrossRef]

20. Calin, G.A.; Croce, C.M. MicroRNA signatures in human cancers. *Nat. Rev. Cancer* **2006**, *6*, 857–866. [CrossRef]

21. Corrà, F.; Agnoletto, C.; Minotti, L.; Baldassari, F.; Volinia, S. The Network of Non-coding RNAs in Cancer Drug Resistance. *Front. Oncol.* **2018**, *8*, 327. [CrossRef]

22. Gelato, K.A.; Shaikhibrahim, Z.; Ocker, M.; Haendler, B. Targeting epigenetic regulators for cancer therapy: Modulation of bromodomain proteins, methyltransferases, demethylases, and microRNAs. *Expert Opin. Ther. Targets* **2016**, *20*, 783–799. [CrossRef] [PubMed]

23. Saini, H.K.; Griffiths-Jones, S.; Enright, A.J. Genomic analysis of human microRNA transcripts. *Proc. Natl. Acad. Sci. USA* **2007**, *104*, 17719–17724. [CrossRef] [PubMed]

24. Krol, J.; Loedige, I.; Filipowicz, W. The widespread regulation of microRNA biogenesis, function and decay. *Nat. Rev. Genet.* **2010**, *11*, 597–610. [CrossRef] [PubMed]

25. Hayes, J.; Peruzzi, P.P.; Lawler, S. MicroRNAs in cancer: Biomarkers, functions and therapy. *Trends Mol. Med.* **2014**, *20*, 460–469. [CrossRef]

26. Ha, M.; Kim, V.N. Regulation of microRNA biogenesis. *Nat. Rev. Mol. Cell Biol.* **2014**, *15*, 509–524. [CrossRef]

27. Broughton, J.P.; Lovci, M.T.; Huang, J.L.; Yeo, G.W.; Pasquinelli, A.E. Pairing beyond the Seed Supports MicroRNA Targeting Specificity. *Mol. Cell* **2016**, *64*, 320–333. [CrossRef]

28. O'Brien, J.; Hayder, H.; Zayed, Y.; Peng, C. Overview of MicroRNA Biogenesis, Mechanisms of Actions, and Circulation. *Front. Endocrinol. (Lausanne)* **2018**, *9*, 402. [CrossRef] [PubMed]

29. Makarova, J.A.; Shkurnikov, M.U.; Wicklein, D.; Lange, T.; Samatov, T.R.; Turchinovich, A.A.; Tonevitsky, A.G. Intracellular and extracellular microRNA: An update on localization and biological role. *Prog. Histochem. Cytochem.* **2016**, *51*, 33–49. [CrossRef] [PubMed]

30. Ling, H.; Fabbri, M.; Calin, G.A. MicroRNAs and other non-coding RNAs as targets for anticancer drug development. *Nat. Rev. Drug Discov.* **2013**, *12*, 847–865. [CrossRef] [PubMed]

31. Oliveira, A.C.; Bovolenta, L.A.; Alves, L.; Figueiredo, L.; Ribeiro, A.O.; Campos, V.F.; Lemke, N.; Pinhal, D. Understanding the Modus Operandi of MicroRNA Regulatory Clusters. *Cells* **2019**, *8*, 1103. [CrossRef] [PubMed]

32. Sun, X.; Li, J.; Sun, Y.; Zhang, Y.; Dong, L.; Shen, C.; Yang, L.; Yang, M.; Li, Y.; Shen, G.; et al. miR-7 reverses the resistance to BRAFi in melanoma by targeting EGFR/IGF-1R/CRAF and inhibiting the MAPK and PI3K/AKT signaling pathways. *Oncotarget* **2016**, *7*, 53558–53570. [CrossRef] [PubMed]

33. Fattore, L.; Ruggiero, C.F.; Pisanu, M.E.; Liguoro, D.; Cerri, A.; Costantini, S.; Capone, F.; Acunzo, M.; Romano, G.; Nigita, G.; et al. Reprogramming miRNAs global expression orchestrates development of drug resistance in BRAF mutated melanoma. *Cell Death Differ.* **2019**, *26*, 1267. [CrossRef] [PubMed]

34. Bennett, P.E.; Bemis, L.; Norris, D.A.; Shellman, Y.G. miR in melanoma development: miRNAs and acquired hallmarks of cancer in melanoma. *Physiol. Genom.* **2013**, *45*, 1049–1059. [CrossRef] [PubMed]

35. Mueller, D.W.; Rehli, M.; Bosserhoff, A.K. miRNA expression profiling in melanocytes and melanoma cell lines reveals miRNAs associated with formation and progression of malignant melanoma. *J. Investig. Dermatol.* **2009**, *129*, 1740–1751. [CrossRef] [PubMed]

36. Wozniak, M.; Mielczarek, A.; Czyz, M. miRNAs in Melanoma: Tumor Suppressors and Oncogenes with Prognostic Potential. *Curr. Med. Chem.* **2016**, *23*, 3136–3153. [CrossRef]

37. Varamo, C.; Occelli, M.; Vivenza, D.; Merlano, M.; Lo Nigro, C. MicroRNAs role as potential biomarkers and key regulators in melanoma. *Genes Chromosomes Cancer* **2017**, *56*, 3–10. [CrossRef]

38. Mirzaei, H.; Gholamin, S.; Shahidsales, S.; Sahebkar, A.; Jaafari, M.R.; Mirzaei, H.R.; Hassanian, S.M.; Avan, A. MicroRNAs as potential diagnostic and prognostic biomarkers in melanoma. *Eur. J. Cancer* **2016**, *53*, 25–32. [CrossRef]

39. Mannavola, F.; Tucci, M.; Felici, C.; Stucci, S.; Silvestris, F. miRNAs in melanoma: A defined role in tumor progression and metastasis. *Expert Rev. Clin. Immunol.* **2016**, *12*, 79–89. [CrossRef] [PubMed]

40. Gajos-Michniewicz, A.; Czyz, M. Role of miRNAs in Melanoma Metastasis. *Cancers* **2019**, *11*, 326. [CrossRef] [PubMed]

41. Valadi, H.; Ekström, K.; Bossios, A.; Sjöstrand, M.; Lee, J.J.; Lötvall, J.O. Exosome-mediated transfer of mRNAs and microRNAs is a novel mechanism of genetic exchange between cells. *Nat. Cell Biol.* **2007**, *9*, 654–659. [CrossRef]

42. Vickers, K.C.; Palmisano, B.T.; Shoucri, B.M.; Shamburek, R.D.; Remaley, A.T. MicroRNAs are transported in plasma and delivered to recipient cells by high-density lipoproteins. *Nat. Cell Biol.* **2011**, *13*, 423–433. [CrossRef]

43. Solé, C.; Tramonti, D.; Schramm, M.; Goicoechea, I.; Armesto, M.; Hernandez, L.I.; Manterola, L.; Fernandez-Mercado, M.; Mujika, K.; Tuneu, A.; et al. The Circulating Transcriptome as a Source of Biomarkers for Melanoma. *Cancers* **2019**, *11*, 70. [CrossRef]

44. Lorusso, C.; De Summa, S.; Pinto, R.; Danza, K.; Tommasi, S. miRNAs as Key Players in the Management of Cutaneous Melanoma. *Cells* **2020**, *9*, 415. [CrossRef] [PubMed]
45. Couts, K.L.; Anderson, E.M.; Gross, M.M.; Sullivan, K.; Ahn, N.G. Oncogenic B-Raf signaling in melanoma cells controls a network of microRNAs with combinatorial functions. *Oncogene* **2013**, *32*, 1959–1970. [CrossRef] [PubMed]
46. Huber, V.; Vallacchi, V.; Fleming, V.; Hu, X.; Cova, A.; Dugo, M.; Shahaj, E.; Sulsenti, R.; Vergani, E.; Filipazzi, P.; et al. Tumor-derived microRNAs induce myeloid suppressor cells and predict immunotherapy resistance in melanoma. *J. Clin. Investig.* **2018**, *128*, 5505–5516. [CrossRef] [PubMed]
47. Zhang, T.; Dutton-Regester, K.; Brown, K.M.; Hayward, N.K. The genomic landscape of cutaneous melanoma. *Pigment Cell Melanoma Res.* **2016**, *29*, 266–283. [CrossRef]
48. Ascierto, P.A.; Kirkwood, J.M.; Grob, J.-J.; Simeone, E.; Grimaldi, A.M.; Maio, M.; Palmieri, G.; Testori, A.; Marincola, F.M.; Mozzillo, N. The role of BRAF V600 mutation in melanoma. *J. Transl. Med.* **2012**, *10*, 85. [CrossRef] [PubMed]
49. Shi, H.; Hugo, W.; Kong, X.; Hong, A.; Koya, R.C.; Moriceau, G.; Chodon, T.; Guo, R.; Johnson, D.B.; Dahlman, K.B.; et al. Acquired resistance and clonal evolution in melanoma during BRAF inhibitor therapy. *Cancer Discov.* **2014**, *4*, 80–93. [CrossRef] [PubMed]
50. Van Allen, E.M.; Wagle, N.; Sucker, A.; Treacy, D.J.; Johannessen, C.M.; Goetz, E.M.; Place, C.S.; Taylor-Weiner, A.; Whittaker, S.; Kryukov, G.V.; et al. The genetic landscape of clinical resistance to RAF inhibition in metastatic melanoma. *Cancer Discov.* **2014**, *4*, 94–109. [CrossRef] [PubMed]
51. Ascierto, P.A.; Grimaldi, A.M.; Anderson, A.C.; Bifulco, C.; Cochran, A.; Garbe, C.; Eggermont, A.M.; Faries, M.; Ferrone, S.; Gershenwald, J.E.; et al. Future perspectives in melanoma research: Meeting report from the "Melanoma Bridge", Napoli, December 5th-8th 2013. *J. Transl. Med.* **2014**, *12*, 277. [CrossRef]
52. Palmieri, G.; Ombra, M.; Colombino, M.; Casula, M.; Sini, M.; Manca, A.; Paliogiannis, P.; Ascierto, P.A.; Cossu, A. Multiple Molecular Pathways in Melanomagenesis: Characterization of Therapeutic Targets. *Front. Oncol.* **2015**, *5*, 183. [CrossRef]
53. Spagnolo, F.; Ghiorzo, P.; Queirolo, P. Overcoming resistance to BRAF inhibition in BRAF-mutated metastatic melanoma. *Oncotarget* **2014**, *5*, 10206–10221. [CrossRef] [PubMed]
54. Ascierto, P.A.; Marincola, F.M.; Atkins, M.B. What's new in melanoma? Combination! *J. Transl. Med.* **2015**, *13*, 213. [CrossRef] [PubMed]
55. Moriceau, G.; Hugo, W.; Hong, A.; Shi, H.; Kong, X.; Yu, C.C.; Koya, R.C.; Samatar, A.A.; Khanlou, N.; Braun, J.; et al. Tunable-combinatorial mechanisms of acquired resistance limit the efficacy of BRAF/MEK cotargeting but result in melanoma drug addiction. *Cancer Cell* **2015**, *27*, 240–256. [CrossRef] [PubMed]
56. Fratangelo, F.; Camerlingo, R.; Carriero, M.V.; Pirozzi, G.; Palmieri, G.; Gentilcore, G.; Ragone, C.; Minopoli, M.; Ascierto, P.A.; Motti, M.L. Effect of ABT-888 on the apoptosis, motility and invasiveness of BRAFi-resistant melanoma cells. *Int. J. Oncol.* **2018**, *53*, 1149–1159. [CrossRef] [PubMed]
57. Lord, C.J.; Ashworth, A. PARP inhibitors: Synthetic lethality in the clinic. *Science* **2017**, *355*, 1152–1158. [CrossRef] [PubMed]
58. Yadav, V.; Burke, T.F.; Huber, L.; Van Horn, R.D.; Zhang, Y.; Buchanan, S.G.; Chan, E.M.; Starling, J.J.; Beckmann, R.P.; Peng, S.-B. The CDK4/6 inhibitor LY2835219 overcomes vemurafenib resistance resulting from MAPK reactivation and cyclin D1 upregulation. *Mol. Cancer Ther.* **2014**, *13*, 2253–2263. [CrossRef]
59. Liu, S.; Tetzlaff, M.T.; Cui, R.; Xu, X. miR-200c inhibits melanoma progression and drug resistance through down-regulation of BMI-1. *Am. J. Pathol.* **2012**, *181*, 1823–1835. [CrossRef]
60. Liu, S.; Tetzlaff, M.T.; Wang, T.; Yang, R.; Xie, L.; Zhang, G.; Krepler, C.; Xiao, M.; Beqiri, M.; Xu, W.; et al. miR-200c/Bmi1 axis and epithelial-mesenchymal transition contribute to acquired resistance to BRAF inhibitor treatment. *Pigment Cell Melanoma Res.* **2015**, *28*, 431–441. [CrossRef]
61. Liu, S.-M.; Lu, J.; Lee, H.-C.; Chung, F.-H.; Ma, N. miR-524-5p suppresses the growth of oncogenic BRAF melanoma by targeting BRAF and ERK2. *Oncotarget* **2014**, *5*, 9444–9459. [CrossRef]
62. Fattore, L.; Mancini, R.; Acunzo, M.; Romano, G.; Laganà, A.; Pisanu, M.E.; Malpicci, D.; Madonna, G.; Mallardo, D.; Capone, M.; et al. miR-579-3p controls melanoma progression and resistance to target therapy. *Proc. Natl. Acad. Sci. USA* **2016**, *113*, E5005–E5013. [CrossRef] [PubMed]
63. Freedman, D.A.; Wu, L.; Levine, A.J. Functions of the MDM2 oncoprotein. *Cell. Mol. Life Sci.* **1999**, *55*, 96–107. [CrossRef]

64. Streicher, K.L.; Zhu, W.; Lehmann, K.P.; Georgantas, R.W.; Morehouse, C.A.; Brohawn, P.; Carrasco, R.A.; Xiao, Z.; Tice, D.A.; Higgs, B.W.; et al. A novel oncogenic role for the miRNA-506-514 cluster in initiating melanocyte transformation and promoting melanoma growth. *Oncogene* **2012**, *31*, 1558–1570. [CrossRef] [PubMed]

65. Stark, M.S.; Bonazzi, V.F.; Boyle, G.M.; Palmer, J.M.; Symmons, J.; Lanagan, C.M.; Schmidt, C.W.; Herington, A.C.; Ballotti, R.; Pollock, P.M.; et al. miR-514a regulates the tumour suppressor NF1 and modulates BRAFi sensitivity in melanoma. *Oncotarget* **2015**, *6*, 17753–17763. [CrossRef] [PubMed]

66. Kim, J.-H.; Ahn, J.-H.; Lee, M. Upregulation of MicroRNA-1246 Is Associated with BRAF Inhibitor Resistance in Melanoma Cells with Mutant BRAF. *Cancer Res. Treat.* **2017**, *49*, 947–959. [CrossRef]

67. Koetz-Ploch, L.; Hanniford, D.; Dolgalev, I.; Sokolova, E.; Zhong, J.; Díaz-Martínez, M.; Bernstein, E.; Darvishian, F.; Flaherty, K.T.; Chapman, P.B.; et al. MicroRNA-125a promotes resistance to BRAF inhibitors through suppression of the intrinsic apoptotic pathway. *Pigment Cell Melanoma Res.* **2017**, *30*, 328–338. [CrossRef]

68. Dror, S.; Sander, L.; Schwartz, H.; Sheinboim, D.; Barzilai, A.; Dishon, Y.; Apcher, S.; Golan, T.; Greenberger, S.; Barshack, I.; et al. Melanoma miRNA trafficking controls tumour primary niche formation. *Nat. Cell Biol.* **2016**, *18*, 1006–1017. [CrossRef]

69. Lunavat, T.R.; Cheng, L.; Kim, D.-K.; Bhadury, J.; Jang, S.C.; Lässer, C.; Sharples, R.A.; López, M.D.; Nilsson, J.; Gho, Y.S.; et al. Small RNA deep sequencing discriminates subsets of extracellular vesicles released by melanoma cells–Evidence of unique microRNA cargos. *RNA Biol.* **2015**, *12*, 810–823. [CrossRef]

70. Lunavat, T.R.; Cheng, L.; Einarsdottir, B.O.; Olofsson Bagge, R.; Veppil Muralidharan, S.; Sharples, R.A.; Lässer, C.; Gho, Y.S.; Hill, A.F.; Nilsson, J.A.; et al. BRAFV600 inhibition alters the microRNA cargo in the vesicular secretome of malignant melanoma cells. *Proc. Natl. Acad. Sci. USA* **2017**, *114*, E5930–E5939. [CrossRef]

71. Díaz-Martínez, M.; Benito-Jardón, L.; Alonso, L.; Koetz-Ploch, L.; Hernando, E.; Teixidó, J. miR-204-5p and miR-211-5p Contribute to BRAF Inhibitor Resistance in Melanoma. *Cancer Res.* **2018**, *78*, 1017–1030. [CrossRef]

72. Kim, M.H.; Kim, J.; Hong, H.; Lee, S.-H.; Lee, J.-K.; Jung, E.; Kim, J. Actin remodeling confers BRAF inhibitor resistance to melanoma cells through YAP/TAZ activation. *EMBO J.* **2016**, *35*, 462–478. [CrossRef]

73. Kim, M.H.; Kim, J. Role of YAP/TAZ transcriptional regulators in resistance to anti-cancer therapies. *Cell. Mol. Life Sci.* **2017**, *74*, 1457–1474. [CrossRef] [PubMed]

74. Choe, M.H.; Yoon, Y.; Kim, J.; Hwang, S.-G.; Han, Y.-H.; Kim, J.-S. miR-550a-3-5p acts as a tumor suppressor and reverses BRAF inhibitor resistance through the direct targeting of YAP. *Cell Death Dis.* **2018**, *9*, 640. [CrossRef]

75. Caporali, S.; Amaro, A.; Levati, L.; Alvino, E.; Lacal, P.M.; Mastroeni, S.; Ruffini, F.; Bonmassar, L.; Antonini Cappellini, G.C.; Felli, N.; et al. miR-126-3p down-regulation contributes to dabrafenib acquired resistance in melanoma by up-regulating ADAM9 and VEGF-A. *J. Exp. Clin. Cancer Res.* **2019**, *38*, 272. [CrossRef]

76. Galore-Haskel, G.; Nemlich, Y.; Greenberg, E.; Ashkenazi, S.; Hakim, M.; Itzhaki, O.; Shoshani, N.; Shapira-Fromer, R.; Ben-Ami, E.; Ofek, E.; et al. A novel immune resistance mechanism of melanoma cells controlled by the ADAR1 enzyme. *Oncotarget* **2015**, *6*, 28999–29015. [CrossRef]

77. Ueda, R.; Kohanbash, G.; Sasaki, K.; Fujita, M.; Zhu, X.; Kastenhuber, E.R.; McDonald, H.A.; Potter, D.M.; Hamilton, R.L.; Lotze, M.T.; et al. Dicer-regulated microRNAs 222 and 339 promote resistance of cancer cells to cytotoxic T-lymphocytes by down-regulation of ICAM-1. *Proc. Natl. Acad. Sci. USA* **2009**, *106*, 10746–10751. [CrossRef] [PubMed]

78. Blank, C.; Brown, I.; Kacha, A.K.; Markiewicz, M.A.; Gajewski, T.F. ICAM-1 contributes to but is not essential for tumor antigen cross-priming and CD8+ T cell-mediated tumor rejection in vivo. *J. Immunol.* **2005**, *174*, 3416–3420. [CrossRef] [PubMed]

79. Kong, D.-H.; Kim, Y.K.; Kim, M.R.; Jang, J.H.; Lee, S. Emerging Roles of Vascular Cell Adhesion Molecule-1 (VCAM-1) in Immunological Disorders and Cancer. *Int. J. Mol. Sci.* **2018**, *19*, 1057. [CrossRef]

80. Tengda, L.; Shuping, L.; Mingli, G.; Jie, G.; Yun, L.; Weiwei, Z.; Anmei, D. Serum exosomal microRNAs as potent circulating biomarkers for melanoma. *Melanoma Res.* **2018**, *28*, 295–303. [CrossRef]

81. Zhang, Y.; Tanno, T.; Kanellopoulou, C. Cancer therapeutic implications of microRNAs in the regulation of immune checkpoint blockade. *ExRNA* **2019**, *1*, 19. [CrossRef]

82. Li, Q.; Johnston, N.; Zheng, X.; Wang, H.; Zhang, X.; Gao, D.; Min, W. miR-28 modulates exhaustive differentiation of T cells through silencing programmed cell death-1 and regulating cytokine secretion. *Oncotarget* **2016**, *7*, 53735–53750. [CrossRef] [PubMed]

83. Martinez-Usatorre, A.; Sempere, L.F.; Carmona, S.J.; Carretero-Iglesia, L.; Monnot, G.; Speiser, D.E.; Rufer, N.; Donda, A.; Zehn, D.; Jandus, C.; et al. MicroRNA-155 Expression Is Enhanced by T-cell Receptor Stimulation Strength and Correlates with Improved Tumor Control in Melanoma. *Cancer Immunol. Res.* **2019**, *7*, 1013–1024. [CrossRef] [PubMed]

84. Schmiedel, D.; Mandelboim, O. NKG2D Ligands-Critical Targets for Cancer Immune Escape and Therapy. *Front Immunol.* **2018**, *9*, 2040. [CrossRef]

85. Heinemann, A.; Zhao, F.; Pechlivanis, S.; Eberle, J.; Steinle, A.; Diederichs, S.; Schadendorf, D.; Paschen, A. Tumor suppressive microRNAs miR-34a/c control cancer cell expression of ULBP2, a stress-induced ligand of the natural killer cell receptor NKG2D. *Cancer Res.* **2012**, *72*, 460–471. [CrossRef]

86. Cobos Jiménez, V.; Bradley, E.J.; Willemsen, A.M.; van Kampen, A.H.C.; Baas, F.; Kootstra, N.A. Next-generation sequencing of microRNAs uncovers expression signatures in polarized macrophages. *Physiol. Genom.* **2014**, *46*, 91–103. [CrossRef]

87. Banerjee, S.; Cui, H.; Xie, N.; Tan, Z.; Yang, S.; Icyuz, M.; Thannickal, V.J.; Abraham, E.; Liu, G. miR-125a-5p regulates differential activation of macrophages and inflammation. *J. Biol. Chem.* **2013**, *288*, 35428–35436. [CrossRef] [PubMed]

88. Lee, H.-M.; Kim, T.S.; Jo, E.-K. MiR-146 and miR-125 in the regulation of innate immunity and inflammation. *BMB Rep.* **2016**, *49*, 311–318. [CrossRef] [PubMed]

89. Sumimoto, H.; Imabayashi, F.; Iwata, T.; Kawakami, Y. The BRAF-MAPK signaling pathway is essential for cancer-immune evasion in human melanoma cells. *J. Exp. Med.* **2006**, *203*, 1651–1656. [CrossRef] [PubMed]

90. Khalili, J.S.; Liu, S.; Rodríguez-Cruz, T.G.; Whittington, M.; Wardell, S.; Liu, C.; Zhang, M.; Cooper, Z.A.; Frederick, D.T.; Li, Y.; et al. Oncogenic BRAF(V600E) promotes stromal cell-mediated immunosuppression via induction of interleukin-1 in melanoma. *Clin. Cancer Res.* **2012**, *18*, 5329–5340. [CrossRef] [PubMed]

91. Jiang, X.; Zhou, J.; Giobbie-Hurder, A.; Wargo, J.; Hodi, F.S. The activation of MAPK in melanoma cells resistant to BRAF inhibition promotes PD-L1 expression that is reversible by MEK and PI3K inhibition. *Clin. Cancer Res.* **2013**, *19*, 598–609. [CrossRef]

92. Vergani, E.; Di Guardo, L.; Dugo, M.; Rigoletto, S.; Tragni, G.; Ruggeri, R.; Perrone, F.; Tamborini, E.; Gloghini, A.; Arienti, F.; et al. Overcoming melanoma resistance to vemurafenib by targeting CCL2-induced miR-34a, miR-100 and miR-125b. *Oncotarget* **2016**, *7*, 4428–4441. [CrossRef] [PubMed]

93. Audrito, V.; Serra, S.; Stingi, A.; Orso, F.; Gaudino, F.; Bologna, C.; Neri, F.; Garaffo, G.; Nassini, R.; Baroni, G.; et al. PD-L1 up-regulation in melanoma increases disease aggressiveness and is mediated through miR-17-5p. *Oncotarget* **2017**, *8*, 15894–15911. [CrossRef] [PubMed]

94. Mitchell, P.S.; Parkin, R.K.; Kroh, E.M.; Fritz, B.R.; Wyman, S.K.; Pogosova-Agadjanyan, E.L.; Peterson, A.; Noteboom, J.; O'Briant, K.C.; Allen, A.; et al. Circulating microRNAs as stable blood-based markers for cancer detection. *Proc. Natl. Acad. Sci. USA* **2008**, *105*, 10513–10518. [CrossRef] [PubMed]

95. Svedman, F.C.; Lohcharoenkal, W.; Bottai, M.; Brage, S.E.; Sonkoly, E.; Hansson, J.; Pivarcsi, A.; Eriksson, H. Extracellular microvesicle microRNAs as predictive biomarkers for targeted therapy in metastastic cutaneous malignant melanoma. *PLoS ONE* **2018**, *13*, e0206942. [CrossRef]

96. Van Laar, R.; Lincoln, M.; Van Laar, B. Development and validation of a plasma-based melanoma biomarker suitable for clinical use. *Br. J. Cancer* **2018**, *118*, 857–866. [CrossRef]

International Journal of
Molecular Sciences

Review

Role of Human Leukocyte Antigen System as A Predictive Biomarker for Checkpoint-Based Immunotherapy in Cancer Patients

Francesco Sabbatino [1,2], Luigi Liguori [3], Giovanna Polcaro [1], Ilaria Salvato [1,4], Gaetano Caramori [4], Francesco A. Salzano [1], Vincenzo Casolaro [1], Cristiana Stellato [1], Jessica Dal Col [1,*] and Stefano Pepe [1,2]

[1] Department of Medicine, Surgery and Dentistry 'Scuola Medica Salernitana', University of Salerno, 84081 Baronissi, Salerno, Italy; fsabbatino@unisa.it (F.S.); gpolcaro@unisa.it (G.P.); il.salvato22@gmail.com (I.S.); frsalzano@unisa.it (F.A.S.); vcasolaro@unisa.it (V.C.); cstellato@unisa.it (C.S.); spepe@unisa.it (S.P.)

[2] Oncology Unit, AOU San Giovanni di Dio e Ruggi D'Aragona, 84131 Salerno, Italy

[3] Department of Clinical Medicine and Surgery, University of Naples "Federico II", 80131 Naples, Italy; luigiliguori1992@gmail.com

[4] Pulmonary Unit, Department of Biomedical Sciences, Dentistry, Morphological and Functional Imaging (BIOMORF), University of Messina, 98125 Messina, Italy; gaetano.caramori@unime.it

* Correspondence: jdalcol@unisa.it; Tel.: +39-08996-5210

Received: 20 August 2020; Accepted: 29 September 2020; Published: 2 October 2020

Abstract: Recent advances in cancer immunotherapy have clearly shown that checkpoint-based immunotherapy is effective in a small subgroup of cancer patients. However, no effective predictive biomarker has been identified so far. The major histocompatibility complex, better known in humans as human leukocyte antigen (HLA), is a very polymorphic gene complex consisting of more than 200 genes. It has a crucial role in activating an appropriate host immune response against pathogens and tumor cells by discriminating self and non-self peptides. Several lines of evidence have shown that down-regulation of expression of HLA class I antigen derived peptide complexes by cancer cells is a mechanism of tumor immune escape and is often associated to poor prognosis in cancer patients. In addition, it has also been shown that HLA class I and II antigen expression, as well as defects in the antigen processing machinery complex, may predict tumor responses in cancer immunotherapy. Nevertheless, the role of HLA in predicting tumor responses to checkpoint-based immunotherapy is still debated. In this review, firstly, we will describe the structure and function of the HLA system. Secondly, we will summarize the HLA defects and their clinical significance in cancer patients. Thirdly, we will review the potential role of the HLA as a predictive biomarker for checkpoint-based immunotherapy in cancer patients. Lastly, we will discuss the potential strategies that may restore HLA function to implement novel therapeutic strategies in cancer patients.

Keywords: major histocompatibility complex (MHC); human leukocyte antigen (HLA); antigen processing machinery (APM) molecules; carcinogenesis; tumor predisposition; biomarker; cancer immunotherapy

---

1. Human Leucocyte Antigen and Antigen Presentation Machinery Molecules: An Overview

1.1. HLA Class I: Structure and Function

The major histocompatibility complex (MHC), better known in humans as human leukocyte antigen (HLA), is a very polymorphic gene complex encoding for cell surface molecules specialized to present and recognize self and non-self peptides [1–8]. HLA complex contains more than 200 identified loci located close together on a 3 Mbp stretch within the short arm of chromosome 6 [6,8,9]. Population surveys have identified several thousands of allelic variants of HLA molecules which mainly affect the nature and composition of their peptide-binding groove, regulating the peptide repertoire presented on the cell membrane [6,8–10]. These allelic variants can be associated with an increased risk of various diseases including cancer [11].

HLA is categorized into three groups on the basis of function and structure: class I, II and III [12,13]. HLA class I molecules are expressed on the surface of nucleated cells, except for germ line and some neuronal cells [14]. HLA class I molecules display on cell membrane peptide fragments derived from endogenously degraded self or non-self proteins to T-cell receptor (TCR) of CD8+ cytotoxic T lymphocytes (CTLs) [7,15–18]. Peptides derived from unmutated (self) proteins are normally ignored by CTLs, whereas those derived from mutated (self) or pathogen (non-self) proteins are recognized and trigger an adaptive immune response [15–17]. Particularly, tumor cells are characterized by mutated genes and aberrantly expressed cellular proteins from which derived tumor specific antigens (TSAs) and tumor-associated antigens (TAAs). Through the presentation of TAAs/TSAs, tumor cells become susceptible to CTL-mediated lysis. Using this immune surveillance system, CTLs eradicate intracellular pathogens and exert potent antitumor activity, eliminating the transformed or infected cells through the adaptive immune response [19]. In addition, HLA class I molecules can also present peptides generated from exogenous proteins, a process known as cross-presentation [14,20]. This process is necessary to recognize and destroy tumor cells as well as viruses that do not readily infect antigen-presenting cells, stimulating naïve T cells into activated CTLs [14,20,21]. Specifically, cross-presentation involves dendritic cells (DCs) which present TAAs/TSAs or pathogen-derived peptides in their HLA class I complex to naïve T cells [22–24]. Extracellular peptide loading on HLA class I complex differs from the canonical way followed by intracellular peptides and it will be discussed below in this review. However, recently, several lines of evidence demonstrated that macrophages are also able to implement a cross-presentation process, subverting the original belief that it is an exclusive characteristic of DCs [22].

Besides mediating an adaptive immune response, HLA class I molecules also play a key role in the innate immune response since they serve as ligands of inhibitory killer cell immunoglobulin-like receptors (KIRs) of Natural Killer (NK) cells [25]. Because the majority of healthy nucleated cells express HLA class I molecules, inhibitory KIRs ensure that NK cells do not attack normal cells which express HLA class I molecules but eliminate infected and tumor cells which may have reduced expression of HLA class I molecules [25,26]. Structurally, HLA class I molecules are heterodimers that consist of two polypeptide chains, alpha (α) heavy chain and β2-microglobulin (β2-m) light chain [27]. The β2-m subunit is not polymorphic and is encoded on human chromosome 15. In contrast, the α chain is polymorphic and is encoded by HLA class I genes, further categorized in HLA-A, -B, and –C, according to the locus of their encoding gene [28–30]. The α heavy chain has three extracellular domains (α 1-3, with α1 being at the N-terminus), a transmembrane region and a C-terminal cytoplasmic tail [30–32]. The only invariant region is the Ig-like α3 domain, essential for non-covalent association with the β2-m light chain [30–32]. The α3-CD8 interaction holds the HLA class I molecules in place, while the TCR binds to α1-α2 and checks the coupled peptide for antigenicity [30,32]. The α1 and α2 domains fold to make up a groove for peptides to bind. Bound peptides are predominantly 8-10 amino acid in length, but longer peptides have also been reported [30,32,33].

Besides HLA-A, B, and C, some other HLA class I molecules are also encoded by non-classical HLA loci. Those include HLA-E, which primarily presents various peptides that are derived from the leader sequence of some HLA class I molecules. It blocks conventional NKs expressing the inhibitory heterodimeric NKG2A/CD94 receptor. Lastly, HLA-F mainly resides intracellularly and rarely reaches the cell surface; HLA-G, plays a role in protecting the fetus from the maternal immune responses [34–39].

1.2. HLA Class I Antigen Processing Machinery Complex and Antigen Presentation

The generation and expression of HLA class I antigen-derived peptide complexes is a multistep process and requires an integral and functional HLA class I antigen processing machinery (APM) [25]. This is constituted by several distinct components, such as the proteasome complex, the ubiquitination system, the transporters associated with antigen processing (TAP)1 and TAP2, the endoplasmic chaperone molecules (calnexin, calreticulin, ERp57, and tapasin), and the Golgi apparatus [25,40–42]. The generation and expression of HLA class I antigen derived peptide complexes and their presentation to naïve CD8+ T cells require four main tasks: (i) peptide generation and trimming; (ii) peptide transport; (iii) assembly of the HLA class I loading complex; and (iv) antigen presentation [25,30,43–48]. Firstly, proteins are targeted for degradation by the covalent attachment of multiple copies of the 76-residue protein ubiquitin to free amino groups of Lys [25,49]. Subsequently, they are transferred to the proteasome, where the catalytic core, called the 20S proteasome, contains α and β subunits. The catalytic core interacts with regulatory particles and creates a physical barrier to regulate access to the gate. The latter has protease catalytic activity [25,50–52]. Three of the 20S proteasome's β subunits $\delta(\beta 1)$, $Z(\beta 5)$, and $MB1(\beta 2)$ may be replaced by the functionally different counterparts low molecular proteins (LMP) as LMP2 (also called $\beta 1i$), LMP7 ($\beta 5i$), and LMP10 ($\beta 2i$), respectively [25,53–55]. Proteasome incorporating LMP2, LMP7 and LMP10 is called immunoproteasomes because it develops under conditions of intensified immune response [25]. The immunoproteasome formation is induced during inflammation by stimulation with type I (α and β) or type II (γ) interferons (IFNs) [25,56,57]. Moreover, the immunoproteasome is involved in other activities such as generation of cytokines as well as regulation of T cell differentiation, survival and function during thymocyte development [25,58,59]. Peptides generated in the proteasome are then actively transported from the cytosol into the endoplasmic reticulum (ER) lumen by TAP [25]. TAP is a heterodimeric complex composed of two half-transporters, TAP1 and TAP2, members of the adenosine triphosphate (ATP)-binding cassette transporter family. This complex forms a transmembrane pore in the ER membrane whose opening and closing depend on ATP binding and hydrolysis, respectively (ATP switch model) [25,60–63]. TAP transports most efficiently peptides of a well-defined length (8–12 residues), while longer peptides can be further trimmed in the ER lumen or, alternatively, can be transported back to the cytosol where they are trimmed by cytosolic peptidases and recycle back to the ER [25,64–70]. Peptides transported into the ER by TAP are loaded onto nascent HLA class I molecules with the assistance of four chaperone proteins: calnexin, the thiol oxidoreductase ERp57, calreticulin, and tapasin [25,71–79]. Specifically, the HLA class I α heavy chain interacts with calnexin, which facilitates its complete folding and, by acting in concert with ERp57, ensures the correct oxidation [25,80,81]. At this point, the conformation of the α heavy chain is recognizable by $\beta 2$-m [25,82]. Their binding triggers the release of calnexin [25,82,83]. The resulting conformational changes give the α heavy chain/$\beta 2$-m heterodimer an "open" form that interacts with calreticulin [25,74]. High affinity peptide binding requires the additional participation of tapasin, which links the complex to nascent HLA class I molecules [25,75,79]. After peptide loading, HLA class I derived peptide complex dissociates from TAP as well as from ER-resident chaperones and clusters at export sites on the ER membrane, where it is selectively recruited into cargo vesicles for transport to the Golgi apparatus and then to the cell membrane [25]. On the membrane, the HLA class I derived peptide complex is extracellularly exposed to be recognized by the TCR of naïve T cells, potentially triggering an adaptive immune response when non self or mutated self antigen derived peptides are expressed [19,25].

During cross-presentation extracellular antigens need to enter into canonical HLA class I route and they can exploit various ways. (i) Extracellular peptides can be directly transferred from infected or tumor cells to the cytosol of DCs through the Gap junctions. (ii) ER components can fuse with endosomal/phagosomal pathway and the exogenous peptides are exported from phagosome into cytosol through the ER-associated protein degradation system. (iii) Recycling HLA class I molecules are loaded with extracellular peptides into recycling endosome. (iv) Exosomes secreted by infected or tumor cells can directly bind to DCs [84,85].

1.3. HLA Class II: Structure and Functions

In contrast to HLA class I molecules, HLA class II molecules are usually present only on professional antigen-presenting cells (APCs) (B cells, macrophages, DCs, Langerhans cells), thymic epithelium and activated (but not resting) T cells [86–88]. In all other nucleated cells, HLA class II antigen expression can be induced by IFN-γ [88–90]. HLA class II molecules promote the switch of naïve T cells into activated T cells by presenting exogenously derived antigen peptides to CD4+ T cells [86–90]. Moreover, HLA class II molecules regulate the functions of B cells, macrophages and T cells [87,88,91]. They are encoded by genes in the HLA-DP, -DQ and -DR loci of the chromosome 6 cluster [6]. HLA class II molecules consist of 2 highly polymorphic polypeptides, the α and β chains [87,92–94]. Only the β2 domain of the β chain is a non-polymorphic region. It constitutes the binding site for the CD4+ T cell co-receptor [87,92–96]. HLA class II molecules have a peptide-binding domain, an Ig-like domain and a transmembrane region with a cytoplasmic tail and are responsible for binding peptides (15-24 amino acids) derived from extracellular sources [87,92–96]. Therefore, HLA class II binds peptides longer than HLA class I and accommodates peptide side chains within its binding pocket. These two features increase HLA class II peptide diversity [97,98].

1.4. HLA Class II Antigen Presentation System: How it Works

Compared to HLA class I, also the HLA class II antigen presentation process is characterized by different tasks. This involves several molecules and protein complexes [99,100]. Firstly, α and β chains are assembled in the ER with the invariant chain (li, CD74), forming the (α/β-li)$_3$ complexes [99–101]. Invariant chain li occupies the peptide binding groove of HLA class II, preventing peptide loading within ER [98,102]. Li targets HLA class II containing vesicles to acidic endosomes. Then, into these acidic endosomes, called MHC class II compartments (MIICs), the li chain undergoes selective proteolytic digestion, forming the class II-associated I chain peptide (CLIP). This peptide occupies the groove of HLA class II dimers [100,103]. Subsequently, CLIP is exchanged by tightly bound peptides derived from proteins degraded into the endosomal pathway [99,100]. HLA-DM molecules are crucial to facilitate this exchange by promoting CLIP removal and stabilizing the peptide free status of HLA class II molecules. Moreover, HLA-DM also catalyzes the release of weakly bound peptide, ensuring that only strong bound peptide HLA class II complexes reach the cell surface [98]. Finally, the HLA class II derived peptides complexes are exposed on APCs [100,103,104].

1.5. HLA Class I and II Transcription Regulation

The transcription of genes encoding for the components of HLA class I and II complex is tightly regulated according with the crucial role of these molecules to obtain an effective adaptive immune response. HLA class I genes, except for HLA-G, contain several conserved *cis*-acting regulatory elements. Specifically, three different elements are important for both constitutive and inducible expression. The first element, called enhancer A, contains a binding-site for the nuclear factor kB (NF-kB). The second one corresponds to an IFN-sensitive response element (ISRE) and allows the binding of IFN Regulatory Factors 1 (IRF1). Lastly, the third one is an SXY module comprising four different boxes: W/S, X1, X2 and Y. Equally, the promoter of β2-m, but not those of other genes involved in antigen processing and presentation such as TAP or LMP, contains all three *cis*-acting regulatory elements in its proximal region [105]. Conversely, HLA class II gene proximal promoters contain only the SXY module which is bound in its X1 box by the regulatory factor X (RFX) complex, which comprises RFX5, RFX-associated ankyrin-containing protein (RFXANK) and RFX-associated protein (RFXAP). The cAMP-responsive element binding protein 1 (CREB1) and the activating transcription factor 1 (ATF1) bind the X2 box; the nuclear transcription factor Y (NFY) complex interacts with the Y box. Instead, the elements interacting with W/S box remains poorly defined [106,107]. The interactors with SXY module of genes belonging to both HLA class I and II are crucial elements in the transcriptional control. Indeed, since the identification in 1993 of the class II trans-activator CIITA [108] and, more recently of the NOD-like receptor 5 (NLRC5) [109], also called class I trans-activator CITA, it has been clearly highlighted that both these NLR proteins miss of a DNA-binding domain. Therefore, both NLRC5/CITA and CIITA need to cooperate with the multiprotein complex that is assembled on the SXY module to exert their transactivation activity (forming CITA- and CIITA enhanceosomes) [110,111]. Different studies exploiting CIITA-deficient mice [112] and several molecular analyses performed in patients affected by bare lymphocyte disease (BLS) with HLA class II deficiency confirmed CIITA as the master regulator of HLA class II expression [92,113,114]. Differently NLRC5/CITA is defined as a key regulator of HLA class I, especially in selected immune cell subsets. Indeed, the generation of NLRC5/CITA knockout mice in three independent studies has allowed to show a retention of HLA class I expression in professional APCs also in the absence of the trans-activator [115–117], suggesting the presence of a compensatory mechanism. These results agreed with previous findings regarding the ability of CIITA to contribute to HLA class I expression control [118]. Moreover, NLRC5/CITA regulates the expression of other genes involved in HLA class I presentation and processing, such as β2-m, LMP2, and TAP1 [109]. Interestingly, the up-regulation of both NLRC5/CITA and CIITA is critical for the efficient induction of HLA class I and II, respectively, by IFN-γ stimulation. The induction of NLRC5/CITA by IFN-γ precedes HLA class I gene expression as well as CIITA transcript levels are induced earlier than HLA class II genes upon IFN-γ stimulation [119].

1.6. HLA Class III: A Poorly Characterized Class

The structure and function of HLA class III molecules are poorly defined. They are not involved in antigen binding but in inflammatory processes. Their gene cluster is present between those of class I and class II molecules and encodes important molecules involved in inflammatory processes including complement components C2 and C4, factor B, tumor necrosis factor (TNF)-α, lymphotoxin, and heat shock proteins [12,120–122].

1.7. Carcinoma Cells as Non-Professional APC: A Novel Role for HLA Class II Complex

As mentioned above, HLA class II is usually express only on APCs' surface, playing a crucial role in CD4+ T cell activation [86–88]. However, several lines of evidence showed that many type of cancer cells can also express MHC class II complex regardless tissue origin [123–130]. So far, the role of tumor specific MHC class II (tsMHC-II) expression remains unclear. Conflicting evidences are reported about how tsMHC-II regulates cancer progression as well as immune checkpoint inhibitor (ICI)-based immunotherapy response [98]. The expression of tsMHC-II has been related to longer progression free survival (PFS) and overall survival (OS) in melanoma and Hodgkin lymphoma patients treated with programmed death cell 1 (PD-1)/programmed death-ligand 1 (PD-L1) monoclonal antibodies (mAbs) [123,125,126]. In contrast, a similar association was not found in melanoma patients treated with a cytotoxic T-lymphocyte-associated protein 4 (CTLA-4) mAb [126]. Two independent studies on breast cancer specimens, evaluating tsMHC-II expression by both immunohistochemistry (IHC) and RNA sequencing demonstrated that tsMHC-II expression positivity correlated with longer disease free survival (DFS) and PFS [124,131]. These results were observed also in advanced-stage serous ovarian cancer [132]. However, further clinical trials are needed to define the role of tsMHC-II expression as a potential biomarker for ICI-based immunotherapy in cancer patients.

2. Defects and Clinical Significance of HLA in Human Cancer

2.1. HLA Class I Molecule Defects

Aberrations in expression of HLA class I derived peptide complex have frequently been observed in several types of cancer both in vivo and in vitro. Their frequency ranges from 0–90%. Depending on tumor types, these defects have been associated with aggressive histopathological features as well as poor survival [133–147]. Most of the defects are caused by genetic or epigenetic mutations as well as by transcriptional or post-translational modifications [145,148,149]. These types of alterations can induce a total loss or down-regulation of HLA class I derived peptide complex as well as selective loss of HLA class I haplotypes or alleles (Figure 1) [150–152].

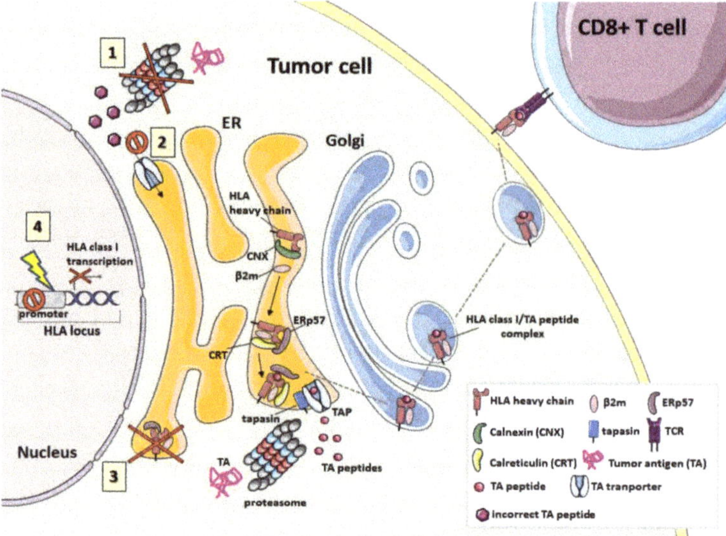

Figure 1. Defects in tumor antigen processing, translocation and loading on HLA class I. Normally tumor antigens (TAs) are degraded by proteasome/immunoproteasome into TA peptides and translocated into the endoplasmic reticulum (ER) through ATP-dependent activation of TAP transporters. Then, different chaperones: tapasin, calnexin (CNX), calreticulin (CRT) and ERp57 form a multimeric complex that provides for the correct assembly of HLA class I and for peptide loading. The lack of TA presentation during tumor development can be determined by different defective mechanisms depicted in the cartoon. **1.** Mutations in genes coding for proteasome subunits or deregulation of their expression implicate an incorrect TA degradation and the production of modified TA peptides. **2.** Mutations in TAP genes, associated with down-regulation of their expression or to their dysfunction, reduce the translocation of TA peptides into the ER. **3.** Defects in the expression of chaperones reduce the stable assembly of the "peptide-free" HLA class I molecule and of the HLA class I molecule-TA peptide complexes inhibiting a correct and efficient TA peptide presentation. **4.** Defects in HLA class I gene expression involve the total loss of these genes or mechanisms that control their transcription resulting in HLA class I molecule down-regulation.

A complete loss of HLA class I derived peptide complex requires two genetic events: a mutation in one copy of the wild-type β2-m and loss of the other non-mutated copy. This phenomenon leads to the loss of heterozygosity (LOH), a genetic abnormality frequently found in malignant cells [153–155]. The mutations in β2-m can range from large deletions to single nucleotide mutations. Both types of alterations in most cases inhibit the translation of β2-m mRNA or abolish the disulfide linkage required for the native structure of β2-m, preventing its binding to HLA class I heavy chains [138,139,153]. Although mutations in β2-m can be randomly distributed, a mutation hot spot located in the CT repeat region of exon 1 has been identified in more than 75% of tumor cells, reflecting an increased genetic instability of this region during malignant transformation. As a result tumor cells present total HLA class I molecule loss since the HLA class I heavy chain-β2-m-peptide complex is not formed and not transported to the cell membrane [156–159]. In contrast, selective HLA class I allospecificity loss requires only one genetic event. This involves mutations of HLA class I allele(s) which inhibit HLA class I molecule transcription or translation. The other allele remains intact and no LOH is required. Loss of one HLA class I haplotype, e.g., HLA-A24, -B56, -Cw7, appears to be frequently caused by loss of segments of the short arm of chromosome 6, where HLA class I genes reside. LOH at chromosome 6 represents a frequent mechanism that contributes to selective HLA haplotype loss in tumors [160,161].

In addition, multiple types of alterations can induce down-regulation of HLA class I molecules [150–152]. Specifically, transcriptional activity of HLA class I heavy chain genes can be suppressed by: (i) the presence of silencer localized at the distal promoter region of HLA class I heavy chain gene [162,163]; (ii) epigenetic mechanisms which alter chromatin structure of the HLA class I heavy chain gene promoters; and (iii) DNA hypermethylation [162,164–166]. It is well known that the constitutive patterns of DNA methylation in solid and hematopoietic human malignancies are characterized by global hypomethylation with concomitant localized hypermethylation of DNA [148,167,168]. Furthermore, an impaired function of one of the APM components can also reduce the expression of HLA class I derived peptide complex [145]. Lastly, down-regulation of HLA class I antigen complex can be caused by alterations in the transcription factors forming the enhanceosome which bind SXY module on HLA class I heavy chain promoters [148,169–172]. Specifically, the expression and function of the NLRC5/CITA trans-activator can be affected by promoter methylation, copy number loss and somatic mutations [173,174]. About 60% of somatic mutations result in the inactivation of NLRC5/CITA [174].

2.2. Proteasome Defects

Alterations of proteasome subunits have been identified by utilizing mAbs that allow semi-quantitative analyses of the constitutive subunits δ, Z and MB1, as well as of the immunoproteasome subunits LMP2, LMP7 and LMP10 [25]. Down-regulation of one of these proteins caused by mutations at coding microsatellites or single nucleotide polymorphisms have been described in several types of tumors including colorectal, bladder, and ovarian carcinomas, as well as in acute myeloid leukemia and melanoma [147,175–179]. As previously described, the proteasome plays a key role in immune regulation. Consequently, inhibition or loss of function in one of the proteasomal components inhibits antigen processing and presentation and modifies the characteristics of processed peptides, decreasing the efficiency of epitope generation and altering tumor cell recognition by naïve T cells [25,169,171].

2.3. Defects in TAP1, TAP2 and Other Chaperones

Among the APM components, TAP genes have been most extensively investigated. At genetic level, mutations in TAP genes, resulting in total protein loss or expression of a non-functional protein, have been described in breast, lung, gastric, colorectal, and cervical carcinomas. Their frequency ranges from 10-84% in the cases analyzed [100,180–184]. TAP abnormalities reduce the translocation of peptides into the ER, resulting in a decreased formation of stable HLA class I derived peptide complexes or expression of "peptide-free" HLA class I molecules [171,172]. Interestingly, TAP-deficient individuals do not succumb to viral infections, suggesting that CD8+ T-cell immunity is sufficiently supported by an increased number of alternative TAP-independent processing pathways [171,172]. Identification of these alternative loading mechanisms into peptide-receptive HLA class I molecules still needs further investigation. It is reported that peptides can walk on multiple different paths before ending up in the grooves of HLA class I molecules [185]. Lastly, a substantial down-regulation in chaperone expression have been also associated to several types of malignancies due to defects in proper loading and assembly of HLA class I molecules, altering their maturation and stability [25,88].

2.4. HLA Class II Defects

Contrasting results have been described about the clinical significance of alterations in HLA class II molecule expression in cancer. Defects in HLA class II pathway, as well as induction of HLA class II molecule expression by non-immune cells have been involved in carcinogenesis [100]. In addition, HLA class II expression by cancer cells has been associated with poor prognosis and disease progression in melanoma and osteosarcomas [186–188]. However, an improved overall survival has been also associated to HLA class II expression by cancer cells in several types of cancer including melanoma, laryngeal, breast, cervical and colorectal cancer [100,188–197]. In various type of cancer (plasmacytoma, small cell lung cancer, and hepatocarcinoma) defects in HLA class II molecules have been associated to CIITA defects. The latter results in a reversible detrimental HLA class II expression that can be restored by CIITA transfection [100,198–200]. Moreover, other HLA class II presentation antigen pathway defects have been also described in Hodgkin's disease cancer cells [100,201].

3. Role of HLA as A Predictive Biomarker for ICI-Based Immunotherapy

3.1. Impact of HLA Class I and II on ICI-Based Immunotherapy In Vivo

As we have described above, HLA class I antigen derived peptide complex is crucial for tumor antigen presentation to naïve T cells. Binding of HLA class I antigen derived peptide complex to the TCR of naïve T cells allows T cell activation and consequently the recognition and the lysis of altered tumor cells [7,15–18]. However, binding of HLA class I derived peptide complex to TCR is not sufficient to activate naïve T cells. Naïve T cell activation requires the interaction between the CD28 family receptors on T cell with their co-stimulatory ligands belonging to B7 family molecules on APCs. Therefore, T cell activation is tightly controlled by co-stimulating or co-inhibiting signaling which are triggered by the interaction between immune checkpoint molecules such as PD-1 and CTLA-4 and their ligands PD-L1 and CD80/CD86 [202–205], expressed on naïve T cells and APCs, respectively. Actually, the interaction between PD-1 and PD-L1 is crucial in the activation phase of T cells as well as in their effector phase, due to PD-L1 expression also on tumor cells. Several lines of evidence, both in vitro and in vivo, have shown that blockade of the co-inhibitory signaling, including PD-1/PD-L1 axis, by mAbs promotes a host immune response against cancer cells by releasing T cells activation [206]. This novel therapeutic approach, called ICI-based immunotherapy, is revolutionizing the treatment of solid tumors [207]. Several clinical trials in various types of malignancies, such as melanoma, head and neck, triple negative breast, lung, kidney and bladder cancer, have demonstrated that administration of mAbs, which inhibit the interaction of immunoregulatory checkpoint molecules, such as CTLA-4 and PD-1, with their ligands CD80, CD86, and PD-L1, can have a major and lasting effect on their clinical course, significantly improving clinical outcomes as compared with standard chemotherapy [208]. However, this type of therapy is effective only in a subgroup of cancer patients, regardless of the tumor type. Therefore, there is an urgent need to identify the mechanisms of resistance as well as predictive biomarkers which may help to select patients who may benefits from this type of therapy [209–211]. Several molecules have been investigated as potential predictive biomarkers of ICI-based immunotherapy [212–215]. Among the postulated escape mechanisms utilized by tumor cells to avoid recognition and destruction by the host's immune system, are defects in the ability of tumor cells to process and present tumor antigens to naïve T cells [216]. This phenomenon is mediated by defects in the expression of HLA class I antigen-tumor antigen derived peptide complexes. Therefore, there has been an interest in investigating whether decreased or complete loss of HLA class I and II molecules as well as defects in the APM molecules might predict the efficacy of ICI-based immunotherapy by impairing naïve T cells activation induced by anti-checkpoint molecules. Several lines of evidence in vivo indicate that HLA class I or II modulation play a major role in the efficacy of ICI-based immunotherapy and various humanized mouse models that reliably reflect the complexity of the human heterogeneous tumour and its TME, have been developed in order to evaluate the potential role of HLA class I and II in predicting the efficacy of ICI-based immunotherapy as well as immune adverse effects for different types of cancer [217]. In the study of Lechner MG et al., six murine solid tumor models (CT26, 4T1, MAD109, RENCA, LLC, and B16) were used to demonstrate that MHC class I expression on tumor cells is an excellent surrogate marker of the overall tumor immunogenicity level as well as a predictor of response to immunotherapy. Specifically, tumor growth rate correlated indirectly with MHC class I expression and overall immunogenicity of the tumor model, with fastest growth in B16, LLC, and MAD109 and slowest growth in CT26, RENCA, and 4T1 [218]. Ashizawa et al.

reported that HLA class I and class II KO NOG mice (NOG dKO) transplanted with human PBMCs and tumor cell lines showed high anticancer effects following a PD-1 antibody treatment [219]. Gettinger et al. functionally demonstrated that loss of HLA class I expression by CRISPR-mediated knock-out of β2-m in an immunocompetent cancer mouse model (A/J mice transplanted with murine lung cancer cell line UN-SCC680AJ) confers resistance to PD-1 blockade and tumour progression [220]. β2-m gene deactivation in a mouse oncogenic TC-1 cell line derived from primary lung epithelial cells has also been shown to lead to negative surface MHC-I expression along with reduced proliferation and tumor rejection. Despite stimulation with IFN-γ, tumour cells were only weakly responsive to combined immunotherapy [221]. In addition to HLA class I expression, it is important to acknowledge that HLA haplotypes have also been shown to correlate with immunotherapy response in vivo. Rangan L et al. described a tumor cell line generated from a naturally occurring tumor in HLA-A*0201/DRB1*0101 (A2/DR1) mouse named SARC-L1 with a very low expression of HLA-A*0201 molecules, absence of HLA-DRB1*0101 and weak but constitutive expression of PD-L1. Histological and genes signature analysis supported the sarcoma origin of this cell line. According to the high frequency of these HLA alleles in the world population, this mouse model gained considerable interest in the field of tumor immunology and it has been used as preclinical tool for the evaluation of antitumor immunotherapies. Interestingly, both HLA-A*0201 and PD-L1 expressions increased on SARC-L1 after IFN-γ exposure in vitro defining this tumor very sensitive to several drugs commonly used to treat sarcoma and susceptible to anti-PD-L1 mAb therapy in vivo [222]. As we have described above, tumor cells might also express tsHLA-II molecules [123–130]. In the majority of studies, tsHLA-II molecule expression by cancer cells is associated with better prognosis, improved response to ICI in humans and increased tumor rejection in mouse models of breast cancer, sarcoma, lung cancer and colon cancer [98]. However, contrasting reports have shown that tsHLA-II or CIITA has no effect or, in some cases, accelerates tumor growth. In a mouse model of lung cancer Mortara L et al. demonstrated that single cell clones derived from a CIITA transduced population grew more aggressively in mice when cell surface MHC-II was highly expressed [223]. These contrasting results are likely to reflect different variables present in the different mouse model and cancer cells utilized, including (i) the ability of TAAs to be presented differentially on MHC class I or II in each model system, (ii) the number of mutations and therefore number of candidate neo-antigens expressed, (iii) the number of tumor cells injected, (iv) the injection site of cancer cells in the mouse, and (v) the mouse strains used that are characterized by different immunological status. An intriguing but underexplored hypothesis is that induction of HLA class I-II molecules may lead to up-regulate immunoinhibitory molecules on tumor-infiltrating lymphocytes (TILs), such as lymphocyte activation gene 3 (LAG-3) that binds HLA class II and negatively regulates cellular proliferation, activation and homeostasis of T cells, in a similar fashion to CTLA-4 and PD-1. This phenomenon has been reported to play a role in regulatory T cell (Treg) suppressive function creating a tolerizing microenvironment for tumor growth. ICIs directed to LAG-3 have been shown to have synergy with PD-1 inhibition in mouse models, suggesting that co-signaling blockade could restore a favorable immune microenvironment that can respond to antigenic stimulation [224].

3.2. HLA Class I and II as Predictive Biomarker for ICI-Based Immunotherapy: Clinical Evidences

Actually, few clinical studies have been investigating the potential role of HLA class I and II antigens in predicting the efficacy of ICI-based immunotherapy and no large clinical cohort analysis of patient population has been performed. Rodig et al. retrospectively evaluated whether HLA proteins confer differential sensitivity to CTLA-4 and PD-1 blockade in pre-treated metastatic melanoma patients. Tumor biopsies were obtained from patients enrolled in two different trials: CheckMate 064 and CheckMate 069. In these trials, patients were treated with the anti-CTLA-4 monoclonal antibody ipilimumab followed by the anti-PD-1 mAb nivolumab, nivolumab followed by ipilimumab, ipilimumab alone, or concurrent nivolumab plus ipilimumab. In this study, Rodig et al. demonstrated that (i) reduced tumor HLA class I molecule expression ($\leq$ 30%) correlated with lack of response to ipilimumab; (ii) HLA class II molecule expression (>1%) correlated with tumor response to nivolumab;

and (iii) among nivolumab plus ipilimumab treated patients, reduced HLA class I molecule expression was not associated with progressive disease as well as a decreased tumor response and overall survival. As a result, HLA class I molecule expression appeared a reliable predictive biomarker of tumor response to anti-CTLA-4 mAb ipilimumab but not to anti-PD-1 mAb nivolumab. In contrast, HLA class II molecule expression might represent a useful predictive biomarker for nivolumab but not ipilimumab therapy [126]. In addition, Chowell et al. retrospectively performed high-resolution HLA class I genotyping of 1535 advanced cancer patients treated with ICI-based immunotherapy. Patients affected by non-small cell lung cancer or melanoma were treated with anti-CTLA-4, anti-PD-1/PD-L1 or combinations of both. Results from this study demonstrated that patients carrying maximal heterozygosity at HLA class I loci (HLA-A, HLA-B, or HLA-C) have an improved overall survival as compared to patients who were homozygous for at least one HLA locus. Moreover, patients carrying HLA-B44 supertype had extended survival, while those carrying HLA-B62 supertype (including HLA-B*15:01) or somatic loss of heterozygosity at HLA class I had poor survival outcomes [225].

4. Restoring HLA Class I Expression as A Novel Therapeutic Strategy for Cancer Immunotherapy

Identification of the molecular aberrations responsible for altered tumor expression of HLA class I derived peptide complexes is crucial for the success of cancer T cell-based immunotherapy as well as for the rational design of novel immunotherapeutic strategies which restore an integral expression of HLA class I derived peptide complexes. Most of the defects of HLA class I derived peptide complex in human cancers are distinguished in "hard" or "soft" lesions [152]. "Hard" lesions are caused by structural gene alterations that induce loss of expression of HLA class I derived peptide complexes. They are reported in about 30–40% of human cancers [226]. LOH and β2-m gene mutations at chromosomes 6 and 15, respectively, represent the major cause of "hard" defects. In addition, a homologous point mutation in codon 67 of the β2-m gene also results in the total loss of HLA class I molecule expression [227,228]. These types of alterations cannot be repaired by any signaling pathway inhibitors as well as chemotherapeutic or immunotherapeutic agents. Del Campo et al. showed that infection of cells carrying β2-m mutations with an adenoviral vector expressing the human β2-m gene caused a total restoration of HLA class I molecule expression [229]. Thus, based on the type of mutated genes, transfection with a wild type gene, such as HLA class I heavy chain or β2-m genes, can potentially restore the expression of HLA class I derived peptide complex [152].

On the other hand, "soft" defects are caused by transcriptional or post-transcriptional modifications of one of HLA class I APM component genes. Activation of pro-tumorigenic pathways or epigenetic modifications which induce a reduced expression of HLA class I APM components are the major causes of "soft" defects [230]. Activation of pro-tumorigenic pathways includes inhibition of Jak/STAT pathway or activation of mitogen-activated protein kinase (MAPK) pathway. Epigenetic modifications include those that cause impairment in gene regulation such as hypermethylation of the HLA-A, B, and C heavy chains, β2m and APM component encoding gene promoter regions, or unbalanced histone acetylation [152,227,228,231]. Lastly, post-transcriptional alterations are induced by aberrant function of micro-RNAs (miRNAs). All these types of alterations can be restored by signaling pathway inhibitors, chemotherapeutic or immunotherapeutic agents (Figure 2). In Table 1, we have summarized the information available in the literature on the main molecules able to restore HLA class I expression in cases associated with "soft" defects.

Table 1. Drugs for restoring MHC class I expression [228,230,232–253].

Name	Target	Combination Therapy	Cancer Type	References
Kinase Inhibitors				
Nimotuzumab Cetuximab Erlotinib	EGFR	IFN-γ	epidermoid carcinoma lung cancer head and neck carcinoma lung adenocarcinoma	Garrido G. et al. 2017 [236] Srivastava et al. 2015 [234] Im et al. 2016 [235]
Trametinib	MEK1/2	IFN-γ	mesothelioma melanoma lung adenocarcinoma colon adenocarcinoma thyroid carcinoma breast cancer head and neck carcinoma	Brea et al. 2016 [232] Loi et al. 2016 [233] Kang et al. 2019 [237]
Vemurafenib	BRAF^{V600E}	IFN-γ IFNα-2b	Melanoma	Sapkota et al. 2013 [238] Sabbatino et al. 2016 [239]
Dabrafenib		Trametinib		Hu-Lieskovan et al. 2015 [240]
Epigenetic Agents				
Vorinostat	HDAC class I, II, and IV	Mithramycin A	Merkel cell carcinoma	Ritter et al. 2017 [241]
			B-cell lymphoma cutaneous T-cell lymphoma acute myeloid leukemia glioma	Chacon et al. 2016 [247] Banik et al. 2019 [228] Sun et al. 2019 [242] Yang et al. 2019 [243] Wang et al. 2020 [253]
Belinostat			peripheral T-cell lymphoma B-cell lymphoma	Banik et al. 2019 [228] Wang et al. 2020 [253]
Panobinostat			multiple myeloma B-cell lymphoma	Banik et al. 2019 [228] Wang et al. 2020 [253]
OKI-179			diffuse large B-cell lymphoma	Wang et al. 2019 [230]

Table 1. *Cont.*

Name	Target	Combination Therapy	Cancer Type	References
Romidepsin	HDAC 1-2		B-cell lymphoma peripheral T cell lymphoma	Banik et al. 2019 [228] Wang et al. 2020 [253]
Tubastatin A	HDAC 6		Melanoma	Woan et al. 2015 [244]
Azacytidine	DNMT		lung carcinoma melanoma	Fonsatti et al. 2007 [246] Chacon et al. 2016 [247]
Guadecitabine			breast cancer	Luo et al. 2018 [248]
Tazemetostat	EZH2		diffuse large B-cell lymphoma	Ennishi et al. 2019 [245]
Chemotherapeutics				
Cisplatin	DNA	Vinorelbine or5-fluorouracil	lung cancer head and neck carcinoma	De Biasi et al. 2014 [249]
Epothilone B Taxol Vinblastine	Microtubules		ovarian cancer	Pellicciotta et al. 2011 [251]
Doxorubicin	Topoisomerase		nasopharyngeal carcinoma	Faè et al. 2016 [250]
Topotecan Etoposide			breast cancer	Wan et al. 2012 [252]

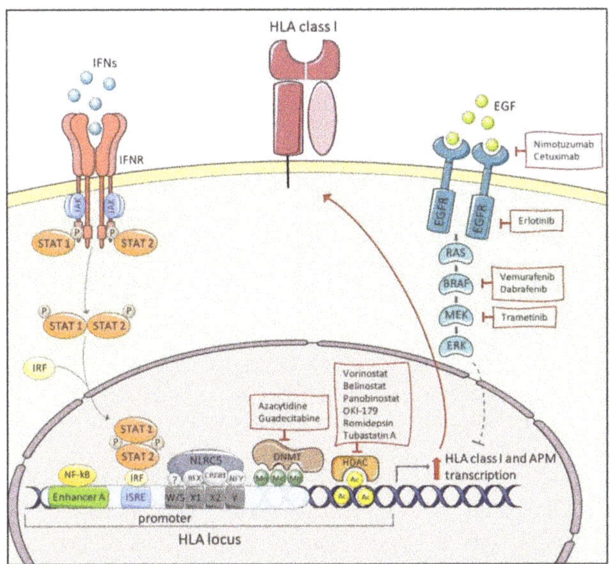

Figure 2. Mechanisms for restoring HLA class I expression. IFN binding to IFNR triggers Jak/STAT transduction pathway. STAT1/STAT2/IRF complex translocates to the nucleus where it binds to ISRE motifs located in HLA promoter region, inducing HLA gene transcription. EGFR and MAPK down-stream pathways suppress HLA class I surface expression. EGFR inhibitors, such as nimotuzumab, cetuximab and erlotinib, BRAF inhibitors, vemurafenib and dabrafenib, and MEK inhibitor trametinib can increase expression of both HLA class I antigens and APM components. DNMT inhibitors (azacytidine and guadecitabine) and HDAC inhibitors (vorinostat, belinostat, panobinostat, OKI-179, romidepsin and tubastatin A) avoid hypermethylation of HLA promoter region and histones hypoacetylation that cause HLA genes silencing. Abbreviations: HLA, human leukocyte antigen; IFNs, interferons; IFNR, interferon receptor; IRF, IFN regulatory factor; ISRE, IFN-sensitive response element; EGF, epidermal growth factor; EGFR, epidermal growth factor receptor; DNMT, DNA methyltransferase; HDAC, histone deacetylase; APM, antigen-processing machinery.

The administration of IFNs might effectively counteract HLA class I down-regulation in cancer cells by boosting the presentation of tumor specific-associated antigens [254–257]. A phase 0 clinical trial showed that systemic administration of IFN-γ increases not only HLA class I expression on tumor cells, but also T-cell infiltration in cold tumors [258] IFN-mediated up-regulation of HLA class I derived peptide complexes occurs via activation of the Jak/STAT pathway which induces the binding of IRFs to ISRE motifs located in HLA class I promoter region [107,254]. In addition, IFN-γ-mediated HLA class I derived peptide complex up-regulation is also associated to an increased histone demethylation and acetylation of APM genes in the MHC locus, particularly of histone H3 in TAP1 promoter locus [259]. Down-regulation of Jak/STAT signal transduction pathway and therefore of HLA class I derived peptide complex is strictly linked to protein kinase activation. Activation of either epithelial growth factor receptor (EGFR) or Ras/MAPK downstream pathway directly suppresses expression of HLA class I derived peptide complex and IFN-induced antigen presentation [232,233]. In this case the combination of IFNs and tyrosine kinase inhibitors (TKIs) can represent a potential therapeutic strategy for recovering HLA class I derived peptide complex expression in cancer cells. EGFR inhibitors such as the monoclonal antibodies nimotuzumab and cetuximab and the tyrosine kinase inhibitor erlotinib have been shown to increase membrane expression of HLA class I APM components in cells with EGFR activation [234–236]. Im and collaborators showed that the sensitivity of lung cancer cells to erlotinib positively correlates with the increase of expression of HLA class I derived peptide complex following IFN-γ treatment [235]. Moreover, the MAPK/ERK kinase (MEK) inhibitor trametinib, in combination with IFN-γ, has been shown to enhance tumor immunogenicity by modulating HLA class I derived peptide complex in different types of malignancies [232], including triple-negative breast cancer [233] and head and neck squamous cell carcinoma [237]. In melanoma patients harboring the BRAF^{V600E} mutations, BRAF inhibitor vemurafenib strengthens the induction of HLA class I antigen expression by both IFN-γ and IFN-α2b [238,239]. Furthermore, this effect was enhanced when BRAF inhibition was combined with the MEK inhibitor trametinib [240].

However, it is important to mention that several tumors develop IFN-γ signaling insensitivity. The lack of response to IFN-γ stimulation results from cellular defects on IFNγR1 receptor or downstream components of the signaling pathway such as Janus kinases (JAK) 1 and 2 [260]. In addition, the absence of STAT1 expression and/or tyrosine-phosphorylation and epigenetic regulation of IRF1 can contribute to the lack of HLA class I expression, restoring in vitro following IFN-γ administration [261,262]. Lastly, given the addiction of IFN-γ-induced HLA class I and II up-regulation by NLRC5/CITA and CIITA, respectively, genetic or functional defects on these two trans-activators abrogate the ability of IFN-γ to boost antigen processing and presentation [106,260].

Inhibition of histone deacetylases (HDACs), enzymes that remove acetyl group from lysines on histones, also promotes the increase of expression of HLA class I APM components. HDAC inhibitors (HDACi) vorinostat, belinostat, and panobinostat, targeting HDAC class I, II, and IV, and romidepsin, a specific HDAC 1 and 2 inhibitor, are currently approved for hematological malignancies [228,230]. Vorinostat, also known as SAHA, in combination with mithramycin A, a Sp1 inhibitor, reverses the histone hypoacetylation which causes APM gene silencing in Merkel cell carcinoma [241]. In addition, vorinostat promotes tumor cell recognition by CTLs in glioma cells [242,243]. In diffuse large B-cell lymphoma, the HDAC class I inhibitor OKI-179 reverts down-regulation of HLA class I derived peptide complex, a typical feature of this hematological cancer [230]. Likewise, the selective inhibition of HDAC6 with tubastatin A improves the immunogenicity of melanoma cells by increasing HLA class I expression [244]. Mutations in EZH2 gene, encoding for a histone-lysine N-methyltransferase, are strictly connected to loss of expression of HLA class I derived peptide complex in large B-cell lymphoma. Treatment with EZH2 inhibitor tazemetostat (EPZ-6438) restores the expression of HLA class I derived peptide complex in EZH2-mutant cell lines [245]. In cases of DNA hypermethylation, therapy with DNA methyltransferase inhibitors (DNMTi) such as azacytidine and decitabine restores the expression of HLA class I derived peptide complex [263]. Azacytidine increased the transcription of APM genes and HLA class I molecule expression in lung carcinoma and melanoma cells [246,247]. Guadecitabine, a novel DNMTi, significantly up-regulated both basal and IFN-γ-dependent expression of HLA class I derived peptide complex in breast cancer cells [248]. In the case of HLA class I APM component down-regulation mediated by aberrant miRNA function, it has been shown that suppression of miR-9 and miR-19 expression restores the expression of HLA class I derived peptide complex [264,265]. Lastly, several chemotherapeutic agents can promote the expression of HLA class I derived peptide complex. In many types of cancer cell lines cisplatin alone or in combination with vinorelbine or 5-fluorouracil [249], doxorubicin [250], the microtubule-destabilizers epothilone B, taxol and vinblastine [251] increase the expression of HLA class I APM components. Moreover, some other chemotherapeutic agents, such as the topoisomerase-I inhibitors topotecan and etoposide, indirectly induce the up-regulation of HLA class I derived peptide complex by stimulating IFN-β autocrine/paracrine signaling of tumor cells [252].

5. Conclusions

In the last decades, important progress has been made pertaining to the knowledge of the structure and the function of HLA class I and II antigen presentation pathways. An improved knowledge of molecular mechanisms underlying HLA class I APM defects will be crucial to better understand the mechanisms determining carcinogenesis, tumor progression and immune escape. Moreover, recent knowledge of the pathways leading to restored HLA expression could be utilized to conceive new cancer therapeutic strategies. The identification of molecular defects resulting in acquired insensitivity to the stimulation with specific adjuvants, such as IFNs, could allow to a better selection of cancer patients who can take advantage from this approach for HLA expression restoring. Because of their crucial role in tumor cell recognition, additional studies are urgently needed to validate the role of expression of HLA class I and II antigens, as well as of APM components as novel potential predictive biomarkers in cancer immunotherapy. Moreover, where HLA class I and II gene expression is not recoverable, the development of different immunotherapeutic approaches able to target native cell surface antigens, and therefore independent from HLA expression, such as chimer antigen receptor (CAR)-T cell or bi-specific T-cell engager antibodies (BiTES) should be taken in consideration.

Author Contributions: Conception and design: F.S. and J.D.C. Writing, review, and/or revision of the manuscript: F.S., L.L., I.S., G.P., G.C., and J.D.C. Study supervision: F.S., V.C., C.S., and S.P. Other (discussed results and implications of findings): F.S., S.P., V.C., F.A.S., and C.S. All authors have read and agreed to the published version of the manuscript.

Funding: The work was supported by Ministero dell'Università e della Ricerca (Progetti di Rilevante Interesse Nazionale (PRIN), 2017, CODICE 2017PHRC8X_003) (to SP) and by "Fondazione con il Sud" (Brains2South 2015-PDR-0224) (to J.D.C.).

Conflicts of Interest: The authors declare no conflict of interest.

Abbreviations

ACT	adoptive cell-transfer
APC	antigen presenting cell
APM	antigen processing machinery
ATF1	activating transcriptor factor 1
ATP	adenosine triphosphate
BiTES	bi-specific T-cell engager antibodies
BLS	bare lymphocyte disease
CAR	chimeric antigen receptor
CIITA	class II trans-activator
CIITA	class I trans activator
CLIP	class II-associated I chain peptide
CNX	calnexin
CREB1	cAMP-responsive element binding protein 1
CRT	calreticulin
CTL	cytotoxic T lymphocyte
CTLA-4	cytotoxic T-lymphocyte antigen 4
DC	dendritic cell
DNMTi	DNA methyltransferase inhibitor
EGFR	epithelial growth factor receptor
ER	endoplasmic reticulum
HDAC	histone deacetylase
HLA	human leukocyte antigen
ICI	immune checkpoint inhibitor
IFN	interferon
IRF	IFN Regulatory Factor
ISRE	IFN-sensitive response element
KIR	killer cell immunoglobulin-like receptor
DFS	disease free survival
JAK	janus kinase
LAG-3	lymphocyte activation gene 3
LMP	low molecular weight protein
LOH	loss of heterozygosity
mAb	monoclonal antibody
MAPK	mitogen-activated protein kinase

MEK	MAPK/ERK kinase
MHC	major histocompatibility complex
MIIC	MHC class II compartment
NF-kB	nuclear factor kB
NFY	nuclear transcriptor factor Y
NK	natural killer
NLRC5	NOD-like receptor 5
OS	overall survival
PD-1	programmed cell death 1
PD-L1	programmed cell death ligand 1
PFS	progression free survival
RFX	regulatory factor X
RFXANK	RFX-associated ankyrin-containing protein
RFXAP	RFX associated protein
TAP	transporters associated with antigen processing
TCR	T-cell receptor
TKI	tyrosine kinase inhibitor
TNF-α	tumor necrosis factor alpha
TA	tumor antigen
TAA	tumor associated antigen
TAM	Type II tumour-associated macrophages
TIL	tumor-infiltrating lymphocyte
TME	tumor microenvironment
Treg	regulatory T cell
TSA	tumor specific antigen
tsMHC-II	tumor specific MHC class II
β2-m	β2-microglobulin

References

1. Zinkernagel, R.M.; Doherty, P.C. Restriction of in vitro T cell-mediated cytotoxicity in lymphocytic choriomeningitis within a syngeneic or semiallogeneic system. *Nature* **1974**, *248*, 701–702. [CrossRef]
2. Townsend, A.R.; Skehel, J.J. Influenza A specific cytotoxic T-cell clones that do not recognize viral glycoproteins. *Nature* **1982**, *300*, 655–657. [CrossRef]
3. Townsend, A.R.; Rothbard, J.; Gotch, F.M.; Bahadur, G.; Wraith, D.; McMichael, A.J. The epitopes of influenza nucleoprotein recognized by cytotoxic T lymphocytes can be defined with short synthetic peptides. *Cell* **1986**, *44*, 959–968. [CrossRef]
4. Bjorkman, P.J.; Saper, M.A.; Samraoui, B.; Bennett, W.S.; Strominger, J.L.; Wiley, D.C. Structure of the human class I histocompatibility antigen, HLA-A2. *Nature* **1987**, *329*, 506–512. [CrossRef]
5. Rötzschke, O.; Falk, K.; Deres, K.; Schild, H.; Norda, M.; Metzger, J.; Jung, G.; Rammensee, H.G. Isolation and analysis of naturally processed viral peptides as recognized by cytotoxic T cells. *Nature* **1990**, *348*, 252–254. [CrossRef]
6. Trowsdale, J. Genomic structure and function in the MHC. *Trends Genet.* **1993**, *9*, 117–122. [CrossRef]
7. Flutter, B.; Gao, B. MHC class I antigen presentation–recently trimmed and well presented. *Cell. Mol. Immunol.* **2004**, *1*, 22–30.
8. Norman, P.J.; Norberg, S.J.; Guethlein, L.A.; Nemat-Gorgani, N.; Royce, T.; Wroblewski, E.E.; Dunn, T.; Mann, T.; Alicata, C.; Hollenbach, J.A.; et al. Sequences of 95 human MHC haplotypes reveal extreme coding variation in genes other than highly polymorphic HLA class I and II. *Genome Res.* **2017**, *27*, 813–823. [CrossRef]
9. Horton, R.; Wilming, L.; Rand, V.; Lovering, R.C.; Bruford, E.A.; Khodiyar, V.K.; Lush, M.J.; Povey, S.; Talbot, C.C.; Wright, M.W.; et al. Gene map of the extended human MHC. *Nat. Rev. Genet.* **2004**, *5*, 889–899. [CrossRef]
10. Vandiedonck, C.; Knight, J.C. The human Major Histocompatibility Complex as a paradigm in genomics research. *Brief. Funct. Genom. Proteom.* **2009**, *8*, 379–394. [CrossRef]

11. Trowsdale, J.; Knight, J.C. Major histocompatibility complex genomics and human disease. *Annu. Rev. Genom. Hum. Genet.* **2013**, *14*, 301–323. [CrossRef] [PubMed]

12. Complete sequence and gene map of a human major histocompatibility complex. The MHC sequencing consortium. *Nature* **1999**, *401*, 921–923. [CrossRef]

13. Milner, C.M.; Campbell, R.D. Genetic organization of the human MHC class III region. *Front. Biosci.* **2001**, *6*, D914–D926. [CrossRef]

14. Neefjes, J.; Jongsma, M.L.M.; Paul, P.; Bakke, O. Towards a systems understanding of MHC class I and MHC class II antigen presentation. *Nat. Rev. Immunol.* **2011**, *11*, 823–836. [CrossRef]

15. Ajitkumar, P.; Geier, S.S.; Kesari, K.V.; Borriello, F.; Nakagawa, M.; Bluestone, J.A.; Saper, M.A.; Wiley, D.C.; Nathenson, S.G. Evidence that multiple residues on both the alpha-helices of the class I MHC molecule are simultaneously recognized by the T cell receptor. *Cell* **1988**, *54*, 47–56. [CrossRef]

16. Garcia, K.C.; Degano, M.; Stanfield, R.L.; Brunmark, A.; Jackson, M.R.; Peterson, P.A.; Teyton, L.; Wilson, I.A. An alphabeta T cell receptor structure at 2.5 A and its orientation in the TCR-MHC complex. *Science* **1996**, *274*, 209–219. [CrossRef]

17. Garboczi, D.N.; Ghosh, P.; Utz, U.; Fan, Q.R.; Biddison, W.E.; Wiley, D.C. Structure of the complex between human T-cell receptor, viral peptide and HLA-A2. *Nature* **1996**, *384*, 134–141. [CrossRef]

18. Blum, J.S.; Wearsch, P.A.; Cresswell, P. Pathways of antigen processing. *Annu. Rev. Immunol.* **2013**, *31*, 443–473. [CrossRef]

19. Farhood, B.; Najafi, M.; Mortezaee, K. CD8+ cytotoxic T lymphocytes in cancer immunotherapy: A review. *J. Cell. Physiol.* **2019**, *234*, 8509–8521. [CrossRef]

20. Cruz, F.M.; Colbert, J.D.; Merino, E.; Kriegsman, B.A.; Rock, K.L. The Biology and Underlying Mechanisms of Cross-Presentation of Exogenous Antigens on MHC-I Molecules. *Annu. Rev. Immunol.* **2017**, *35*, 149–176. [CrossRef]

21. Rock, K.L.; Shen, L. Cross-presentation: Underlying mechanisms and role in immune surveillance. *Immunol. Rev.* **2005**, *207*, 166–183. [CrossRef] [PubMed]

22. Muntjewerff, E.M.; Meesters, L.D.; Van den Bogaart, G. Antigen Cross-Presentation by Macrophages. *Front. Immunol.* **2020**, *11*. [CrossRef] [PubMed]

23. Bevan, M.J. Minor H antigens introduced on H-2 different stimulating cells cross-react at the cytotoxic T cell level during in vivo priming. *J. Immunol.* **1976**, *117*, 2233–2238.

24. Embgenbroich, M.; Burgdorf, S. Current Concepts of Antigen Cross-Presentation. *Front. Immunol.* **2018**, *9*. [CrossRef]

25. Leone, P.; Shin, E.-C.; Perosa, F.; Vacca, A.; Dammacco, F.; Racanelli, V. MHC class I antigen processing and presenting machinery: Organization, function, and defects in tumor cells. *J. Natl. Cancer Inst.* **2013**, *105*, 1172–1187. [CrossRef] [PubMed]

26. Thielens, A.; Vivier, E.; Romagné, F. NK cell MHC class I specific receptors (KIR): From biology to clinical intervention. *Curr. Opin. Immunol.* **2012**, *24*, 239–245. [CrossRef]

27. Wieczorek, M.; Abualrous, E.T.; Sticht, J.; Álvaro-Benito, M.; Stolzenberg, S.; Noé, F.; Freund, C. Major Histocompatibility Complex (MHC) Class I and MHC Class II Proteins: Conformational Plasticity in Antigen Presentation. *Front. Immunol.* **2017**, *8*, 292. [CrossRef]

28. Cunningham, B.A. Structure and significance of beta2-microglobulin. *Fed. Proc.* **1976**, *35*, 1171–1176.

29. Cunningham, B.A.; Berggård, I. Structure, evolution and significance of beta2-microglobulin. *Transplant. Rev.* **1974**, *21*, 3–14. [CrossRef]

30. Madden, D.R. The three-dimensional structure of peptide-MHC complexes. *Annu. Rev. Immunol.* **1995**, *13*, 587–622. [CrossRef]

31. Wilson, I.A.; Fremont, D.H. Structural analysis of MHC class I molecules with bound peptide antigens. *Semin. Immunol.* **1993**, *5*, 75–80. [CrossRef] [PubMed]

32. Persson, K.; Schneider, G. Three-dimensional structures of MHC class I-peptide complexes: Implications for peptide recognition. *Arch. Immunol. Ther. Exp.* **2000**, *48*, 135–142.

33. Mage, M.G.; Dolan, M.A.; Wang, R.; Boyd, L.F.; Revilleza, M.J.; Robinson, H.; Natarajan, K.; Myers, N.B.; Hansen, T.H.; Margulies, D.H. The peptide-receptive transition state of MHC class I molecules: Insight from structure and molecular dynamics. *J. Immunol.* **2012**, *189*, 1391–1399. [CrossRef] [PubMed]

34. Braud, V.M.; Allan, D.S.; O'Callaghan, C.A.; Söderström, K.; D'Andrea, A.; Ogg, G.S.; Lazetic, S.; Young, N.T.; Bell, J.I.; Phillips, J.H.; et al. HLA-E binds to natural killer cell receptors CD94/NKG2A, B and C. *Nature* **1998**, *391*, 795–799. [CrossRef] [PubMed]

35. Carosella, E.D.; HoWangYin, K.-Y.; Favier, B.; LeMaoult, J. HLA-G-dependent suppressor cells: Diverse by nature, function, and significance. *Hum. Immunol.* **2008**, *69*, 700–707. [CrossRef]

36. Lepin, E.J.; Bastin, J.M.; Allan, D.S.; Roncador, G.; Braud, V.M.; Mason, D.Y.; Van der Merwe, P.A.; McMichael, A.J.; Bell, J.I.; Powis, S.H.; et al. Functional characterization of HLA-F and binding of HLA-F tetramers to ILT2 and ILT4 receptors. *Eur. J. Immunol.* **2000**, *30*, 3552–3561. [CrossRef]

37. Persson, G.; Jørgensen, N.; Nilsson, L.L.; Andersen, L.H.J.; Hviid, T.V.F. A role for both HLA-F and HLA-G in reproduction and during pregnancy? *Hum. Immunol.* **2020**, *81*, 127–133. [CrossRef] [PubMed]

38. Hackmon, R.; Pinnaduwage, L.; Zhang, J.; Lye, S.J.; Geraghty, D.E.; Dunk, C.E. Definitive class I human leukocyte antigen expression in gestational placentation: HLA-F, HLA-E, HLA-C, and HLA-G in extravillous trophoblast invasion on placentation, pregnancy, and parturition. *Am. J. Reprod. Immunol.* **2017**, *77*. [CrossRef]

39. Joosten, S.A.; Sullivan, L.C.; Ottenhoff, T.H.M. Characteristics of HLA-E Restricted T-Cell Responses and Their Role in Infectious Diseases. *J. Immunol. Res.* **2016**, *2016*, 2695396. [CrossRef]

40. Raghavan, M.; Del Cid, N.; Rizvi, S.M.; Peters, L.R. MHC class I assembly: Out and about. *Trends Immunol.* **2008**, *29*, 436–443. [CrossRef]

41. Paulsson, K.M. Evolutionary and functional perspectives of the major histocompatibility complex class I antigen-processing machinery. *Cell. Mol. Life Sci.* **2004**, *61*, 2446–2460. [CrossRef]

42. Wearsch, P.A.; Cresswell, P. The quality control of MHC class I peptide loading. *Curr. Opin. Cell Biol.* **2008**, *20*, 624–631. [CrossRef] [PubMed]

43. Antoniou, A.N.; Powis, S.J.; Elliott, T. Assembly and export of MHC class I peptide ligands. *Curr. Opin. Immunol.* **2003**, *15*, 75–81. [CrossRef]

44. Jensen, P.E. Recent advances in antigen processing and presentation. *Nat. Immunol.* **2007**, *8*, 1041–1048. [CrossRef] [PubMed]

45. Kloetzel, P.M. Antigen processing by the proteasome. *Nat. Rev. Mol. Cell Biol.* **2001**, *2*, 179–187. [CrossRef]

46. Sant, A.; Yewdell, J. Antigen processing and recognition. *Curr. Opin. Immunol.* **2003**, *15*, 66–68. [CrossRef]

47. Schölz, C.; Tampé, R. The peptide-loading complex—antigen translocation and MHC class I loading. *Biol. Chem.* **2009**, *390*, 783–794. [CrossRef]

48. Kotsias, F.; Cebrian, I.; Alloatti, A. Antigen processing and presentation. *Int. Rev. Cell Mol. Biol* **2019**, *348*, 69–121. [CrossRef]

49. Koegl, M.; Hoppe, T.; Schlenker, S.; Ulrich, H.D.; Mayer, T.U.; Jentsch, S. A novel ubiquitination factor, E4, is involved in multiubiquitin chain assembly. *Cell* **1999**, *96*, 635–644. [CrossRef]

50. Goldberg, A.L. Functions of the proteasome: The lysis at the end of the tunnel. *Science* **1995**, *268*, 522–523. [CrossRef]

51. Kloetzel, P.-M. The proteasome and MHC class I antigen processing. *Biochim. Biophys. Acta* **2004**, *1695*, 225–233. [CrossRef]

52. Rechsteiner, M.; Hoffman, L.; Dubiel, W. The multicatalytic and 26 S proteases. *J. Biol. Chem.* **1993**, *268*, 6065–6068.

53. Griffin, T.A.; Nandi, D.; Cruz, M.; Fehling, H.J.; Kaer, L.V.; Monaco, J.J.; Colbert, R.A. Immunoproteasome assembly: Cooperative incorporation of interferon gamma (IFN-gamma)-inducible subunits. *J. Exp. Med.* **1998**, *187*, 97–104. [CrossRef]

54. Groettrup, M.; Standera, S.; Stohwasser, R.; Kloetzel, P.M. The subunits MECL-1 and LMP2 are mutually required for incorporation into the 20S proteasome. *Proc. Natl. Acad. Sci. USA* **1997**, *94*, 8970–8975. [CrossRef]

55. Nandi, D.; Woodward, E.; Ginsburg, D.B.; Monaco, J.J. Intermediates in the formation of mouse 20S proteasomes: Implications for the assembly of precursor beta subunits. *EMBO J.* **1997**, *16*, 5363–5375. [CrossRef]

56. Aki, M.; Shimbara, N.; Takashina, M.; Akiyama, K.; Kagawa, S.; Tamura, T.; Tanahashi, N.; Yoshimura, T.; Tanaka, K.; Ichihara, A. Interferon-gamma induces different subunit organizations and functional diversity of proteasomes. *J. Biochem.* **1994**, *115*, 257–269. [CrossRef]

57. Shin, E.-C.; Seifert, U.; Kato, T.; Rice, C.M.; Feinstone, S.M.; Kloetzel, P.-M.; Rehermann, B. Virus-induced type I IFN stimulates generation of immunoproteasomes at the site of infection. *J. Clin. Investig.* **2006**, *116*, 3006–3014. [CrossRef]

58. Chen, W.; Norbury, C.C.; Cho, Y.; Yewdell, J.W.; Bennink, J.R. Immunoproteasomes shape immunodominance hierarchies of antiviral CD8(+) T cells at the levels of T cell repertoire and presentation of viral antigens. *J. Exp. Med.* **2001**, *193*, 1319–1326. [CrossRef]

59. Melnikova, V.I.; Sharova, N.P.; Maslova, E.V.; Voronova, S.N.; Zakharova, L.A. Ontogenesis of rat immune system: Proteasome expression in different cell populations of the developing thymus. *Cell. Immunol.* **2010**, *266*, 83–89. [CrossRef]

60. Schmitt, L.; Tampé, R. Structure and mechanism of ABC transporters. *Curr. Opin. Struct. Biol.* **2002**, *12*, 754–760. [CrossRef]

61. Arora, S.; Lapinski, P.E.; Raghavan, M. Use of chimeric proteins to investigate the role of transporter associated with antigen processing (TAP) structural domains in peptide binding and translocation. *Proc. Natl. Acad. Sci. USA* **2001**, *98*, 7241–7246. [CrossRef]

62. Higgins, C.F.; Linton, K.J. The ATP switch model for ABC transporters. *Nat. Struct. Mol. Biol.* **2004**, *11*, 918–926. [CrossRef] [PubMed]

63. Parcej, D.; Tampé, R. ABC proteins in antigen translocation and viral inhibition. *Nat. Chem. Biol.* **2010**, *6*, 572–580. [CrossRef]

64. Herget, M.; Baldauf, C.; Schölz, C.; Parcej, D.; Wiesmüller, K.-H.; Tampé, R.; Abele, R.; Bordignon, E. Conformation of peptides bound to the transporter associated with antigen processing (TAP). *Proc. Natl. Acad. Sci. USA* **2011**, *108*, 1349–1354. [CrossRef]

65. Koopmann, J.O.; Post, M.; Neefjes, J.J.; Hämmerling, G.J.; Momburg, F. Translocation of long peptides by transporters associated with antigen processing (TAP). *Eur. J. Immunol.* **1996**, *26*, 1720–1728. [CrossRef]

66. Momburg, F.; Roelse, J.; Howard, J.C.; Butcher, G.W.; Hämmerling, G.J.; Neefjes, J.J. Selectivity of MHC-encoded peptide transporters from human, mouse and rat. *Nature* **1994**, *367*, 648–651. [CrossRef]

67. Momburg, F.; Roelse, J.; Hämmerling, G.J.; Neefjes, J.J. Peptide size selection by the major histocompatibility complex-encoded peptide transporter. *J. Exp. Med.* **1994**, *179*, 1613–1623. [CrossRef]

68. Momburg, F.; Hämmerling, G.J. Generation and TAP-mediated transport of peptides for major histocompatibility complex class I molecules. *Adv. Immunol.* **1998**, *68*, 191–256. [CrossRef]

69. Schumacher, T.N.; Kantesaria, D.V.; Heemels, M.T.; Ashton-Rickardt, P.G.; Shepherd, J.C.; Fruh, K.; Yang, Y.; Peterson, P.A.; Tonegawa, S.; Ploegh, H.L. Peptide length and sequence specificity of the mouse TAP1/TAP2 translocator. *J. Exp. Med.* **1994**, *179*, 533–540. [CrossRef]

70. Roelse, J.; Grommé, M.; Momburg, F.; Hämmerling, G.; Neefjes, J. Trimming of TAP-translocated peptides in the endoplasmic reticulum and in the cytosol during recycling. *J. Exp. Med.* **1994**, *180*, 1591–1597. [CrossRef]

71. Diedrich, G.; Bangia, N.; Pan, M.; Cresswell, P. A role for calnexin in the assembly of the MHC class I loading complex in the endoplasmic reticulum. *J. Immunol.* **2001**, *166*, 1703–1709. [CrossRef]

72. Hughes, E.A.; Cresswell, P. The thiol oxidoreductase ERp57 is a component of the MHC class I peptide-loading complex. *Curr. Biol.* **1998**, *8*, 709–712. [CrossRef]

73. Morrice, N.A.; Powis, S.J. A role for the thiol-dependent reductase ERp57 in the assembly of MHC class I molecules. *Curr. Biol.* **1998**, *8*, 713–716. [CrossRef]

74. Sadasivan, B.; Lehner, P.J.; Ortmann, B.; Spies, T.; Cresswell, P. Roles for calreticulin and a novel glycoprotein, tapasin, in the interaction of MHC class I molecules with TAP. *Immunity* **1996**, *5*, 103–114. [CrossRef]

75. Momburg, F.; Tan, P. Tapasin-the keystone of the loading complex optimizing peptide binding by MHC class I molecules in the endoplasmic reticulum. *Mol. Immunol.* **2002**, *39*, 217–233. [CrossRef]

76. Ortmann, B.; Copeman, J.; Lehner, P.J.; Sadasivan, B.; Herberg, J.A.; Grandea, A.G.; Riddell, S.R.; Tampé, R.; Spies, T.; Trowsdale, J.; et al. A critical role for tapasin in the assembly and function of multimeric MHC class I-TAP complexes. *Science* **1997**, *277*, 1306–1309. [CrossRef]

77. Dick, T.P.; Bangia, N.; Peaper, D.R.; Cresswell, P. Disulfide bond isomerization and the assembly of MHC class I-peptide complexes. *Immunity* **2002**, *16*, 87–98. [CrossRef]

78. Ortmann, B.; Androlewicz, M.J.; Cresswell, P. MHC class I/beta 2-microglobulin complexes associate with TAP transporters before peptide binding. *Nature* **1994**, *368*, 864–867. [CrossRef]

79. Wright, C.A.; Kozik, P.; Zacharias, M.; Springer, S. Tapasin and other chaperones: Models of the MHC class I loading complex. *Biol. Chem.* **2004**, *385*, 763–778. [CrossRef]

80. Ellgaard, L.; Helenius, A. ER quality control: Towards an understanding at the molecular level. *Curr. Opin. Cell Biol.* **2001**, *13*, 431–437. [CrossRef]

81. Oliver, J.D.; Roderick, H.L.; Llewellyn, D.H.; High, S. ERp57 functions as a subunit of specific complexes formed with the ER lectins calreticulin and calnexin. *Mol. Biol. Cell* **1999**, *10*, 2573–2582. [CrossRef] [PubMed]

82. Degen, E.; Cohen-Doyle, M.F.; Williams, D.B. Efficient dissociation of the p88 chaperone from major histocompatibility complex class I molecules requires both beta 2-microglobulin and peptide. *J. Exp. Med.* **1992**, *175*, 1653–1661. [CrossRef] [PubMed]

83. Nössner, E.; Parham, P. Species-specific differences in chaperone interaction of human and mouse major histocompatibility complex class I molecules. *J. Exp. Med.* **1995**, *181*, 327–337. [CrossRef] [PubMed]

84. Groothuis, T.A.M.; Neefjes, J. The many roads to cross-presentation. *J. Exp. Med.* **2005**, *202*, 1313–1318. [CrossRef]

85. Neefjes, J.; Sadaka, C. Into the Intracellular Logistics of Cross-Presentation. *Front. Immunol.* **2012**, *3*. [CrossRef]

86. Lechler, R.I. Structure-function relationships of MHC class II molecules. *Immunol. Suppl.* **1988**, *1*, 25–26.

87. Stern, L.J.; Potolicchio, I.; Santambrogio, L. MHC class II compartment subtypes: Structure and function. *Curr. Opin. Immunol.* **2006**, *18*, 64–69. [CrossRef]

88. Rock, K.L.; Reits, E.; Neefjes, J. Present Yourself! By MHC Class I and MHC Class II Molecules. *Trends Immunol.* **2016**, *37*, 724–737. [CrossRef]

89. Chang, C.H.; Flavell, R.A. Class II transactivator regulates the expression of multiple genes involved in antigen presentation. *J. Exp. Med.* **1995**, *181*, 765–767. [CrossRef]

90. Collins, T.; Korman, A.J.; Wake, C.T.; Boss, J.M.; Kappes, D.J.; Fiers, W.; Ault, K.A.; Gimbrone, M.A.; Strominger, J.L.; Pober, J.S. Immune interferon activates multiple class II major histocompatibility complex genes and the associated invariant chain gene in human endothelial cells and dermal fibroblasts. *Proc. Natl. Acad. Sci. USA* **1984**, *81*, 4917–4921. [CrossRef]

91. Holling, T.M.; Schooten, E.; Van Den Elsen, P.J. Function and regulation of MHC class II molecules in T-lymphocytes: Of mice and men. *Hum. Immunol.* **2004**, *65*, 282–290. [CrossRef] [PubMed]

92. Reith, W.; LeibundGut-Landmann, S.; Waldburger, J.-M. Regulation of MHC class II gene expression by the class II transactivator. *Nat. Rev. Immunol.* **2005**, *5*, 793–806. [CrossRef] [PubMed]

93. Ting, J.P.-Y.; Trowsdale, J. Genetic control of MHC class II expression. *Cell* **2002**, *109* (Suppl. S21–S33). [CrossRef]

94. Boss, J.M.; Jensen, P.E. Transcriptional regulation of the MHC class II antigen presentation pathway. *Curr. Opin. Immunol.* **2003**, *15*, 105–111. [CrossRef]

95. Neumann, J.; Koch, N. Assembly of major histocompatibility complex class II subunits with invariant chain. *FEBS Lett.* **2005**, *579*, 6055–6059. [CrossRef]

96. Harton, J.; Jin, L.; Hahn, A.; Drake, J. Immunological Functions of the Membrane Proximal Region of MHC Class II Molecules. *F1000Res* **2016**, *5*. [CrossRef]

97. Arnold, P.Y.; La Gruta, N.L.; Miller, T.; Vignali, K.M.; Adams, P.S.; Woodland, D.L.; Vignali, D.A.A. The majority of immunogenic epitopes generate CD4+ T cells that are dependent on MHC class II-bound peptide-flanking residues. *J. Immunol.* **2002**, *169*, 739–749. [CrossRef]

98. Axelrod, M.L.; Cook, R.S.; Johnson, D.B.; Balko, J.M. Biological Consequences of MHC-II Expression by Tumor Cells in Cancer. *Clin. Cancer Res.* **2019**, *25*, 2392–2402. [CrossRef]

99. Pieters, J. MHC class II restricted antigen presentation. *Curr. Opin. Immunol.* **1997**, *9*, 89–96. [CrossRef]

100. Seliger, B.; Maeurer, M.J.; Ferrone, S. Antigen-processing machinery breakdown and tumor growth. *Immunol. Today* **2000**, *21*, 455–464. [CrossRef]

101. Cresswell, P. Invariant chain structure and MHC class II function. *Cell* **1996**, *84*, 505–507. [CrossRef]

102. Busch, R.; Cloutier, I.; Sékaly, R.P.; Hämmerling, G.J. Invariant chain protects class II histocompatibility antigens from binding intact polypeptides in the endoplasmic reticulum. *EMBO J.* **1996**, *15*, 418–428. [CrossRef]

103. Denzin, L.K.; Cresswell, P. HLA-DM induces CLIP dissociation from MHC class II alpha beta dimers and facilitates peptide loading. *Cell* **1995**, *82*, 155–165. [CrossRef]

104. Kropshofer, H.; Arndt, S.O.; Moldenhauer, G.; Hämmerling, G.J.; Vogt, A.B. HLA-DM acts as a molecular chaperone and rescues empty HLA-DR molecules at lysosomal pH. *Immunity* **1997**, *6*, 293–302. [CrossRef]

105. Van Den Elsen, P.J. Expression Regulation of Major Histocompatibility Complex Class I and Class II Encoding Genes. *Front. Immunol.* **2011**, *2*. [CrossRef] [PubMed]

106. Kobayashi, K.S.; Van den Elsen, P.J. NLRC5: A key regulator of MHC class I-dependent immune responses. *Nat. Rev. Immunol.* **2012**, *12*, 813–820. [CrossRef]

107. Jongsma, M.L.M.; Guarda, G.; Spaapen, R.M. The regulatory network behind MHC class I expression. *Mol. Immunol.* **2019**, *113*, 16–21. [CrossRef] [PubMed]

108. Steimle, V.; Otten, L.A.; Zufferey, M.; Mach, B. Complementation cloning of an MHC class II transactivator mutated in hereditary MHC class II deficiency (or bare lymphocyte syndrome). *Cell* **1993**, *75*, 135–146. [CrossRef]

109. Meissner, T.B.; Liu, Y.-J.; Lee, K.-H.; Li, A.; Biswas, A.; Van Eggermond, M.C.J.A.; Van den Elsen, P.J.; Kobayashi, K.S. NLRC5 cooperates with the RFX transcription factor complex to induce MHC class I gene expression. *J. Immunol.* **2012**, *188*, 4951–4958. [CrossRef]

110. Drozina, G.; Kohoutek, J.; Jabrane-Ferrat, N.; Peterlin, B.M. Expression of MHC II genes. *Curr. Top. Microbiol. Immunol.* **2005**, *290*, 147–170. [CrossRef]

111. Gobin, S.J.; Van Zutphen, M.; Westerheide, S.D.; Boss, J.M.; Van den Elsen, P.J. The MHC-specific enhanceosome and its role in MHC class I and beta(2)-microglobulin gene transactivation. *J. Immunol.* **2001**, *167*, 5175–5184. [CrossRef] [PubMed]

112. Harton, J.A.; Ting, J.P. Class II transactivator: Mastering the art of major histocompatibility complex expression. *Mol. Cell. Biol.* **2000**, *20*, 6185–6194. [CrossRef] [PubMed]

113. Reith, W.; Mach, B. The bare lymphocyte syndrome and the regulation of MHC expression. *Annu. Rev. Immunol.* **2001**, *19*, 331–373. [CrossRef] [PubMed]

114. Krawczyk, M.; Reith, W. Regulation of MHC class II expression, a unique regulatory system identified by the study of a primary immunodeficiency disease. *Tissue Antigens* **2006**, *67*, 183–197. [CrossRef]

115. Robbins, G.R.; Truax, A.D.; Davis, B.K.; Zhang, L.; Brickey, W.J.; Ting, J.P.-Y. Regulation of Class I Major Histocompatibility Complex (MHC) by Nucleotide-binding Domain, Leucine-rich Repeat-containing (NLR) Proteins. *J. Biol. Chem.* **2012**, *287*, 24294–24303. [CrossRef]

116. Biswas, A.; Meissner, T.B.; Kawai, T.; Kobayashi, K.S. Cutting edge: Impaired MHC class I expression in mice deficient for Nlrc5/class I transactivator. *J. Immunol.* **2012**, *189*, 516–520. [CrossRef]

117. Ludigs, K.; Jandus, C.; Utzschneider, D.T.; Staehli, F.; Bessoles, S.; Dang, A.T.; Rota, G.; Castro, W.; Zehn, D.; Vivier, E.; et al. NLRC5 shields T lymphocytes from NK-cell-mediated elimination under inflammatory conditions. *Nat. Commun.* **2016**, *7*, 10554. [CrossRef]

118. Martin, B.K.; Chin, K.C.; Olsen, J.C.; Skinner, C.A.; Dey, A.; Ozato, K.; Ting, J.P. Induction of MHC class I expression by the MHC class II transactivator CIITA. *Immunity* **1997**, *6*, 591–600. [CrossRef]

119. Meissner, T.B.; Li, A.; Kobayashi, K.S. NLRC5: A newly discovered MHC class I transactivator (CITA). *Microbes Infect.* **2012**, *14*, 477–484. [CrossRef]

120. Gruen, J.R.; Weissman, S.M. Human MHC class III and IV genes and disease associations. *Front. Biosci.* **2001**, *6*, D960–D972. [CrossRef]

121. Deakin, J.E.; Papenfuss, A.T.; Belov, K.; Cross, J.G.R.; Coggill, P.; Palmer, S.; Sims, S.; Speed, T.P.; Beck, S.; Graves, J.A.M. Evolution and comparative analysis of the MHC Class III inflammatory region. *BMC Genom.* **2006**, *7*, 281. [CrossRef] [PubMed]

122. Xie, T.; Rowen, L.; Aguado, B.; Ahearn, M.E.; Madan, A.; Qin, S.; Campbell, R.D.; Hood, L. Analysis of the gene-dense major histocompatibility complex class III region and its comparison to mouse. *Genome Res.* **2003**, *13*, 2621–2636. [CrossRef] [PubMed]

123. Johnson, D.B.; Estrada, M.V.; Salgado, R.; Sanchez, V.; Doxie, D.B.; Opalenik, S.R.; Vilgelm, A.E.; Feld, E.; Johnson, A.S.; Greenplate, A.R.; et al. Melanoma-specific MHC-II expression represents a tumour-autonomous phenotype and predicts response to anti-PD-1/PD-L1 therapy. *Nat. Commun.* **2016**, *7*, 10582. [CrossRef] [PubMed]

124. Park, I.A.; Hwang, S.-H.; Song, I.H.; Heo, S.-H.; Kim, Y.-A.; Bang, W.S.; Park, H.S.; Lee, M.; Gong, G.; Lee, H.J. Expression of the MHC class II in triple-negative breast cancer is associated with tumor-infiltrating lymphocytes and interferon signaling. *PLoS ONE* **2017**, *12*, e0182786. [CrossRef] [PubMed]

125. Roemer, M.G.M.; Redd, R.A.; Cader, F.Z.; Pak, C.J.; Abdelrahman, S.; Ouyang, J.; Sasse, S.; Younes, A.; Fanale, M.; Santoro, A.; et al. Major Histocompatibility Complex Class II and Programmed Death Ligand 1 Expression Predict Outcome After Programmed Death 1 Blockade in Classic Hodgkin Lymphoma. *J. Clin. Oncol.* **2018**, *36*, 942–950. [CrossRef]

126. Rodig, S.J.; Gusenleitner, D.; Jackson, D.G.; Gjini, E.; Giobbie-Hurder, A.; Jin, C.; Chang, H.; Lovitch, S.B.; Horak, C.; Weber, J.S.; et al. MHC proteins confer differential sensitivity to CTLA-4 and PD-1 blockade in untreated metastatic melanoma. *Sci. Transl. Med.* **2018**, *10*. [CrossRef]

127. Seliger, B.; Kloor, M.; Ferrone, S. HLA class II antigen-processing pathway in tumors: Molecular defects and clinical relevance. *Oncoimmunology* **2017**, *6*. [CrossRef]

128. Oldford, S.A.; Robb, J.D.; Watson, P.H.; Drover, S. HLA-DRB alleles are differentially expressed by tumor cells in breast carcinoma. *Int. J. Cancer* **2004**, *112*, 399–406. [CrossRef]

129. Bustin, S.A.; Li, S.R.; Phillips, S.; Dorudi, S. Expression of HLA class II in colorectal cancer: Evidence for enhanced immunogenicity of microsatellite-instability-positive tumours. *Tumour Biol.* **2001**, *22*, 294–298. [CrossRef]

130. Younger, A.R.; Amria, S.; Jeffrey, W.A.; Mahdy, A.E.M.; Goldstein, O.G.; Norris, J.S.; Haque, A. HLA class II antigen presentation by prostate cancer cells. *Prostate Cancer Prostatic Dis.* **2008**, *11*, 334–341. [CrossRef]

131. Forero, A.; Li, Y.; Chen, D.; Grizzle, W.E.; Updike, K.L.; Merz, N.D.; Downs-Kelly, E.; Burwell, T.C.; Vaklavas, C.; Buchsbaum, D.J.; et al. Expression of the MHC Class II Pathway in Triple-Negative Breast Cancer Tumor Cells Is Associated with a Good Prognosis and Infiltrating Lymphocytes. *Cancer Immunol. Res.* **2016**, *4*, 390–399. [CrossRef] [PubMed]

132. Callahan, M.J.; Nagymanyoki, Z.; Bonome, T.; Johnson, M.E.; Litkouhi, B.; Sullivan, E.H.; Hirsch, M.S.; Matulonis, U.A.; Liu, J.; Birrer, M.J.; et al. Increased HLA-DMB expression in the tumor epithelium is associated with increased CTL infiltration and improved prognosis in advanced-stage serous ovarian cancer. *Clin. Cancer Res.* **2008**, *14*, 7667–7673. [CrossRef] [PubMed]

133. Dierssen, J.W.F.; De Miranda, N.F.C.C.; Ferrone, S.; Van Puijenbroek, M.; Cornelisse, C.J.; Fleuren, G.J.; Van Wezel, T.; Morreau, H. HNPCC versus sporadic microsatellite-unstable colon cancers follow different routes toward loss of HLA class I expression. *BMC Cancer* **2007**, *7*, 33. [CrossRef] [PubMed]

134. De Miranda, N.F.C.C.; Nielsen, M.; Pereira, D.; Van Puijenbroek, M.; Vasen, H.F.; Hes, F.J.; Van Wezel, T.; Morreau, H. MUTYH-associated polyposis carcinomas frequently lose HLA class I expression—A common event amongst DNA-repair-deficient colorectal cancers. *J. Pathol.* **2009**, *219*, 69–76. [CrossRef]

135. Bicknell, D.C.; Kaklamanis, L.; Hampson, R.; Bodmer, W.F.; Karran, P. Selection for beta 2-microglobulin mutation in mismatch repair-defective colorectal carcinomas. *Curr. Biol.* **1996**, *6*, 1695–1697. [CrossRef]

136. Cabrera, T.; Collado, A.; Fernandez, M.A.; Ferron, A.; Sancho, J.; Ruiz-Cabello, F.; Garrido, F. High frequency of altered HLA class I phenotypes in invasive colorectal carcinomas. *Tissue Antigens* **1998**, *52*, 114–123. [CrossRef]

137. Cabrera, T.; Maleno, I.; Collado, A.; Lopez Nevot, M.A.; Tait, B.D.; Garrido, F. Analysis of HLA class I alterations in tumors: Choosing a strategy based on known patterns of underlying molecular mechanisms. *Tissue Antigens* **2007**, *69* (Suppl. S1), 264–268. [CrossRef]

138. Benitez, R.; Godelaine, D.; Lopez-Nevot, M.A.; Brasseur, F.; Jiménez, P.; Marchand, M.; Oliva, M.R.; Van Baren, N.; Cabrera, T.; Andry, G.; et al. Mutations of the beta2-microglobulin gene result in a lack of HLA class I molecules on melanoma cells of two patients immunized with MAGE peptides. *Tissue Antigens* **1998**, *52*, 520–529. [CrossRef]

139. Hicklin, D.J.; Wang, Z.; Arienti, F.; Rivoltini, L.; Parmiani, G.; Ferrone, S. beta2-Microglobulin mutations, HLA class I antigen loss, and tumor progression in melanoma. *J. Clin. Investig.* **1998**, *101*, 2720–2729. [CrossRef]

140. Tang, Q.; Zhang, J.; Qi, B.; Shen, C.; Xie, W. Downregulation of HLA class I molecules in primary oral squamous cell carcinomas and cell lines. *Arch. Med. Res.* **2009**, *40*, 256–263. [CrossRef]

141. Maleno, I.; López-Nevot, M.A.; Cabrera, T.; Salinero, J.; Garrido, F. Multiple mechanisms generate HLA class I altered phenotypes in laryngeal carcinomas: High frequency of HLA haplotype loss associated with loss of heterozygosity in chromosome region 6p21. *Cancer Immunol. Immunother.* **2002**, *51*, 389–396. [CrossRef] [PubMed]

142. Cabrera, T.; Angustias Fernandez, M.; Sierra, A.; Garrido, A.; Herruzo, A.; Escobedo, A.; Fabra, A.; Garrido, F. High frequency of altered HLA class I phenotypes in invasive breast carcinomas. *Hum. Immunol.* **1996**, *50*, 127–134. [CrossRef]

143. Qifeng, S.; Bo, C.; Xingtao, J.; Chuanliang, P.; Xiaogang, Z. Methylation of the promoter of human leukocyte antigen class I in human esophageal squamous cell carcinoma and its histopathological characteristics. *J. Thorac. Cardiovasc. Surg.* **2011**, *141*, 808–814. [CrossRef] [PubMed]

144. Facoetti, A.; Nano, R.; Zelini, P.; Morbini, P.; Benericetti, E.; Ceroni, M.; Campoli, M.; Ferrone, S. Human leukocyte antigen and antigen processing machinery component defects in astrocytic tumors. *Clin. Cancer Res.* **2005**, *11*, 8304–8311. [CrossRef]

145. Cai, L.; Michelakos, T.; Yamada, T.; Fan, S.; Wang, X.; Schwab, J.H.; Ferrone, C.R.; Ferrone, S. Defective HLA class I antigen processing machinery in cancer. *Cancer Immunol. Immunother.* **2018**, *67*, 999–1009. [CrossRef] [PubMed]

146. Garrido, F.; Algarra, I. MHC antigens and tumor escape from immune surveillance. *Adv. Cancer Res.* **2001**, *83*, 117–158. [CrossRef]

147. Kageshita, T.; Hirai, S.; Ono, T.; Hicklin, D.J.; Ferrone, S. Down-regulation of HLA class I antigen-processing molecules in malignant melanoma: Association with disease progression. *Am. J. Pathol.* **1999**, *154*, 745–754. [CrossRef]

148. Campoli, M.; Ferrone, S. HLA antigen changes in malignant cells: Epigenetic mechanisms and biologic significance. *Oncogene* **2008**, *27*, 5869–5885. [CrossRef]

149. Chang, C.-C.; Pirozzi, G.; Wen, S.-H.; Chung, I.-H.; Chiu, B.-L.; Errico, S.; Luongo, M.; Lombardi, M.L.; Ferrone, S. Multiple structural and epigenetic defects in the human leukocyte antigen class I antigen presentation pathway in a recurrent metastatic melanoma following immunotherapy. *J. Biol. Chem.* **2015**, *290*, 26562–26575. [CrossRef]

150. Garrido, F.; Cabrera, T.; Concha, A.; Glew, S.; Ruiz-Cabello, F.; Stern, P.L. Natural history of HLA expression during tumour development. *Immunol. Today* **1993**, *14*, 491–499. [CrossRef]

151. Ljunggren, H.G.; Kärre, K. In search of the "missing self": MHC molecules and NK cell recognition. *Immunol. Today* **1990**, *11*, 237–244. [CrossRef]

152. Garrido, F.; Cabrera, T.; Aptsiauri, N. "Hard" and "soft" lesions underlying the HLA class I alterations in cancer cells: Implications for immunotherapy. *Int. J. Cancer* **2010**, *127*, 249–256. [CrossRef] [PubMed]

153. Paschen, A.; Arens, N.; Sucker, A.; Greulich-Bode, K.M.; Fonsatti, E.; Gloghini, A.; Striegel, S.; Schwinn, N.; Carbone, A.; Hildenbrand, R.; et al. The coincidence of chromosome 15 aberrations and beta2-microglobulin gene mutations is causative for the total loss of human leukocyte antigen class I expression in melanoma. *Clin. Cancer Res.* **2006**, *12*, 3297–3305. [CrossRef] [PubMed]

154. Del Campo, A.B.; Aptsiauri, N.; Méndez, R.; Zinchenko, S.; Vales, A.; Paschen, A.; Ward, S.; Ruiz-Cabello, F.; González-Aseguinolaza, G.; Garrido, F. Efficient recovery of HLA class I expression in human tumor cells after beta2-microglobulin gene transfer using adenoviral vector: Implications for cancer immunotherapy. *Scand. J. Immunol.* **2009**, *70*, 125–135. [CrossRef]

155. Maleno, I.; Aptsiauri, N.; Cabrera, T.; Gallego, A.; Paschen, A.; López-Nevot, M.A.; Garrido, F. Frequent loss of heterozygosity in the β2-microglobulin region of chromosome 15 in primary human tumors. *Immunogenetics* **2011**, *63*, 65–71. [CrossRef]

156. Pérez, B.; Benitez, R.; Fernández, M.A.; Oliva, M.R.; Soto, J.L.; Serrano, S.; López Nevot, M.A.; Garrido, F. A new beta 2 microglobulin mutation found in a melanoma tumor cell line. *Tissue Antigens* **1999**, *53*, 569–572. [CrossRef]

157. Chang, C.-C.; Campoli, M.; Restifo, N.P.; Wang, X.; Ferrone, S. Immune selection of hot-spot beta 2-microglobulin gene mutations, HLA-A2 allospecificity loss, and antigen-processing machinery component down-regulation in melanoma cells derived from recurrent metastases following immunotherapy. *J. Immunol.* **2005**, *174*, 1462–1471. [CrossRef]

158. Bernal, M.; Ruiz-Cabello, F.; Concha, A.; Paschen, A.; Garrido, F. Implication of the β2-microglobulin gene in the generation of tumor escape phenotypes. *Cancer Immunol. Immunother.* **2012**, *61*, 1359–1371. [CrossRef]

159. Garrido, F.; Romero, I.; Aptsiauri, N.; Garcia-Lora, A.M. Generation of MHC class I diversity in primary tumors and selection of the malignant phenotype. *Int. J. Cancer* **2016**, *138*, 271–280. [CrossRef]

160. Browning, M.; Petronzelli, F.; Bicknell, D.; Krausa, P.; Rowan, A.; Tonks, S.; Murray, N.; Bodmer, J.; Bodmer, W. Mechanisms of loss of HLA class I expression on colorectal tumor cells. *Tissue Antigens* **1996**, *47*, 364–371. [CrossRef]

161. Sette, A.; Chesnut, R.; Fikes, J. HLA expression in cancer: Implications for T cell-based immunotherapy. *Immunogenetics* **2001**, *53*, 255–263. [CrossRef] [PubMed]

162. Khleif, S. (Ed.) Cancer Treatment and Research. In *Tumor Immunology and Cancer Vaccines*; Springer: New York, NY, USA, 2005; ISBN 978-1-4020-8119-4.

163. Murphy, C.; Nikodem, D.; Howcroft, K.; Weissman, J.D.; Singer, D.S. Active repression of major histocompatibility complex class I genes in a human neuroblastoma cell line. *J. Biol. Chem.* **1996**, *271*, 30992–30999. [CrossRef] [PubMed]

164. Nie, Y.; Yang, G.; Song, Y.; Zhao, X.; So, C.; Liao, J.; Wang, L.D.; Yang, C.S. DNA hypermethylation is a mechanism for loss of expression of the HLA class I genes in human esophageal squamous cell carcinomas. *Carcinogenesis* **2001**, *22*, 1615–1623. [CrossRef] [PubMed]

165. Coral, S.; Sigalotti, L.; Gasparollo, A.; Cattarossi, I.; Visintin, A.; Cattelan, A.; Altomonte, M.; Maio, M. Prolonged upregulation of the expression of HLA class I antigens and costimulatory molecules on melanoma cells treated with 5-aza-2'-deoxycytidine (5-AZA-CdR). *J. Immunother.* **1999**, *22*, 16–24. [CrossRef] [PubMed]

166. Coral, S.; Sigalotti, L.; Altomonte, M.; Engelsberg, A.; Colizzi, F.; Cattarossi, I.; Maraskovsky, E.; Jager, E.; Seliger, B.; Maio, M. 5-aza-2'-deoxycytidine-induced expression of functional cancer testis antigens in human renal cell carcinoma: Immunotherapeutic implications. *Clin. Cancer Res.* **2002**, *8*, 2690–2695. [PubMed]

167. Esteller, M. Epigenetics provides a new generation of oncogenes and tumour-suppressor genes. *Br. J. Cancer* **2006**, *94*, 179–183. [CrossRef]

168. Lettini, A.A.; Guidoboni, M.; Fonsatti, E.; Anzalone, L.; Cortini, E.; Maio, M. Epigenetic remodelling of DNA in cancer. *Histol. Histopathol.* **2007**, *22*, 1413–1424. [CrossRef]

169. Ferris, R.L.; Hunt, J.L.; Ferrone, S. Human leukocyte antigen (HLA) class I defects in head and neck cancer: Molecular mechanisms and clinical significance. *Immunol. Res.* **2005**, *33*, 113–133. [CrossRef]

170. Ferrone, S.; Marincola, F.M. Loss of HLA class I antigens by melanoma cells: Molecular mechanisms, functional significance and clinical relevance. *Immunol. Today* **1995**, *16*, 487–494. [CrossRef]

171. Concha-Benavente, F.; Srivastava, R.; Ferrone, S.; Ferris, R.L. Immunological and clinical significance of HLA class I antigen processing machinery component defects in malignant cells. *Oral Oncol.* **2016**, *58*, 52–58. [CrossRef]

172. Seliger, B.; Ferrone, S. HLA Class I Antigen Processing Machinery Defects in Cancer Cells-Frequency, Functional Significance, and Clinical Relevance with Special Emphasis on Their Role in T Cell-Based Immunotherapy of Malignant Disease. *Methods Mol. Biol.* **2020**, *2055*, 325–350. [CrossRef]

173. Yoshihama, S.; Roszik, J.; Downs, I.; Meissner, T.B.; Vijayan, S.; Chapuy, B.; Sidiq, T.; Shipp, M.A.; Lizee, G.A.; Kobayashi, K.S. NLRC5/MHC class I transactivator is a target for immune evasion in cancer. *Proc. Natl. Acad. Sci. USA* **2016**, *113*, 5999–6004. [CrossRef]

174. Yoshihama, S.; Vijayan, S.; Sidiq, T.; Kobayashi, K.S. NLRC5/CITA: A key player in cancer immune surveillance. *Trends Cancer* **2017**, *3*, 28–38. [CrossRef] [PubMed]

175. Kloor, M.; Becker, C.; Benner, A.; Woerner, S.M.; Gebert, J.; Ferrone, S.; Von Knebel Doeberitz, M. Immunoselective pressure and human leukocyte antigen class I antigen machinery defects in microsatellite unstable colorectal cancers. *Cancer Res.* **2005**, *65*, 6418–6424. [CrossRef]

176. Cathro, H.P.; Smolkin, M.E.; Theodorescu, D.; Jo, V.Y.; Ferrone, S.; Frierson, H.F. Relationship between HLA class I antigen processing machinery component expression and the clinicopathologic characteristics of bladder carcinomas. *Cancer Immunol. Immunother.* **2010**, *59*, 465–472. [CrossRef]

177. Leffers, N.; Gooden, M.J.M.; Mokhova, A.A.; Kast, W.M.; Boezen, H.M.; Ten Hoor, K.A.; Hollema, H.; Daemen, T.; Van der Zee, A.G.J.; Nijman, H.W. Down-regulation of proteasomal subunit MB1 is an independent predictor of improved survival in ovarian cancer. *Gynecol. Oncol.* **2009**, *113*, 256–263. [CrossRef]

178. Hoves, S.; Aigner, M.; Pfeiffer, C.; Laumer, M.; Obermann, E.C.; Mackensen, A. In situ analysis of the antigen-processing machinery in acute myeloid leukaemic blasts by tissue microarray. *Leukemia* **2009**, *23*, 877–885. [CrossRef]

179. Kamarashev, J.; Ferrone, S.; Seifert, B.; Böni, R.; Nestle, F.O.; Burg, G.; Dummer, R. TAP1 down-regulation in primary melanoma lesions: An independent marker of poor prognosis. *Int. J. Cancer* **2001**, *95*, 23–28. [CrossRef]

180. Hicklin, D.J.; Marincola, F.M.; Ferrone, S. HLA class I antigen downregulation in human cancers: T-cell immunotherapy revives an old story. *Mol. Med. Today* **1999**, *5*, 178–186. [CrossRef]

181. Vitale, M.; Rezzani, R.; Rodella, L.; Zauli, G.; Grigolato, P.; Cadei, M.; Hicklin, D.J.; Ferrone, S. HLA class I antigen and transporter associated with antigen processing (TAP1 and TAP2) down-regulation in high-grade primary breast carcinoma lesions. *Cancer Res.* **1998**, *58*, 737–742.

182. Atkins, D.; Breuckmann, A.; Schmahl, G.E.; Binner, P.; Ferrone, S.; Krummenauer, F.; Störkel, S.; Seliger, B. MHC class I antigen processing pathway defects, ras mutations and disease stage in colorectal carcinoma. *Int. J. Cancer* **2004**, *109*, 265–273. [CrossRef] [PubMed]

183. Hirata, T.; Yamamoto, H.; Taniguchi, H.; Horiuchi, S.; Oki, M.; Adachi, Y.; Imai, K.; Shinomura, Y. Characterization of the immune escape phenotype of human gastric cancers with and without high-frequency microsatellite instability. *J. Pathol.* **2007**, *211*, 516–523. [CrossRef] [PubMed]

184. Chen, H.L.; Gabrilovich, D.; Tampé, R.; Girgis, K.R.; Nadaf, S.; Carbone, D.P. A functionally defective allele of TAP1 results in loss of MHC class I antigen presentation in a human lung cancer. *Nat. Genet.* **1996**, *13*, 210–213. [CrossRef] [PubMed]

185. Oliveira, C.C.; Van Hall, T. Alternative Antigen Processing for MHC Class I: Multiple Roads Lead to Rome. *Front. Immunol.* **2015**, *6*, 298. [CrossRef]

186. Van Duinen, S.G.; Ruiter, D.J.; Broecker, E.B.; Van der Velde, E.A.; Sorg, C.; Welvaart, K.; Ferrone, S. Level of HLA antigens in locoregional metastases and clinical course of the disease in patients with melanoma. *Cancer Res.* **1988**, *48*, 1019–1025.

187. Zaloudik, J.; Moore, M.; Ghosh, A.K.; Mechl, Z.; Rejthar, A. DNA content and MHC class II antigen expression in malignant melanoma: Clinical course. *J. Clin. Pathol.* **1988**, *41*, 1078–1084. [CrossRef]

188. Trieb, K.; Lechleitner, T.; Lang, S.; Windhager, R.; Kotz, R.; Dirnhofer, S. Evaluation of HLA-DR expression and T-lymphocyte infiltration in osteosarcoma. *Pathol. Res. Pract.* **1998**, *194*, 679–684. [CrossRef]

189. Moretti, S.; Pinzi, C.; Berti, E.; Spallanzani, A.; Chiarugi, A.; Boddi, V.; Reali, U.M.; Giannotti, B. In situ expression of transforming growth factor beta is associated with melanoma progression and correlates with Ki67, HLA-DR and beta 3 integrin expression. *Melanoma Res.* **1997**, *7*, 313–321. [CrossRef]

190. Diederichsen, A.C.; Stenholm, A.C.; Kronborg, O.; Fenger, C.; Jensenius, J.C.; Zeuthen, J.; Christensen, P.B.; Kristensen, T.; Ostenhom, A.C. Flow cytometric investigation of immune-response-related surface molecules on human colorectal cancers. *Int. J. Cancer* **1998**, *79*, 283–287. [CrossRef]

191. Hilders, C.G.; Houbiers, J.G.; Van Ravenswaay Claasen, H.H.; Veldhuizen, R.W.; Fleuren, G.J. Association between HLA-expression and infiltration of immune cells in cervical carcinoma. *Lab. Investig.* **1993**, *69*, 651–659.

192. Coleman, N.; Stanley, M.A. Analysis of HLA-DR expression on keratinocytes in cervical neoplasia. *Int. J. Cancer* **1994**, *56*, 314–319. [CrossRef] [PubMed]

193. Cromme, F.V.; Meijer, C.J.; Snijders, P.J.; Uyterlinde, A.; Kenemans, P.; Helmerhorst, T.; Stern, P.L.; Van den Brule, A.J.; Walboomers, J.M. Analysis of MHC class I and II expression in relation to presence of HPV genotypes in premalignant and malignant cervical lesions. *Br. J. Cancer* **1993**, *67*, 1372–1380. [CrossRef] [PubMed]

194. Jackson, P.A.; Green, M.A.; Marks, C.G.; King, R.J.; Hubbard, R.; Cook, M.G. Lymphocyte subset infiltration patterns and HLA antigen status in colorectal carcinomas and adenomas. *Gut* **1996**, *38*, 85–89. [CrossRef] [PubMed]

195. Matsushita, K.; Takenouchi, T.; Kobayashi, S.; Hayashi, H.; Okuyama, K.; Ochiai, T.; Mikata, A.; Isono, K. HLA-DR antigen expression in colorectal carcinomas: Influence of expression by IFN-gamma in situ and its association with tumour progression. *Br. J. Cancer* **1996**, *73*, 644–648. [CrossRef] [PubMed]

196. Norheim Andersen, S.; Breivik, J.; Løvig, T.; Meling, G.I.; Gaudernack, G.; Clausen, O.P.; Schjölberg, A.; Fausa, O.; Langmark, F.; Lund, E.; et al. K-ras mutations and HLA-DR expression in large bowel adenomas. *Br. J. Cancer* **1996**, *74*, 99–108. [CrossRef]

197. Anichini, A.; Mortarini, R.; Nonaka, D.; Molla, A.; Vegetti, C.; Montaldi, E.; Wang, X.; Ferrone, S. Association of Antigen-Processing Machinery and HLA Antigen Phenotype of Melanoma Cells with Survival in American Joint Committee on Cancer Stage III and IV Melanoma Patients. *Cancer Res.* **2006**, *66*, 6405–6411. [CrossRef]

198. Sartoris, S.; Valle, M.T.; Barbaro, A.L.; Tosi, G.; Cestari, T.; D'Agostino, A.; Megiovanni, A.M.; Manca, F.; Accolla, R.S. HLA class II expression in uninducible hepatocarcinoma cells after transfection of AIR-1 gene product CIITA: Acquisition of antigen processing and presentation capacity. *J. Immunol.* **1998**, *161*, 814–820.

199. Yazawa, T.; Kamma, H.; Fujiwara, M.; Matsui, M.; Horiguchi, H.; Satoh, H.; Fujimoto, M.; Yokoyama, K.; Ogata, T. Lack of class II transactivator causes severe deficiency of HLA-DR expression in small cell lung cancer. *J. Pathol.* **1999**, *187*, 191–199. [CrossRef]

200. Mach, B.; Steimle, V.; Martinez-Soria, E.; Reith, W. Regulation of MHC class II genes: Lessons from a disease. *Annu. Rev. Immunol.* **1996**, *14*, 301–331. [CrossRef]

201. Bosshart, H.; Jarrett, R.F. Deficient major histocompatibility complex class II antigen presentation in a subset of Hodgkin's disease tumor cells. *Blood* **1998**, *92*, 2252–2259. [CrossRef]

202. Fife, B.T.; Bluestone, J.A. Control of peripheral T-cell tolerance and autoimmunity via the CTLA-4 and PD-1 pathways. *Immunol. Rev.* **2008**, *224*, 166–182. [CrossRef]

203. Mueller, D.L.; Jenkins, M.K.; Schwartz, R.H. Clonal expansion versus functional clonal inactivation: A costimulatory signalling pathway determines the outcome of T cell antigen receptor occupancy. *Annu. Rev. Immunol.* **1989**, *7*, 445–480. [CrossRef] [PubMed]

204. Schneider, H.; Downey, J.; Smith, A.; Zinselmeyer, B.H.; Rush, C.; Brewer, J.M.; Wei, B.; Hogg, N.; Garside, P.; Rudd, C.E. Reversal of the TCR stop signal by CTLA-4. *Science* **2006**, *313*, 1972–1975. [CrossRef] [PubMed]

205. Nishimura, H.; Honjo, T. PD-1: An inhibitory immunoreceptor involved in peripheral tolerance. *Trends Immunol.* **2001**, *22*, 265–268. [CrossRef]

206. Ribas, A. Releasing the Brakes on Cancer Immunotherapy. *N. Engl. J. Med.* **2015**, *373*, 1490–1492. [CrossRef] [PubMed]

207. Webb, E.S.; Liu, P.; Baleeiro, R.; Lemoine, N.R.; Yuan, M.; Wang, Y.-H. Immune checkpoint inhibitors in cancer therapy. *J. Biomed. Res.* **2018**, *32*, 317–326. [CrossRef]

208. Keung, E.Z.; Wargo, J.A. The Current Landscape of Immune Checkpoint Inhibition for Solid Malignancies. *Surg. Oncol. Clin. N. Am.* **2019**, *28*, 369–386. [CrossRef]

209. Pitt, J.M.; Vétizou, M.; Daillère, R.; Roberti, M.P.; Yamazaki, T.; Routy, B.; Lepage, P.; Boneca, I.G.; Chamaillard, M.; Kroemer, G.; et al. Resistance Mechanisms to Immune-Checkpoint Blockade in Cancer: Tumor-Intrinsic and -Extrinsic Factors. *Immunity* **2016**, *44*, 1255–1269. [CrossRef]

210. Restifo, N.P.; Smyth, M.J.; Snyder, A. Acquired resistance to immunotherapy and future challenges. *Nat. Rev. Cancer* **2016**, *16*, 121–126. [CrossRef]

211. Jenkins, R.W.; Barbie, D.A.; Flaherty, K.T. Mechanisms of resistance to immune checkpoint inhibitors. *Br. J. Cancer* **2018**, *118*, 9–16. [CrossRef]

212. Meng, X.; Huang, Z.; Teng, F.; Xing, L.; Yu, J. Predictive biomarkers in PD-1/PD-L1 checkpoint blockade immunotherapy. *Cancer Treat. Rev.* **2015**, *41*, 868–876. [CrossRef] [PubMed]

213. Maleki Vareki, S.; Garrigós, C.; Duran, I. Biomarkers of response to PD-1/PD-L1 inhibition. *Crit. Rev. Oncol. Hematol.* **2017**, *116*, 116–124. [CrossRef] [PubMed]

214. Manson, G.; Norwood, J.; Marabelle, A.; Kohrt, H.; Houot, R. Biomarkers associated with checkpoint inhibitors. *Ann. Oncol.* **2016**, *27*, 1199–1206. [CrossRef] [PubMed]

215. Gibney, G.T.; Weiner, L.M.; Atkins, M.B. Predictive biomarkers for checkpoint inhibitor-based immunotherapy. *Lancet Oncol.* **2016**, *17*, e542–e551. [CrossRef]

216. Zaretsky, J.M.; Garcia-Diaz, A.; Shin, D.S.; Escuin-Ordinas, H.; Hugo, W.; Hu-Lieskovan, S.; Torrejon, D.Y.; Abril-Rodriguez, G.; Sandoval, S.; Barthly, L.; et al. Mutations Associated with Acquired Resistance to PD-1 Blockade in Melanoma. *N. Engl. J. Med.* **2016**, *375*, 819–829. [CrossRef]

217. Kametani, Y.; Ohno, Y.; Ohshima, S.; Tsuda, B.; Yasuda, A.; Seki, T.; Ito, R.; Tokuda, Y. Humanized Mice as an Effective Evaluation System for Peptide Vaccines and Immune Checkpoint Inhibitors. *Int. J. Mol. Sci.* **2019**, *20*, 6337. [CrossRef]

218. Lechner, M.G.; Karimi, S.S.; Barry-Holson, K.; Angell, T.E.; Murphy, K.A.; Church, C.H.; Ohlfest, J.R.; Hu, P.; Epstein, A.L. Immunogenicity of murine solid tumor models as a defining feature of in vivo behavior and response to immunotherapy. *J. Immunother.* **2013**, *36*, 477–489. [CrossRef]

219. Ashizawa, T.; Iizuka, A.; Nonomura, C.; Kondou, R.; Maeda, C.; Miyata, H.; Sugino, T.; Mitsuya, K.; Hayashi, N.; Nakasu, Y.; et al. Antitumor Effect of Programmed Death-1 (PD-1) Blockade in Humanized the NOG-MHC Double Knockout Mouse. *Clin. Cancer Res.* **2017**, *23*, 149–158. [CrossRef]

220. Gettinger, S.; Choi, J.; Hastings, K.; Truini, A.; Datar, I.; Sowell, R.; Wurtz, A.; Dong, W.; Cai, G.; Melnick, M.A.; et al. Impaired HLA Class I Antigen Processing and Presentation as a Mechanism of Acquired Resistance to Immune Checkpoint Inhibitors in Lung Cancer. *Cancer Discov.* **2017**, *7*, 1420–1435. [CrossRef]

221. Lhotakova, K.; Grzelak, A.; Polakova, I.; Vackova, J.; Smahel, M. Establishment and characterization of a mouse tumor cell line with irreversible downregulation of MHC class I molecules. *Oncol. Rep.* **2019**, *42*, 2826–2835. [CrossRef]

222. Rangan, L.; Galaine, J.; Boidot, R.; Hamieh, M.; Dosset, M.; Francoual, J.; Beziaud, L.; Pallandre, J.-R.; Lauret Marie Joseph, E.; Asgarova, A.; et al. Identification of a novel PD-L1 positive solid tumor transplantable in HLA-A*0201/DRB1*0101 transgenic mice. *Oncotarget* **2017**, *8*, 48959–48971. [CrossRef] [PubMed]

223. Mortara, L.; Castellani, P.; Meazza, R.; Tosi, G.; De Lerma Barbaro, A.; Procopio, F.A.; Comes, A.; Zardi, L.; Ferrini, S.; Accolla, R.S. CIITA-induced MHC class II expression in mammary adenocarcinoma leads to a Th1 polarization of the tumor microenvironment, tumor rejection, and specific antitumor memory. *Clin. Cancer Res.* **2006**, *12*, 3435–3443. [CrossRef] [PubMed]

224. Murciano-Goroff, Y.R.; Warner, A.B.; Wolchok, J.D. The future of cancer immunotherapy: Microenvironment-targeting combinations. *Cell Res.* **2020**, *30*, 507–519. [CrossRef] [PubMed]

225. Chowell, D.; Morris, L.G.T.; Grigg, C.M.; Weber, J.K.; Samstein, R.M.; Makarov, V.; Kuo, F.; Kendall, S.M.; Requena, D.; Riaz, N.; et al. Patient HLA class I genotype influences cancer response to checkpoint blockade immunotherapy. *Science* **2018**, *359*, 582–587. [CrossRef] [PubMed]

226. Garrido, F.; Aptsiauri, N.; Doorduijn, E.M.; Garcia Lora, A.M.; Van Hall, T. The urgent need to recover MHC class I in cancers for effective immunotherapy. *Curr. Opin. Immunol.* **2016**, *39*, 44–51. [CrossRef]

227. Garrido, F.; Aptsiauri, N. Cancer immune escape: MHC expression in primary tumours versus metastases. *Immunology* **2019**, *158*, 255–266. [CrossRef]

228. Banik, D.; Moufarrij, S.; Villagra, A. Immunoepigenetics Combination Therapies: An Overview of the Role of HDACs in Cancer Immunotherapy. *Int. J. Mol. Sci.* **2019**, *20*, 2241. [CrossRef]

229. Del Campo, A.B.; Carretero, J.; Muñoz, J.A.; Zinchenko, S.; Ruiz-Cabello, F.; González-Aseguinolaza, G.; Garrido, F.; Aptsiauri, N. Adenovirus expressing β2-microglobulin recovers HLA class I expression and antitumor immunity by increasing T-cell recognition. *Cancer Gene Ther.* **2014**, *21*, 317–332. [CrossRef]

230. Wang, X.; Waschke, B.C.; Woolaver, R.A.; Chen, Z.; Zhang, G.; Piscopio, A.D.; Liu, X.; Wang, J.H. Histone Deacetylase Inhibition Sensitizes PD1 Blockade-Resistant B-cell Lymphomas. *Cancer Immunol. Res.* **2019**, *7*, 1318–1331. [CrossRef]

231. Ramsuran, V.; Kulkarni, S.; O'huigin, C.; Yuki, Y.; Augusto, D.G.; Gao, X.; Carrington, M. Epigenetic regulation of differential HLA-A allelic expression levels. *Hum. Mol. Genet.* **2015**, *24*, 4268–4275. [CrossRef]

232. Brea, E.J.; Oh, C.Y.; Manchado, E.; Budhu, S.; Gejman, R.S.; Mo, G.; Mondello, P.; Han, J.E.; Jarvis, C.A.; Ulmert, D.; et al. Kinase Regulation of Human MHC Class I Molecule Expression on Cancer Cells. *Cancer Immunol. Res.* **2016**, *4*, 936–947. [CrossRef] [PubMed]

233. Loi, S.; Dushyanthen, S.; Beavis, P.A.; Salgado, R.; Denkert, C.; Savas, P.; Combs, S.; Rimm, D.L.; Giltnane, J.M.; Estrada, M.V.; et al. RAS/MAPK Activation Is Associated with Reduced Tumor-Infiltrating Lymphocytes in Triple-Negative Breast Cancer: Therapeutic Cooperation Between MEK and PD-1/PD-L1 Immune Checkpoint Inhibitors. *Clin. Cancer Res.* **2016**, *22*, 1499–1509. [CrossRef] [PubMed]

234. Srivastava, P.K. Neoepitopes of Cancers: Looking Back, Looking Ahead. *Cancer Immunol. Res.* **2015**, *3*, 969–977. [CrossRef] [PubMed]

235. Im, J.S.; Herrmann, A.C.; Bernatchez, C.; Haymaker, C.; Molldrem, J.J.; Hong, W.K.; Perez-Soler, R. Immune-Modulation by Epidermal Growth Factor Receptor Inhibitors: Implication on Anti-Tumor Immunity in Lung Cancer. *PLoS ONE* **2016**, *11*, e0160004. [CrossRef]

236. Garrido, G.; Rabasa, A.; Garrido, C.; Chao, L.; Garrido, F.; García-Lora, Á.M.; Sánchez-Ramírez, B. Upregulation of HLA Class I Expression on Tumor Cells by the Anti-EGFR Antibody Nimotuzumab. *Front. Pharmacol.* **2017**, *8*, 595. [CrossRef]

237. Kang, S.-H.; Keam, B.; Ahn, Y.-O.; Park, H.-R.; Kim, M.; Kim, T.M.; Kim, D.-W.; Heo, D.S. Inhibition of MEK with trametinib enhances the efficacy of anti-PD-L1 inhibitor by regulating anti-tumor immunity in head and neck squamous cell carcinoma. *Oncoimmunology* **2019**, *8*, e1515057. [CrossRef]

238. Sapkota, B.; Hill, C.E.; Pollack, B.P. Vemurafenib enhances MHC induction in BRAFV600E homozygous melanoma cells. *Oncoimmunology* **2013**, *2*, e22890. [CrossRef]

239. Sabbatino, F.; Wang, Y.; Scognamiglio, G.; Favoino, E.; Feldman, S.A.; Villani, V.; Flaherty, K.T.; Nota, S.; Giannarelli, D.; Simeone, E.; et al. Antitumor Activity of BRAF Inhibitor and IFNα Combination in BRAF-Mutant Melanoma. *J. Natl. Cancer Inst.* **2016**, *108*. [CrossRef]

240. Hu-Lieskovan, S.; Mok, S.; Homet Moreno, B.; Tsoi, J.; Robert, L.; Goedert, L.; Pinheiro, E.M.; Koya, R.C.; Graeber, T.G.; Comin-Anduix, B.; et al. Improved antitumor activity of immunotherapy with BRAF and MEK inhibitors in BRAF(V600E) melanoma. *Sci. Transl. Med.* **2015**, *7*, 279ra41. [CrossRef]

241. Ritter, C.; Fan, K.; Paschen, A.; Reker Hardrup, S.; Ferrone, S.; Nghiem, P.; Ugurel, S.; Schrama, D.; Becker, J.C. Epigenetic priming restores the HLA class-I antigen processing machinery expression in Merkel cell carcinoma. *Sci. Rep.* **2017**, *7*, 2290. [CrossRef]

242. Sun, T.; Li, Y.; Yang, W.; Wu, H.; Li, X.; Huang, Y.; Zhou, Y.; Du, Z. Histone deacetylase inhibition up-regulates MHC class I to facilitate cytotoxic T lymphocyte-mediated tumor cell killing in glioma cells. *J. Cancer* **2019**, *10*, 5638–5645. [CrossRef] [PubMed]

243. Yang, W.; Li, Y.; Gao, R.; Xiu, Z.; Sun, T. MHC class I dysfunction of glioma stem cells escapes from CTL-mediated immune response via activation of Wnt/β-catenin signaling pathway. *Oncogene* **2020**, *39*, 1098–1111. [CrossRef] [PubMed]

244. Woan, K.V.; Lienlaf, M.; Perez-Villaroel, P.; Lee, C.; Cheng, F.; Knox, T.; Woods, D.M.; Barrios, K.; Powers, J.; Sahakian, E.; et al. Targeting histone deacetylase 6 mediates a dual anti-melanoma effect: Enhanced antitumor immunity and impaired cell proliferation. *Mol. Oncol.* **2015**, *9*, 1447–1457. [CrossRef] [PubMed]

245. Ennishi, D.; Takata, K.; Béguelin, W.; Duns, G.; Mottok, A.; Farinha, P.; Bashashati, A.; Saberi, S.; Boyle, M.; Meissner, B.; et al. Molecular and Genetic Characterization of MHC Deficiency Identifies EZH2 as Therapeutic Target for Enhancing Immune Recognition. *Cancer Discov.* **2019**, *9*, 546–563. [CrossRef]

246. Fonsatti, E.; Nicolay, H.J.M.; Sigalotti, L.; Calabrò, L.; Pezzani, L.; Colizzi, F.; Altomonte, M.; Guidoboni, M.; Marincola, F.M.; Maio, M. Functional up-regulation of human leukocyte antigen class I antigens expression by 5-aza-2′-deoxycytidine in cutaneous melanoma: Immunotherapeutic implications. *Clin. Cancer Res.* **2007**, *13*, 3333–3338. [CrossRef]

247. Chacon, J.A.; Schutsky, K.; Powell, D.J. The Impact of Chemotherapy, Radiation and Epigenetic Modifiers in Cancer Cell Expression of Immune Inhibitory and Stimulatory Molecules and Anti-Tumor Efficacy. *Vaccines* **2016**, *4*, 43. [CrossRef]

248. Luo, N.; Nixon, M.J.; Gonzalez-Ericsson, P.I.; Sanchez, V.; Opalenik, S.R.; Li, H.; Zahnow, C.A.; Nickels, M.L.; Liu, F.; Tantawy, M.N.; et al. DNA methyltransferase inhibition upregulates MHC-I to potentiate cytotoxic T lymphocyte responses in breast cancer. *Nat. Commun.* **2018**, *9*, 248. [CrossRef]

249. De Biasi, A.R.; Villena-Vargas, J.; Adusumilli, P.S. Cisplatin-induced antitumor immunomodulation: A review of preclinical and clinical evidence. *Clin. Cancer Res.* **2014**, *20*, 5384–5391. [CrossRef]

250. Faè, D.A.; Martorelli, D.; Mastorci, K.; Muraro, E.; Dal Col, J.; Franchin, G.; Barzan, L.; Comaro, E.; Vaccher, E.; Rosato, A.; et al. Broadening Specificity and Enhancing Cytotoxicity of Adoptive T Cells for Nasopharyngeal Carcinoma Immunotherapy. *Cancer Immunol. Res.* **2016**, *4*, 431–440. [CrossRef]

251. Pellicciotta, I.; Yang, C.-P.H.; Goldberg, G.L.; Shahabi, S. Epothilone B enhances Class I HLA and HLA-A2 surface molecule expression in ovarian cancer cells. *Gynecol. Oncol.* **2011**, *122*, 625–631. [CrossRef]

252. Wan, S.; Pestka, S.; Jubin, R.G.; Lyu, Y.L.; Tsai, Y.-C.; Liu, L.F. Chemotherapeutics and radiation stimulate MHC class I expression through elevated interferon-beta signaling in breast cancer cells. *PLoS ONE* **2012**, *7*, e32542. [CrossRef] [PubMed]

253. Wang, X.; Waschke, B.C.; Woolaver, R.A.; Chen, S.M.Y.; Chen, Z.; Wang, J.H. HDAC inhibitors overcome immunotherapy resistance in B-cell lymphoma. *Protein Cell* **2020**. [CrossRef] [PubMed]

254. Zhou, F. Molecular mechanisms of IFN-gamma to up-regulate MHC class I antigen processing and presentation. *Int. Rev. Immunol.* **2009**, *28*, 239–260. [CrossRef] [PubMed]

255. De Charette, M.; Marabelle, A.; Houot, R. Turning tumour cells into antigen presenting cells: The next step to improve cancer immunotherapy? *Eur. J. Cancer* **2016**, *68*, 134–147. [CrossRef] [PubMed]

256. Medrano, R.F.V.; Hunger, A.; Mendonça, S.A.; Barbuto, J.A.M.; Strauss, B.E. Immunomodulatory and antitumor effects of type I interferons and their application in cancer therapy. *Oncotarget* **2017**, *8*, 71249–71284. [CrossRef]

257. Selinger, E.; Reiniš, M. Epigenetic View on Interferon γ Signalling in Tumour Cells. *Folia Biol. (Praha)* **2018**, *64*, 125–136.

258. Zhang, S.; Kohli, K.; Black, R.G.; Yao, L.; Spadinger, S.M.; He, Q.; Pillarisetty, V.G.; Cranmer, L.D.; Van Tine, B.A.; Yee, C.; et al. Systemic Interferon-γ Increases MHC Class I Expression and T-cell Infiltration in Cold Tumors: Results of a Phase 0 Clinical Trial. *Cancer Immunol. Res.* **2019**, *7*, 1237–1243. [CrossRef]

259. Vlková, V.; Štěpánek, I.; Hrušková, V.; Šenigl, F.; Mayerová, V.; Šrámek, M.; Šímová, J.; Bieblová, J.; Indrová, M.; Hejhal, T.; et al. Epigenetic regulations in the IFNγ signalling pathway: IFNγ-mediated MHC class I upregulation on tumour cells is associated with DNA demethylation of antigen-presenting machinery genes. *Oncotarget* **2014**, *5*, 6923–6935. [CrossRef]

260. Castro, F.; Cardoso, A.P.; Gonçalves, R.M.; Serre, K.; Oliveira, M.J. Interferon-Gamma at the Crossroads of Tumor Immune Surveillance or Evasion. *Front. Immunol.* **2018**, *9*. [CrossRef]

261. Rodríguez, T.; Méndez, R.; Del Campo, A.; Jiménez, P.; Aptsiauri, N.; Garrido, F.; Ruiz-Cabello, F. Distinct mechanisms of loss of IFN-gamma mediated HLA class I inducibility in two melanoma cell lines. *BMC Cancer* **2007**, *7*, 34. [CrossRef]

262. Sucker, A.; Zhao, F.; Pieper, N.; Heeke, C.; Maltaner, R.; Stadtler, N.; Real, B.; Bielefeld, N.; Howe, S.; Weide, B.; et al. Acquired IFNγ resistance impairs anti-tumor immunity and gives rise to T-cell-resistant melanoma lesions. *Nat. Commun.* **2017**, *8*, 15440. [CrossRef] [PubMed]

263. Aspeslagh, S.; Morel, D.; Soria, J.-C.; Postel-Vinay, S. Epigenetic modifiers as new immunomodulatory therapies in solid tumours. *Ann. Oncol.* **2018**, *29*, 812–824. [CrossRef] [PubMed]

264. Gao, F.; Zhao, Z.-L.; Zhao, W.-T.; Fan, Q.-R.; Wang, S.-C.; Li, J.; Zhang, Y.-Q.; Shi, J.-W.; Lin, X.-L.; Yang, S.; et al. miR-9 modulates the expression of interferon-regulated genes and MHC class I molecules in human nasopharyngeal carcinoma cells. *Biochem. Biophys. Res. Commun.* **2013**, *431*, 610–616. [CrossRef]

265. Li, J.; Lin, T.-Y.; Chen, L.; Liu, Y.; Dian, M.-J.; Hao, W.-C.; Lin, X.-L.; Li, X.-Y.; Li, Y.-L.; Lian, M.; et al. miR-19 regulates the expression of interferon-induced genes and MHC class I genes in human cancer cells. *Int. J. Med. Sci.* **2020**, *17*, 953–964. [CrossRef] [PubMed]

International Journal of
Molecular Sciences

Article

Transcriptional Profiling of Advanced Urothelial Cancer Predicts Prognosis and Response to Immunotherapy

Seung-Woo Baek [1,2,†], In-Hwan Jang [1,2,†], Seon-Kyu Kim [2,3], Jong-Kil Nam [4], Sun-Hee Leem [5] and In-Sun Chu [1,2,*]

[1] Genome Editing Research Center, Korea Research Institute of Bioscience and Biotechnology (KRIBB), Daejeon 34141, Korea; baek@kribb.re.kr (S.-W.B.); jangih@kribb.re.kr (I.-H.J.)
[2] Department of Bioinformatics, KRIBB School of Bioscience, Korea University of Science and Technology, Daejeon 34113, Korea; seonkyu@kribb.re.kr
[3] Personalized Genomic Medicine Research Center, KRIBB, Daejeon 34141, Korea
[4] Department of Urology, Research Institute for Convergence of Biochemical Science and Technology, Pusan National University Yangsan Hospital, Yangsan 50612, Korea; tuff-kil@hanmail.net
[5] Department of Biological Science, Dong-A University, Busan 49315, Korea; shleem@dau.ac.kr
* Correspondence: chu@kribb.re.kr; Tel.: +82-42-879-8520
† These authors contributed equally to this work.

Received: 31 January 2020; Accepted: 6 March 2020; Published: 8 March 2020

Abstract: Recent investigations reported that some subtypes from the Lund or The Cancer Genome Atlas (TCGA) classifications were most responsive to PD-L1 inhibitor treatment. However, the association between previously reported subtypes and immune checkpoint inhibitor (ICI) therapy responsiveness has been insufficiently explored. Despite these contributions, the ability to predict the clinical applicability of immune checkpoint inhibitor therapy in patients remains a major challenge. Here, we aimed to re-classify distinct subtypes focusing on ICI responsiveness using gene expression profiling in the IMvigor 210 cohort (n = 298). Based on the hierarchical clustering analysis, we divided advanced urothelial cancer patients into three subgroups. To confirm a prognostic impact, we performed survival analysis and estimated the prognostic value in the IMvigor 210 and TCGA cohort. The activation of CD8$^+$ T effector cells was common for patients of classes 2 and 3 in the TCGA and IMvigor 210 cohort. Survival analysis showed that patients of class 3 in the TCGA cohort had a poor prognosis, while patients of class 3 showed considerably prolonged survival in the IMvigor 210 cohort. One of the distinct characteristics of patients in class 3 is the inactivation of the TGFβ and YAP/TAZ pathways and activation of the cell cycle and DNA replication and DNA damage (DDR). Based on our identified transcriptional patterns and the clinical outcomes of advanced urothelial cancer patients, we constructed a schematic summary. When comparing clinical and transcriptome data, patients with downregulation of the TGFβ and YAP/TAZ pathways and upregulation of the cell cycle and DDR may be more responsive to ICI therapy.

Keywords: bladder cancer; immune checkpoint inhibitor; CD8$^+$ T effector cells

1. Introduction

Bladder cancer is the sixth most common malignant disease. In 2019, 80,470 new cases of bladder cancer were diagnosed, and 17,670 deaths were due to bladder cancer in the United States [1]. Bladder cancer is generally divided into pathological subtypes: non-muscle-invasive bladder cancer (NMIBC) and muscle-invasive bladder cancer (MIBC). Cisplatin-based chemotherapy followed by radical cystectomy is the standard of care in previously untreated patients with MIBC. However, patients who relapse after cisplatin-based chemotherapy experience a very poor prognosis [2]. Since cisplatin-based chemotherapy also has the limitation of drug resistance, it is necessary to provide a variety of treatments such as immunotherapy. Previous investigations of immunotherapy have opened a new frontier in the treatment of advanced urothelial cancer [3,4]. Although the response rates are moderately high, it is promising that responsive patients experience durable disease management. Unlike conventional cisplatin-based chemotherapy, immunotherapy enhances the patient's own immune environment and can be combined with conventional therapies to produce additive effects [5].

Recently, immunotherapy studies have focused on a way to improve efficacy in individual patients. Numerous studies reported several molecular subtypes in bladder cancer, including immunotherapy-associated subgroups, the genomically unstable (GU) subtype of the Lund classification and the neuronal subtype in The Cancer Genome Atlas (TCGA) classification [6–10]. PD-L1 protein expression on immune cells, tumor mutation burden (TMB), and the TGFβ pathway have been shown to correlate with the clinical outcome of immune checkpoint inhibitor (ICI) therapy for advanced urothelial cancer [11]. The previous molecular subtypes of advanced urothelial cancer are not a classification system directly related to immunotherapy. In the existing classification systems, only some patients with specific subtypes of the Lund and TCGA were identified to be responsive to immunotherapy. Furthermore, overall immune system activities such as transcriptional activities of CD8$^+$ T effector (T$_{eff}$) cell have not been fully elucidated in bladder cancer. To provide ICI therapy to more patients with advanced urothelial cancer, we wanted to explore both various clinical outcomes and a new subset related to ICI therapy that contains many limiting factors.

In this study, we identified a gene expression signature from the IMvigor 210 cohort [11] revealing distinct three molecular subgroups showing different clinical characteristics and core biological pathways in advanced urothelial cancer patients. To validate a signature, we applied the signature into the TCGA and Lund cohorts and confirmed similar characteristics. By performing a survival analysis, we confirmed that the patients who could potentially benefit from anti-PD-L1 treatment actually represented a difference in the IMvigor 210 and TCGA cohorts.

2. Results

2.1. Discovery of Distinct Three Subtypes and Clinical Characteristics

To select patients with a high response to ICI therapy, we performed unsupervised clustering analysis using gene expression profiling from bladder cancer patients treated with the PD-L1 inhibitor atezolizumab (the IMvigor 210 trial [11]). Based on the hierarchical clustering analysis of the expression patterns of 2366 genes correlated with the IMvigor 210 cohort, we then divided advanced urothelial cancer patients into three subgroups (Figure S1). From the clustering analysis results, we chose 24 genes associated with the three subgroups and identified a transcriptional pattern according to these genes. In addition, we identified clinical characteristics such as PD-L1 expression on immune cells (IC PD-L1), PD-L1 expression on tumor cells (TC PD-L1), tumor mutation burden (TMB), and Lund and TCGA subtypes related with the three subgroups. We identified that IC PD-L1 and TC PD-L1 were increased in classes 2 and 3. Furthermore, we identified that TMB was highest in class 3. Importantly, we confirmed that many patients with the neuronal subtype from the TCGA classification were included in class 3 and also a subset of patients with the GU subtype from the Lund classification was involved in class 3 (Figure 1A).

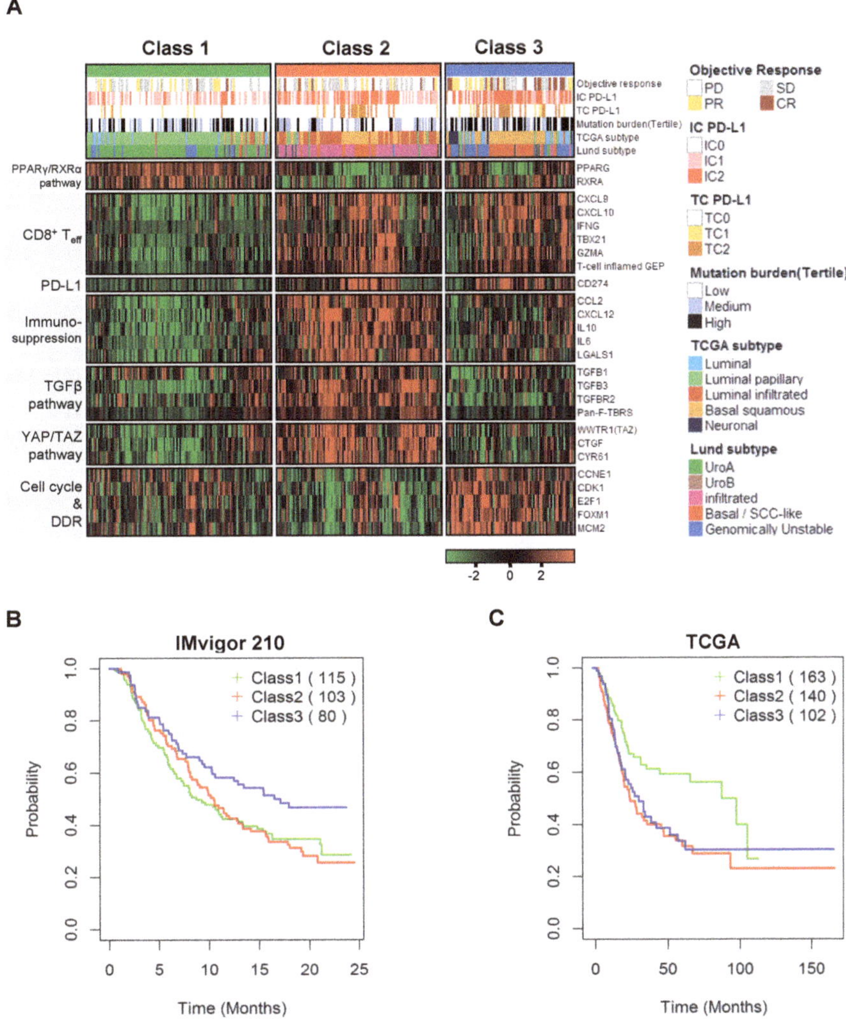

Figure 1. Core biological pathways associated with immune checkpoint inhibitor (ICI) therapy in the IMvigor 210 cohort and survival analyses. (**A**) Heat map of immunotherapy-associated clinical and biological features. **On top**, samples are ordered according to gene expression patterns. Gene signatures, including the pan-fibroblast TGFβ response signature (pan-F-TBRS) and the T cell inflamed gene expression profile (GEP) scores, were selected to explore the correlation between expression patterns and other relevant biological processes. Gene expression levels and signatures such as the Pan-F-TBRS and GEP were ordered and grouped by pathway. The coloring in the heat map reflects relatively high (red) and low (green) expression (Z score) levels; the same representation is used for high and low gene signatures. T_{eff}, T effector. (**B**) Overall survival in the IMvigor 210 cohort ($p = 0.04$ by the log-rank test). (**C**) Overall survival in the TCGA cohort ($p = 0.001$ by the log-rank test).

2.2. Biological Insight into the Newly Identified Subtypes

Next, we investigated core biological pathways that are known to play major roles in the immune system. Immune cell infiltration is controlled by activated PPARγ/RXRα, which inhibits the host immune response by suppressing the expression and secretion of inflammatory cytokines [12].

The genes involved in the PPARγ/RXRα pathway (e.g., *PPARG* and *RXRA*) were upregulated in class 1. According to a recent report, urothelial cancer patients with *FGFR3* mutations had lower immune cell infiltration and lower TGFβ signals than patients without *FGFR3* mutations [13]. We identified that *FGFR3* mutations were enriched in class 1 (Figure S2). On the other hand, the expression of CD8$^+$ T effector cell-related genes (e.g., *CXCL9, CXCL10, IFNG, TBX21,* and *GZMA*) and PD-L1 (*CD274*) was relatively upregulated in classes 2 and 3. In addition, the T-cell-inflamed gene expression profile (GEP), which was correlated with a clinical benefit in a clinical study of pembrolizumab [14], was activated in classes 2 and 3. Exceptionally, in patients with the neuronal subtype from the TCGA classification and some of the GU subtype from the Lund classification in class 3, CD8$^+$ T effector cell-related genes and the GEP were relatively downregulated. The expression of immune-suppression-related genes (e.g., *CCL2, CXCL12, IL10, IL6,* and *LGALS1*) was also upregulated in class 2. Furthermore, TGFβ pathway genes (e.g., *TGFB1, TGFB3,* and *TGFBR2*) and pan-fibroblast TGFβ response signature (pan-F-TBRS) scores were upregulated in class 2 but downregulated in class 3, consistent with a previous report that TGFβ attenuates the response to PD-L1 inhibitors [11]. Multiple cancer-associated signaling networks engage in regulatory crosstalk with the YAP/TAZ pathway, which has been reported to functionally interact with the TGFβ pathway [15]. The activation of YAP/TAZ-pathway-related genes (e.g., *WWTR1, CTGF,* and *CYR61*) also supported the activation of the TGFβ pathway and immune suppression in class 2. Notably, *CTGF*, a major target gene of the YAP/TAZ pathway that is associated with angiogenesis, epithelial-mesenchymal transition and wound healing, was differentially expressed between classes 2 and 3. Additionally, the cell cycle and DNA replication and DNA damage (DDR) genes (e.g., *CCNE1, CDK1, E2F1, FOXM1,* and *MCM2*) were upregulated in class 3. These results support significant differences in clinical characteristics and core biological pathways between the three subtypes.

2.3. Prognostic Impact Based on Unsupervised Clustering Analysis

To investigate the utility of the three molecular subtypes, we performed survival analysis and estimated the prognostic value by a log-rank test. As a result, we identified that patients with activated CD8$^+$ T effector cells in classes 2 and 3 showed slightly prolonged survival after treatment with the PD-L1 inhibitor. Importantly, patients in class 3 had better survival than those in the other subgroups (Figure 1B). Interestingly, patients in classes 2 and 3 had poorer prognoses than those in class 1 before treatment with the PD-L1 inhibitor in the TCGA cohort (Figure S2 and Figure 1C). These results suggest that patients with poor prognoses in class 3 exhibited prolonged survival after treatment with a PD-L1 inhibitor.

2.4. Comparison of Clinical Outcomes in the Three Subgroups

PD-L1 protein expression on immune cells, which correlated with the activation of CD8$^+$ T effector cells, was present in high scores over the classes 2 and 3 (Figure 2A). The results indicated that both PD-L1 protein expression and gene expression were largely related. When the objective response rate (ORR) was compared among these subgroups, class 3 exhibited a higher response rate than the other classes (Figure 2B). For a comparison with previously reported subtypes, we investigated the distribution of the Lund and TCGA subtypes in each subgroup [6,7]. The GU subtype of the Lund classification was mostly distributed in classes 1 and 3. When comparing the ORR of the GU subtypes across classes, interestingly, the complete response rate was significantly higher in class 3 than in class 1 (Figure 2C), implying that the activation of CD8$^+$ T effector cell-related genes may play an important role in the response to ICI therapy beyond the GU subtype. The basal/SCC-like subtype of the Lund classification was divided into classes 2 and 3, whereas most of the patients with the infiltrated subtype were included in class 2 (Figure 2D). We also observed that the neuronal subtype, known as the most responsive subtype of the TCGA classification [10], was classified into class 3 (Figure 2E). These results indicate that patients who would be the most responsive to ICI therapy, including those with the neuronal, GU, or basal/SCC-like subtypes, could be re-stratified extensively by our classification. In addition, TMB was also significantly higher in class 3 than in other classes (Figure 2F). To validate the characteristics of the three subgroups, we applied our transcriptional patterns to other muscle-invasive bladder cancer patient cohorts. Similar gene expression patterns and TMB values were observed in the TCGA cohort (Figure S2 and Figure 2G). We also observed consistent biological characteristics in the Lund cohort (Figure S3). The gene expression patterns from the validation cohorts were also related to the activation of the cell cycle and the DDR and TGFβ and YAP/TAZ pathway genes, such as *FOXM1*, *TGFBR2*, and *CTGF* (Figure 1A, Figure 3 and Figure S2).

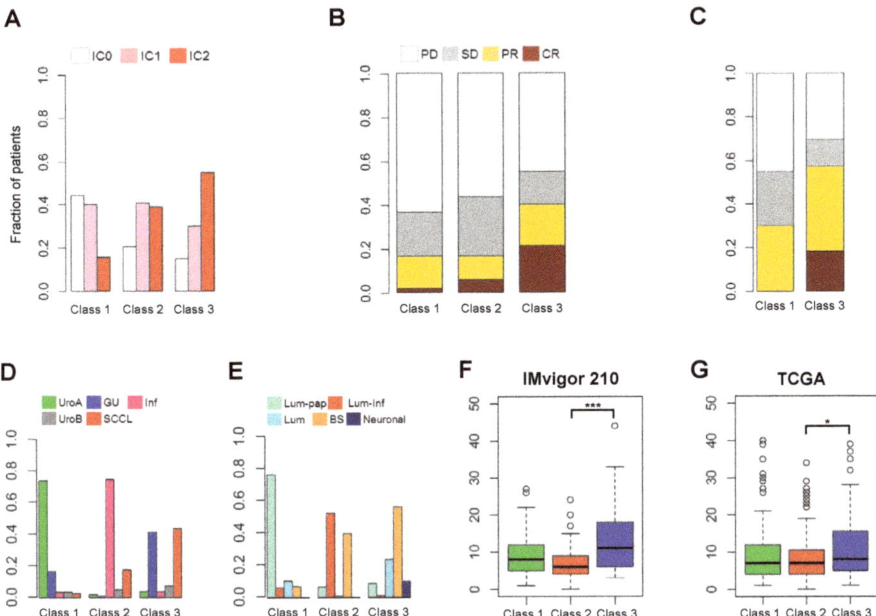

Figure 2. Clinical and biological characteristics of Figure 1A. (**A**) Distribution of PD-L1 protein expression levels on immune cells in each class (p = 0.0003 by the chi-squared test). (**B**) Objective response rate stratified by the three subgroups (p = 2.714×10^{-5} by the chi-squared test). (**C**) Comparison of the objective response rate between class 1 and class 3 in the GU subtype of the Lund classification (p = 0.046 by the two-tailed Fisher's exact test). PD, progressive disease; SD, stable disease; PR, partial response; CR, complete response. (**D**) Distribution of subtypes of the Lund classification in each subgroup (p < 2.2×10^{-16} by the chi-squared test). UroA, urothelial-like A; GU, genomically unstable; Inf, infiltrated; UroB, urothelial-like B; SCCL, squamous cell carcinoma-like. (**E**) Distribution of subtypes of the TCGA classification in each subgroup (p < 4.6×10^{-51} by the two-tailed Fisher's exact test). Lum-pap, luminal-papillary; Lum-inf, luminal-infiltrated; Lum, luminal; BS, basal squamous. (**F**) Reported tumor mutation burden (TMB) classified by the three subgroups (p = 1.83×10^{-8} by the two-sample t-test). (**G**) Reported TMB, classified by the three subgroups in the TCGA cohort (p = 0.012 by the two-sample t-test; class 2 vs. class 3). * p < 0.05, *** p < 0.001. Clinical and biological characteristics of Figure 1A. (**A**) Distribution of PD-L1 protein expression levels on immune cells in each class (p = 0.0003 by the chi-squared test). (**B**) Objective response rate stratified by the three subgroups (p = 2.714×10^{-5} by the chi-squared test). (**C**) Comparison of the objective response rate between class 1 and class 3 in the GU subtype of the Lund classification (p = 0.046 by the two-tailed Fisher's exact test). PD, progressive disease; SD, stable disease; PR, partial response; CR, complete response. (**D**) Distribution of subtypes of the Lund classification in each subgroup (p < 2.2×10^{-16} by the chi-squared test). UroA, urothelial-like A; GU, genomically unstable; Inf, infiltrated; UroB, urothelial-like B; SCCL, squamous cell carcinoma-like. (**E**) Distribution of subtypes of the TCGA classification in each subgroup

$(p < 4.6 \times 10^{-51}$ by the two-tailed Fisher's exact test). Lum-pap, luminal-papillary; Lum-inf, luminal-infiltrated; Lum, luminal; BS, basal squamous. (**F**) Reported tumor mutation burden (TMB) classified by the three subgroups $(p = 1.83 \times 10^{-8}$ by the two-sample *t*-test). (**G**) Reported TMB, classified by the three subgroups in the TCGA cohort $(p = 0.012$ by the two-sample *t*-test; class 2 vs. class 3). * $p < 0.05$, *** $p < 0.001$.

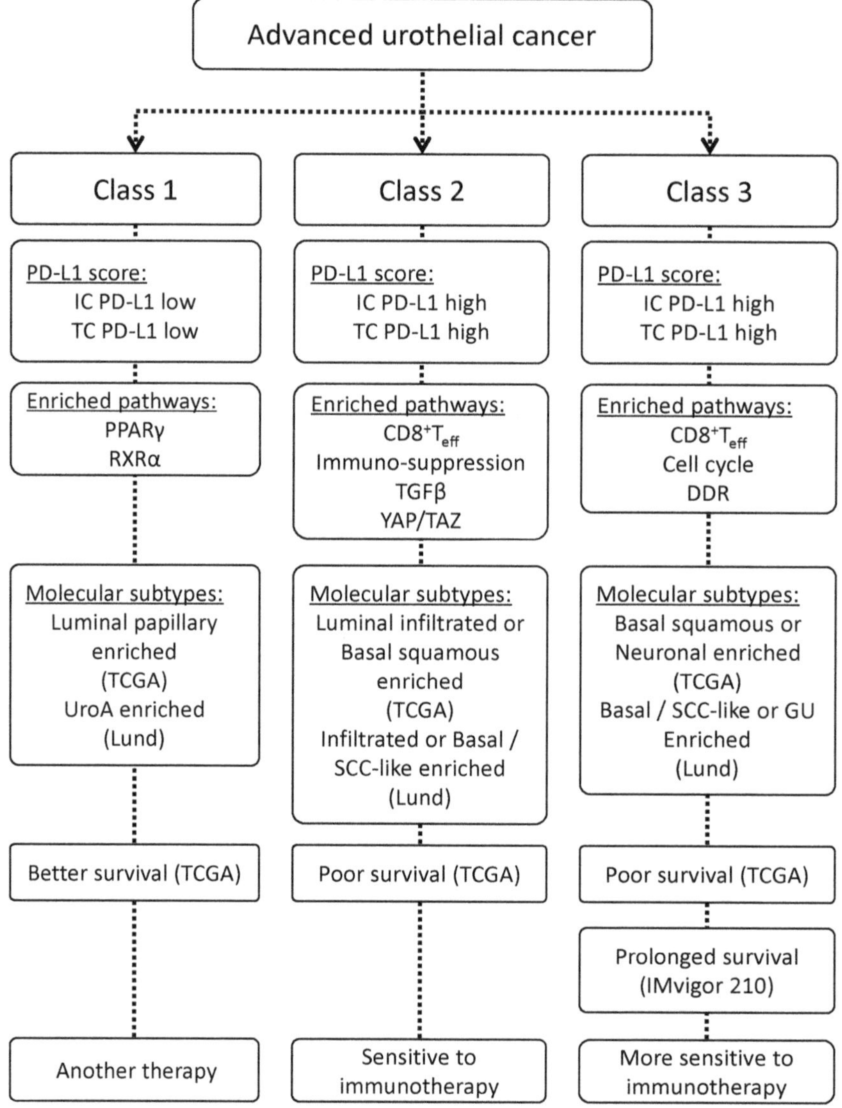

Figure 3. Schematic diagram of the characteristics of advanced urothelial cancer. T_{eff}, T effector; DDR, DNA replication and DNA damage response; IC PD-L1, PD-L1 expression on immune cells; TC PD-L1, PD-L1 expression on tumor cell; SCC, squamous cell carcinoma; GU, genomically unstable.

2.5. Schematic Diagram of the Characteristics of Advanced Urothelial Cancer

Based on our identified transcriptional patterns and the clinical outcomes of advanced urothelial bladder cancer patients, we constructed a schematic summary (Figure 3). In the TCGA cohort, the overall survival rate of the class 1 patients was significantly higher than that of the patients in classes 2 and 3. In addition, the IC PD-L1 and TC PD-L1 scores were relatively low in the class 1 patients. In contrast, for the patients in classes 2 and 3, the overall survival rate was significantly lower than that of the class 1 patients, and the IC PD-L1 and TC PD-L1 scores were relatively high. On the other hand, we suspect that the class 2 and class 3 patients had similar characteristics, but we observed a significant difference in biological pathways. In particular, a relative difference in the YAP/TAZ pathway has not yet been mentioned with other immunotherapies for advanced urothelial cancer. Taken together, these observations suggest that the high-risk patients in class 3 are the most likely to respond favorably to anti-PD-L1 treatment.

3. Discussion

Advanced urothelial cancer is clinically heterogeneous and exhibits poor outcomes. Using multiple bladder cancer patient cohorts, we carried out transcriptional profiling analyses, which identified a signature of distinct prognostic subtypes of advanced urothelial cancer. The signature showed therapeutic relevance in that those patients with enriched CD8[+] T effector cell-related genes benefit from ICI therapy. Interestingly, among these individuals, patients with inactivation of the TGFβ and YAP/TAZ pathways and activation of the cell cycle and DDR were more responsive to ICI therapy than patients without these traits.

Recently, considerable effort has been devoted to elucidating the molecular characteristics of bladder cancer [6–9]. It has been reported that the GU subtype of the Lund classification and the neuronal subtype of the TCGA classification respond best to anti-PD-L1 treatment [10,11]. Despite these contributions, predicting clinically relevant patients responsive to ICI therapy remains a major challenge. We tried to directly compare the survival rates between the TCGA and the IMvigor 210 cohort. Through the results, beyond the previously known subtypes, we also tried to contribute to precisely selecting the patients who would be most responsive to treatment by introducing subtypes that reflect clinical characteristics and core biological pathways.

The most interesting finding of our study was the relative difference in the enriched biological pathways between our subtypes. In class 1, we confirmed that immune cell infiltration was controlled by activated PPARγ/RXRα, which inhibited host immune systems [12]. In recently updated data from TCGA, these patients showed enrichment of *FGFR3* mutations. Bladder cancer patients with *FGFR3* mutations have been associated with lower immune cell infiltration and lower TGFβ signals than patients without *FGFR3* mutations [13]. Patients with *FGFR* mutations or fusions may be less likely to have a response to immunotherapy than those without such alterations. The pan-FGFR inhibitor erdafitinib had a measurable benefit in patients with advanced urothelial carcinoma with *FGFR* alteration [16]. Therefore, we suggest that immunotherapy is not suitable for patients in class 1. In class 2, we identified an enrichment of CD8[+] T effector cell-related genes. One of the most distinct characteristics was the co-activation of the TGFβ and YAP/TAZ pathways. Multiple cancer-associated signaling networks engage in regulatory crosstalk with the YAP/TAZ pathway, which has been reported to functionally interact with the TGFβ pathway. YAP/TAZ expression in immune cells, including T cells, B cells, and macrophages, regulates the differentiation and functionality of immune cells, which are important for tumor immunity [17]. Notably, CTGF, a major target gene that is associated with immune suppression and epithelial-mesenchymal transition, was differentially expressed. In class 3, we also identified enrichment of CD8[+] T effector cells. However, the patients in class 3 showed inactivation of the TGFβ and YAP/TAZ pathways and activation of the cell cycle and the DDR. Cell cycle and DDR regulatory genes, which are significantly associated with TMB, play an important role in selecting patients with high response rates to ICI therapy.

However, the unclear relationship between DDR gene alterations and expression and response to immunotherapy remains a challenge. DDR gene alterations are independently associated with the response to PD-1/PD-L1 inhibitors in patients with advanced urothelial cancer [18]. Future studies using next-generation sequencing technologies will continue to uncover associations between mutation- or expression-based changes in tumor DNA repair pathway function and response to immunotherapy [19].

Through further analysis, we confirmed that patients with activated $CD8^+$ T effector cells in class 2 and class 3 showed slightly prolonged survival after treatment with the PD-L1 inhibitor. In class 3, in particular, we identified that patients showed considerably prolonged survival after treatment. These patients also included a subset of patients with the GU subtype in the Lund classification and all the neuronal subtypes in the TCGA classification.

In conclusion, to effectively select patients who will respond to ICI therapy, we suggest that many aspects should be considered, including predefined subtypes, clinical characteristics, and core biological pathways. It is clear that the combined use of multiple markers can improve the performance of ICI therapy compared to a single marker. Our investigations will contribute to the development of predictive markers and therapeutic options.

4. Materials and Methods

4.1. Patients and Gene Expression Data

RNA-seq datasets from 348 patients with bladder cancer were obtained from the IMvigor 210 dataset [11]. Among the 348 patients, 298 patients who had received immunotherapy were used as the discovery cohort (The IMvigor 210 cohort, $n = 298$). Gene expression datasets from patients with bladder cancer from The Cancer Genome Atlas (TCGA, $n = 407$) and the Lund cohort (GSE83586, $n = 307$) were used as the validation cohorts [6,7]. Fragments per kilobase of transcript per million mapped reads (FPKM) values were calculated from sequence read count data in the IMvigor 210 dataset. All gene expression data were transformed to a log2 scale and normalized by quantile normalization. Clinical data were obtained from the supplementary information of the corresponding literature.

4.2. Gene Expression Analysis

For the IMvigor 210 cohort, we selected 2366 genes with FPKM values that were detected with confidence (FPKM > 1) and exhibited at least a two-fold difference relative to the median value in greater than 30% of the samples. To classify patients into three groups, we used a by-hierarchical clustering algorithm using the centered correlation coefficient as the measure of similarity and centroid linkage clustering. To develop the prediction model, we used Prediction Analysis for Microarrays (PAM) and 1659 selected genes (R-package: PAMR). To explore significantly enriched functions, we performed gene ontology (GO) enrichment analysis using the DAVID tool (http://david.ncifcrf.gov) with significance criteria (FDR < 0.01). To integrate previous gene sets with our signature, we standardized a pan-fibroblast TGFβ response signature (pan-F-TBRS) and T-cell-inflamed GEP score [11,14]. Hierarchical clustering analysis was conducted using Gene Cluster 3.0 and visualized using TreeView^{TM}.

4.3. Statistical Analysis and Data Visualization

We estimated patient prognosis using Kaplan–Meier plots and the log-rank test. The significance of the distribution of subtypes and comparisons of objective responses were estimated using Fisher's exact test. The significance of the distribution of IC PD-L1 protein expression levels was estimated using the chi-squared test. The reported tumor mutation burden (TMB) was estimated using the two-sample t-test. All statistical analyses were performed in the R 3.6.1 language environment (http://www.r-project.org).

4.4. Data Availability

IMvigor 210 data and clinical information were obtained from the IMvigor210CoreBiologies R package [11]. The Cancer Genome Atlas (TCGA) and the Lund datasets were obtained through Cancer Browser (https://xenabrowser.net) and Gene Expression Omnibus (GEO), respectively, with the accession number GSE83586.

5. Conclusions

We identified three molecular subtypes of advanced urothelial cancer that reflect clinical and biological features and consider the TCGA and Lund classifications. When comparing clinical and transcriptome data, patients with downregulation of the TGFβ and YAP/TAZ pathways and upregulation of the cell cycle, DNA replication and DNA damage response showed significantly prolonged survival. Because only a subset of patients benefits from immune checkpoint inhibitor therapy, our investigations will contribute to the development of predictive markers and therapeutic options.

Supplementary Materials: The following are available online at http://www.mdpi.com/1422-0067/21/5/1850/s1, Figure S1: Hierarchical clustering analysis of gene expression data from the IMvigor 210 cohort. Three subgroups of patients and four distinct subsets of genes were revealed from unsupervised clustering analysis. The genes were grouped as G1, G2, G3, and G4. G1 was highly enriched in genes involved in the immune response and significantly highly expressed in classes 2 and 3. G2 was highly enriched in genes involved in angiogenesis, collagen fibril organization, and wound healing associated with the immunosuppressive reaction, and it was recently shown that this group attenuates the immune reaction towards the tumor via the TGFβ pathway. G3 was highly enriched in cell cycle-, histone-, or DNA repair-associated genes, implying that the high responsiveness to PD-L1 blockade of class 3 may be mediated by these genes. G4, predominantly expressed in class 1, was enriched in the metabolic process and FGFR3 pathway genes; Figure S2: Validation of the TCGA cohort (n=407). Heat map of the selected gene list associated with Figure 1. Samples are ordered according to the TCGA subtypes in each subgroup; Figure S3: Validation of the Lund cohort (n=307). Heat map of the selected gene set in the Lund cohort. Samples are ordered according to the three subgroups associated with Figure 1.

Author Contributions: I.-S.C., S.-H.L. and J.-K.N. conceived and designed the project; S.-W.B., I.-H.J., S.-K.K. and I.-S.C. collected and assembled the data; S.-W.B. and I.-H.J. analyzed the data; S.-W.B., I.-H.J., S.-K.K. and I.-S.C. drafted the manuscript; All authors proofread and approved the final version of the manuscript.

Funding: This research was supported by a National Research Foundation of Korea (NRF) grant NRF-2017R1A2B2007302 funded by the Korean government (MSIP) and a grant from the KRIBB Research Initiative Program.

Conflicts of Interest: The authors declare no conflict of interest. The funders had no role in the design of the study; in the collection, analyses, or interpretation of data; in the writing of the manuscript, or in the decision to publish the results

References

1. Siegel, R.L.; Miller, K.D.; Jemal, A. Cancer statistics, 2019. *CA Cancer J. Clin.* **2019**, *69*, 7–34. [CrossRef] [PubMed]

2. Sanli, O.; Dobruch, J.; Knowles, M.A.; Burger, M.; Alemozaffar, M.; Nielsen, M.E.; Lotan, Y. Bladder cancer. *Nat. Rev. Dis. Prim* **2017**, *3*, 17022. [CrossRef] [PubMed]

3. Rosenberg, J.E.; Hoffman-Censits, J.; Powles, T.; van der Heijden, M.S.; Balar, A.V.; Necchi, A.; Dawson, N.; O'Donnell, P.H.; Balmanoukian, A.; Loriot, Y.; et al. Atezolizumab in patients with locally advanced and metastatic urothelial carcinoma who have progressed following treatment with platinum-based chemotherapy: A single-arm, multicentre, phase 2 trial. *Lancet* **2016**, *387*, 1909–1920. [CrossRef]

4. Balar, A.V.; Galsky, M.D.; Rosenberg, J.E.; Powles, T.; Petrylak, D.P.; Bellmunt, J.; Loriot, Y.; Necchi, A.; Hoffman-Censits, J.; Perez-Gracia, J.L.; et al. Atezolizumab as first-line treatment in cisplatin-ineligible patients with locally advanced and metastatic urothelial carcinoma: A single-arm, multicentre, phase 2 trial. *Lancet* **2017**, *389*, 67–76. [CrossRef]

5. Fakhrejahani, F.; Tomita, Y.; Maj-Hes, A.; Trepel, J.B.; De Santis, M.; Apolo, A.B. Immunotherapies for bladder cancer: A new hope. *Curr. Opin. Urol.* **2015**, *25*, 586–596. [CrossRef] [PubMed]

6.	Sjodahl, G.; Eriksson, P.; Liedberg, F.; Hoglund, M. Molecular classification of urothelial carcinoma: Global mRNA classification versus tumour-cell phenotype classification. *J. Pathol.* **2017**, *242*, 113–125. [CrossRef] [PubMed]

7.	Robertson, A.G.; Kim, J.; Al-Ahmadie, H.; Bellmunt, J.; Guo, G.; Cherniack, A.D.; Hinoue, T.; Laird, P.W.; Hoadley, K.A.; Akbani, R.; et al. Comprehensive Molecular Characterization of Muscle-Invasive Bladder Cancer. *Cell* **2018**, *174*, 1033. [CrossRef] [PubMed]

8.	Song, B.N.; Kim, S.K.; Mun, J.Y.; Choi, Y.D.; Leem, S.H.; Chu, I.S. Identification of an immunotherapy-responsive molecular subtype of bladder cancer. *EBioMedicine* **2019**, *50*, 238–245. [CrossRef] [PubMed]

9.	Kamoun, A.; de Reynies, A.; Allory, Y.; Sjodahl, G.; Robertson, A.G.; Seiler, R.; Hoadley, K.A.; Groeneveld, C.S.; Al-Ahmadie, H.; Choi, W.; et al. A Consensus Molecular Classification of Muscle-invasive Bladder Cancer. *Eur. Urol.* **2019**. [CrossRef] [PubMed]

10.	Kim, J.; Kwiatkowski, D.; McConkey, D.J.; Meeks, J.J.; Freeman, S.S.; Bellmunt, J.; Getz, G.; Lerner, S.P. The Cancer Genome Atlas Expression Subtypes Stratify Response to Checkpoint Inhibition in Advanced Urothelial Cancer and Identify a Subset of Patients with High Survival Probability. *Eur. Urol.* **2019**, *75*, 961–964. [CrossRef] [PubMed]

11.	Mariathasan, S.; Turley, S.J.; Nickles, D.; Castiglioni, A.; Yuen, K.; Wang, Y.; Kadel, E.E., III; Koeppen, H.; Astarita, J.L.; Cubas, R.; et al. TGFβ attenuates tumour response to PD-L1 blockade by contributing to exclusion of T cells. *Nature* **2018**, *554*, 544–548. [CrossRef] [PubMed]

12.	Korpal, M.; Puyang, X.; Jeremy Wu, Z.; Seiler, R.; Furman, C.; Oo, H.Z.; Seiler, M.; Irwin, S.; Subramanian, V.; Julie Joshi, J.; et al. Evasion of immunosurveillance by genomic alterations of PPARgamma/RXRalpha in bladder cancer. *Nat. Commun.* **2017**, *8*, 103. [CrossRef] [PubMed]

13.	Wang, L.; Gong, Y.; Saci, A.; Szabo, P.M.; Martini, A.; Necchi, A.; Siefker-Radtke, A.; Pal, S.; Plimack, E.R.; Sfakianos, J.P.; et al. Fibroblast Growth Factor Receptor 3 Alterations and Response to PD-1/PD-L1 Blockade in Patients with Metastatic Urothelial Cancer. *Eur. Urol.* **2019**. [CrossRef] [PubMed]

14.	Ayers, M.; Lunceford, J.; Nebozhyn, M.; Murphy, E.; Loboda, A.; Kaufman, D.R.; Albright, A.; Cheng, J.D.; Kang, S.P.; Shankaran, V.; et al. IFN-γ–related mRNA profile predicts clinical response to PD-1 blockade. *J. Clin. Investig.* **2017**, *127*, 2930–2940. [CrossRef] [PubMed]

15.	Harvey, K.F.; Zhang, X.; Thomas, D.M. The Hippo pathway and human cancer. *Nat. Rev. Cancer* **2013**, *13*, 246–257. [CrossRef] [PubMed]

16.	Loriot, Y.; Necchi, A.; Park, S.H.; Garcia-Donas, J.; Huddart, R.; Burgess, E.; Fleming, M.; Rezazadeh, A.; Mellado, B.; Varlamov, S.; et al. Erdafitinib in Locally Advanced or Metastatic Urothelial Carcinoma. *N. Engl. J. Med.* **2019**, *381*, 338–348. [CrossRef] [PubMed]

17.	Pan, Z.; Tian, Y.; Cao, C.; Niu, G. The Emerging Role of YAP/TAZ in Tumor Immunity. *Mol. Cancer Res.* **2019**, *17*, 1777–1786. [CrossRef] [PubMed]

18.	Teo, M.Y.; Seier, K.; Ostrovnaya, I.; Regazzi, A.M.; Kania, B.E.; Moran, M.M.; Cipolla, C.K.; Bluth, M.J.; Chaim, J.; Al-Ahmadie, H.; et al. Alterations in DNA Damage Response and Repair Genes as Potential Marker of Clinical Benefit From PD-1/PD-L1 Blockade in Advanced Urothelial Cancers. *J. Clin. Oncol.* **2018**, *36*, 1685–1694. [CrossRef] [PubMed]

19.	Mouw, K.W.; Goldberg, M.S.; Konstantinopoulos, P.A.; D'Andrea, A.D. DNA Damage and Repair Biomarkers of Immunotherapy Response. *Cancer Discov.* **2017**, *7*, 675–693. [CrossRef] [PubMed]

 International Journal of *Molecular Sciences*

Article

NLRP3 as Putative Marker of Ipilimumab-Induced Cardiotoxicity in the Presence of Hyperglycemia in Estrogen-Responsive and Triple-Negative Breast Cancer Cells

Vincenzo Quagliariello [1,*], Michelino De Laurentiis [2], Stefania Cocco [2], Giuseppina Rea [3], Annamaria Bonelli [1], Antonietta Caronna [1], Maria Cristina Lombari [1], Gabriele Conforti [1], Massimiliano Berretta [4], Gerardo Botti [5] and Nicola Maurea [1,*]

[1] Division of Cardiology, Istituto Nazionale Tumori- IRCCS- Fondazione G. Pascale, 80131 Napoli, Italy; a.bonelli@istitutotumori.na.it (A.B.); a.caronna@istitutotumori.na.it (A.C.); m.lombari@istitutotumori.na.it (M.C.L.); g.conforti@istitutotumori.na.it (G.C.)

[2] Breast Unit, Istituto Nazionale Tumori- IRCCS- Fondazione G. Pascale, 80131 Napoli, Italy; m.delaurentiis@istitutotumori.na.it (M.D.L.); s.cocco@istitutotumori.na.it (S.C.)

[3] UOC Bersagli Molecolari del Microambiente, Istituto Nazionale Tumori, IRCCS Fondazione G. Pascale, 80131 Naples, Italy; pina.rea@hotmail.it

[4] Department of MedicalOncology-Centro di Riferimento Oncologico di Aviano (CRO), IRCCS, 33081 Aviano, Italy; mberretta@gmail.com

[5] Scientific Direction, Istituto Nazionale Tumori- IRCCS- Fondazione G. Pascale, 80131 Napoli, Italy; g.botti@istitutotumori.na.it

* Correspondence: quagliariello.enzo@gmail.com (V.Q.); n.maurea@istitutotumori.na.it (N.M.)

Received: 9 September 2020; Accepted: 20 October 2020; Published: 21 October 2020

Abstract: Hyperglycemia, obesity and metabolic syndrome are negative prognostic factors in breast cancer patients. Immune checkpoint inhibitors (ICIs) have revolutionized cancer treatment, achieving unprecedented efficacy in multiple malignancies. However, ICIs are associated with immune-related adverse events involving cardiotoxicity. We aimed to study if hyperglycemia could affect ipilimumab-induced anticancer efficacy and enhance its cardiotoxicity. Human cardiomyocytes and estrogen-responsive and triple-negative breast cancer cells (MCF-7 and MDA-MB-231 cell lines) were exposed to ipilimumab under high glucose (25 mM); low glucose (5.5 mM); high glucose and co-administration of SGLT-2 inhibitor (empagliflozin); shifting from high glucose to low glucose. Study of cell viability and the expression of new putative biomarkers of cardiotoxicity and resistance to ICIs (NLRP3, MyD88, cytokines) were quantified through ELISA (Cayman Chemical) methods. Hyperglycemia during treatment with ipilimumab increased cardiotoxicity and reduced mortality of breast cancer cells in a manner that is sensitive to NLRP3. Notably, treatment with ipilimumab and empagliflozin under high glucose or shifting from high glucose to low glucose reduced significantly the magnitude of the effects, increasing responsiveness to ipilimumab and reducing cardiotoxicity. To our knowledge, this is the first evidence that hyperglycemia exacerbates ipilimumab-induced cardiotoxicity and decreases its anticancer efficacy in MCF-7 and MDA-MB-231 cells. This study sets the stage for further tests on other breast cancer cell lines and primary cardiomyocytes and for preclinical trials in mice aimed to decrease glucose through nutritional interventions or administration of gliflozines during treatment with ipilimumab.

Keywords: hyperglycemia; cardioncology; nivolumab; breast cancer; cytokines; cardiotoxicity

---

1. Introduction

Immune checkpoint inhibitors (ICIs) improved overall survival in cancer patients both as monotherapy or combined with chemotherapies for primary and metastatic cancer patients [1,2]. The family of ICIs involves anti-PD-1 (called nivolumab and pembrolizumab), anti-PD-L1 (called atezolizumab, avelumab, and durvalumab) and anti-CTLA-4 antibodies (called ipilimumab and tremelimumab) [1,3,4]. The combinatorial strategies of ICIs are currently under study in metastatic cancer patients, aimed to reduce immune-resistance of cancer cells, enhancing their apoptosis and necrosis [5]. However, ICIs showed several autoimmune or inflammatory side effects, collectively termed immune-related adverse events, including diabetes, chronic inflammatory bowel diseases, thyroiditis and cardiovascular diseases [6]. Cardiovascular immune-related adverse events involved myocarditis [7]; its pathogenesis is based on lymphocytic infiltration in myocardial tissue and a direct/indirect interaction with cardiomyocytes expressing PD-1/PDL-1 and other immune-sensitive antigens [7,8]. The prevalence of myocarditis ranged from 0.06% to 2.4%, with a higher risk in combination immunotherapy [8]. Other cardiovascular diseases induced by ICIs involves pericardial disease, vasculitis, Takotsubo syndrome, destabilization of atherosclerotic lesions, venous thromboembolism, and conduction abnormalities [9]. Patients with diabetes have increased risk of breast, liver, bladder, pancreatic, colorectal, endometrial and prostate cancers [10,11]. Hyperglycemia is a well-recognized prognostic factor in patients with several chronic diseases like myocardial injuries and cancer [12,13]. Hyperglycemia increases the prevalence and mortality of cancer patients [14]. Causes of cancer risks in diabetic patients involved hyperglycemia, hyperinsulinemia, insulin resistance, distorted insulin-like growth factor-1 (IGF-1) pathway, oxidative stress, enhanced inflammatory processes and aberrant sex hormone production [15,16]. Joshi et al. [17] pointed out that hyperglycemia could provide nutrients for the rapid proliferation of malignant tumor cells, thereby accelerating the process of tumor cells. Hou et al. [18] reported that high-concentration glucose (25 mM) significantly increased the proliferation of breast cancer cells compared to low-concentration glucose (5 mM). Hyperglycemia accelerates the progression of tumor, increases the proliferation, migration and invasion of cancer cells [19]. Recent findings reported that hyperglycemia could increase anticancer-induced cardiotoxicity [20] through involvement of AMPK, mitochondrial proteins, reactive oxygen species and pro-inflammatory cytokines involved in pro-fibrogenic and pro-apoptotic signaling [21,22].

NLRP3 is a new prognostic marker in oncology and acts as a key player in immune-related events involving bacterial and viral infection as well as autoimmune diseases [23,24]. NLRP3 inflammasome activation increases gene expression of IL-1 and IL-6, thereby enhancing production of hs-CRP [23]. Recently, NLRP3 inflammasome was proposed as a new biomarker of cardiovascular diseases and predictor of hospitalization and death for myocardial injuries [25,26]. MyD88 complex (called myddosome) is another protein regulator of death-like signals in human cells [27]; it is a putative marker of the incidence and prognosis of cardiovascular diseases and cancer [28]. However, to the best of our knowledge, no studies have analyzed the effects of high glucose or low glucose on breast cancer responsiveness to ICIs and their cardiotoxicity. To date, only few studies correlated NLRP3 and MyD88 with hyperglycemia damages, like diabetes-induced endothelial inflammation and atherosclerosis [26]. Considering the high prevalence of breast cancer in women with diabetes [29,30], we studied if hyperglycemia could exacerbate ipilimumab-induced cardiotoxicity and decreases its anticancer efficacy in human breast cancer cells (estrogen responsive and triple negative cells) and verified the involvement of NLRP3 and MyD88 in these processes. Moreover, we highlighted the effects of the low glucose or Sodium glucose co-transporter 2 inhibitor (SGLT-2i), called empagliflozin, on the reduction of magnitude of the pro-inflammatory effects mediated by hyperglycemia on cancer cells and cardiomyocytes.

2. Results

2.1. CTLA-4 Expression in Human Breast Cancer Cells

Firstly, we investigated the intracellular and surface expression of CTLA-4 in breast cancer cell lines by Fluorescence-activated cell sorting (FACS) analysis. As expected and reported in the literature, CTLA-4 expression in the breast cancer cell lines was detectable; the higher expression was seen in MDA-MB-231 cells (Figure 1A) compared to MCF-7 (Figure 1B). Moreover, the intracellular expression was generally higher than the surface expression and these data are in line with the literature, confirming the interesting putative role of CTLA-4-related pathway in the breast cancer cell metabolism.

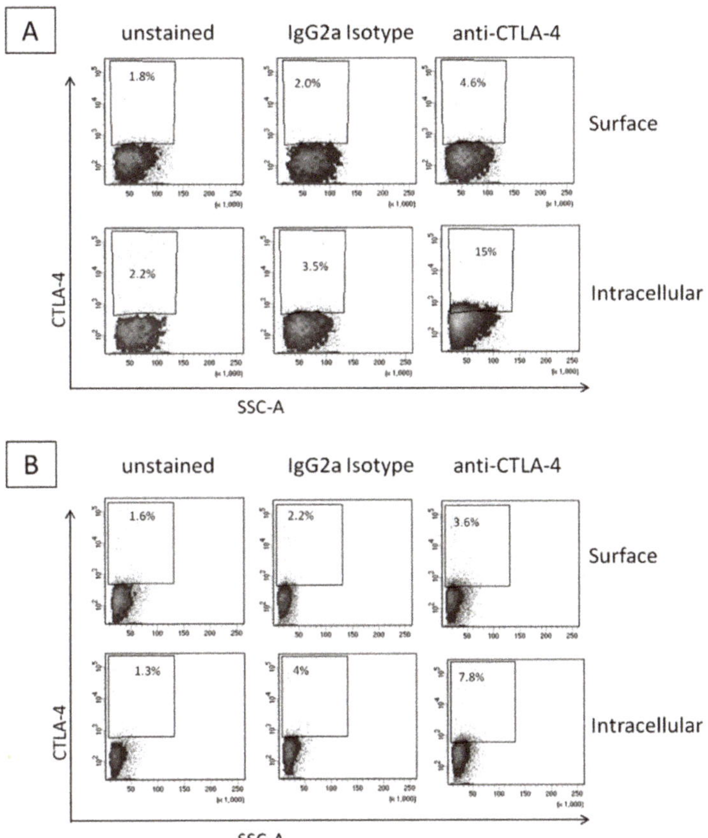

Figure 1. Flow-cytometric analysis of CTLA-4 in human breast cancer cells MDA-MB-231 (**A**) and MCF-7 (**B**). MDA-MB-231 and MCF-7 were stained on their surface or intracellularly with the designated antibodies. IgG2a Isotype corresponds to the staining with a negative class-matched control antibody. Results are expressed as percentage of stained cells.

2.2. Glucose Reduces Ipilimumab-Related Anticancer Activities and Increases its Cardiotoxicity

We studied how high glucose could affect the anticancer properties and cardiotoxicity induced by ipilimumab in a co-culture of cardiomyocytes or breast cancer cells with human peripheral blood mononuclear cells (hPBMCs) [19]. We found that sensitivity to ipilimumab was reduced by 25 mM glucose compared to 5.5 mM glucose (Figure 2A,B). Shifting from a high glucose to a low glucose as well as the treatment with empagliflozin ameliorated breast cancer cell responsiveness to ipilimumab (Figure 2A,B for MCF-7 and MDA-MB-231 cells, respectively). Notably, triple negative breast cancer cells showed more sensitivity to ipilimumab than MCF-7 cells. A different behavior was seen in cardiomyocytes co-cultured with PBMCs (Figure 2C): hyperglicemia increased significantly ipilimumab-induced toxicity than hypoglicemia (paired *t*-test $p < 0.001$, $n = 3$); administration of empagliflozin during high glucose and shifting from high glucose to low glucose reduced the magnitude of the effects. These results indicated that hyperglicemia significantly influenced the cytotoxicity of ipilimumab in breast cancer cells and cardiomyocytes; low glucose and exposure to empagliflozin under hyperglicemia increases the anticancer efficacy of the CTLA-4 blocking agent in breast cancer cells and reduces cytotoxicity.

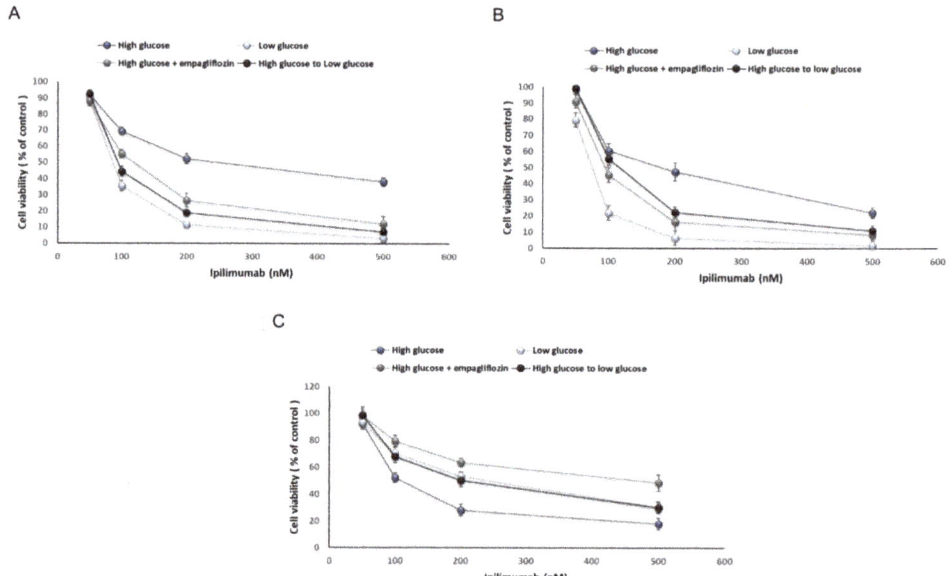

Figure 2. Cell viability of MCF-7 (**A**) and MDA-MB-231 (**B**) cells after 72 h of incubation with ipilimumab under different condition (high glucose; low glucose; high glucose + empagliflozin at 500 nM; switch high glucose to low glucose); (**C**) Cell viability of AC16 cells after 72 h of incubation with ipilimumab under different condition (high glucose; low glucose; high glucose + empagliflozin at 500 nM; shifting from a high glucose to low glucose). Error bars depict means ± SD ($n = 3$). Statistical analysis was performed using paired *t*-test.

2.3. Glucose Increases Leukotriene-Mediated Cardiotoxicity of Ipilimumab

To evaluate the effects of hyperglicemia on lipid metabolism transduction signal pathways during ipilimumab exposure in cancer cells and cardiomyocytes, we quantified the production of leukotrienes B4 (Figure 3). After incubation with ipilimumab under hyperglicemia, MCF-7 cells increased production of leukotrienes compared to low-glucose (125.6 ± 7.4 vs. 43,3 ± 5.5 pg/mg of protein, paired *t*-test $p < 0.001$, $n = 3$) (Figure 3A); shifting from high glucose to low glucose (73.5 ± 6.1 vs. 125.6 ± 7.4 pg/mg of protein, paired *t*-test $p < 0.001$, $n = 3$), as well as the treatment with empagliflozin under hyperglicemic conditions (53.3 ± 3.3 vs. 125.6 ± 7.4 pg/mg of protein, paired *t*-test $p < 0.001$, $n = 3$) reduced significantly the production of leukotrienes indicating anti-inflammatory effects (Figure 3A). A different picture was seen in MDA-MB-231 cells (Figure 3B); after incubation with ipilimumab under hyperglicemia, triple negative cells increased production of leukotrienes compared to low-glucose (154.5 ± 8.3 vs. 53,6 ± 3.4 pg/mg of protein, paired *t*-test $p < 0.001$, $n = 3$) (Figure 3A);

shifting from high glucose to low glucose (89.9 ± 8.2 vs. 154.5 ± 8.3 pg/mg of protein, paired *t*-test $p < 0.001$, $n = 3$), as well as the treatment with empagliflozin under hyperglicemic condition (80.5 ± 7.6 vs. 154.5 ± 8.3 pg/mg of protein, paired *t*-test $p < 0.001$, $n = 3$) reduced significantly the production of leukotrienes indicating anti-inflammatory effects (Figure 3B). Human cardiomyocytes exposed to ipilimumab under hyperglicemic conditions (74.2 ± 7.4 vs. 27.2 ±5.4 pg/mg of protein, paired *t*-test $p < 0.001$, $n = 3$) increased the production of leukotrienes and these effects were partially reduced after a change to low-glucose (46.6 ± 6.1 pg/mg of protein) and treatment with empagliflozin (29.9 ± 3.3 pg/mg of protein) (Figure 2B).

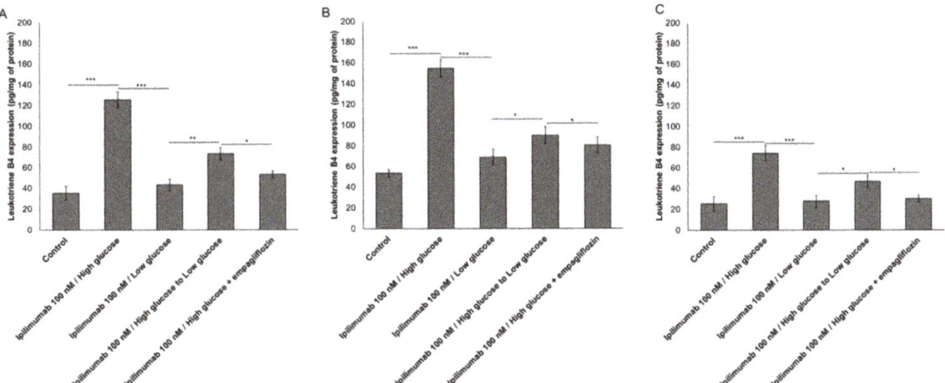

Figure 3. Leukotrienes type B4 production by MCF-7 (**A**) and MDA-MB-231 (**B**) cells, treated with ipilimumab mAb for 24 h, in the presence of human peripheral blood mononuclear cells (hPBMCs) under different condition (high glucose; low glucose; high glucose + empagliflozin at 50 nm; shifting from a high glucose to low glucose). Untreated or treated cells with an unrelated control IgG (control) were used as negative controls; (**C**) Leukotrienes type B4 production by AC-16 cells, treated with ipilimumab mAb for 24 h, in the presence of hPBMCs under different condition (high glucose; low glucose; high glucose + empagliflozin at 500 nM; shifting from a high glucose to low glucose). Untreated or treated cells with an unrelated control IgG (control) were used as negative controls. Error bars depict means ± SD ($n = 3$). Statistical analysis was performed using paired *t*-test. *** $p < 0.001$.** $p < 0.01$.* $p < 0.05$.

2.4. Hyperglycemia Have Pro-Oxidative Effects during Treatment with Ipilimumab

Reactive oxygen species (ROS) overproduction induces several cellular damages and activates pro-inflammatory pathways in myocytes and cancer cells [31,32]. In breast cancer cells, incubation with ipilimumab increased ROS production compared to untreated cells (Figure 4A,B). Under low glucose or after treatment with empagliflozin, surprisingly, ROS production was increased (Figure 4A,B). A different picture was seen in cardiomyocytes (Figure 4C); in fact, treatment with ipilimumab under low glucose or during empagliflozin partially reduced ROS production compared with myocytes grown under hyperglicemic conditions (Figure 4C). These effects were confirmed through the quantification of malondialdeyde (MDA) as a marker of lipid peroxidation [33] that was increased significantly in MCF-7 (Figure 4D) and MDA-MB-231 cells (Figure 4E) and reduced in cardyomyocytes (Figure 4F) under low glucose.

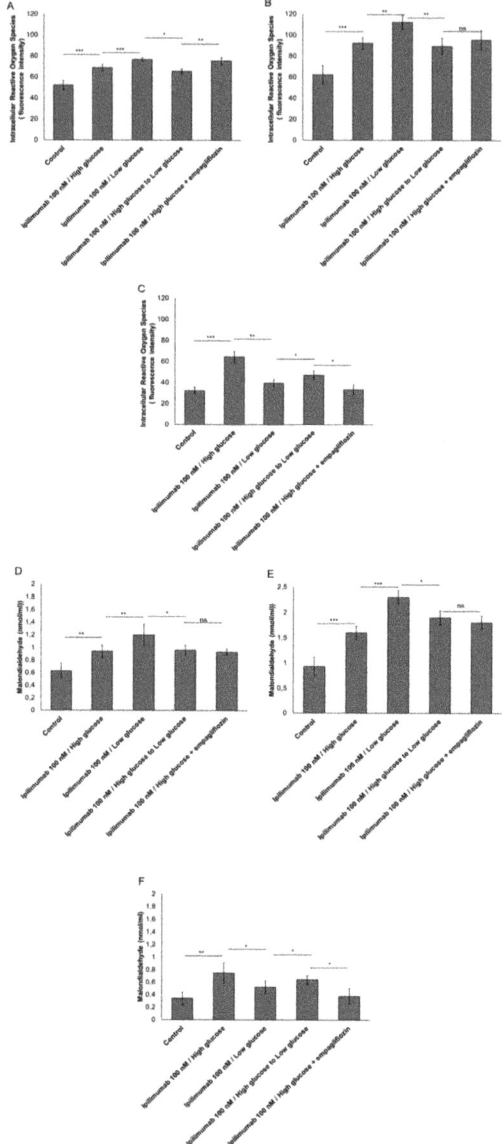

Figure 4. Intracellular Reactive Oxygen Species (ROS) and Malondialdeyde (MDA) quantification in MCF-7 cells (**A,D**) and MDA-MB-231 (**B,E**) cells treated with ipilimumab mAb in the presence of hPBMCs under different condition (high glucose; low glucose; high glucose + empagliflozin at 50 nm; shifting from a high glucose to low glucose). Untreated or treated cells with an unrelated control IgG (control) were used as negative controls; (**C,F**) Intracellular Reactive Oxygen Species (ROS) and Malondialdeyde (MDA) quantification in AC-16 cells, treated with ipilimumab mAb for 24 h, in the presence of hPBMCs under different conditions (high glucose; low glucose; high glucose + empagliflozin at 500 nM; shifting from a high glucose to low glucose). Untreated or treated cells with an unrelated control IgG (control) were used as negative controls. Error bars depict means ± SD ($n = 3$). Statistical analysis was performed using paired t-test. *** $p < 0.001$. ** $p < 0.01$. * $p < 0.05$.

2.5. p65-NF-κB is Overexpressed under Hyperglicemic Condition

As shown in Figure 5A,B, p65-NF-κB expression was increased by 2.8 and 3.4 times in MCF-7 and MDA-MB-231 cells, respectively, under high glucose and exposure to ipilimumab. This trend was reduced by shifting high glucose to low glucose (-0.9 ± 0.13 for MCF-7 cells; -1.2 ± 0.11 for MDA-MB-231 cells; paired *t*-test $p < 0.001$, $n = 3$ for both) and after administration of empagliflozin (-1.4 ± 0.008 for MCF-7 cells; -1.6 ± 0.03 for MDA-MB-231 cells; paired *t*-test $p < 0.001$, $n = 3$ for both). Additionally, cardiomyocytes exposed to ipilimumab under high glucose increased by 2.3 times the expression of p65-NF-κB compared with untreated cells and shifting from high glucose to low glucose reduced the magnitude of the effects (Figure 5C). These effects indicate anti-inflammatory properties of hypoglicemia and treatment with empagliflozin during treatment with ipilimumab.

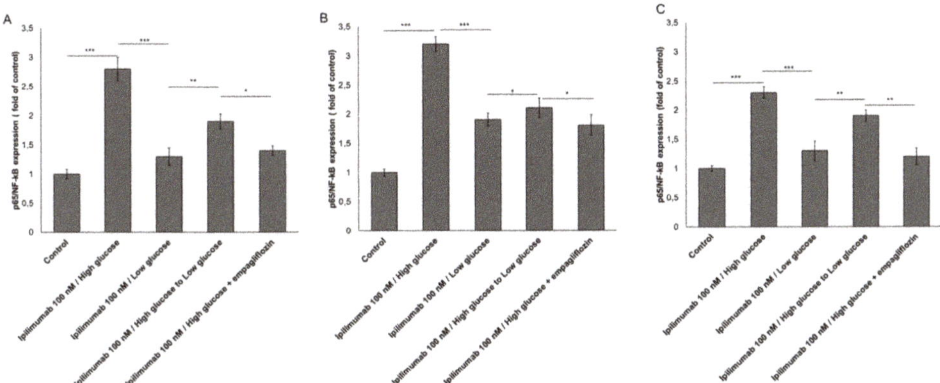

Figure 5. p65/NF-kB expression(fold of control) expression in MCF-7 (**A**), MDA-MB-231 (**B**) and AC-16 (**C**) cells, treated with ipilimumab mAb in the presence of hPBMCs under different condition (high glucose; low glucose; high glucose + empagliflozin at 500 nM; shifting from a high glucose to low glucose). Untreated or treated cells with an unrelated control IgG (control) were used as negative controls. Error bars depict means $\pm$ SD ($n = 3$). Statistical analysis was performed using paired *t*-test. *** $p < 0.001$. ** $p < 0.01$. * $p < 0.05$.

2.6. NLRP3 and MYD88 Expression Are Key Mediators of Hyperglicemia-Mediated Effects in Human Breast Cancer Cells and Cardiomyocytes

We investigated on NLRP3 and MyD88 as key prognostic factors of dendrimental effects of hyperglicema in ipilimumab-induced cardiotoxicity and anticancer effects. In MCF-7 cells, NLRP3 (3.8 ± 0.3 vs. 1.65 ± 0.2, (fold of control) paired *t*-test $p < 0.001$, $n = 3$) (Figure 6A) and MyD88 (2.5 ± 0.3 vs. 1.3 ± 0.1, (fold of control) paired *t*-test $p < 0.001$, $n = 3$) (Figure 6D) were clearly overexpressed under high glucose compared with low glucose. In MDA-MB-231 cells, NLRP3 (4.5 ± 0.14 vs. 2.4 ± 0.11, (fold of control) paired *t*-test $p < 0.001$, $n = 3$) (Figure 6B) and MyD88 (3.3 ± 0.21 vs. 1.7 ± 0.12, (fold of control) paired *t*-test $p < 0.001$, $n = 3$) (Figure 6E) were clearly overexpressed under high glucose compared with low glucose. These effects were reversible by shifting from high glucose to low glucose (Figure 6). Lower levels of NLRP3 and MyD88 protein in high glucose cells treated with empagliflozin were also seen (Figure 6). To assess if NLRP3 controls the sensitivity of high glucose cells to ipilimumab, MCF-7 (Figure 6G,H) and MDA-MB-231 cells (Figure 6I,L) were treated with the anti CTLA-4 antibody in the presence (or absence) of OLT1177 (a selective NLRP3 inhibitor). Treatment with OLT1177 significantly increased responsiveness to ipilimumab in both hyperglicemic and hypoglicemic conditions (Figure 6). A similar behavior was seen in cardiomyocytes: hyperglycemia increased expression of NLRP3 and MyD88 in a way that is sensitive to empagliflozin or shifting from high glucose to low glucose (Figure 6C,F). Selective inhibition of NLRP3 decreases significantly the cardiotoxicity of ipilimumab under high glucose (Figure 6M,N).

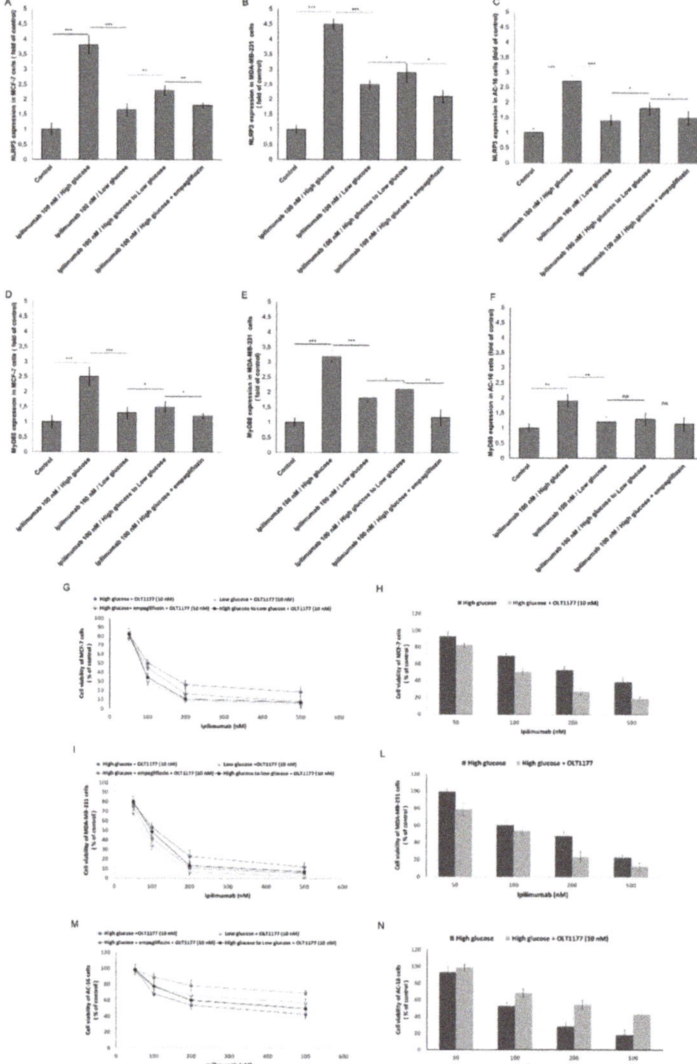

Figure 6. NLRP3 (fold of control) expression in MCF-7 (**A**), MDA-MB-231 (**B**), and AC-16 cells (**C**), treated with ipilimumab mAb in the presence of hPBMCs under different condition (high glucose; low glucose; high glucose + empagliflozin at 500 nM; shifting from a high glucose to low glucose). MyD88 (fold of control) expression in MCF-7 (**D**), MDA-MB-231 (**E**), and AC-16 cells (**F**), treated with ipilimumab in the presence of hPBMCs under different condition (high glucose; low glucose; high glucose + empagliflozin at 50 nm; shifting from a high glucose to low glucose). Cell viability of MCF-7 (**G,H**), MDA-MB-231 (**I,L**), and AC-16 cells (**M,N**) under high glucose (with or without empagliflozin), low glucose, shifting from high glucose to low glucose and always exposed to ipilimumab and NLRP3 selective inhibitor OLT-1177. For all experiments, untreated or treated cells with an unrelated control IgG (control) were used as negative controls. Error bars depict means ± SD (n = 3). Statistical analysis was performed using paired t-test. *** p < 0.001. ** p < 0.01. * p < 0.05.

2.7. NLRP3 Staining in Breast Cancer Cells and Cardiomyocytes

NLRP3 plays a key role in the pathogenesis of breast cancer and heart failure. Based on the considerable changes in NLRP3 expression in MCF-7, MDA-MB-231, and AC-16 cells under high glucose and low glucose, we analyzed cellular staining of NLRP3 through a confocal laser scanning microscope (Figure 7). Breast cancer cells (Figure 7B,N) and cardiomyocytes (Figure 7G) under high glucose and exposed to ipilimumab showed a considerably higher amount of NLRP3 (green signals) than the control (Figure 7A,F,M). Treatment with empagliflozin under high glucose (Figure 7E,L,Q), shifting from high glucose to low glucose (Figure 7D,I,P) and growth in low glucose (Figure 7C,H,O), decreased significantly NLRP3 staining in cell cytoplasm indicating anti-inflammatory effects. These results corroborated the quantitative data described in Figure 6.

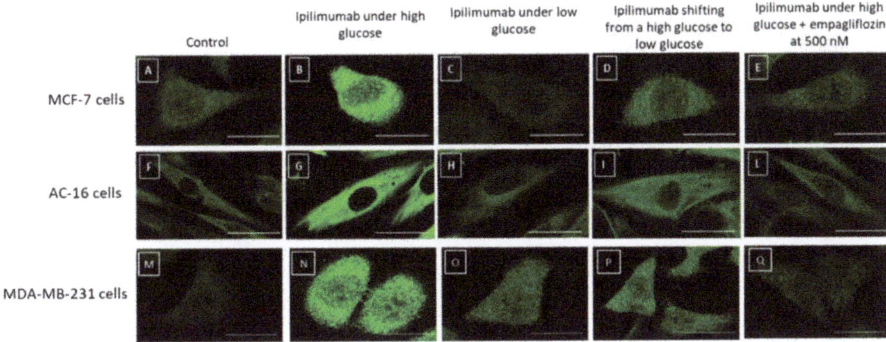

Figure 7. NLRP3 straining (green signals) in MCF-7 (**A–E**), AC-16 (**F–I,L**), and MDA-MB-231 cells (**M–Q**) treated with ipilimumab under high glucose (**B,G,N**); low glucose (**C,H,O**); shifting from a high glucose to low glucose (**D,I,P**), and high glucose + empagliflozin at 500 nM (**E,L,Q**). Untreated cells with an unrelated control IgG (control) were used as negative controls (**A,F,M**). Scale bar: 50 μm.

2.8. Pro-Inflamamtory Cytokines and Growth Factors Are Overexpressed during Hyperglycemia

NLRP3 and MyD88 are primary activators of cytokine storm in human cells in response to pro-inflammatory stimuli, as well as bacterial and viral infection [34]. We investigated cytokines and growth factors during exposure to ipilimumab under hyperglycemia, hypoglycemia, exposure to empagliflozin (under high glucose), and after shifting from high glucose to low glucose. Effectively, under hyperglycemia, compared to low glucose, MCF-7 cells exposed to ipilimumab (Figure 8A) overexpressed IL-1β (245.5 ± 11.5 vs. 156.6 ± 3.4 pg/mg of protein, paired *t*-test $p < 0.001$, $n = 3$), IL-6 (173.3 ± 3.3 vs. 96.6 ± 6.7 pg/mg of protein, paired *t*-test $p < 0.001$, $n = 3$), PDGF (121.1 ± 8.3 vs. 72.3 ± 3.5 pg/mg of protein, paired *t*-test $p < 0.001$, $n = 3$), VEGF (182.5 ± 5.5 vs. 95.3 ± 3.8 pg/mg of protein, paired *t*-test $p < 0.001$, $n = 3$), and TGF-β (165.5 ± 8.5 vs. 97.7 ± 5.3 pg/mg of protein, paired *t*-test $p < 0.001$ $n = 3$). A slightly different picture was seen for triple negative breast cancer cells (Figure 8B): under hyperglycemia, compared to low glucose, MDA-MB-231 cells exposed to ipilimumab overexpressed IL-1β (321.1 ± 10.8 vs. 188.4 ± 7.6 pg/mg of protein, paired *t*-test $p < 0.001$, $n = 3$), IL-6 (256.6 ± 4.5 vs. 102.8 ± 7.2 pg/mg of protein, paired *t*-test $p < 0.001$, $n = 3$), PDGF (168.8 ± 8.9 vs. 78.9 ± 12.2 pg/mg of protein, paired *t*-test $p < 0.001$, $n = 3$), VEGF (234.4 ± 10.3 vs. 112.5 ± 11.2 pg/mg of protein, paired *t*-test $p < 0.001$, $n = 3$), and TGF-β (188.8 ± 5.6 vs. 121.1 ± 8.9 pg/mg of protein, paired *t*-test $p < 0.001$ $n = 3$). For both, shifting from high glucose to low glucose or being treated with empagliflozin under high glucose reduced significantly the expression of cytokines and growth factors compared with cells exposed to high glucose (Figure 8A,B). Instead, cardiomyocytes exposed to ipilimumab under high glucose (Figure 8C) overexpressed IL-1β (154.4 ± 6.7 vs. 106.6 ± 5.8 pg/mg of protein, paired *t*-test $p < 0.001$, $n = 3$), IL-6 (112.2 ± 5.6 vs. 57.7 ± 6.7 pg/mg of protein, paired *t*-test $p < 0.001$, $n = 3$), PDGF (107.7 ± 7.2 vs. 62.1 ± 5.5 pg/mg of protein, paired *t*-test $p < 0.001$, $n = 3$),

VEGF (56.7 ± 7.7 vs. 28.7 ± 6.4 pg/mg of protein, paired *t*-test $p < 0.001$, $n = 3$), and TGF-β (72.3 ± 4.3 vs. 44.7 ± 7.8 pg/mg of protein, paired *t*-test $p < 0.001$, $n = 3$), compared with low glucose cells. Also in this case, under low glucose or after treatment with empagliflozin, the increase rates of cytokines and growth factors were significantly reduced.

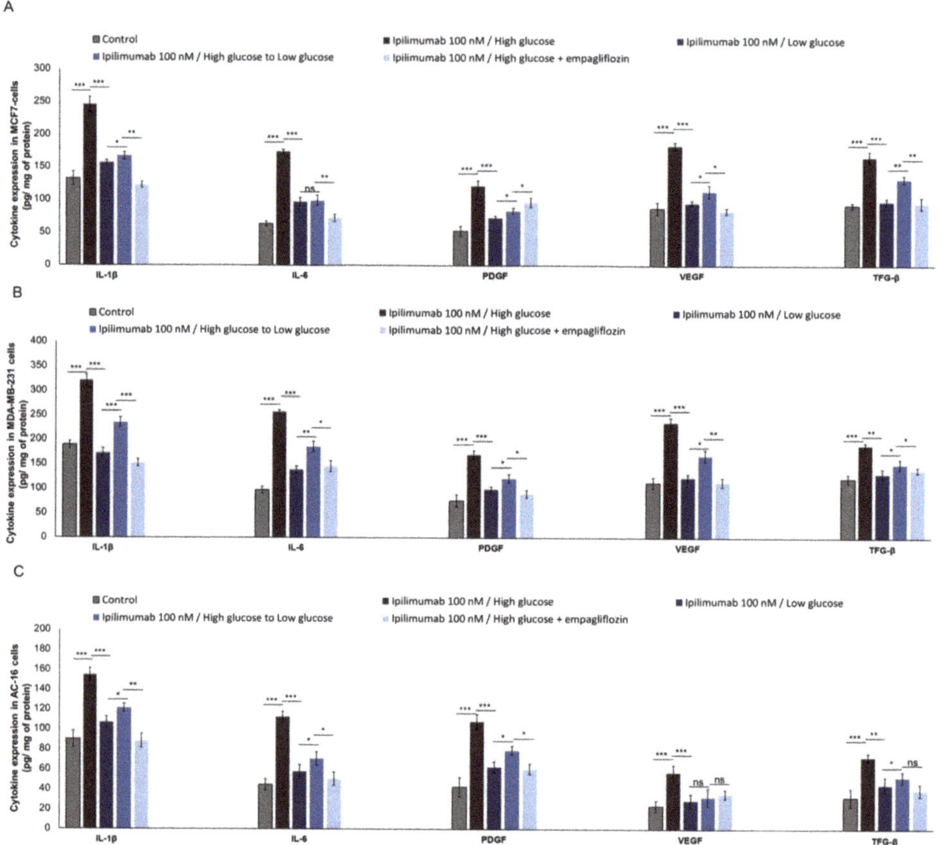

Figure 8. Expression of IL-1, IL-6, PDGF, VEGF, and TGF in MCF-7 (**A**), MDA-MB-231 (**B**), and AC-16 cells (**C**). Cells were treated with ipilimumab mAb for 24 h, in the presence of hPBMCs under different condition (high glucose; low glucose; shifting from a high glucose to low glucose; high glucose + empagliflozin at 500 nm). Error bars depict means ± SD ($n = 3$). Statistical analysis was performed using paired *t*-test. *** $p < 0.001$. ** $p < 0.01$. * $p < 0.05$.

3. Discussion

In this study, we demonstrated that hyperglycemia reduces ipilimumab-related anticancer functions and enhances its cardiotoxicity in cellular models through mechanisms mediated by MyD88 and NLRP3 signaling. More specifically, our findings provide a proof of principle that hyperglycemia increased cytokine storm in human breast cancer cells and cardiomyocytes placing the conditions for cardiotoxic and immune-resistance phenomena. Several studies demonstrated that hyperglycemia or diabetes increased both incidence and recurrence of cancer (breast, liver, prostate, brain, and pancreas) [35]. Chronic or intermittent hyperglycemia is associated with the development of diabetic complications like cardiovascular diseases and chronic inflammatory diseases. Several signaling pathways could be modified during high glucose, for example the increase of inoxidative stress and the overproduction of advanced glycation end products (AGEs) associated with pro-inflammatory cytokines leading to cellular death or chemo-resistance [36,37]. However, the signaling pathways directly triggered by hyperglycemia appear to have a pivotal role in diabetic complications due to the production of reactive oxygen species (ROS) and lipid peroxides [38]. However, surprisingly, in co-cultures of breast cancer cells and hPBMCs, low glucose increases ROS production (Figure 3) compared to high glucose; these effects could be explained by the production of ROS following the cytotoxic effect of hPBMCs against cancer cells [39]; T-cell mediated cytotoxicity involves granzyme B and other pro-apoptotic and pro-oxidizing factors. [39] However, here we have not studied the type of cell death induced by high glucose or the amount of granzyme B secreted by hPBMCs. Further studies on the cell death mechanism will be performed.

Hyperglycemia and Amadori products induce the overexpression of ERK/MAPK leading to the production of ROS and pro-inflammatory and immune-modulating interleukins like interleukin-1 [40]. Interleukin 1 is a key factor in the progression of cancer, cardiovascular diseases, and mortality [41]. Recent cardiovascular outcome trials (CVOTs) demonstrated that pharmacological inhibition of interleukin-1 by subcutaneous administration of the anti IL-1 antibody (i.e., canakimumab) reduced the risk of adverse cardiac events and mortality [42]. Other studies showed that patients treated with canakimumab had reduced cancer-related mortality compared with untreated patients [43]. Results of this study focus on a direct link between hyperglycemia and the pro-inflammatory cytokines/growth factors involved in cancer survival and cardiotoxicity during the CTLA4 blocking agent ipilimumab; these effects are reversible by shifting from high glucose to low glucose or through concomitant treatment with an anti SGLT-2 drug (empagliflozin). Recent studies in cancer-bearing mice demonstrated the chemo-preventive and anti-inflammatory properties of calorie-restriction [44]. In another study, low glucose was associated with higher response to tamoxifen in breast cancer patients [19]. Clinical trials on calorie restriction or hypoglycemia and responsiveness to ICIs are scarce: one retrospective cohort study performed in 55 cancer patients demonstrated trending improvements in overall and progression-free survival in participants with metastatic malignant melanoma who receive a hypoglycemic drug (metformin) in combination with ipilimumab/nivolumab and/or pembrolizumab [45].

Another trial is currently under investigation to combine metformin with ICIs in non-small cell lung cancer patients (NCT03048500) [46]. Other trials in non-cancer [47] and cancer [48,49] patients suggests that natural flavonoids with hypoglycemic properties (i.e., resveratrol) improve T-cell function and increase responsiveness to anti-cancer drugs. Clearly, there is a need for larger retrospective analyses and multi-center prospective studies aimed to evaluate the potential benefits anti-hyperglycemic agents or calorie restriction mimetic strategies (such as hydroxycitrate, metformin, and complementary and alternative medicines) combined with ICIs or conventional anticancer drugs [50–52].

MyD88 is a molecular complex involved in regulation of the immune system, cardiovascular, and cancer metabolism [27]. Patients with viral myocarditis have a high expression of MyD88, CD3+ lymphocytes, and collagen fibers (markers of fibrosis) in myocardial tissue [27]. MyD88 is essential for stimulating resistance to paclitaxel, doxorubicin, and tamoxifen [28]; however, its roles in the ICI-related resistance of breast cancer cells remain unclear. Here, we hypothesize that MyD88, directly associated with hyperglycemia, could be directly involved in the cardiotoxicity and anticancer functions of ipilimumab. Notably, shifting from hyperglycemia to hypoglycemia, as well as the treatment with empagliflozin, reduced the magnitude of these effects and expression of MyD88 both in MCF7, MDA-MB-231 cells, and AC16 cells.

Inflammasomes are multiprotein complexes regulating pro-inflammatory factors including IL-1β and IL-18. IL-18 induces programmed cell death protein 1-dependent immunosuppression in cancer [53], and IL-1β is one of the most important pro-inflammatory mediators involved in immune resistance [54]. The nucleotide-binding oligomerization domain-like receptors (NOD-like receptors) family pyrin domain, containing 3 (NLRP3), is the most widely studied inflammasome. A previous study demonstrated that activated NLRP3 increases IL-18 in patients with lymphoma [55].

NLRP3 inflammasome could represent a novel potential target for the treatment of breast cancer [56]. In a recent preclinical study, the pharmacological inhibition of NLRP3 through miRNA reduced the tumor growth and the immune-resistance in breast cancer-bearing mice through ASC/IL-1/IL-18 pathways; these data provide new clinical insights for breast cancer management [57]. Notably, NLRP3 is associated with myocardial injuries, atherosclerosis, and diabetes mellitus [57]; high glucose stimulates NLRP3 expression thorough inhibition of its ubiquitinization in human cells [57]; therefore, we can speculate that hyperglycemia could enhance cardiotoxicity and responsiveness to ipilimumab through the NLRP3 complex. Therefore, NLRP3 could became a predictive marker of immune-resistance and cardiotoxicity to ipilimumab. Notably, high glucose breast cancer cells and cardiomyocytes exposed to ipilimumab increased NLRP3 inflammasome expression. Administration of the NLRP3 selective inhibitor (OLT1177) increased responsiveness to ipilimumab in breast cancer cells and reduced cytotoxicity in AC16 cells. Interleukin-6 is another key player in ICI-induced cardiotoxicity; it is an independent predictor of diabetes and cardiovascular diseases [58]. Adipocytes and macrophages are the major sources of IL-6 in patients with metabolic syndrome and obesity [58]. Notably, IL-6 has immunosuppressive properties in colorectal cancer cells through the recruitment of immune-suppressive cells and reduction of T cell infiltration in cancer tissues [59]. Inhibition of IL-6 enhanced the efficacy of anti-PD-L1 antibodies in colorectal cancer providing a novel strategy to overcome anti-PD-L1 resistance [60]. Here, we have shown that high glucose increased IL-6 and IL-1 expression in breast cancer cells and cardiomyocytes exposed to ipilimumab; notably, also in this case, shifting from high glucose to low glucose and the treatment with empagliflozin reduced significantly IL-6 expression, providing new insight on the putative protective role of low glucose in IL-6 mediated immune-suppression and cardiotoxicity.

Based on this scenario, overexpression of NLRP3/MyD88 and cytokines during hyperglycemia in breast cancer cells and cardiomyocytes could be a key player in cardiotoxicity and resistance to ipilimumab. There are some limitations of this study: firstly, human leukocyte antigen (HLA)-class I molecules on tumor cells have been regarded as crucial sites where cytotoxic T lymphocytes can recognize tumor-specific antigens. HLA mismatch could be a possible limitation of co-cultures of tumor cells with hPMBCs, despite co-culture models often being used and recommended in preliminary cell-lymphocytes interaction studies, as reported in the literature [61–64]. Although many immortalized cell lines may have a reduced HLA expression [65–67], it is conceivable that some cancer cells induce an HLA mismatch when co-incubated with hPBMCs, being from different donors. However, in line with other similar papers published in the literature [61–64], for all tested combinations (in our case controls and treatments with ipilimumab under high glucose and low glucose) the differences between groups are still due to the treatment conditions. Moreover, different open discussion papers are available about the role, for example, of HLA-E in co-cultures of cancer cells and PBMCs [68]. Interestingly, a recent study Tristan Courau et al. [68] points to another resistance mechanism used by tumor cells that try to evade immune recognition. This is illustrated by HLA-E upregulation on tumor cells upon spheroid infiltration, associated with NKG2A increase in infiltrating immune cells. NKG2A-HLA-E pathway has already been described as a potential inhibitor of antitumor immune responses. A deep study of the role of high glucose in HLA-E expression in co-cultures of breast cancer cells and PBMCs should be also investigated. Another limitation of the study is the lack of a deeper analysis of metabolome in breast cancer cells and cardiomyocytes under high glucose or low glucose. Other studies will be performed during treatments with other ICIs, including anti-PD-1 and anti PDL-1 blocking agents. Moreover, preclinical studies will be made on mice models following a hyperglycemic/hypoglycemic diet, in order to corroborate cellular results described herein. The effects seen herein are principally due to the effects of ipilimumab on hPBMCs and not against cancer cells or cardiomyocytes. To date, the mechanisms and key players of ipilimumab-induced myocardial injuries are not completely understood [69]. Immune cells uptake and infiltration in heart tissues were always seen in human histological studies with high amounts of CD4+/CD8+ T lymphocytes and CD68+ cells [70]; this interaction involves some chemokines like Interleukin 1, 6, and 8 and chemokines that increase the granzyme B-mediated cytotoxicity driving cardiac injury [71]. Treatment with ipilmumab and other ICIs increase the cancer cell recognition of lymphocytes and their release of cytotoxic degranulation markers perforin and granzyme B [72]. According to the literature, CTLA-4 is also expressed in a subsample of breast cancer cells and their role in cancer cell responsiveness to ICIs is actually unknown [73]. It is possible that isolated CTLA-4 positive breast cancer cells could have a different responsiveness to ipilimumab or hyperglycemia; further studies will be performed on isolated CTLA-4 + breast cancer cells, studying their metabolism and their responsiveness to glucose and growth factors.

High glucose mediates NLRP3 inflammasome activation via upregulation of E74-like transcription factor (ELF3) expression [74]. Microtubule affinity regulating kinase 4 (MARK4) plays a crucial role in the regulation of NLRP3 inflammasome activation, which leads to the generation of IL-1β [74]. High glucose increases NLRP3 activation, probably involved in ICIs-induced resistance and cardiotoxicity; in fact, recently, a mechanism was identified whereby CD8+ T cell activation in response to PD-1 blockade induced a NLRP3 inflammasome signaling cascade that ultimately led to the recruitment of granulocytic myeloid-derived suppressor cells (MDSCs) into tumor tissues, thereby dampening the resulting antitumor immune response [75]. Genetic and pharmacologic inhibition of NLRP3 enhances the efficacy of anti–PD-1 Ab immunotherapy [76]. We hypothesize that high glucose increases NLRP3 expression thereby reducing the ipilimumab-mediated cytotoxic efficacy against MCF-7 and MDA-MB-231 cells. To our knowledge, this is the first evidence that hyperglycemia exacerbates ipilimumab-induced cardiotoxicity and decreases its anticancer efficacy in MCF-7 and MDA-MB-231 cells. Moreover, to clarify whether hypoglycemia can increase responsiveness to ipilimumab against ER+ and triple negative breast cancer and reduces cardiovascular side effects, further studies in breast cancer-bearing mice are also required.

4. Materials and Methods

4.1. Cell Culture

AC16 human cardiomyocytes were purchased from American Type Culture Collection (ATCC®, LGC Standards, Teddington, UK) and cultured in Gibco® Dulbecco's modified Eagle's medium (Gibco, Milan, Italy): Nutrient mixture F-12 (DMEM/F12) supplemented with 10% fetal bovine serum (FBS) (HyClone™, GE Healthcare Life Sciences, Milan, Italy) and Penicillin-Streptomycin (100 U/mL, Gibco®, Milan, Italy). MCF-7 human breast cancer cells (ERα+, PR+, HER2-) were cultured in Dulbecco's modified Eagle's medium (DMEM) supplemented with 10% fetal bovine serum (FBS), 2 mM glutamine, 100 units/mL penicillin and 100 units/mL streptomycin. Triple negative MDA-MB-231 (ATCC® HTB-26™) cells were grown in ATCC-formulated Leibovitz's L-15 Medium supplemented with 10% fetal bovine serum (FBS) (HyClone™, GE Healthcare Life Sciences, Milan, Italy) and Penic illin-Streptomycin (100 U/mL, Gibco®, Milan, Italy). Cell cultures were maintained in a humidified atmosphere of 95% air and 5% CO_2 at 37 °C.

4.2. Co-Cultures

To test the biological effects of hyperglycemia on breast cancer and cardiac metabolism, we used co-cultures of MCF-7 and MDA-MB-231 breast cancer cells or human cardiomyocytes (AC16) (that do not express CTLA-4) with human peripheral blood mononuclear cells (hPBMCs). Cancer cells and cardiomyocytes were plated in 96-well flat-bottom plates at the density of 150,000 cells/well for 16 h. hPBMCs, a population of immune cells consisting of T cells, B cells, natural killer cells, dendritic cells, and monocytes [77] from healthy donors, were added at effector: target ratio 5:1 in the absence or presence of ipilimumab as described previously [78,79]; in fact, hPBMCs are conventionally used in cellular experiments of responsiveness to PD-1, PDL-1, or CTLA-4 blocking agents [80–82]. Co-culture of MCF7 or MDA-MB-231 cells or AC16 cells with hPBMCs mimics, respectively, the immune cell infiltration in breast cancer tissue (the immune infiltration of tumors is closely related to clinical outcomes in breast cancer patients [83,84]), as well as in myocardial tissue (patients treated with CTLA-4 blocking agents developed severe/fatal myocarditis which was likely a result of the lymphocytic infiltration of lymphocytes [85,86]).

4.3. CTLA-4 Expression in Breast Cancer Cells through Flow Cytometry

For the cell-surface marker, MDA-MB-231 and MCF-7 cells (100.000 cells/tube) were harvested and stained with anti-human CD152 (CTLA-4) (BD Bioscience, San Diego, CA, USA) and LIVE/DEAD fixable Aqua (Thermo Fisher, Milan, Italy, at 4 °C for 30 min in cell stain Buffer (BSA 0.2%) (BD Pharmingen™, San Diego, CA, USA). Cells were then washed twice and resuspended in 200 μL of in cell stain Buffer. To avoid non-specific anti-CTLA-4 antibody binding to the Fc portion of the receptor, the cell suspension was pre-treated with FcR Blocking Reagent, human (130-059-901 MiltenyiBiotec, GmbH, Bergisch Gladbach, Germany), and then stained with the anti-CTLA-4 antibody. As negative control, we stained cells with a negative class-matched control antibody (IgG2a isotype). For intracellular staining, cells were fixed and permeabilized using intracellular buffer set BD Cytofix/Cytoperm™ (BD Bioscience, San Jose, CA, USA) according to the manufacturer's instructions and stained with anti-human CD152 at ice for 30 min in permeabilization buffer. The samples were washed and resuspended in 200 μL of cell stain Buffer (BSA0.2%) (BD Pharmingen™, San Diego, CA, USA) cells were washed twice with permeabilization buffer. A minimum of 100.000 events for each sample were collected by FACS ARIA III (Becton Dickinson, Mountain View, CA, USA) and data were analyzed using FACSDiva™ 8.0 Software (BD Bioscience, San Diego, CA, USA).

4.4. Cell Viability

To test the effects of hyperglycemia on cellular mitochondrial viability, we work with co-cultures of MCF-7 or MDA-MB-231 cells or human cardiomyocytes (AC16) with hPBMCs, the cells were plated in 96-well flat-bottom plates at the density of 150,000 cells/well for 16 h. Human Peripheral Blood Mononuclear Cells (hPBMCs) were added at an effector:target ratio of 5:1 in the absence or presence of ipilimumab at 50, 100, 200 and 500 nM and incubated for 72 h at 37 °C, as described previously [78,79]. Controls included target cells incubated in the absence of effector cells or in the presence of anti CTLA-4 antibody. Notably, cells in co-culture were grown in 5.5 mM glucose, corresponding to normal fasting glucose levels in humans, or in 25 mM glucose resembling hyperglycemia in humans, following well established protocols [19]; moreover, as a control, only during the hyperglycemic condition, cells were co-incubated with ipilimumab (100 nM, [78]) and empagliflozin (500 nM, [87]), an oral antidiabetic agent with cardioprotective properties. After treatments, lymphocytes were removed and adherent cells were washed three times with PBS at pH 7.4 and incubated with 100 μL of an MTT solution (0.5 mg/mL in cell culture medium) for 4 h at 37 °C. Absorbance readings were acquired at a wavelength of 450 nm with the Tecan Infinite M200 plate-reader (Tecan Life Sciences Home, Männedorf, Switzerland) using I-control software. Relative cell viability (%) was calculated with the following formula: [A]test/[A]control × 100, where "[A]test" is the absorbance of the test sample, and "[A]control" is the absorbance of the control cells incubated solely in culture medium. After the evaluation of cell cytotoxicity, we measured the total protein content using the Pierce Micro BCA protein assay kit (Thermo Fisher, Milan, Italy). Briefly, the cells were washed with ice-cold PBS, and incubated for 15 min in 150 μL cell lysis buffer (0.5% *v/v* Triton X-100 in PBS) that included 150 μL of the Micro BCA protein assay kit reagent (prepared according to the manufacturer's instructions). Absorbance at 562 nm was measured on a plate reader. Cytotoxicity measurements were normalized by the amount of total protein content in each well.

4.5. Expression of Leukotriene B4 (LTB4)

Co-cultures of cardiomyocytes/hPBMCs and breast cancer cells/hPBMCs were untreated (control) or treated with ipilimumab (100 nM) under high glucose, low glucose, shifting from high glucose to low glucose, or treated with empagliflozin (500 nM) under high glucose for 12 h. After treatments, leukotriene B4 ((5S,12R)-dihydroxy-6,14Z-8,10E-eicosatetraenoic acid) expression in cell lysates was determined through ELISA (Cayman Chemical, Ann Arbor, MI, USA) following the supplier's instructions [78]; data were expressed as pg of leukotriene B4/mg of cell proteins calculated by QuantiPro Assay (Biorad, Milan, Italy).

4.6. Reactive Oxygen Species and Lipid Peroxidation

Lipid peroxidation is a key player in heart failure and chemo-resistance phenomena. Co-cultures of cardiomyocytes/hPBMCs and breast cancer cells/hPBMCs were untreated (control) or treated with ipilimumab (100 nM) for 6 h under high glucose, low glucose, shifting from high glucose to low glucose, or treated with empagliflozin (500 nM) under high glucose. After treatments, cells were washed three times with cold PBS, harvested with 0.25% v/v Trypsin, and centrifuged at 1000× g for 10 min. The supernatant was discarded and the cell pellet sonicated in cold PBS. After a centrifugation step at 800× g for 5 min, we quantified malondialdehyde (MDA) by using a commercial kit with a spectrophotometer according to the manufacturer's protocols (Sigma Aldrich, Milan, Italy) [79]. We measured the protein content of the cell homogenates using the Micro BCA protein assay kit (Pierce, Thermo Fisher, Milan, Italy) according to the kit instructions.

4.7. p65-NF-κB Expression

Co-cultures of cardiomyocytes/hPBMCs and breast cancer cells/hPBMCs were untreated (control) or treated with ipilimumab (100 nM) under high glucose, low glucose, shifting from high glucose to low glucose or treated with empagliflozin (500 nM) under high glucose for 12 h. After treatments, cells were harvested and lysed in lyses buffer (50 mMTris-HCl, pH 7.4, 1 mM EDTA, 100 mM NaCl, 20 mM NaF, 3 mM Na_3VO_4, 1 mM PMSF, and protease inhibitor cocktail). Lysates were then centrifuged, the supernatants were collected and analyzed using the TransAM NF-κB p65 transcription factor assay kit (Active Motif, Carlsbad, CA, USA), according to the manufacturer's recommendations. NF-κB complexes were captured by binding to a consensus 5′-GGGACTTTCC-3′ oligonucleotide immobilized on a 96-well plate. Bound NF-κB was quantified by incubating with anti-p65 primary antibody followed by horseradish peroxidase (HRP)-conjugated goat anti-rabbit IgG and spectrophotometric detection at a wavelength of 450 nm using a microplate spectrofluorometer. Data were expressed as the percentage of p65/NF-κB DNA binding relative to control (untreated) cells.

4.8. NLRP3 and MyD88: Key Mediators of Hyperglycemia-Induced Cardiotoxicity

Co-cultures of cardiomyocytes/hPBMCs and breast cancer cells/hPBMCs were untreated (control) or treated with ipilimumab (100 nM) under high glucose, low glucose, shifting from high glucose to low glucose, or treated with empagliflozin (500 nM) under high glucose for 12 h. After treatments, cells were harvested and lysed in lyses buffer (50 mMTris-HCl, pH 7.4, 1 mM EDTA, 100 mM NaCl, 20 mM NaF, 3mM Na_3VO_4, 1 mM PMSF, and protease inhibitor cocktail). Lysates were then centrifuged, the supernatants were collected and submitted to the ELISA protocol for quantification of MyD88 (Human MyD88 ELISA Kit (ab171341), Abcam, Milan, Italy) and NLRP3 (Human NLRP3 ELISA Kit (OKEH03368), Aviva Systems Biology, San Diego, CA, USA) [88,89]. Briefly, an antibody against NLRP3 or MyD88 was pre-coated onto a 96-wellplate (12 × 8 Well Strips) and blocked. Standards or test samples were added to the wells and incubated for 1h. After washing, a biotinylated detector antibody specific to NLRP3 or MyD88 was added, incubated and followed by washing for 30 s. Avidin-Peroxidase Conjugate was then added, incubated, and unbound conjugate was washed away. An enzymatic reaction was produced through the addition of TMB substrate which is catalyzed by HRP generating a blue color product that changes yellow after adding acidic stop solution. The density of yellow coloration read by absorbance at 450 nm was quantitatively proportional to the amount of sample NLRP3 or MYD88 captured in well. For human MyD88 ELISA, the sensitivity was <10 pg/mL and range of detection was 156 pg/mL–10,000 pg/mL; for human NLRP3 ELISA assay, the sensitivity was <0.078 ng/mL and range of detection was 0.156–10 ng/mL. Moreover, we utilized pharmacological inhibitor to identify the major signaling pathway involved in hyperglicemia-related cancer sensitivity and the cardiotoxicity of ipilimumab. To this aim, we pre-incubated MCF-7, MDA-MB.231, and AC-16 cells with NLRP3 inhibitor (OLT1177) at 100 nM [90] during incubation with ipilimumab and performed cell viability assay as described in paragraph 4.4.

4.9. Confocal Laser Scanning Microscope

Cardiomyocytes and human breast cancer cells cultured as described previously under standard conditions at 37 °C in a humidified 5% CO_2 atmosphere, were seeded in a 24-well plate (5000 cells/well) and allowed to grow for 24 h. Cells were then untreated (control) or treated with ipilimumab (100 nM) under high glucose, low glucose, shifting from high glucose to low glucose, or treated with empagliflozin (500 nM) under high glucose for 12 h. Cells were then fixed with 2.5% glutaraldehyde in PBS at room temperature for 20 min, rinsed with PBS, and permeabilzed with 0.1% Triton X-100 for 5 min. NLRP3 was stained by incubation with a primary antibody against NLRP3 (Life Span Bio Sciences) for 1 h under gentle stirring, followed by an anti-human NLRP3 polyclonal antibody (Life Span BioSciences, Seattle, WA, USA). The detection was performed by the addition of a Goat Anti-Rabbit IgG H&L (FITC) (ab6717, Abcam, Milan, Italy) for 1 h, under gentle stirring. Using a Confocal Microscope (C1 Nikon) equipped with a EZ-C1 Software for data acquisition and 60× oil immersion objective, NLRP3 was imaged through excitation/emission at 488/515 nm.

4.9.1. Cytokines and Growth Factors Assay

The expression of IL-1β, IL-6, PDGF, FGF, VEGF, and TGF-βin cardiomyocytes and breast cancer cells was performed through the ELISA method, as described elsewhere [78]. Co-cultures of cardiomyocytes/hPBMCs and breast cancer cells/hPBMCs were untreated (control) or treated with ipilimumab (100 nM) under high glucose, low glucose, shifting from high glucose to low glucose, or treated with empagliflozin (500 nM) under high glucose for 12 h. Culture supernatants were centrifuged to pellet any detached cells and measured using the appropriate ELISA kits according to the manufacturer's instructions (Sigma Aldrich, Milan, Italy). The sensitivity of this method was below 10 (pg/mL), and the assay accurately detected cytokines in the range of 1–32,000 pg/mL.

4.9.2. Statistical Analysis

All cell-based assays were performed in triplicates ($n = 3$) and results are presented as mean ± Standard Deviation (SD). To compare cell culture conditions, a paired-*t* test was used through the use of Sigmaplot software (Systat Software Inc., San Jose, CA, USA). $p < 0.05$ was considered to indicate a statistically significant difference.

5. Conclusions

Hyperglycemia is a prognostic marker in oncology; glucose-related damages increased cancer cell metabolism, chemo, and immune resistance [91]; in fact, integrating glycometabolism targeting (aimed to reduce glucose in cancer cells) and immunotherapy seems to be a rational strategy for improving overall survival in cancer patients [92]. This study reveals that hyperglycemia during treatment with ipilimumab, a CTLA-4 blocking agent, increased cardiotoxicity and reduced mortality of MCF-7 and MDA-MB-231 cells in a manner that is sensitive to NLRP3. Therefore, NLRP3 could became a valid biomarker of ipilimumab-induced cardiotoxicity under hypoglycemia; pharmacological inhibition of NLRP3, through clinically available drugs [90,93], safe in humans, currently studied for therapy of acute gout and arthritis could be an effective therapeutic strategy aimed at improving anticancer responsiveness to ipilimumab and reducing its cardiovascular side effects. Notably, cells treated with ipilimumab and SGLT-2 inhibitor (empagliflozin) under high glucose or shifting from high glucose to low glucose reduced significantly the magnitude of the effects. Further studies will be done in other breast cancer cell lines and cardiomyocytes (i.e., primary ventricular cardiomyocytes) exposed to CTLA-4 blocking agents as well as other immune check point inhibitors (PD-1 or PDL-1 blocking agents). This study set the stage to preclinical trials in mice models aimed to decrease glucose through nutritional intervention or through treatment with gliflozines during therapy with ipilimumab.

Author Contributions: Conceptualization, V.Q., M.D.L., and N.M.; methodology, V.Q., S.C., and G.R.; software, M.B., S.C., and G.R.; validation, A.B., M.C.L., and N.M.; formal analysis, A.C., M.B.; investigation, G.C., A.B.; writing—original draft preparation, V.Q.; writing—review and editing, N.M., G.B.; visualization, A.B., M.C.L., A.C., and G.G.; supervision, N.M., G.B.; funding acquisition, G.B., N.M. All authors have read and agreed to the published version of the manuscript.

Funding: This work was funded by a "Ricerca Corrente" grant from the Italian Ministry of Health. "Cardiotossicità dei trattamenti antineoplastici: identificazione precoce e strategie di cardioprotezione" Project code: M1/5-C.

Acknowledgments: We thank Grimaldi Immacolata, working in the Division of Cardiology, Istituto Nazionale Tumori, IRCCS, Fondazione G. Pascale of Naples, for effective research support activities.

Conflicts of Interest: The authors declare no conflict of interest.

Abbreviations

ICIs	Immune checkpoint inhibitors
SGLT-2	Sodium-glucose cotransporter type-2
NLRP3	NOD-like receptor family pyrin domain, containing 3
MyD88	Myddosome type 88
PD1	Programmed cell death protein 1
PDL1	Programmed Death-Ligand 1
CTLA4	Cytotoxic T-Lymphocyte Antigen 4
AMPK	5′ AMP-activated protein kinase
hs-CRP	High sensitivity C-reactive protein
ROS	Reactive oxygen species
MDA	Malondialdeyde
p65-NF-κB	Nuclear factor kappa-light-chain-enhancer of activated B cells
ILs	Interleukins
PDGF	Platelet-Derived Growth Factor
VEGF	Vascular endothelia growth factor
TGF-β	Transforming growth factor beta
AGEs	Advanced glycation end products
CVOTs	Cardiovascular outcome trials
VEGF	Vascular endothelia growth factor
TGF-β	Transforming growth factor beta
AGEs	Advanced glycation end products

References

1. The Lancet Oncology. Immunotherapy: Hype and hope. *Lancet Oncol.* **2018**, *19*, 845. [CrossRef]
2. Haslam, A.; Prasad, V. Estimation of the Percentage of US Patients With Cancer Who Are Eligible for and Respond to Checkpoint Inhibitor Immunotherapy Drugs. *JAMA Netw. Open* **2019**, *2*, e192535. [CrossRef] [PubMed]
3. Salmaninejad, A.; Valilou, S.F.; Shabgah, A.G.; Aslani, S.; Alimardani, M.; Pasdar, A.; Sahebkar, A. PD-1/PD-L1 pathway: Basic biology and role in cancer immunotherapy. *J. Cell. Physiol.* **2019**, *234*, 16824–16837. [CrossRef] [PubMed]
4. Seidel, J.A.; Otsuka, A.; Kabashima, K. Anti-PD-1 and Anti-CTLA-4 Therapies in Cancer: Mechanisms of Action, Efficacy, and Limitations. *Front. Oncol.* **2018**, *8*, 86. [CrossRef]
5. Larkin, J.; Chiarion-Sileni, V.; Gonzalez, R.; Grob, J.J.; Rutkowski, P.; Lao, C.D.; Cowey, C.L.; Schadendorf, D.; Wagstaff, J.; Dummer, R.; et al. Five-Year Survival with Combined Nivolumab and Ipilimumab in Advanced Melanoma. *N. Engl. J. Med.* **2019**, *381*, 1535–1546. [CrossRef] [PubMed]
6. Martins, F.; Sofiya, L.; Sykiotis, G.P.; Lamine, F.; Maillard, M.; Fraga, M.; Shabafrouz, K.; Ribi, C.; Cairoli, A.; Guex-Crosier, Y.; et al. Adverse effects of immune-checkpoint inhibitors: Epidemiology, management and surveillance. *Nat. Rev. Clin. Oncol.* **2019**, *16*, 563–580. [CrossRef] [PubMed]

7. Zhou, Y.W.; Zhu, Y.J.; Wang, M.N.; Xie, Y.; Chen, C.Y.; Zhang, T.; Xia, F.; Ding, Z.Y.; Liu, J.Y. Immune Checkpoint Inhibitor-Associated Cardiotoxicity: Current Understanding on Its Mechanism, Diagnosis and Management. *Front. Pharmacol.* **2019**, *10*, 1350. [CrossRef]

8. Ganatra, S.; Neilan, T.G. Immune Checkpoint Inhibitor-Associated Myocarditis. *Oncologist* **2018**, *23*, 879–886. [CrossRef]

9. Michel, L.; Rassaf, T.; Totzeck, M. Cardiotoxicity from immune checkpoint inhibitors. *Int. J. Cardiol. Heart Vasc.* **2019**, *25*, 100420. [CrossRef]

10. Ottaiano, A.; Nappi, A.; Tafuto, S.; Nasti, G.; De Divitiis, C.; Romano, C.; Cassata, A.; Casaretti, R.; Silvestro, L.; Avallone, A.; et al. Diabetes and Body Mass Index Are Associated with Neuropathy and Prognosis in Colon Cancer Patients Treated with Capecitabine and Oxaliplatin Adjuvant Chemotherapy. *Oncology* **2016**, *90*, 36–42. [CrossRef]

11. Noto, H.; Goto, A.; Tsujimoto, T.; Osame, K.; Noda, M. Latest insights into the risk of cancer in diabetes. *J. Diabetes Investig.* **2013**, *4*, 225–232. [CrossRef] [PubMed]

12. Ryu, T.Y.; Park, J.; Scherer, P.E. Hyperglycemia as a risk factor for cancer progression. *Diabetes Metab. J.* **2014**, *38*, 330–336. [CrossRef] [PubMed]

13. Habib, S.L.; Rojna, M. Diabetes and risk of cancer. *ISRN Oncol.* **2013**, *2013*, 583786. [CrossRef]

14. Gapstur, S.M.; Gann, P.H.; Lowe, W.; Liu, K.; Colangelo, L.; Dyer, A. Abnormal glucose metabolism and pancreatic cancer mortality. *JAMA* **2000**, *283*, 2552–2558. [CrossRef] [PubMed]

15. Leroith, D.; Scheinman, E.J.; Bitton-Worms, K. The Role of Insulin and Insulin-like Growth Factors in the Increased Risk of Cancer in Diabetes. *Rambam Maimonides Med. J.* **2011**, *2*, e0043. [CrossRef]

16. Bowers, L.W.; Rossi, E.L.; O'Flanagan, C.H.; deGraffenried, L.A.; Hursting, S.D. The Role of the Insulin/IGF System in Cancer: Lessons Learned from Clinical Trials and the Energy Balance-Cancer Link. *Front. Endocrinol. (Lausanne)* **2015**, *6*, 77. [CrossRef]

17. Joshi, S.; Liu, M.; Turner, N. Diabetes and its link with cancer: Providing the fuel and spark to launch an aggressive growth regime. *Biomed Res. Int.* **2015**, *2015*, 390863. [CrossRef]

18. Hou, Y.; Zhou, M.; Xie, J.; Chao, P.; Feng, Q.; Wu, J. High glucose levels promote the proliferation of breast cancer cells through GTPases. *Breast Cancer (Dove Med. Press)* **2017**, *9*, 429–436. [CrossRef]

19. Ambrosio, M.R.; D'Esposito, V.; Costa, V.; Liguoro, D.; Collina, F.; Cantile, M.; Prevete, N.; Passaro, C.; Mosca, G.; De Laurentiis, M.; et al. Glucose impairs tamoxifen responsiveness modulating connective tissue growth factor in breast cancer cells. *Oncotarget* **2017**, *8*, 109000–109017. [CrossRef]

20. Zhuang, X.D.; Hu, X.; Long, M.; Dong, X.B.; Liu, D.H.; Liao, X.X. Exogenous hydrogen sulfide alleviates high glucose-induced cardiotoxicity via inhibition of leptin signaling in H9c2 cells. *Mol. Cell. Biochem.* **2014**, *391*, 147–155. [CrossRef]

21. Bell, D.S.H.; Goncalves, E. Heart failure in the patient with diabetes: Epidemiology, aetiology, prognosis, therapy and the effect of glucose-lowering medications. *Diabetes Obes. Metab.* **2019**, *21*, 1277–1290. [CrossRef] [PubMed]

22. Russo, I.; Frangogiannis, N.G. Diabetes-associated cardiac fibrosis: Cellular effectors, molecular mechanisms and therapeutic opportunities. *J. Mol. Cell. Cardiol.* **2016**, *90*, 84–93. [CrossRef] [PubMed]

23. Mangan, M.S.J.; Olhava, E.J.; Roush, W.R.; Seidel, H.M.; Glick, G.D.; Latz, E. Targeting the NLRP3 inflammasome in inflammatory diseases. *Nat. Rev. Drug Discov.* **2018**, *17*, 588–606. [CrossRef]

24. Kelley, N.; Jeltema, D.; Duan, Y.; He, Y. The NLRP3 Inflammasome: An Overview of Mechanisms of Activation and Regulation. *Int. J. Mol. Sci.* **2019**, *20*, 3328. [CrossRef] [PubMed]

25. Xue, Y.; Du, H.D.; Tang, D.; Zhang, D.; Zhou, J.; Zhai, C.W.; Yuan, C.C.; Hsueh, C.Y.; Li, S.J.; Heng, Y.; et al. Correlation Between the NLRP3 Inflammasome and the Prognosis of Patients With LSCC. *Front. Oncol.* **2019**, *9*, 588. [CrossRef] [PubMed]

26. Luo, B.; Huang, F.; Liu, Y.; Liang, Y.; Wei, Z.; Ke, H.; Zeng, Z.; Huang, W.; He, Y. NLRP3 Inflammasome as a Molecular Marker in Diabetic Cardiomyopathy. *Front. Physiol.* **2017**, *8*, 519. [CrossRef]

27. Feng, Y.; Zou, L.; Zhang, M.; Li, Y.; Chen, C.; Chao, W. MyD88 and Trif signaling play distinct roles in cardiac dysfunction and mortality during endotoxin shock and polymicrobial sepsis. *Anesthesiology* **2011**, *115*, 555–567. [CrossRef] [PubMed]

28. Blyszczuk, P.; Kania, G.; Dieterle, T.; Marty, R.R.; Valaperti, A.; Berthonneche, C.; Pedrazzini, T.; Berger, C.T.; Dirnhofer, S.; Matter, C.M.; et al. Myeloid differentiation factor-88/interleukin-1 signaling controls cardiac fibrosis and heart failure progression in inflammatory dilated cardiomyopathy. *Circ. Res.* **2009**, *105*, 912–920. [CrossRef]

29. Thi, H.T.H.; Hong, S. Inflammasome as a Therapeutic Target for Cancer Prevention and Treatment. *J. Cancer Prev.* **2017**, *22*, 62–73. [CrossRef]

30. Boyle, P.; Boniol, M.; Koechlin, A.; Robertson, C.; Valentini, F.; Coppens, K.; Fairley, L.L.; Boniol, M.; Zheng, T.; Zhang, Y.; et al. Diabetes and breast cancer risk: A meta-analysis. *Br. J. Cancer* **2012**, *107*, 1608–1617. [CrossRef]

31. Zinman, B.; Wanner, C.; Lachin, J.M.; Fitchett, D.; Bluhmki, E.; Hantel, S.; Mattheus, M.; Devins, T.; Johansen, O.E.; Woerle, H.J.; et al. Empagliflozin, Cardiovascular Outcomes, and Mortality in Type 2 Diabetes. *N. Engl. J. Med.* **2015**, *373*, 2117–2128. [CrossRef]

32. Quagliariello, V.; Coppola, C.; Mita, D.G.; Piscopo, G.; Iaffaioli, R.V.; Botti, G.; Maurea, N. Low doses of Bisphenol A have pro-inflammatory and pro-oxidant effects, stimulate lipid peroxidation and increase the cardiotoxicity of Doxorubicin in cardiomyoblasts. *Environ. Toxicol. Pharmacol.* **2019**, *69*, 1–8. [CrossRef] [PubMed]

33. Ayala, A.; Muñoz, M.F.; Argüelles, S. Lipid peroxidation: Production, metabolism, and signaling mechanisms of malondialdehyde and 4-hydroxy-2-nonenal. *Oxid. Med. Cell. Longev.* **2014**, *2014*, 360438. [CrossRef] [PubMed]

34. Rodriguez, A.E.; Bogart, C.; Gilbert, C.M.; McCullers, J.A.; Smith, A.M., Kanneganti, T.D., Lupfer, C.R. Enhanced IL-1β production is mediated by a TLR2-MYD88-NLRP3 signaling axis during coinfection with influenza A virus and Streptococcus pneumoniae. *PLoS ONE* **2019**, *14*, e0212236. [CrossRef] [PubMed]

35. Ramteke, P.; Deb, A.; Shepal, V.; Bhat, M.K. Hyperglycemia Associated Metabolic and Molecular Alterations in Cancer Risk, Progression, Treatment, and Mortality. *Cancers (Basel)* **2019**, *11*, 1402. [CrossRef]

36. Hartog, J.W.; Voors, A.A.; Bakker, S.J.; Smit, A.J.; van Veldhuisen, D.J. Advanced glycation end-products (AGEs) and heart failure: Pathophysiology and clinical implications. *Eur. J. Heart Fail.* **2007**, *9*, 1146–1155. [CrossRef]

37. Kenny, H.C.; Abel, E.D. Heart Failure in Type 2 Diabetes Mellitus. *Circ. Res.* **2019**, *124*, 121–141. [CrossRef] [PubMed]

38. Kaludercic, N.; Di Lisa, F. Mitochondrial ROS Formation in the Pathogenesis of Diabetic Cardiomyopathy. *Front. Cardiovasc. Med.* **2020**, *7*, 12. [CrossRef] [PubMed]

39. Duan, X.; Chan, C.; Han, W.; Guo, N.; Weichselbaum, R.R.; Lin, W. Immunostimulatorynanomedicines synergize with checkpoint blockade immunotherapy to eradicate colorectal tumors. *Nat. Commun.* **2019**, *10*, 1899. [CrossRef]

40. Busik, J.V.; Mohr, S.; Grant, M.B. Hyperglycemia-induced reactive oxygen species toxicity to endothelial cells is dependent on paracrine mediators. *Diabetes* **2008**, *57*, 1952–1965. [CrossRef]

41. Mantovani, A.; Barajon, I.; Garlanda, C. IL-1 and IL-1 regulatory pathways in cancer progression and therapy. *Immunol. Rev.* **2018**, *281*, 57–61. [CrossRef] [PubMed]

42. Ridker, P.M.; Everett, B.M.; Thuren, T.; MacFadyen, J.G.; Chang, W.H.; Ballantyne, C.; Fonseca, F.; Nicolau, J.; Koenig, W.; Anker, S.D.; et al. Antiinflammatory Therapy with Canakinumab for Atherosclerotic Disease. *N. Engl. J. Med.* **2017**, *377*, 1119–1131. [CrossRef] [PubMed]

43. Schenk, K.M.; Reuss, J.E.; Choquette, K.; Spira, A.I. A review of canakinumab and its therapeutic potential for non-small cell lung cancer. *Anticancer Drugs* **2019**, *30*, 879–885. [CrossRef] [PubMed]

44. Brandhorst, S.; Longo, V.D. Fasting and Caloric Restriction in Cancer Prevention and Treatment. *Recent Results Cancer Res.* **2016**, *207*, 241–266. [CrossRef] [PubMed]

45. Afzal, M.Z.; Mercado, R.R.; Shirai, K. Efficacy of metformin in combination with immune checkpoint inhibitors (anti-PD-1/anti-CTLA-4) in metastatic malignant melanoma. *J. Immunother. Cancer* **2018**, *6*, 64. [CrossRef]

46. Nivolumab and Metformin Hydrochloride in Treating Patients with Stage III-IV Non-small Cell Lung Cancer That Cannot Be Removed by Surgery. Available online: https://clinicaltrials.gov/ct2/show/NCT03048500 (accessed on 9 February 2017).

47. Malaguarnera, L. Influence of Resveratrol on the Immune Response. *Nutrients* **2019**, *11*, 946. [CrossRef]

48. Honari, M.; Shafabakhsh, R.; Reiter, R.J.; Mirzaei, H.; Asemi, Z. Resveratrol is a promising agent for colorectal cancer prevention and treatment: Focus on molecular mechanisms. *Cancer Cell Int.* **2019**, *19*, 180. [CrossRef]

49. Berretta, M.; Bignucolo, A.; Di Francia, R.; Comello, F.; Facchini, G.; Ceccarelli, M.; Iaffaioli, R.V.; Quagliariello, V.; Maurea, N. Resveratrol in Cancer Patients: From Bench to Bedside. *Int. J. Mol. Sci.* **2020**, *21*, 2945. [CrossRef]

50. Barbieri, A.; Quagliariello, V.; Del Vecchio, V.; Falco, M.; Luciano, A.; Amruthraj, N.J.; Nasti, G.; Ottaiano, A.; Berretta, M.; Iaffaioli, R.V.; et al. Anticancer and Anti-Inflammatory Properties of Ganodermalucidum Extract Effects on Melanoma and Triple-Negative Breast Cancer Treatment. *Nutrients* **2017**, *9*, 210. [CrossRef]

51. Berretta, M.; Della Pepa, C.; Tralongo, P.; Fulvi, A.; Martellotta, F.; Lleshi, A.; Nasti, G.; Fisichella, R.; Romano, C.; De Divitiis, C.; et al. Use of Complementary and Alternative Medicine (CAM) in cancer patients: An Italian multicenter survey. *Oncotarget* **2017**, *8*, 24401–24414. [CrossRef]

52. Lestuzzi, C.; Bearz, A.; Lafaras, C.; Gralec, R.; Cervesato, E.; Tomkowski, W.; DeBiasio, M.; Viel, E.; Bishiniotis, T.; Platogiannis, D.N.; et al. Neoplastic pericardial disease in lung cancer: Impact on outcomes of different treatment strategies. A multicenter study. *Lung Cancer* **2011**, *72*, 340–347. [CrossRef] [PubMed]

53. Terme, M.; Ullrich, E.; Aymeric, L.; Meinhardt, K.; Desbois, M.; Delahaye, N.; Viaud, S.; Ryffel, B.; Yagita, H.; Kaplanski, G.; et al. IL-18 induces PD-1-dependent immunosuppression in cancer. *Cancer Res.* **2011**, *71*, 5393–5399. [CrossRef] [PubMed]

54. Kaplanov, I.; Carmi, Y.; Kornetsky, R.; Shemesh, A.; Shurin, G.V.; Shurin, M.R.; Dinarello, C.A.; Voronov, E.; Apte, R.N. Blocking IL-1β reverses the immunosuppression in mouse breast cancer and synergizes with anti-PD-1 for tumor abrogation. *Proc. Natl. Acad. Sci. USA* **2019**, *116*, 1361–1369. [CrossRef]

55. Zhao, X.; Zhang, C.; Hua, M.; Wang, R.; Zhong, C.; Yu, J.; Han, F.; He, N.; Zhao, Y.; Liu, G.; et al. NLRP3 inflammasome activation plays a carcinogenic role through effector cytokine IL-18 in lymphoma. *Oncotarget* **2017**, *8*, 108571–108583. [CrossRef] [PubMed]

56. Lee, H.E.; Lee, J.Y.; Yang, G.; Kang, H.C.; Cho, Y.Y.; Lee, H.S.; Lee, J.Y. Inhibition of NLRP3 inflammasome in tumor microenvironment leads to suppression of metastatic potential of cancer cells. *Sci. Rep.* **2019**, *9*, 12277. [CrossRef]

57. Zhang, L.; Li, H.; Zang, Y.; Wang, F. NLRP3 inflammasome inactivation driven by miR-223-3p reduces tumor growth and increases anticancer immunity in breast cancer. *Mol. Med. Rep.* **2019**, *19*, 2180–2188. [CrossRef]

58. Qu, D.; Liu, J.; Lau, C.W.; Huang, Y. IL-6 in diabetes and cardiovascular complications. *Br. J. Pharmacol.* **2014**, *171*, 3595–3603. [CrossRef]

59. Waldner, M.J.; Foersch, S.; Neurath, M.F. Interleukin-6–a key regulator of colorectal cancer development. *Int. J. Biol. Sci.* **2012**, *8*, 1248–1253. [CrossRef]

60. Li, S.; Tian, J.; Zhang, H.; Zhou, S.; Wang, X.; Zhang, L.; Yang, J.; Zhang, Z.; Ji, Z. Down-regulating IL-6/GP130 targets improved the anti-tumor effects of 5-fluorouracil in colon cancer. *Apoptosis* **2018**, *23*, 356–374. [CrossRef]

61. Babini, G.; Morini, J.; Barbieri, S.; Baiocco, G.; Ivaldi, G.B.; Liotta, M.; Tabarelli de Fatis, P.; Ottolenghi, A. A Co-culture Method to Investigate the Crosstalk Between X-ray Irradiated Caco-2 Cells and PBMC. *J. Vis. Exp.* **2018**, *131*, e56908. [CrossRef]

62. Chang, D.H.; Rutledge, J.R.; Patel, A.A.; Heerdt, B.G.; Augenlicht, L.H.; Korst, R.J. The effect of lung cancer on cytokine expression in peripheral blood mononuclear cells. *PLoS ONE* **2013**, *8*, e64456. [CrossRef] [PubMed]

63. Passariello, M.; Camorani, S.; Vetrei, C.; Ricci, S.; Cerchia, L.; De Lorenzo, C. Ipilimumab and Its Derived EGFR Aptamer-Based Conjugate Induce Efficient NK Cell Activation against Cancer Cells. *Cancers (Basel)* **2020**, *12*, 331. [CrossRef] [PubMed]

64. Zheng, Y.; Fang, Y.C.; Li, J. PD-L1 expression levels on tumor cells affect their immunosuppressive activity. *Oncol. Lett.* **2019**, *18*, 5399–5407. [CrossRef] [PubMed]

65. Pandha, H.; Rigg, A.; John, J.; Lemoine, N. Loss of expression of antigen-presenting molecules in human pancreatic cancer and pancreatic cancer cell lines. *Clin. Exp. Immunol.* **2007**, *148*, 127–135. [CrossRef]

66. Kaklamanis, L.; Leek, R.; Koukourakis, M.; Gatter, K.C.; Harris, A.L. Loss of transporter in antigen processing 1 transport protein and major histocompatibility complex class I molecules in metastatic versus primary breast cancer. *Cancer Res.* **1995**, *55*, 5191–5194.

67. Gudmundsdóttir, I.; GunnlaugurJónasson, J.; Sigurdsson, H.; Olafsdóttir, K.; Tryggvadóttir, L.; Ogmundsdóttir, H.M. Altered expression of HLA class I antigens in breast cancer: Association with prognosis. *Int. J. Cancer* **2000**, *89*, 500–505. [CrossRef]

68. Courau, T.; Bonnereau, J.; Chicoteau, J.; Bottois, H.; Remark, R.; Assante Miranda, L.; Toubert, A.; Blery, M.; Aparicio, T.; Allez, M.; et al. Cocultures of human colorectal tumor spheroids with immune cells reveal the therapeutic potential of MICA/B and NKG2A targeting for cancer treatment. *J. Immunother. Cancer* **2019**, *7*, 74. [CrossRef]

69. Salem, J.E.; Manouchehri, A.; Moey, M. Cardiovascular toxicities associated with immune checkpoint inhibitors: An observational, retrospective, pharmacovigilance study. *Lancet Oncol.* **2018**, *19*, 1579–1589. [CrossRef]

70. Tomoaia, R.; Beyer, R.S.; Pop, D.; Minciună, I.A.; Dădârlat-Pop, A. Fatal association of fulminant myocarditis and rhabdomyolysis after immune checkpoint blockade. *Eur. J. Cancer* **2020**, *132*, 224–227. [CrossRef]

71. Martin Huertas, R.; Saavedra Serrano, C.; Perna, C.; Ferrer Gómez, A.; Alonso Gordoa, T. Cardiac toxicity of immune-checkpoint inhibitors: A clinical case of nivolumab-induced myocarditis and review of the evidence and new challenges. *Cancer Manag. Res.* **2019**, *11*, 4541–4548. [CrossRef]

72. Kitano, S.; Tsuji, T.; Liu, C.; Hirschhorn-Cymerman, D.; Kyi, C.; Mu, Z.; Allison, J.P.; Gnjatic, S.; Yuan, J.D.; Wolchok, J.D. Enhancement of tumor-reactive cytotoxic CD4+ T cell responses after ipilimumab treatment in four advanced melanoma patients. *Cancer Immunol. Res.* **2013**, *1*, 235–244. [CrossRef] [PubMed]

73. Lan, G.; Li, J.; Wen, Q.; Lin, L.; Chen, L.; Chen, L.; Chen, X. Cytotoxic T lymphocyte associated antigen 4 expression predicts poor prognosis in luminal B HER2-negative breast cancer. *Oncol. Lett.* **2018**, *15*, 5093–5097. [CrossRef] [PubMed]

74. Wang, J.; Shen, X.; Liu, J.; Chen, W.; Wu, F.; Wu, W.; Meng, Z.; Zhu, M.; Miao, C. High glucose mediates NLRP3 inflammasome activation via upregulation of ELF3 expression. *Cell Death Dis.* **2020**, *11*, 383. [CrossRef] [PubMed]

75. Khaled, Y.S.; Ammori, B.J.; Elkord, E. Increased levels of granulocytic myeloid-derived suppressor cells in peripheral blood and tumour tissue of pancreatic cancer patients. *J. Immunol. Res.* **2014**, *2014*, 879897. [CrossRef]

76. Theivanthiran, B.; Evans, K.S.; DeVito, N.C.; Plebanek, M.; Sturdivant, M.; Wachsmuth, L.P.; Salama, A.K.; Kang, Y.; Hsu, D.; Balko, J.M.; et al. A tumor-intrinsic PD-L1/NLRP3 inflammasome signaling pathway drives resistance to anti-PD-1 immunotherapy. *J. Clin. Investig.* **2020**, *130*, 2570–2586. [CrossRef]

77. Haudek-Prinz, V.J.; Klepeisz, P.; Slany, A.; Griss, J.; Meshcheryakova, A.; Paulitschke, V.; Mitulovic, G.; Stöckl, J.; Gerner, C. Proteome signatures of inflammatory activated primary human peripheral blood mononuclear cells. *J. Proteom.* **2012**, *76*, 150–162. [CrossRef]

78. Quagliariello, V.; Passariello, M.; Coppola, C.; Rea, D.; Barbieri, A.; Scherillo, M.; Monti, M.G.; Iaffaioli, R.V.; De Laurentiis, M.; Ascierto, P.A.; et al. Cardiotoxicity and pro-inflammatoryeffects of the immune checkpoint inhibitorPembrolizumabassociated to Trastuzumab. *Int. J. Cardiol.* **2019**, *292*, 171–179. [CrossRef]

79. Quagliariello, V.; Vecchione, R.; Coppola, C.; Di Cicco, C.; De Capua, A.; Piscopo, G.; Paciello, R.; Narciso, V.; Formisano, C.; Taglialatela-Scafati, O.; et al. Cardioprotective Effects of Nanoemulsions Loaded with Anti-Inflammatory Nutraceuticals against Doxorubicin-Induced Cardiotoxicity. *Nutrients* **2018**, *10*, 1304. [CrossRef]

80. Passariello, M.; D'Alise, A.M.; Esposito, A.; Vetrei, C.; Froechlich, G.; Scarselli, E.; Nicosia, A.; De Lorenzo, C. Novel Human Anti-PD-L1 mAbs Inhibit Immune-Independent Tumor Cell Growth and PD-L1 Associated Intracellular Signalling. *Sci. Rep.* **2019**, *9*, 13125. [CrossRef]

81. Theodoro, T.R.; Matos, L.L.; Cavalheiro, R.P.; Justo, G.Z.; Nader, H.B.; Pinhal, M.A.S. Crosstalk between tumor cells and lymphocytes modulates heparanase expression. *J. Transl. Med.* **2019**, *17*, 103. [CrossRef]

82. Laurent, S.; Queirolo, P.; Boero, S.; Salvi, S.; Piccioli, P.; Boccardo, S.; Minghelli, S.; Morabito, A.; Fontana, V.; Pietra, G.; et al. The engagement of CTLA-4 on primary melanoma cell lines induces antibody-dependent cellular cytotoxicity and TNF-α production. *J. Transl. Med.* **2013**, *11*, 108. [CrossRef] [PubMed]

83. Zhang, S.C.; Hu, Z.Q.; Long, J.H.; Zhu, G.M.; Wang, Y.; Jia, Y.; Zhou, J.; Ouyang, Y.; Zeng, Z. Clinical Implications of Tumor-Infiltrating Immune Cells in Breast Cancer. *J. Cancer* **2019**, *10*, 6175–6184. [CrossRef] [PubMed]

84. Stanton, S.E.; Disis, M.L. Clinical significance of tumor-infiltrating lymphocytes in breast cancer. *J. Immunother. Cancer* **2016**, *4*, 59. [CrossRef]

85. Tivol, E.A.; Borriello, F.; Schweitzer, A.N.; Lynch, W.P.; Bluestone, J.A.; Sharpe, A.H. Loss of CTLA-4 leads to massive lymphoproliferation and fatal multiorgan tissue destruction, revealing a critical negative regulatory role of CTLA-4. *Immunity* **1995**, *3*, 541–547. [CrossRef]

86. Johnson, D.B.; Balko, J.M.; Compton, M.L.; Chalkias, S.; Gorham, J.; Xu, Y.; Hicks, M.; Puzanov, I.; Alexander, M.R.; Bloomer, T.L.; et al. Fulminant myocarditis with combination immune checkpoint blockade. *N. Engl. J. Med.* **2016**, *375*, 1749–1755. [CrossRef] [PubMed]

87. Andreadou, I.; Efentakis, P.; Balafas, E.; Togliatto, G.; Davos, C.H.; Varela, A.; Dimitriou, C.A.; Nikolaou, P.E.; Maratou, E.; Lambadiari, V.; et al. Empagliflozin Limits Myocardial Infarction in Vivo and Cell Death in Vitro: Role of STAT3, Mitochondria, and Redox Aspects. *Front. Physiol.* **2017**, *8*, 1077. [CrossRef]

88. Lebreton, F.; Berishvili, E.; Parnaud, G.; Rouget, C.; Bosco, D.; Berney, T.; Lavallard, V. NLRP3 inflammasome is expressed and regulated in human islets. *Cell Death Dis.* **2018**, *9*, 726. [CrossRef]

89. Yamamoto, T.; Tsutsumi, N.; Tochio, H.; Ohnishi, H.; Kubota, K.; Kato, Z.; Shirakawa, M.; Kondo, N. Functional assessment of the mutational effects of human IRAK4 and MyD88 genes. *Mol. Immunol.* **2014**, *58*, 66–76. [CrossRef]

90. Marchetti, C.; Swartzwelter, B.; Gamboni, F.; Neff, C.P.; Richter, K.; Azam, T.; Carta, S.; Tengesdal, I.; Nemkov, T.; D'Alessandro, A.; et al. OLT1177, a β-sulfonyl nitrile compound, safe in humans, inhibits the NLRP3 inflammasome and reverses the metabolic cost of inflammation. *Proc. Natl. Acad. Sci. USA* **2018**, *115*, E1530–E1539. [CrossRef]

91. Pitt, J.M.; Vétizou, M.; Daillère, R.; Roberti, M.P.; Yamazaki, T.; Routy, B.; Lepage, P.; Boneca, I.G.; Chamaillard, M.; Kroemer, G.; et al. Resistance Mechanisms to Immune-Checkpoint Blockade in Cancer: Tumor-Intrinsic and -Extrinsic Factors. *Immunity* **2016**, *44*, 1255–1269. [CrossRef]

92. Afonso, J.; Santos, L.L.; Longatto-Filho, A.; Baltazar, F. Competitive glucose metabolism as a target to boost bladder cancer immunotherapy. *Nat. Rev. Urol.* **2020**, *17*, 77–106. [CrossRef] [PubMed]

93. Toldo, S.; Abbate, A. The NLRP3 inflammasome in acute myocardial infarction. *Nat. Rev. Cardiol.* **2018**, *15*, 203–214. [CrossRef] [PubMed]

Publisher's Note: MDPI stays neutral with regard to jurisdictional claims in published maps and institutional affiliations.

International Journal of
Molecular Sciences

Review

NAA10 as a New Prognostic Marker for Cancer Progression

Sun Myung Kim [1,†], **Eunyoung Ha** [2,†], **Jinyoung Kim** [3], **Chiheum Cho** [1], **So-Jin Shin** [1,*] **and Ji Hae Seo** [2,*]

[1] Department of Gynecology and Obstetrics and Institute for Cancer Research, School of Medicine, Keimyung University, Daegu 42601, Korea; sun7@snu.ac.kr (S.M.K.); c0035@dsmc.or.kr (C.C.)

[2] Department of Biochemistry, School of Medicine, Keimyung University, Daegu 42601, Korea; eyha@dsmc.or.kr

[3] Department of Internal Medicine, School of Medicine, Keimyung University, Daegu 42601, Korea; takgu@dsmc.or.kr

* Correspondence: sjshinhope2014@kmu.ac.kr (S.-J.S.); seojh@kmu.ac.kr (J.H.S.); Tel.: +82-53-258-7807 (S.-J.S.); +82-53-258-7436 (J.H.S.)

† These authors contributed equally to this work.

Received: 5 October 2020; Accepted: 26 October 2020; Published: 28 October 2020

Abstract: N-α-acetyltransferase 10 (NAA10) is an acetyltransferase that acetylates both N-terminal amino acid and internal lysine residues of proteins. NAA10 is a crucial player to regulate cell proliferation, migration, differentiation, apoptosis, and autophagy. Recently, mounting evidence presented the overexpression of NAA10 in various types of cancer, including liver, bone, lung, breast, colon, and prostate cancers, and demonstrated a correlation of overexpressed NAA10 with vascular invasion and metastasis, thereby affecting overall survival rates of cancer patients and recurrence of diseases. This evidence all points NAA10 toward a promising biomarker for cancer prognosis. Here we summarize the current knowledge regarding the biological functions of NAA10 in cancer progression and provide the potential usage of NAA10 as a prognostic marker for cancer progression.

Keywords: acetyltransferase; biomarker; cancer prognosis; NAA10

---

1. Introduction

Cancer is currently the second leading cause of death worldwide following heart disease [1]. With the rapidly advancing biomedical technologies, discovery of biomarkers for early detection and progress of cancer has been the focus of intense research [2].

N-terminal acetyltransferase (NAT) is an acetyltransferase that targets the N-terminal α-amino group of nascent proteins. NAA10, a catalytic subunit of NatA, also functions as lysine acetyltransferase (KAT) that acetylates internal lysine residues of proteins [3]. Accumulating evidence demonstrated that NAA10, both via NAT and KAT activities, plays key roles in regulating tumorigenesis processes such as cellular proliferation, apoptosis, migration, and autophagy. Evidence also demonstrated that NAA10 is highly upregulated in various malignancies, including breast, bone, colorectal, liver, lung, and prostate cancers, and that expression level of NAA10 is correlated with the cancer progression, an implication for possibility of NAA10 as a cancer biomarker [4–9]. Based on above referenced studies, this review discusses the possibility of NAA10 as a prognostic cancer biomarker and describes the biological functions of NAA10 in cancer progression.

2. NATs

Most proteins undergo one or more types of modifications to form stable structure and/or maintain catalytic activities. More than 200 different types of protein modification, ranging from small chemical modifications—acetylation and phosphorylation—to small molecule bindings—ubiquitination and sumoylation—occur in the cell [10,11]. N-terminal acetylation (Nt-acetylation) catalyzed by NAT, is the attachment of an acetyl group from acetyl-CoA to the α-amino group of N-terminal residues of newly synthesized polypeptides (Figure 1A). It is a representative process of co-translational protein modifications in eukaryotes and affects more than 80% of all human proteins [12]. Nt-acetylation of proteins regulates various cellular events, including protein–protein interaction, subcellular localization, aggregation and folding, and protein turnover [13–16]. Along with Nt-acetylation, acetylation of the ε-amino group of an internal lysine residue of protein, mediated by KATs, also frequently occurs in the cell (Figure 1B) [17].

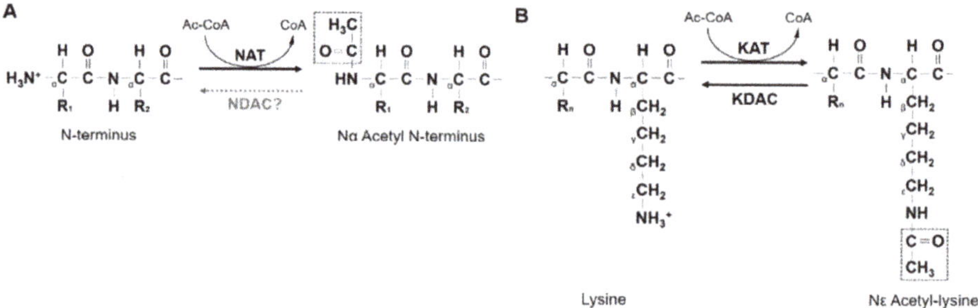

Figure 1. N-terminal and internal lysine acetylation by NAA10. (**A**) NAA10 acetylates the α-amino group of N-terminal residue of new peptides. An acetyl group (dotted rectangle) is transferred from acetyl-CoA to a free α-amino group at the N-terminal. (**B**) NAA10 also acetylates the ε-amino group of an internal lysine residue of the protein. The acetylated lysine can be deacetylated by the lysine deacetylase (KDAC), whereas N-terminal deacetylases (NDACs) have not been reported yet. NAT, N-terminal acetyltransferase; KAT, lysine acetyltransferase.

To date, NAT family in eukaryotes comprise eight isoforms, from NatA to NatH. NatA, B, C, and E are composed of a unique catalytic subunit and auxiliary subunits and the others have one catalytic subunit. Subcellular localization of NATs varies from cytosolic and ribosome-associated (NatA–NatE) areas to Golgi membrane (NatF), organelle lumen (NatG), and cytosolic but non-ribosomal (NatH) area [3]. The substrate specificities of NAT complexes toward different proteins are determined by the identity of the first two amino acids. NatA, a major NAT that acetylates about 40% of the human proteome, is composed of a catalytic subunit NAA10 and an auxiliary subunit NAA15. NatA co-translationally acetylates the N-terminus of small amino acids (Ala, Cys, Ser, Gly, Thr, and Val) that are exposed after methionine cleavage by a methionine aminopeptidase [3,18].

3. NAA10

NAA10 is a catalytic subunit of the NatA complex that acetylates the N-terminus of proteins following aminopeptidase mediated methionine cleavage [19]. NAA10 is a human orthologous gene of arrest-defective 1 (ARD1), which was first identified in *Saccharomyces cerevisiae* by Whiteway and Szostak in 1985 [20]. In humans, NAA10 is located on chromosome Xq28 and composed of eight exons, which is highly conserved across organisms from yeasts to mammals [21].

NAA10 acetylates the ε-amino group of lysine residues of the substrate proteins. The substrates of NAA10 include androgen receptor (AR), Runt-related transcription factor 2 (Runx2), heat shock protein 70 (Hsp70), and phosphoglycerate kinase 1 (PGK1) [22–25]. DePaolo et al. reported the function of NAA10 in the tumorigenesis of prostate cancer. They demonstrate that NAA10 acetylates AR at Lys618, an event that dissociates AR from AR-HSP90 complex and translocates AR into nucleus where AR activates its target gene expressions and stimulates the androgen-dependent tumorigenesis [22]. NAA10 also regulates bone formation. Yoon et al. reported that NAA10 acetylates Lys225 residue of

Runx2, a modification that inhibits Runx2-mediated gene transcription and regulates the differentiation of osteoblast differentiation in the bone [23]. More recently, Seo et al. revealed a role of NAA10 in the stress response. They unraveled that NAA10 acetylates the Lys77 residue of Hsp70 in response to cellular stress and this modification is attributable to the maintenance of protein homeostasis and cell survival under the stress conditions [24]. Based on the evidence reported for decades, it has been established that NAA10 plays important roles to regulate cell proliferation, differentiation, and survival by acetylating its target proteins.

With regard to the molecular structure, NAA10 consists of 235 amino acids and N-terminal region of NAA10 is critical to form the NatA complex with NAA15 [26]. Acetyltransferase domain is located between amino acids 45 and 130, containing an acetyl-CoA binding site (Figure 2). Intriguingly, NAA10 is found both in the cytosol and in the nucleus, whereas NAA15 is localized only in the cytosol, implying a unique function of NAA10 in the nucleus independent of NAA15 [26]. Park et al. showed that NAA10 has a nuclear localization signal (NLS) in ATD and that NLS-deleted NAA10, which cannot enter the nucleus, resulted in cell morphological changes and growth impairment, demonstrating a critical role of NAA10 translocation into nucleus in cell cycle progression [27]. On the basis of its KAT activity, NAA10 could possibly be involved in epigenetic regulation of gene expression. Another NAT family, including NAA20, NAA30, NAA40, and NAA50, are also observed both in the cytosol and in the nucleus, however, their translocations or cellular functions in nucleus have not yet been investigated [28].

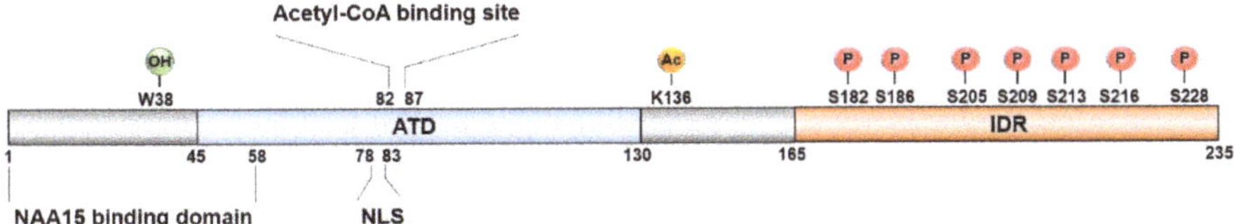

Figure 2. Domain structure of NAA10. The human NAA10 protein consists of 235 amino acids. Acetyltransferase domain (ATD) is located between amino acids 45 and 130, containing acetyl-CoA binding site and nuclear localization signal (NLS). N-terminal region of NAA10 is critical for the binding with NAA15 and C-terminal region is predicted as an intrinsically disordered region (IDR). NAA10 has several post-translational modification sites for phosphorylation, acetylation, and hydroxylation. Ac, acetylation; OH, hydroxylation; P, phosphorylation.

In contrast to N-terminal region of NAA10, C-terminal region is predicted as a highly intrinsically disordered region (IDR) which lacks a fixed three-dimensional structure, allowing NAA10 to interact with different proteins with different consequences, implying that structural and functional properties of NAA10 might be decided by diverse protein modifications and formation of protein complexes, which is essential to conduct a proper function in a specific cell type under a certain condition [29].

Indeed, previous studies reported several post-translational modifications of NAA10 that occur under the specific condition and regulate the enzymatic activity and cellular function of NAA10. Autoacetylation of NAA10 at K136 residue, which is essential for enzymatic activity of NAA10, is rapidly stimulated by anticancer drug treatment, leading to the activation of cellular stress response and the protection of cancer cells against cell death [30]. Under glutamine deprivation or hypoxic condition, mTOR-mediated NAA10 phosphorylation at S228 residue is downregulated, resulting in the activation of an acetyltransferase activity of NAA10 and the promotion of protective autophagy processes in cancer cells [25,31]. Phosphorylation also occurs at S209 residue of NAA10 by IKKβ, which induces proteosomal degradation of NAA10 [32]. Under normoxic condition, NAA10 is hydroxylated by factor inhibiting HIF (FIH) at W38 residue, leading to opening the gate at the catalytic pocket of NAA10 that allows the lysine acetylation of substrate protein [33]. In addition to these sites, bioinformatics data (Uniprot: P41227) uncovered that NAA10 has at least additional five post-translational modification sites in C-terminal region of NAA10, including S182, S186, S205, S213, and S216 (Figure 2). Furthermore,

NAA10 is also observed to be cleaved after anticancer drug treatment, although the biological meaning of this event needs to be elucidated [26].

Previous studies also support the importance of interacting protein of NAA10 for its functional properties. In vitro, purified NAA10 recombinant is prone to aggregate then easily loses its catalytic activity, suggesting that stability and enzymatic activity of NAA10 protein rely on its interaction with other proteins [34]. Crystal structure analysis of NAA10 revealed that binding of NAA10 to NAA15 induces a conformational change and increases the NAT activity of NatA complex [35]. Enzymatic and physical properties of NatA complex are also regulated by the interaction with an intrinsically disordered Huntingtin yeast two-hybrid protein K (HYPK) [36]. All these results suggest that the physiological roles of NAA10 could be dynamically and transiently regulated by protein modifications and interactions over time and space, which should not be overlooked to understand the physiological roles of NAA10 in cancer.

4. Expression of NAA10 in Cancer

N-terminal acetylation occurs co-translationally over 80% of the human proteome. Due to its essential role in protein synthesis, NAA10 is expressed in a wide range of cell types. N-terminal acetylation of newly synthesized proteins might be especially important in rapidly dividing cells. On the basis of this reason, it is reasonable to expect that the role of NAA10 is critical for cancer cells. Accordingly, the expressions of NAA10 are highly upregulated in various types of cancer tissues compared to its adjutant normal tissues.

4.1. Breast Cancer (BCa)

NAA10 has been implicated in the oncogenesis of BCa. BCa-derived tissues exhibited elevated expressions of NAA10. Wang et al. analyzed 356 clinical breast specimens and confirmed that the expression levels of NAA10 in cancerous tissues were upregulated compared with those in non-cancerous tissues [4]. They also demonstrated that higher expression level of NAA10 is correlated with the degree of cancer invasiveness and metastasis, an implication that NAA10 may be a potential prognostic biomarker for monitoring the progress of BCa. Of note, contrary to Wang et al., Kuo et al. reported a possible function of NAA10 as a tumor suppressor in BCa. They showed that high expression of NAA10 was associated with better clinical outcomes for patients with BCa [37]. This issue will be discussed further later in this review.

4.2. Lung Cancer (LCa)

Overexpression of NAA10 in LCa has been reported. Lee et al. reported upregulated expression of NAA10 in LCa and its correlation with the cancer progression [8]. They also showed the oncogenic effect of NAA10 in vitro. They proved that NAA10 interacts with DNA methyltransferase 1 (DNMT1) thereby silencing E-cadherin, a tumor suppressor gene. Hua et al. reported a contradicting result that showed downregulated expression of NAA10 in LCa, proposing tumor-suppressive effect of NAA10 [38].

4.3. Hepatocellular Carcinoma (HCC)

HCC, the most frequent representative malignancy of the liver, is the second leading cause of cancer-related deaths in East Asia and the third in western countries [39]. Similar to many other malignancies, the overexpression of NAA10 has been observed in HCC. Shim et al. investigated the role and clinical involvement of NAA10 in HCC development [7]. They measured intratumoral NAA10 mRNA levels in patient-derived HCC tissues and found that the high transcription level of NAA10 mRNA was closely related to HCC progression. Lee et al. added the same finding by showing proportionally increased levels of NAA10 from low grade dysplasia to HCC and a correlation between NAA10 expression and HCC progression [19].

4.4. Colorectal Cancer (CRC)

CRC, a malignancy of the inner lining of the colon or rectum, is the third leading cause of cancer-related mortality worldwide [40]. Similar to the results above, increased expressions of NAA10 have been observed in CRC. Ren et al. showed increased mRNA and protein expressions of NAA10 in CRC tissues [41]. Jiang et al. and Yang et al. reported high expression levels of NAA10 in patients with CRC. These results suggest the potential role of NAA10 as a prognostic biomarker for CRC [6,42].

4.5. Osteosarcoma

NAA10 is known to take part in embryogenesis and regulate bone formation. Consequently, its dysregulation causes severe developmental defects and bone cancer [43,44]. A recent report showed an association of the higher transcription level of NAA10 with several types of bone cancer [5].

4.6. Oral Squamous Cell Carcinoma (OSCC)

OSCC is a major malignancy in the oral capacity, comprising approximately 90% of all oral neoplasms [45]. Few studies reported the role of NAA10 in OSCC. Zeng et al. studied the involvement of NAA10 in the OSCC and found that NAA10 expression levels were highly increased in 98 out of 124 OSCC specimens [46].

4.7. Prostate Cancer (PCa)

PCa, one of the most common malignancies in men, gives rise to the second leading cause of cancer-related deaths worldwide [47]. The progression of PCa is highly dependent on AR signaling, a pathway that is essential for the development and function of the prostate gland [48]. Patients with PCa show significantly elevated expression levels of AR. Additionally, many studies showed that AR is causative in the pathogenesis and the progression of PCa [49]. Of a couple of post-translational modifications that modulate the activities of AR, acetylation activates AR and AR dependent signaling pathway ultimately leading to the development and the progression of PCa [50]. Given the function of NAA10 in the pathogenesis of various cancers, Wang et al. showed that NAA10 expressions were highly increased in PCa [9]. They conducted immunohistochemistry (IHC) on 64 PCa tissues and found substantially higher levels of NAA10 in PCa tissues than in adjacent normal tissues [9]. Furthermore, DePaolo et al. confirmed that NAA10 forms NAA10-Hsp90-AR complex and acetylates Lys618 residue of AR. This acetylation allows AR to translocate into the nucleus and sequentially stimulates the expression of AR target genes required for the progression of PCa [22]. These results suggest the possible usage of NAA10 as a novel biomarker for PCa.

5. NAA10 as a Prognostic Marker

Innumerable studies have demonstrated that the expression level of NAA10 in tumor tissues is closely correlated with disease progression and clinical outcomes of various cancers (Table 1).

Table 1. Clinical outcome in cancer tissues overexpressing NAA10.

Prognosis	Cancer Type	Clinical Outcome	Reference
Overall Survival	Breast cancer	Low	Wang et al., 2011 [4]
		High	Kuo et al., 2010 [37]
		High	Zeng et al., 2014 [51]
	Colon cancer	Low	Jiang et al., 2010 [6]
	HCC	Low	Lee et al., 2018 [19]
	Lung cancer	Low	Lee et al., 2010 [8]
		High	Hua et al., 2011 [38]
	OSCC	High	Zeng et al., 2016 [46]
	Osteosarcoma	Low	Chien et al., 2018 [5]
Invasiveness	HCC	MVI	Shim et al., 2012 [7]
Metastasis	Breast cancer	High	Wang et al., 2011 [4]
		Low	Kuo et al., 2010 [37]
		Low	Zeng et al., 2011 [51]
	Lung cancer	Low	Hua et al., 2011 [38]
	OSCC	Low	Zeng et al., 2016 [46]
	Osteosarcoma	High	Chien et al., 2018 [5]
Recurrence	Breast cancer	High	Wang et al., 2011 [4]
	HCC	High	Lee et al., 2018 [19]
	OSCC	Low	Zeng et al., 2016 [46]

HCC, hepatocellular carcinoma; OSCC, oral squamous carcinoma; MVI, microvascular invasion.

5.1. NAA10 and Cancer Survival

5.1.1. BCa

NAA10 appeared to be a promising biomarker for assessment of prognosis after postoperative chemotherapy. Wang et al. showed that NAA10 was positive up to 50/82 (61.0%) at primary diagnosis in breast invasive ductal carcinomas (IDC) specimens. Of 50 patients with positive NAA10, 29 patients recurred and underwent a second surgery [4]. From these observations, they suggested that the elevated NAA10 protein is associated with poor prognosis in BCa patients, and thus that NAA10 may be utilized for the prediction of prognosis. Caution should be taken since the data are controversial. Kuo et al. found higher levels of NAA10 transcript in BCa tissues from longer relapse-free surviving patients compared with those from shorter relapse-free surviving patients [37]. In addition, Zeng et al. showed that highly expressed NAA10 was associated with better survival rate in BCa patients [51]. These reports suggest that NAA10 expression in BCa correlates positively with cancer survival.

5.1.2. LCa

Lee et al. showed that NAA10 overexpression is associated with poor survival of LCa [8]. They performed IHC of NAA10 expression from 90 patients with lung adenocarcinoma and found that 48 patients with high levels of NAA10 showed poorer survival rates compared to 42 patients with low levels of NAA10. These results indicate that NAA10 was overexpressed in more than half of 90 LCa tissues and that NAA10 may play a role in the progression of LCa. However, as in the case of BCa, the other study has presented an opposite result regarding the oncogenic role of NAA10. Hua et al. reported that the higher expression of NAA10 is correlated with better survival of female patients with adenocarcinoma, suggesting the suppressive role of NAA10 in LCa progression [38].

5.1.3. HCC

HCC is characterized by high mortality and poor survival rate. Lee et al. showed that NAA10 is highly upregulated in HCC tissues and that NAA10 overexpression was associated with microvascular invasion, poor tumor differentiation, and poor survival rate [19].

5.1.4. CRC

Jiang et al. demonstrated that, of 106 patients with high expression of NAA10, 74 (69.8%) died of cancer-related causes, as opposed to 7 (41.1%) out of 17 patients with low NAA10 [6]. They revealed a significantly shortened overall survival in patients with NAA10 positive expressions.

5.1.5. Osteosarcoma

Osteosarcoma patients with more than average levels of NAA10 expression showed significantly shorter overall survival, an implication that enhanced NAA10 expression is associated with poor prognosis [5].

5.1.6. OSCC

A study that analyzed the survival of OSCC showed that patients with positive NAA10 expression had a better overall survival rate than those with negative NAA10 expression, a statement that high level of NAA10 in OSCC is correlated with the better prognosis [46].

5.2. Invasion and Metastasis

5.2.1. BCa

As previously mentioned, Wang et al. highlighted the correlation of NAA10 expression with BCa progression [4]. They showed that the level of NAA10 protein in breast carcinoma patients was distinctly related to lymph node metastasis, with 94.0% (47/50) of metastatic tumor showing increased expression, as compared to 6.0% (3/50) of non-metastatic tumors. Conversely, other studies have emphasized that NAA10 expression levels are negatively associated with lymph node metastasis in BCa patients [37,51]. Kuo et al. found that NAA10 transcription levels were higher in patients with fewer lymph node metastases than in those with more lymph node metastases [37]. Zeng et al. also showed that NAA10 levels were higher in patients with fewer lymph node metastases [51].

5.2.2. LCa

There are no reports of whether high levels of NAA10 expression are positively correlated with LCa invasion and metastasis. However, Hua et al. found lower expression of NAA10 in malignancies with lymph node metastasis compared to non-lymph node metastasis [38]. They showed significantly decreased levels of NAA10 in 13 out of 15 lung cancer patients with positive lymph node metastasis compared with those with negative lymph node metastasis, suggesting the suppressive role of NAA10 in the tumor metastasis.

5.2.3. HCC

The high level of intratumoral NAA10 mRNA has been implicated in the process of microvascular invasion in HCC patients. Shim et al. investigated the role and clinical involvement of NAA10 in the development of HCC [7]. They measured intratumoral NAA10 mRNA levels in patient-derived HCC specimens and observed that patients with higher expression level of NAA10 showed more frequent microvascular invasions than patients with lower expression levels of NAA10.

5.2.4. Osteosarcoma

Chien et al. observed higher expression level of NAA10 in osteosarcoma patients with metastasis compared with those without metastasis [5]. They showed that NAA10 interacts with matrix metalloproteinase-2 (MMP-2) via its acetyltransferase domain and in turn promotes the stabilization of MMP-2 protein, leading to the increases in cell invasion and tumor metastasis.

5.2.5. OSCC

NAA10 was proposed as a tumor suppressor in OSCC. Higher expression levels of NAA10 negatively correlated with lymph node metastasis in OSCC [46].

5.3. Recurrence

Not sufficient studies as to the correlation between the expression level of NAA10 and cancer recurrence have been conducted. Wang et al. reported recurrence after postoperative chemotherapy in 29 breast cancer patients with high levels of NAA10 [4]. The expression levels of NAA10 in HCC were positively correlated with tumor recurrence [19]. However, inverse correlations were revealed between NAA10 expression and the recurrence of OSCS [46].

6. NAA10 in Tumorigenesis

As mentioned above, several research groups have demonstrated that the expression levels of NAA10 in cancer tissues are considerably correlated with the progression, metastasis, and the survival. However, the roles of NAA10 in cancer tissues seem to be very complicated because the overexpression of NAA10 is closely associated with poor outcomes in HCC and CRC, but with better outcomes in BCa and OSCC. Mechanistic studies have revealed diverse molecular mechanisms of NAA10 and defined NAA10 as a double-faced player, an oncoprotein, and a tumor suppressor. As mentioned before, NAA10 has an intrinsically disordered structure, suggesting the existence of numerous kinds of protein complexes that are functionally different and specific for space–time context. NAA10 is reported to have 95 kinds of physical interactions in databases (Uniprot: P41227), however, the physiological meaning and regulation of these interactions are mostly unknown. This review introduces a part of them based on the functional analyses by biochemical and cellular approaches (Table 2).

Table 2. The role of NAA10 and its regulatory effects on target proteins in cancer progression.

Role	Target Protein	Function	Effect of NAA10	Activity	Reference
Oncoprotein	β-catenin	Proliferation	Acetylation	Activated	Lim et al., 2016 [6]
	Cdc25A	Proliferation	Acetylation	Activated	Lozada et al., 2016 [52]
	AR	Proliferation	Acetylation	Activated	DePaolo et al., 2016 [22]
	PGK1	Autophagy	Acetylation at K388	Activated	Qian et al., 2017 [25]
	Hsp70	Apoptosis	Acetylation at K77	Activated	Park et al., 2017 [53]
	DNMT1	Migration	Interaction	Activated	Lee et al., 2010 [8]
	AuA	Migration	Acetylation at K75/K125	Activated	Vo et al., 2017 [54]
	MMP-2	Migration	Stabilization	Activated	Chien et al., 2018 [5]
Tumor suppressor	TSC2	Autophagy	Acetylation	Activated	Kuo et al., 2010 [37]
	MLCK	Migration	Acetylation at K608	Inactivated	Shin et al., 2009 [55]
	PIX	Migration	Interaction	Inactivated	Hua et al., 2011 [38]
	STAT5α	Migration	Interaction	Inactivated	Zeng et al., 2014 [51]

6.1. NAA10 as an Oncoprotein

The oncogenic properties of NAA10 have been firmly established. Overexpression of NAA10 is associated with microvascular invasion, lymph node metastasis, and lower survival rates in various malignancies, including BCa, LCa, HCC, and osteosarcoma [17]. These findings surely imply the role of NAA10 as an oncoprotein.

Molecular mechanistic studies have revealed that NAA10-mediated protein acetylation plays a crucial role in regulating cellular events that are significant for cancer development, such as cell cycle progression, cell death, migration, and autophagy. In LCa cells, NAA10 acetylates and activates β-catenin, an event that enhances the expression of cyclin D1 protein and then leads to the uncontrolled cell proliferation [56]. NAA10 acetylates and stabilizes phosphatase Cdc25A [52]. The stabilized Cdc25A dephosphorylates cyclin-Cdk complexes and consequently promotes cell proliferation. Furthermore, NAA10 suppresses the tumor suppressor gene E-cadherin by regulating DNMT1 in LCa cells [8]. NAA10 recruited DNMT1 to the E-cadherin promoter independent of acetyltransferase activity, suggesting acetylation-independent oncogenic potential of NAA10. Recently, Vo et al. showed that NAA10-mediated acetylation of aurora kinase A (AuA) promotes proliferation and migration of BCa cells, indicating the probable role of NAA10 in cancer development [54]. As for apoptosis, Park et al. described cancer cell survival mechanisms mediated by Hsp70 acetylation under stress conditions [53]. NAA10-mediated Hsp70 acetylation inhibits cell death by preventing apoptotic protease-activating factor-1 (Apaf-1) and apoptosis-inducing factor (AIF)-controlled apoptotic processes. For the autophagy process involved in brain tumor progression, Qian et al. showed that NAA10 acetylates Lys388 residue of PGK1 under glutamine deprivation [25]. Acetylated PGK1 binds to and phosphorylates Beclin1 and induces in glioblastomas. These results suggest that NAA10 could be a promising candidate for cancer biomarker.

6.2. NAA10 as a Tumor Suppressor

As previously described, NAA10 can also act as a tumor suppressor in various malignancies, including BCa, LCa, and OSCC [37,38,46]. In those malignancies, higher expression levels of NAA10 are correlated with better clinical outcomes; better survival, smaller tumor volume, and lower rates of lymph node metastasis.

New insights into the molecular mechanisms of how NAA10 reduces tumor proliferation and metastasis have been discovered. Kuo et al. identified a tumor suppressive activity of NAA10 in BCa, in which NAA10 retarded cancer cell growth and stimulated autophagy by inhibiting mTOR signaling [37]. Regarding the cellular motility, NAA10 directly acetylates the Lys608 residue of myosin light chain kinase (MLCK) and inhibits MLCK activity required for cell migration and expansion [55]. Hua et al. have depicted the molecular mechanism by which NAA10 interrupts cancer cell metastasis in LCa [38]. They reported that NAA10 binds to p21-activated kinase-interacting exchange factor (PIX) to interrupt its downstream Rac1/Cdc42 pathway. Consequently, blockade of the Rac1/Cdc42 pathway results in the inhibition of cancer cell metastasis. This process does not appear to be related to the acetyltransferase activity of NAA10. In BCa, NAA10 inhibits cancer cell migration by blocking the signal transducer and activator of transcription 5α (STAT5α), regardless of acetyltransferase activity [51].

Whether NAA10 retains the opposite functions in different cancer types or under distinct conditions requires further study. Additionally, it is worthwhile to investigate the functions of NAA10, be it as an oncoprotein or a tumor suppressor, that could be rendered by cancer specific microenvironment.

7. Other NATs in Cancer

In addition to NAA10, other NATs play roles in cancer progression. NatB, NatC, and NatD have been reported as essential components for cancer cell proliferation and survival. The NatB subunits are overexpressed in HCC, and this upregulation is associated with microscopic vascular invasion [57]. Neri et al. showed that NatB silencing blocks tumor formation and proliferation possibly via NatB-mediated Nt-acetylation of cyclin-dependent kinase 2 and tropomyosin [57]. Overexpression of the NatC catalytic subunit NAA30 increases cancer cell viability [58]. Conversely, knockdown of NAA30 induces p53-dependent apoptosis, disruption of mitochondrial function, and reduction of tumorigenic features in cancer cells [59–61]. NatD catalytic subunit NAA40 is required for the survival of human colon cancer cells and its depletion induces apoptotic cell death [62]. Ju et al. reported that NatD is frequently upregulated in primary LCa and its expression level correlates with enhanced invasiveness and poor clinical outcomes [63]. They indicated that NatD is a crucial epigenetic modulator of cell invasion during LCa progression. Recently, NatH (NAA80)-mediated actin Nt-acetylation has been reported to regulate cytoskeleton assembly and cell motility and to act as a potential inhibitor of cancer cell motility [64].

8. Conclusions and Perspectives

With NAT and KAT activities, NAA10 is a multifunctional protein involved in various cellular activities required for proliferation, differentiation, autophagy, and apoptosis. NAA10 has been found to play crucial roles in tumorigenesis and its upregulated expression has been observed in several cancer tissues, including BCa, LCa, CRC, HCC, PCa, and osteosarcoma [4–9]. NAA10 overexpression in these cancer tissues has been found to be significantly correlated with microvascular invasion, lymph node metastasis, survival rate, and recurrence. These oncogenic properties of NAA10 suggest the potential usage of NAA10 as a biomarker for cancer diagnosis and prognosis. However, contrary to the oncogenic roles of NAA10, some studies have reported that NAA10 acts as a tumor suppressor in LCa, BCa, and OSCC [37,38,46,51]. These reports indicate that NAA10 expression is closely associated with improved clinical outcomes, such as longer survival times, smaller tumor size, and less lymph node metastasis.

These contradictory effects of NAA10 may be due to the different signaling pathways in diverse cancer types or the acetylation of different substrates under the cancer specific microenvironment conditions. NAA10 might appear in various kinds of molecular forms in cancer tissues and its physiological roles should be specified depending on the specific molecular form in certain cell type and condition. Therefore, further studies are necessary to investigate the distinguishable functions of NAA10 depending on the protein interactions and modifications specific for cellular context and metabolic stimulation. For instance, new technology, such as single cell analysis, would be helpful to characterize the different functions of NAA10 over space and time in heterogeneous cancer tissues.

Most of current studies for NAA10 introduced in this review measured the expression level of NAA10 using IHC or mRNA analysis. However, one thing that should not be overlooked is that NAA10 is an enzyme. Analysis for its gene expression or protein amount might not reflect the specific characteristics of NAA10 as an acetyltransferase enzyme. Many previous studies support that NAT and KAT activities of NAA10 are essential for its biological roles and that NAA10 function is stimulated by protein modifications rather than gene expression [65]. Therefore, the development of new methods detecting enzyme activity or different modified forms of NAA10 in cancer tissues could be helpful to provide more accurate information.

Since the NAT and KAT activities of NAA10 stimulate cancer development, identifying the substrates and upstream regulators of NAA10 is also important to elucidate its roles in tumor progression. Several studies have elucidated the molecular mechanisms that regulate the expression or catalytic activity of NAA10. Yang et al. showed that miRNA-342-5p and miR-608 inhibit CRC tumorigenesis by targeting NAA10 mRNA for degradation [42]. In addition, phosphorylation appears to negatively regulate the stability or enzymatic activity of NAA10. Phosphorylation of Ser209 and Ser228 residues decreased the stability and activity of NAA10 [25,32]. In contrast to phosphorylation, acetylation of K136 residue enhance the KAT activity of NAA10, promoting proliferation and survival of cancer cells [24,30]. These findings may provide a clue for discovering inhibitors or activators of NAA10 that can be useful to improve therapeutic treatments for cancer patients.

In this review, we provide evidence that NAA10 can be used as a novel target and a potential biomarker for the development of more advanced cancer therapeutics. If used in combination with other biomarkers, NAA10 could provide more precise and accurate therapeutic modalities for cancer patients. Future studies should focus on clarifying the precise roles of NAA10 in tumorigenesis and identifying a new molecular control system of NAA10. In addition, the role of NAA10 in other cancers not mentioned in this review should be explored.

Author Contributions: S.M.K., E.H., S.-J.S., and J.H.S. conceptualized and designed the study. S.M.K., E.H., and J.H.S. performed the literature analysis and wrote the original manuscript draft. J.K., C.C., and S.-J.S. reviewed and edited the final draft. All authors have read and agreed to the published version of the manuscript.

Funding: This work was supported by the Basic Science Research Program through the National Research Foundation of Korea (NRF) funded by the Ministry of Education (NRF-2016R1A6A1A03011325, NRF- 2020R1I1A3071851) and the Korean Government (MSIT) (NRF-2019R1C1C1005855).

Conflicts of Interest: The authors declare no conflict of interest.

Abbreviations

AIF	Apoptosis-inducing factor
Apaf-1	Apoptotic protease-activating factor-1
AR	Androgen receptor
ARD1	Arrest-defective 1
ATD	Acetyltransferase domain
AuA	Aurora kinase A
BCa	Breast cancer
CRC	Colorectal cancer
DNMT1	DNA methyltransferase 1
FIH	Factor Inhibiting HIF
HCC	Hepatocellular carcinoma
Hsp70	Heat shock protein 70
HYPK	Huntingtin yeast two-hybrid protein K
IDC	Invasive ductal carcinoma
IDR	Intrinsically disordered region
IKKβ	IκB kinase β
KAT	Lysine acetyltransferase
KDAC	Lysine deacetylase
LCa	Lung cancer
MLCK	Myosin light chain kinase
MMP-2	Matrix metalloproteinase-2
MVI	Microvascular invasion
NAA10	N-α-acetyltransferase 10
NAT	N-terminal acetyltransferase
OSCC	Oral squamous cell carcinoma
PGK1	Phosphoglycerate kinase 1
PIX	p21-activated kinase-interacting exchange factor
PCa	Prostate cancer
Runx2	Runt-related transcription factor 2
STAT5α	Signal transducer and activator of transcription 5α

References

1. Heron, M. Deaths: Leading causes for 2016. *Natl. Vital Stat. Rep.* **2018**, *67*, 1–77. [PubMed]
2. Koncina, E.; Haan, S.; Rauh, S.; Letellier, E. Prognostic and predictive molecular biomarkers for colorectal cancer: Updates and challenges. *Cancers* **2020**, *12*, 319. [CrossRef] [PubMed]
3. Aksnes, H.; Ree, R.; Arnesen, T. Co-translational, post-translational, and non-catalytic roles of N-Terminal acetyltransferases. *Mol. Cell* **2019**, *73*, 1097–1114. [CrossRef] [PubMed]
4. Wang, Z.H.; Gong, J.L.; Yu, M.; Yang, H.; Lai, J.H.; Ma, M.X.; Wu, H.; Li, L.; Tan, D.Y. Up-regulation of human arrest-defective 1 protein is correlated with metastatic phenotype and poor prognosis in breast cancer. *Asian Pac. J. Cancer Prev.* **2011**, *12*, 1973–1977. [PubMed]
5. Chien, M.H.; Lee, W.J.; Yang, Y.C.; Tan, P.; Pan, K.F.; Liu, Y.C.; Tsai, H.C.; Hsu, C.H.; Wen, Y.C.; Hsiao, M.; et al. N-alpha-acetyltransferase 10 protein promotes metastasis by stabilizing matrix metalloproteinase-2 protein in human osteosarcomas. *Cancer Lett.* **2018**, *433*, 86–98. [CrossRef]

6. Jiang, B.; Ren, T.; Dong, B.; Qu, L.; Jin, G.; Li, J.; Qu, H.; Meng, L.; Liu, C.; Wu, J.; et al. Peptide mimic isolated by autoantibody reveals human arrest defective 1 overexpression is associated with poor prognosis for colon cancer patients. *Am. J. Pathol.* **2010**, *177*, 1095–1103. [CrossRef]

7. Shim, J.H.; Chung, Y.H.; Kim, J.A.; Lee, D.; Kim, K.M.; Lim, Y.S.; Lee, H.C.; Lee, Y.S.; Yu, E.; Lee, Y.J. Clinical implications of arrest-defective protein 1 expression in hepatocellular carcinoma: A novel predictor of microvascular invasion. *Dig. Dis.* **2012**, *30*, 603–608. [CrossRef]

8. Lee, C.F.; Ou, D.S.; Lee, S.B.; Chang, L.H.; Lin, R.K.; Li, Y.S.; Upadhyay, A.K.; Cheng, X.; Wang, Y.C.; Hsu, H.S.; et al. hNaa10p contributes to tumorigenesis by facilitating DNMT1-mediated tumor suppressor gene silencing. *J. Clin. Investig.* **2010**, *120*, 2920–2930. [CrossRef]

9. Wang, Z.; Wang, Z.; Guo, J.; Li, Y.; Bavarva, J.H.; Qian, C.; Brahimi-Horn, M.C.; Tan, D.; Liu, W. Inactivation of androgen-induced regulator ARD1 inhibits androgen receptor acetylation and prostate tumorigenesis. *Proc. Natl. Acad. Sci. USA* **2012**, *109*, 3053–3058. [CrossRef]

10. Minguez, P.; Letunic, I.; Parca, L.; Bork, P. PTMcode: A database of known and predicted functional associations between post-translational modifications in proteins. *Nucleic Acids Res.* **2013**, *41*, D306–D311. [CrossRef]

11. Minguez, P.; Parca, L.; Diella, F.; Mende, D.R.; Kumar, R.; Helmer-Citterich, M.; Gavin, A.C.; van Noort, V.; Bork, P. Deciphering a global network of functionally associated post-translational modifications. *Mol. Syst. Biol.* **2012**, *8*, 599–612. [CrossRef]

12. Ree, R.; Varland, S.; Arnesen, T. Spotlight on protein N-terminal acetylation. *Exp. Mol. Med.* **2018**, *50*, 1–13. [CrossRef] [PubMed]

13. Arnaudo, N.; Fernandez, I.S.; McLaughlin, S.H.; Peak-Chew, S.Y.; Rhodes, D.; Martino, F. The N-terminal acetylation of Sir3 stabilizes its binding to the nucleosome core particle. *Nat. Struct. Mol. Biol.* **2013**, *20*, 1119–1121. [CrossRef] [PubMed]

14. Aksnes, H.; Hole, K.; Arnesen, T. Molecular, cellular, and physiological significance of N-terminal acetylation. *Int. Rev. Cell Mol. Biol.* **2015**, *316*, 267–305.

15. Raychaudhuri, S.; Sinha, M.; Mukhopadhyay, D.; Bhattacharyya, N.P. HYPK, a Huntingtin interacting protein, reduces aggregates and apoptosis induced by N-terminal Huntingtin with 40 glutamines in Neuro2a cells and exhibits chaperone-like activity. *Hum. Mol. Genet.* **2008**, *17*, 240–255. [CrossRef] [PubMed]

16. Hwang, C.S.; Shemorry, A.; Varshavsky, A. N-terminal acetylation of cellular proteins creates specific degradation signals. *Science* **2010**, *327*, 973–977. [CrossRef]

17. Chaudhary, P.; Ha, E.; Vo, T.T.L.; Seo, J.H. Diverse roles of arrest defective 1 in cancer development. *Arch. Pharm. Res.* **2019**, *42*, 1040–1051. [CrossRef]

18. Lee, K.E.; Heo, J.E.; Kim, J.M.; Hwang, C.S. N-Terminal acetylation-targeted N-End rule proteolytic system: The Ac/N-End rule pathway. *Mol. Cells* **2016**, *39*, 169–178.

19. Lee, D.; Jang, M.K.; Seo, J.H.; Ryu, S.H.; Kim, J.A.; Chung, Y.H. ARD1/NAA10 in hepatocellular carcinoma: Pathways and clinical implications. *Exp. Mol. Med.* **2018**, *50*, 1–12. [CrossRef]

20. Whiteway, M.; Szostak, J.W. The ARD1 gene of yeast functions in the switch between the mitotic cell cycle and alternative developmental pathways. *Cell* **1985**, *43*, 483–492. [CrossRef]

21. Arnesen, T.; Van Damme, P.; Polevoda, B.; Helsens, K.; Evjenth, R.; Colaert, N.; Varhaug, J.E.; Vandekerckhove, J.; Lillehaug, J.R.; Sherman, F.; et al. Proteomics analyses reveal the evolutionary conservation and divergence of N-terminal acetyltransferases from yeast and humans. *Proc. Natl. Acad. Sci. USA* **2009**, *106*, 8157–8162. [CrossRef] [PubMed]

22. DePaolo, J.S.; Wang, Z.; Guo, J.; Zhang, G.; Qian, C.; Zhang, H.; Zabaleta, J.; Liu, W. Acetylation of androgen receptor by ARD1 promotes dissociation from HSP90 complex and prostate tumorigenesis. *Oncotarget* **2016**, *7*, 71417–71428. [CrossRef]

23. Yoon, H.; Kim, H.L.; Chun, Y.S.; Shin, D.H.; Lee, K.H.; Shin, C.S.; Lee, D.Y.; Kim, H.H.; Lee, Z.H.; Ryoo, H.M.; et al. NAA10 controls osteoblast differentiation and bone formation as a feedback regulator of Runx2. *Nat. Commun.* **2014**, *5*, 5176. [CrossRef]

24. Seo, J.H.; Park, J.H.; Lee, E.J.; Vo, T.T.; Choi, H.; Kim, J.Y.; Jang, J.K.; Wee, H.J.; Lee, H.S.; Jang, S.H.; et al. ARD1-mediated Hsp70 acetylation balances stress-induced protein refolding and degradation. *Nat. Commun.* **2016**, *7*, 12882. [CrossRef]

25. Qian, X.; Li, X.; Lu, Z. Protein kinase activity of the glycolytic enzyme PGK1 regulates autophagy to promote tumorigenesis. *Autophagy* **2017**, *13*, 1246–1247. [CrossRef]

26. Arnesen, T.; Anderson, D.; Baldersheim, C.; Lanotte, M.; Varhaug, J.E.; Lillehaug, J.R. Identification and characterization of the human ARD1-NATH protein acetyltransferase complex. *Biochem. J.* **2005**, *386*, 433–443. [CrossRef] [PubMed]

27. Park, J.H.; Seo, J.H.; Wee, H.J.; Vo, T.T.; Lee, E.J.; Choi, H.; Cha, J.H.; Ahn, B.J.; Shin, M.W.; Bae, S.J.; et al. Nuclear translocation of hARD1 contributes to proper cell cycle progression. *PLoS ONE* **2014**, *9*, e105185. [CrossRef] [PubMed]

28. Aksnes, H.; Van Damme, P.; Goris, M.; Starheim, K.K.; Marie, M.; Stove, S.I.; Hoel, C.; Kalvik, T.V.; Hole, K.; Glomnes, N.; et al. An organellar nalpha-acetyltransferase, naa60, acetylates cytosolic N termini of transmembrane proteins and maintains Golgi integrity. *Cell Rep.* **2015**, *10*, 1362–1374. [CrossRef] [PubMed]

29. Wright, P.E.; Dyson, H.J. Intrinsically disordered proteins in cellular signalling and regulation. *Nat. Rev. Mol. Cell Biol.* **2015**, *16*, 18–29. [CrossRef]

30. Seo, J.H.; Cha, J.H.; Park, J.H.; Jeong, C.H.; Park, Z.Y.; Lee, H.S.; Oh, S.H.; Kang, J.H.; Suh, S.W.; Kim, K.H.; et al. Arrest defective 1 autoacetylation is a critical step in its ability to stimulate cancer cell proliferation. *Cancer Res.* **2010**, *70*, 4422–4432. [CrossRef]

31. Qian, X.; Li, X.; Cai, Q.; Zhang, C.; Yu, Q.; Jiang, Y.; Lee, J.H.; Hawke, D.; Wang, Y.; Xia, Y.; et al. Phosphoglycerate kinase 1 phosphorylates beclin1 to induce autophagy. *Mol. Cell* **2017**, *65*, 917–931. [CrossRef] [PubMed]

32. Kuo, H.P.; Lee, D.F.; Xia, W.; Lai, C.C.; Li, L.Y.; Hung, M.C. Phosphorylation of ARD1 by IKKbeta contributes to its destabilization and degradation. *Biochem. Biophys. Res. Commun.* **2009**, *389*, 156–161. [CrossRef]

33. Kang, J.; Chun, Y.S.; Huh, J.; Park, J.W. FIH permits NAA10 to catalyze the oxygen-dependent lysyl-acetylation of HIF-1α. *Redox Biol.* **2018**, *19*, 364–374. [CrossRef] [PubMed]

34. Vo, T.T.L.; Park, J.H.; Lee, E.J.; Nguyen, Y.T.K.; Han, B.W.; Nguyen, H.T.T.; Mun, K.C.; Ha, E.; Kwon, T.K.; Kim, K.W.; et al. Characterization of lysine acetyltransferase activity of recombinant human ARD1/NAA10. *Molecules* **2020**, *25*, 588. [CrossRef]

35. Liszczak, G.; Arnesen, T.; Marmorstein, R. Structure of a ternary Naa50p (NAT5/SAN) N-terminal acetyltransferase complex reveals the molecular basis for substrate-specific acetylation. *J. Biol. Chem.* **2011**, *286*, 37002–37010. [CrossRef] [PubMed]

36. Gottlieb, L.; Marmorstein, R. Structure of human NatA and its regulation by the huntingtin interacting protein HYPK. *Structure* **2018**, *26*, 925–935. [CrossRef] [PubMed]

37. Kuo, H.P.; Lee, D.F.; Chen, C.T.; Liu, M.; Chou, C.K.; Lee, H.J.; Du, Y.; Xie, X.; Wei, Y.; Xia, W.; et al. ARD1 stabilization of TSC2 suppresses tumorigenesis through the mTOR signaling pathway. *Sci. Signal.* **2010**, *3*, ra9. [CrossRef] [PubMed]

38. Hua, K.T.; Tan, C.T.; Johansson, G.; Lee, J.M.; Yang, P.W.; Lu, H.Y.; Chen, C.K.; Su, J.L.; Chen, P.B.; Wu, Y.L.; et al. N-alpha-acetyltransferase 10 protein suppresses cancer cell metastasis by binding PIX proteins and inhibiting Cdc42/Rac1 activity. *Cancer Cell* **2011**, *19*, 218–231. [CrossRef]

39. Bray, F.; Ferlay, J.; Soerjomataram, I.; Siegel, R.L.; Torre, L.A.; Jemal, A. Global cancer statistics 2018: GLOBOCAN estimates of incidence and mortality worldwide for 36 cancers in 185 countries. *CA Cancer J. Clin.* **2018**, *68*, 394–424. [CrossRef]

40. Siegel, R.L.; Miller, K.D.; Goding Sauer, A.; Fedewa, S.A.; Butterly, L.F.; Anderson, J.C.; Cercek, A.; Smith, R.A.; Jemal, A. Colorectal cancer statistics, 2020. *CA Cancer J. Clin.* **2020**, *70*, 145–164. [CrossRef]

41. Ren, T.; Jiang, B.; Jin, G.; Li, J.; Dong, B.; Zhang, J.; Meng, L.; Wu, J.; Shou, C. Generation of novel monoclonal antibodies and their application for detecting ARD1 expression in colorectal cancer. *Cancer Lett.* **2008**, *264*, 83–92. [CrossRef] [PubMed]

42. Yang, H.; Li, Q.; Niu, J.; Li, B.; Jiang, D.; Wan, Z.; Yang, Q.; Jiang, F.; Wei, P.; Bai, S. microRNA-342-5p and miR-608 inhibit colon cancer tumorigenesis by targeting NAA10. *Oncotarget* **2016**, *7*, 2709–2720. [CrossRef]

43. Dorfel, M.J.; Lyon, G.J. The biological functions of Naa10—From amino-terminal acetylation to human disease. *Gene* **2015**, *567*, 103–131. [CrossRef] [PubMed]

44. Lee, M.N.; Kweon, H.Y.; Oh, G.T. N-α-acetyltransferase 10 (NAA10) in development: The role of NAA10. *Exp. Mol. Med.* **2018**, *50*, 1–11. [CrossRef]

45. Markopoulos, A.K. Current aspects on oral squamous cell carcinoma. *Open Dent. J.* **2012**, *6*, 126–130. [CrossRef]

46. Zeng, Y.; Zheng, J.; Zhao, J.; Jia, P.R.; Yang, Y.; Yang, G.J.; Ma, J.F.; Gu, Y.Q.; Xu, J. High expression of Naa10p associates with lymph node metastasis and predicts favorable prognosis of oral squamous cell carcinoma. *Tumour. Biol.* **2016**, *37*, 6719–6728. [CrossRef] [PubMed]

47. Merriel, S.W.D.; Funston, G.; Hamilton, W. Prostate cancer in primary care. *Adv. Ther.* **2018**, *35*, 1285–1294. [CrossRef] [PubMed]

48. Lonergan, P.E.; Tindall, D.J. Androgen receptor signaling in prostate cancer development and progression. *J. Carcinog.* **2011**, *10*, 20.

49. Zhou, Y.; Bolton, E.C.; Jones, J.O. Androgens and androgen receptor signaling in prostate tumorigenesis. *J. Mol. Endocrinol.* **2015**, *54*, R15–R29. [CrossRef]

50. Lavery, D.N.; Bevan, C.L. Androgen receptor signalling in prostate cancer: The functional consequences of acetylation. *J. Biomed. Biotechnol.* **2011**, *2011*, 862125. [CrossRef]

51. Zeng, Y.; Min, L.; Han, Y.; Meng, L.; Liu, C.; Xie, Y.; Dong, B.; Wang, L.; Jiang, B.; Xu, H.; et al. Inhibition of STAT5a by Naa10p contributes to decreased breast cancer metastasis. *Carcinogenesis* **2014**, *35*, 2244–2253. [CrossRef] [PubMed]

52. Lozada, E.M.; Andrysik, Z.; Yin, M.; Redilla, N.; Rice, K.; Stambrook, P.J. Acetylation and deacetylation of Cdc25A constitutes a novel mechanism for modulating Cdc25A functions with implications for cancer. *Oncotarget* **2016**, *7*, 20425–20439. [CrossRef]

53. Park, Y.H.; Seo, J.H.; Park, J.H.; Lee, H.S.; Kim, K.W. Hsp70 acetylation prevents caspase-dependent/independent apoptosis and autophagic cell death in cancer cells. *Int. J. Oncol.* **2017**, *51*, 573–578. [CrossRef]

54. Vo, T.T.L.; Park, J.H.; Seo, J.H.; Lee, E.J.; Choi, H.; Bae, S.J.; Le, H.; An, S.; Lee, H.S.; Wee, H.J.; et al. ARD1-mediated aurora kinase A acetylation promotes cell proliferation and migration. *Oncotarget* **2017**, *8*, 57216–57230. [CrossRef] [PubMed]

55. Shin, D.H.; Chun, Y.S.; Lee, K.H.; Shin, H.W.; Park, J.W. Arrest defective-1 controls tumor cell behavior by acetylating myosin light chain kinase. *PLoS ONE* **2009**, *4*, e7451. [CrossRef] [PubMed]

56. Lim, J.H.; Park, J.W.; Chun, Y.S. Human arrest defective 1 acetylates and activates beta-catenin, promoting lung cancer cell proliferation. *Cancer Res.* **2006**, *66*, 10677–10682. [CrossRef]

57. Neri, L.; Lasa, M.; Elosegui-Artola, A.; D'Avola, D.; Carte, B.; Gazquez, C.; Alve, S.; Roca-Cusachs, P.; Inarrairaegui, M.; Herrero, J.; et al. NatB-mediated protein N-alpha-terminal acetylation is a potential therapeutic target in hepatocellular carcinoma. *Oncotarget* **2017**, *8*, 40967–40981. [CrossRef]

58. Varland, S.; Myklebust, L.M.; Goksoyr, S.O.; Glomnes, N.; Torsvik, J.; Varhaug, J.E.; Arnesen, T. Identification of an alternatively spliced nuclear isoform of human N-terminal acetyltransferase Naa30. *Gene* **2018**, *644*, 27–37. [CrossRef]

59. Starheim, K.K.; Gromyko, D.; Evjenth, R.; Ryningen, A.; Varhaug, J.E.; Lillehaug, J.R.; Arnesen, T. Knockdown of human N α-terminal acetyltransferase complex C leads to p53-dependent apoptosis and aberrant human Arl8b localization. *Mol. Cell. Biol.* **2009**, *29*, 3569–3581. [CrossRef]

60. Van Damme, P.; Kalvik, T.V.; Starheim, K.K.; Jonckheere, V.; Myklebust, L.M.; Menschaert, G.; Varhaug, J.E.; Gevaert, K.; Arnesen, T. A role for human N-alpha acetyltransferase 30 (Naa30) in maintaining mitochondrial integrity. *Mol. Cell. Proteom.* **2016**, *15*, 3361–3372. [CrossRef]

61. Mughal, A.A.; Grieg, Z.; Skjellegrind, H.; Fayzullin, A.; Lamkhannat, M.; Joel, M.; Ahmed, M.S.; Murrell, W.; Vik-Mo, E.O.; Langmoen, I.A.; et al. Knockdown of NAT12/NAA30 reduces tumorigenic features of glioblastoma-initiating cells. *Mol. Cancer* **2015**, *14*, 160. [CrossRef]

62. Pavlou, D.; Kirmizis, A. Depletion of histone N-terminal-acetyltransferase Naa40 induces p53-independent apoptosis in colorectal cancer cells via the mitochondrial pathway. *Apoptosis* **2016**, *21*, 298–311. [CrossRef] [PubMed]

63. Ju, J.; Chen, A.; Deng, Y.; Liu, M.; Wang, Y.; Wang, Y.; Nie, M.; Wang, C.; Ding, H.; Yao, B.; et al. NatD promotes lung cancer progression by preventing histone H4 serine phosphorylation to activate Slug expression. *Nat. Commun.* **2017**, *8*, 928. [CrossRef]

64. Drazic, A.; Aksnes, H.; Marie, M.; Boczkowska, M.; Varland, S.; Timmerman, E.; Foyn, H.; Glomnes, N.; Rebowski, G.; Impens, F.; et al. NAA80 is actin's N-terminal acetyltransferase and regulates cytoskeleton assembly and cell motility. *Proc. Natl. Acad. Sci. USA* **2018**, *115*, 4399–4404. [CrossRef]

65. Vo, T.T.L.; Jeong, C.H.; Lee, S.; Kim, K.W.; Ha, E.; Seo, J.H. Versatility of ARD1/NAA10-mediated protein lysine acetylation. *Exp. Mol. Med.* **2018**, *50*, 1–13. [CrossRef] [PubMed]

Publisher's Note: MDPI stays neutral with regard to jurisdictional claims in published maps and institutional affiliations.

International Journal of
Molecular Sciences

Review

Head and Neck Paragangliomas—A Genetic Overview

Anna Majewska [1,*], **Bartłomiej Budny** [2], **Katarzyna Ziemnicka** [2], **Marek Ruchała** [2] and **Małgorzata Wierzbicka** [1]

[1] Department of Otolaryngology, Head and Neck Surgery, Poznan University of Medical Sciences, 60-355 Poznań, Poland; otosk2@gmail.com
[2] Department of Endocrinology, Metabolism and Internal Diseases, Poznan University of Medical Sciences, 60-355 Poznań, Poland; bbudny@ump.edu.pl (B.B.); kaziem@ump.edu.pl (K.Z.); mruchala@ump.edu.pl (M.R.)
* Correspondence: majewska.anna@spsk2.pl

Received: 25 September 2020; Accepted: 14 October 2020; Published: 16 October 2020

Abstract: Pheochromocytomas (PCC) and paragangliomas (PGL) are rare neuroendocrine tumors. Head and neck paragangliomas (HNPGL) can be categorized into carotid body tumors, which are the most common, as well as jugular, tympanic, and vagal paraganglioma. A review of the current literature was conducted to consolidate knowledge concerning PGL mutations, familial occurrence, and the practical application of this information. Available scientific databases were searched using the keywords head and neck paraganglioma and genetics, and 274 articles in PubMed and 1183 in ScienceDirect were found. From these articles, those concerning genetic changes in HNPGLs were selected. The aim of this review is to describe the known genetic changes and their practical applications. We found that the etiology of the tumors in question is based on genetic changes in the form of either germinal or somatic mutations. 40% of PCC and PGL have a predisposing germline mutation (including *VHL, SDHB, SDHD, RET, NF1, THEM127, MAX, SDHC, SDHA, SDHAF2, HIF2A, HRAS, KIF1B, PHD2,* and *FH*). Approximately 25–30% of cases are due to somatic mutations, such as *RET, VHL, NF1, MAX,* and *HIF2A*. The tumors were divided into three main clusters by the Cancer Genome Atlas (TCGA); namely, the pseudohypoxia group, the Wnt signaling group, and the kinase signaling group. The review also discusses genetic syndromes, epigenetic changes, and new testing technologies such as next-generation sequencing (NGS).

Keywords: pheochromocytoma; paraganglioma; head and neck neoplasms; head and neck tumors; genetic syndromes; mutations

---

1. Introduction

Pheochromocytomas (PCC) and paragangliomas (PGL) are rare neuroendocrine tumors originating from either adrenomedullary chromaffin cells (PCCs); sympathetic ganglia of the thorax (T-PGL); or abdominal (A-PGL), pelvic, or parasympathetic ganglia in the head and neck (HNPGL) [1,2]. They are referred to collectively as PPGL. PCCs typically secrete one or more than one catecholamine: epinephrine, norepinephrine, and dopamine [1], while PGLs in most cases are non-secretory [1,3–5]. PCC represent 80% to 85% of chromaffin-cell tumors, and PGL represent 15% to 20% [6]. These tumors are characteristically well-vascularized and typically benign; nonetheless, roughly 10–15% may metastasize to the lungs, bone, liver, and lymph nodes. They most frequently occur between the third and sixth decades of life and present more commonly in women [7]. HNPGL can be categorized into carotid body tumors, which are the most common, as well as jugular, tympanic, and vagal paraganglioma. Other rare locations include the larynx, thyroid gland, parathyroid gland, nose, paranasal sinuses, parotid gland, or orbit [8]. PGL have also been described in the urogenital system, in the spermatic cord in particular [9]. Clinical symptoms vary according to the location and size of the tumor. Carotid body tumors typically produce a painless, slow-growing neck mass [10,11] that may eventually cause dysphagia and cranial nerve disorders. In contrast, pulsatile tinnitus and conductive hearing loss are characteristic of tympanic paraganglioma [12].

Neuroendocrine tumors show the highest degree of heritability in all neoplasms (approximately 40–50%) [13–17]. The first reports of the familial occurrence of PGL date from 1933, when carotid paragangliomas were first described by Chase [18,19]. In recent years, it has been confirmed that more than one-third of these tumors are genetically determined [20]. Today, the planning of further treatment considers family history, the extent and location of the tumor, its genetic origin, and the molecular pathways involved, especially as genetic testing becomes increasingly available and consistently improves the efficacy of therapy [3]. When a mutation is detected in a susceptibility gene such as *VHL, SDH,* or the recently discovered *MDH2,* a search for common co-occurring tumors is indicated [20,21]. Mutation in the SDHB subunit is also associated with the risk for malignancy and worse prognosis [3,10,22,23]. In 50% of patients with metastatic disease, a mutation in the *SDHB* gene was found. In the remaining 50% of cases, the genetic factors of the malignancy are still unidentified [23]. With this knowledge, genetic testing of PGL and the testing of first-degree family members should be routinely implemented to diagnose low-grade tumors [24]. Therefore, we aim to comprehend and conclude the most recent knowledge surrounding mutations in PGL, family occurrence, and their practical application based on the current literature and the paradigm of diagnostics.

2. Results

The outcomes are presented in the form of a literature review, structured by thematic subsections concerning the classification of head and neck paragangliomas with regard to genetic and molecular changes (based on 21 papers), as well as elucidation of genetic syndromes (based on 19 publications). Moreover, the review presents new methods as they pertain to the investigation of these tumors, such as investigation of epigenetic patterns or the application of new advanced molecular tools like next-generation sequencing (NGS) (based on five publications).

The details concerning the content of the presented articles (materials, methods, and conclusions) are presented in Table 1.

Table 1. The table includes details concerning the content of the presented articles (authors, year of publication, number of patients in the study, reported genes, and most significant findings). Only data from original papers are included; no reviews are considered.

Author, Year	No. of Patients	Genes	Findings
Niemann et al. (2001) [25]	Five patients with histologically proven paraganglioma (single family members) and one patient (of this family) with imaging findings consistent with a PGL. 33 family members were clinically unaffected.	*SDHC* gene location	The disease locus in PGL3 was determined to be located at 1q21-q23.
Mannelli et al. (2009) [26]	501 patients with PCC and/or PGL 160 patients under 50 years of age whose DNA sequencing results revealed wild-type *RET, VHL, SDHB, SDHC,* and *SDHD* were subsequently analyzed for genomic rearrangements involving the *VHL* gene or one of the *SDH* genes.	*RET VHL SDHD SDHB SDHC* Genomic rearrangements (total deletion of the *SDHD* gene)	Detection of germinal mutations (such as *VHL, RET, NF1, SDHB, SDHC* and *SDHD*) in 32.1% of cases. From 100% in patients with associated lesions to 11.6% in patients with a single tumor. Genomic rearrangements were found in two of 160 patients (1.2%), both involving total deletion of the *SDHD* gene.
Bayley et al. (2010) [27]	443 patients with apparently sporadic PCC/PGL who did not have mutations in *SDHD, SDHC,* or *SDHB*. Examination of a Spanish family with HNPGL presenting with a young age of onset.	*SDHAF2*	No germinal (315 patients) or somatic (128 patients) mutations, and no germinal deletions of the *SDHAF2* gene were found. After pedigree analysis of a Spanish family with HNPGL a pathogenic mutation in *SDHAF2* was found that resulted in an amino acid substitution (p.Gly78Arg). The same mutation was noted previously in a Dutch kindred.
Kunst et al. (2011) [28]	57 family members.	*SDHAF2*	Establishing a correlation between HNPGL occurrence (based on phenotypic analysis) and *SDHAF2* mutation. The mutation carriers showed early onset of the disease and high levels of multifocality.
Casey et al. (2014) [29]	31 patients with confirmed PCC/PGL.	*TMEM127 SDHAF2 RET*	The occurrence of *TMEM127, SDHAF2* and *RET* mutations was found in patients without indications for genetic testing based on phenotypic evaluation.
Fishbein et al. (2015) [23]	Stage 1: whole exome sequencing on a discovery set of 21 patients with PCC/PGL. Stage 2: targeted sequencing of a separate validation set of 103 patients withPCC/PGL.	*NF1 ATRX*	Mutations in *NF1* were detected in 42% of tumors. In 28% of *SDHB*-related tumors, deleterious variants of *ATRX* were found (PP119F1 p.W2275* and PP098F2 p.R2197H). ATRX protein was not detected in tumor cells by immunohistochemistry. The study found somatic mutation of *ATRX* in 12.6% of cases; 30% of them had truncating mutations and 69% missense mutations, classified as deleterious.
Luchetti et al. (2015) [30]	85 patients: PCC 60, PGL 5, HNPGL 20.	*HRAS BRAF*	Missense mutation was found in six cases (PCC = 6/60, PGL = 0/5, and HNPGL = 0/20) in *HRAS* in the hotspot region of codon 13 and 61. In one case of PCC, an activating *BRAF* mutation was found. In two patients a missense mutation was identified in the tetramerization domain of TP53 protein.
Fishbein et al. (2017) [31]	173 patients with PCCs/PGLs.	*SDHB, RET, WHL, NF1, SDHD, MAX EGLN1 (PHD2), TMEM127, CSDE1, HRAS, EPAS1, MAML3, BRAF, NGFR*	27% of patients had germinal mutations (including *SDHB* 9%, *RET* 6%, *VHL* 4%, and *NF1* 3%). *SDHD, MAX, EGLN1 (PHD2),* and *TMEM127* mutations were found in less than 2% each. *CSDE1* was identified as a somatically mutated driver gene complementary to the other four known drivers (*HRAS, RET, EPAS1,* and *NF1*). *MAML3, BRAF, NGFR,* and *NF1* fusion genes were discovered.

Table 1. *Cont.*

Author, Year	No. of Patients	Genes	Findings
Bausch et al. (2017) [32]	972 unrelated patients without mutations in the classic PCC/PGL associated genes.	SDHA, TMEM127, MAX, SDHAF2	Six percent of patients were mutation carriers (including *SDHA, TMEM127, MAX,* and *SDHAF2*). 91% of patients had familial, multiple, extra-adrenal, and/or malignant tumors and/or had younger age of onset. Extra-adrenal tumors occurred in 48% of mutation carriers and in 79% of carriers with HNPGL.
Chen et al. (2017) [22]	37 patients with HNPGLs.	SDHD SDHB SDHAF2	*SDHD* gene mutations were found in: the Chinese founder mutation (c.3G>C, p.Met1Ile) in six cases, a missense mutation (c.284T>C, p.L95P) in one case, an in-frame deletion (c.278–280delATT, p.Y93S) in one case. A missense *SDHB* mutation (c.647A>G) and a nonsense *SDHAF2* mutation (c.130C>T, p.Gln44Ter) were found in two cases. Frequent methylation was observed in six of the TSGs tested (*HIC1, DcR1, DcR2, DR4, DR5,* and *CASPS8*). Four of them (*HIC1, DcR1, DcR2* and *CASPS8*) showed more frequent mutations in *SDH*-associated HNPGL than in non-mutated ones.
Calsina et al. (2018) [21]	830 patients with PPGLs, negative for the main PPGL driver genes.	MDH2	Twelve heterozygous variants of *MDH2* were found (five of the 12 were missense (41.7%), one synonymous (8.3%), four were located in the intronic region (33.3%), one was an in-frame deletion (8.3%), and one affected a donor splice-site (8.3%). Five of these were unreported variants. The study showed the functional impact of two variants (p.Arg104Gly and p.Lys314del) and suggests altered molecular function of two variants (p.Val160Met and p.Ala256Thr).
Ding et al. (2019) [33]	23 cases of multiple HNPGL.	SDHD, SDHB, SDHC, SDHAF2, VHL, RET	Family 1: 12 *SDHD* mutations (8 bilateral carotid body tumor (CBT) with 1 bilateral malignant CBT) Family 2: 3 *SDHD* mutations (1 bilateral CBT, 2 unilateral CBT) Family 3: 2 cases of *SDHD* mutations (vagus PGL and pheochromocytoma) Other patients: sporadic manifestations (5 cases *SDHD* gene mutation, 1 case *RET* gene mutation). Two novel mutations were found: c.387–393del7 mutation of *SDHD* gene and c.3247A>G mutation of *RET* gene. More frequent occurrence of *SDHD* mutations was found in patients and family members with multiple HNPGL.

2.1. Classification Based on the Genetic and Molecular Background

Germinal mutations occur in the germ line and are passed on to all cells of the developing body [34]. A germline predisposing mutation is found in approximately 40% of PCCs and PGLs in one of at least 12 genes (*VHL, SDHB, SDHD, RET, NF1, THEM127, MAX, SDHC, SDHA, SDHAF2, HIF2A, HRAS, KIF1B, PHD2, FH*). The second type of genetic alteration is classified as somatic. These occur later in life, affecting only a single cell of a particular tissue, and give rise to the development of a specific neoplasm. Somatic mutations of *RET, VHL, NF1, MAX*, and *HIF2A* account for 25–30% of these tumors [13,16,23,32,33,35,36].

PGLs are classified into three clusters by the Cancer Genome Atlas (TCGA) on the basis of molecular, cytogenetic abnormalities, and specific single-nucleotide causative mutations, which led to the development of PPGLs. Moreover, contributing genes are grouped according to their biological activity—namely, the pseudohypoxia group, the Wnt signaling group, and the kinase signaling group. This division into groups with different clinical, imaging, molecular, and biochemical features allows for the personalization of patient care as well as the development of new screening and treatment guidelines [14,35,37,38].

The pseudohypoxia group can be further divided into two subgroups. The first comprises tricarboxylic acid cycle (TCA)-related factors concerning 10–15% of PPGLs. This group includes germline mutations in succinate dehydrogenase subunits *SDHA, SDHB, SDHC, SDHD* or *SDHAF2* (SDHx)—succinate dehydrogenase complex assembly factor 2, and FH (a second enzyme in the TCA cycle). The second subgroup encompasses *VHL/EPAS1*-related genes and accounts for 15–20% of PPGLs [14,35,37–40].

Activation of hypoxia inducible factors (HIFs) is a mutual characteristic for this cluster. HIFs are released in physiological response to cellular hypoxia. A pseudo-hypoxic state is caused by the presence of abnormal, mutated *VHL, SDH, EGLN1*, and *HIF2A* genes. The effect of this is constant activation of HIF pathways in the cell despite normal oxygen levels. This condition causes epigenetic changes in HIF target genes, which affects many processes including proliferation, angiogenesis, migration, apoptosis, and invasion. These events may all contribute to PPGL formation [19,35,38,41–44].

The Wnt signaling cluster is another group that are, in particular, triggered by somatic mutations in the *CSDE1* gene or somatic gene fusions which affect the *MAML3* gene. This results in the activation of Wnt and Hedgehog signaling pathways. Patients with sporadic PPGLs (5–10% of all PPGLs) are grouped here. Many developmental processes such as proliferation, cell polarity, adhesion, or differentiation are regulated by the Wnt pathway. As a result, these tumors are considered more aggressive, recur significantly, and are often prone to metastases [14,31,37–39,45].

The kinase signaling cluster (50–60% of PPGLs) includes germline or somatic mutations in *RET, NF1, MAX, HRAS*, and *TMEM127* genes [14,37]. The RAS/MAPK and PI3/AKT signaling pathways are enabled due to *RET* proto-oncogene activation or *NF1* tumor suppressor inactivation, resulting in tumor formation. In contrast, *TMEM127* mutations trigger the mTOR pathways. Another mechanism includes deactivation of the *MAX* suppressor gene, causing an abnormally elevated expression of cofactor *MYC* (proto-oncogene), resulting in the formation of PPGLs [14,38–41,43,44].

Several genetic syndromes are associated with PPGL: Multiple endocrine neoplasia type 2 (MEN2), Neurofbromatosis type 1 (NF1), Von Hippel–Lindau (VHL) disease, and Hereditary paraganglioma syndrome (PGL 1, PGL2, PGL3 and PGL4) [46,47].

HNPGL are very rare in NF1, MEN 2, and VHL patients. Rather, they display a predisposition toward the development of PCCs.

2.2. Genetic Syndromes

HNPGL are a solid manifestation in hereditary paraganglioma syndromes. They are caused by mutations in the succinate dehydrogenase (SDH) complex, which is necessary for the mitochondrial electron transport chain and ATP generation. This compound is composed of four subunits (A-D) with

SDHAF2 stabilizing the entire complex. Subunits B, C, and D are strongly correlated with PCCs and PGLs [8,12,14,26,35,42,46–48].

PGL1 syndrome is an autosomal dominant disease linked to HNPGLs. It is correlated with inactivating mutations of the *SDHD* gene localized on chromosome 11q23. PCCs and sympathetic PGLs occur in 40% of cases, and bilateral or multifocal tumors are present in approximately 74% of patients. Though these tumors are typically not malignant, they have a tendency toward recurrence [14]. *SDHD* mutations are also associated with maternal genomic imprinting. Tumors are more likely to develop in children if the father is affected or a mutation carrier himself. If the mutation is inherited from the mother, it is inactivated but still genetically transmitted [8,12,15,35,41,47,49].

PGL4 syndrome also arises from a mutation with an autosomal dominant mode of inheritance, is responsible for inactivating the *SDHB* gene located on 11p35. In this condition, the following symptoms are reported: sympathetic extra-adrenal PGLs, PCCs, and HNPGLs. In up to 70% of all cases of PGL4 syndrome, the tumors are malignant [13]. PGLs typically produce catecholamines such as dopamine and norepinephrine, and only 10% of *SDHB* mutated tumors are biochemically silent; however, the clinical consequences are generally the result of significant mass effect rather than catecholamine excess. Typical tumor localizations include the abdomen and the mediastinum. The *SDHB* gene mutation increases the risk of renal cell carcinoma, gastrointestinal stromal tumor (GIST), and breast and papillary thyroid carcinoma, and while patients with metastatic disease should be routinely tested for the presence of the predisposing *SDHB* mutation, there are no guidelines regarding the screening of asymptomatic *SDHx* gene mutation carriers. Experts do suggest annual biochemical screening for PCC/PGLs from between the ages of five and 10, as well as full-body MRI screening for all associated tumor types every 2–5 years [8,12,14,35,41,47,49].

PGL3 syndrome is caused by an *SDHC* gene mutation located on 1q21-q23 and is inherited in an autosomal dominant pattern. PGL3 is associated with the occurrence of benign HNPGL, sympathetic PGL, and PCC and is typically multifocal. Metastases of these tumors is exceedingly rare [8,25,42,47,49,50].

Mutations in the *SDHAF2* gene have also been recently reported. *SDHAF2* mutation results in a rare type of familial paraganglioma syndrome that leads to HNPGL, but only in the children of a father who is a carrier of the defective gene. This syndrome is transmitted in an autosomal dominant manner, and usually manifests in the third decade of life. Genetic screening of *SDHAF2* mutation is crucial in patients with HNPGL with suspicious family history, young age of onset, or multiple tumors and have already tested negative for *SDHB*, *SDHC*, and *SDHD* mutations [27–29,37,46,47].

2.3. Epigenetic Patterns in HNPGL

Epigenetic changes are gene modifications that do not change the DNA sequence but affect gene activity. Most often the changes include methylation—the addition of a methyl group to the DNA strand—which results in the switching off or silencing of the gene and subsequent altered protein production. Other types of epigenetic modification include acetylation, phosphorylation, ubiquitylation, and sumoylation. Some of these changes can be inherited [51]. However, the most frequent of all epigenetic markers in DNA is cytosine methylation. This change in the human genome is referred to as "CpG methylation" or "DNA methylation" [52]. Inactivation of tumor-suppressor genes (TSGs) is caused by overall DNA hypomethylation and hypermethylation of CpG islands located in the closest vicinity of the promoter. Tumorigenesis of HNPGL is not yet fully explained, and the search for new genetic as well as epigenetic changes is ongoing.

In a study by Chen et al. [22], the methylation status of a panel of TSGs (*p16, HIC1, DcR1, DcR2, DR4, DR5, CASP8, HSP47, MGMT,* and *RASSF1A*) has been determined and compared in HNPGLs with and without SDH mutations. A correlation between the methylation index (MI) and the presence of germline mutations was observed. Six out of 10 TSGs showed frequent methylation: HIC1 and those involved in the apoptosis pathway DcR1, DcR2, DR4, DR5, and CASPS8. More frequent methylation in SDH-related HNPGLs compared to non-mutated analogues was observed in four analyzed TSGs (CASPS8, HIC1, DcR1, and DcR2).

2.4. Next-Generation Sequencing (NGS)

Most of the studies conducted as of today have utilized conventional Sanger sequencing. Next-generation sequencing (NGS), in contrast to Sanger sequencing, enables broader and more accurate sequencing, leading to the detection of mutations in multiple genes. This technology allows for sample multiplication and also increases capacity and effectiveness, as well as reducing costs. Therefore, the use of NGS could provide the opportunity to test all patients at risk, rather than just a few selected targets [36]. It may provide a better understanding of the crucial role of the mutations acquired on various level of disease development, as well as those underlying the carcinogenesis of HNPGLs [53]. Luchetti et al. [30] analyzed 50 "mutation hotspot" variants in PCC and PGL using NGS in 20 patients with HNPGL and 85 patients with PPGL. The authors identified mutations in *HRAS* (7.1%), and *BRAF* (1.2%) as well as for *TP53* in 2.35% of cases. In the group of PPGL tumors with identified hereditary mutations (21 cases), *HRAS, BRAF,* and *TP53* genes were not mutated. It was concluded that the occurrence of *HRAS/BRAF* mutations predominates in sporadic PPGL (8.9%) but is inconsequential for inherited PPGL.

3. Materials and Methods

This study assumes a review of world scientific literature. An online search was conducted using the scientific databases PubMed and ScienceDirect applying the key words head and neck paraganglioma and genetics. The first resulting article in PubMed dated from 1981 and from 1996 in ScienceDirect. Over the last 10 years, the number of articles on the subject has doubled. While this review considers articles from the last 20 years, over 85% of them were published in the last 10 years. Detailed data concerning the number of articles in each year are presented in Figure 1.

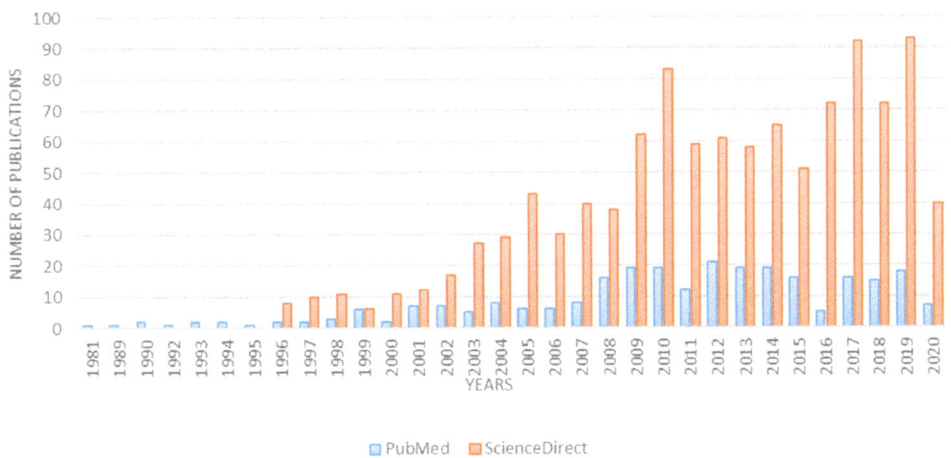

Figure 1. Number of publications according to key words head and neck paraganglioma and genetics.

In total, 274 articles containing the indicated keywords were found in PubMed and 1183 in ScienceDirect. Of these, only those from the last 20 years reporting genetic changes in head and neck paraganglioma were selected.

4. Conclusions

The conclusions of this review are based on the entire overview of the literature and may prove useful for the improvement of diagnostic and therapeutic schemes surrounding PCCs and PGLs. According to the article "Recommendations for Somatic and Germline Genetic Testing of Single Pheochromocytoma and Paraganglioma" [54], the study of germline DNA should be prioritized in head and neck paraganglioma and thorax paraganglioma. A strong recommendation for genetic testing—somatic as well as germline mutations, regardless of the age at diagnosis—is indicated. It is also strongly recommended even in patients with a negative family history, especially if the lesions occur at a young age and are multifocal [55]. Genetic testing is very effective for predicting the incidence of metastatic tumors. Numerous authors [3,56,57] have demonstrated the variability in the *SDHB* gene, which leads to metastatic disease in 40% or more of patients. An agreement in the literature on the selection of mutations in HNPGL has been drawn, and encompasses the following genes: *SDHA, SDHB, SDHD, SDHAF2, SDHC, SDHB, VHL, FH, RET*. These should be routinely determined in PGL patients. Different combinations of these genes should be tested depending on the availability of a tumor sample or the performance of SDHB-immunohistochemistry (SDHB-IHC).

To conclude, the diagnostic schedule in PGL should include the collection of clinical data including epidemiology, family history concerning neoplasms, the course of the disease (e.g., tumor growth rate), and/or its relapses. Radiological evaluation of the tumor consisting of imaging and angiography (assessment of tumor size, vascularization, localization, position relative to other structures, presence of metastases) should also be considered. Furthermore, in light of the expanding knowledge of the genetic basis of this disease, genetic testing concerning causative alterations has become increasingly important. A multidisciplinary team consisting of an ENT specialist, a radiologist, an endocrinologist, a nuclear medicine physician, and a geneticist can qualify the patient on the grounds of such information for further treatment and the management of follow-up.

Author Contributions: All authors have read and agreed to the published version of the manuscript.

Funding: This research was funded by the National Science Centre, Poland, grant number 2019/33/N/NZ5/00872. The APC was funded by the National Science Centre, Poland, grant number 2019/33/N/NZ5/00872.

Acknowledgments: The authors thank Jadzia Tin-Tsen Chou for language revision.

References

1. Lenders, J.W.M.; Duh, Q.-Y.; Eisenhofer, G.; Gimenez-Roqueplo, A.-P.; Grebe, S.K.G.; Murad, M.H.; Naruse, M.; Pacak, K.; Young, W.F. Pheochromocytoma and Paraganglioma: An Endocrine Society Clinical Practice Guideline. *J. Clin. Endocrinol. Metab.* **2014**, *99*, 1915–1942. [CrossRef] [PubMed]

2. Koopman, K.; Gaal, J.; De Krijger, R.R. Pheochromocytomas and Paragangliomas: New Developments with Regard to Classification, Genetics, and Cell of Origin. *Cancers* **2019**, *11*, 1070. [CrossRef] [PubMed]

3. Asban, A.; Kluijfhout, W.P.; Drake, F.T.; Beninato, T.; Wang, E.; Chomsky-Higgins, K.; Shen, W.T.; Gosnell, J.E.; Suh, I.; Duh, Q.-Y. Trends of genetic screening in patients with pheochromocytoma and paraganglioma: 15-year experience in a high-volume tertiary referral center. *J. Surg. Oncol.* **2018**, *117*, 1217–1222. [CrossRef] [PubMed]

4. Fishbein, L. Pheochromocytoma and Paraganglioma: Genetics, Diagnosis, and Treatment. *Hematol. Clin.* **2016**, *30*, 135–150. [CrossRef]

5. Burnichon, N.; Rohmer, V.; Amar, L.; Herman, P.; Leboulleux, S.; Darrouzet, V.; Niccoli, P.; Gaillard, D.; Chabrier, G.; Chabolle, F.; et al. The Succinate Dehydrogenase Genetic Testing in a Large Prospective Series of Patients with Paragangliomas. *J. Clin. Endocrinol. Metab.* **2009**, *94*, 2817–2827. [CrossRef]

6. Lenders, J.W.M.; Eisenhofer, G.; Mannelli, M.; Pacak, K. Phaeochromocytoma. *Lancet* **2005**, *366*, 665–675. [CrossRef]

7. Boedeker, C.C. Paragangliomas and paraganglioma syndromes. *GMS Curr. Top. Otorhinolaryngol. Head Neck Surg.* **2011**, *10*. [CrossRef]

8. Fishbein, L.; Nathanson, K.L. Pheochromocytoma and paraganglioma: Understanding the complexities of the genetic background. *Cancer Genet.* **2012**, *205*, 1–11. [CrossRef]

9. Sun, D.; Yang, P.; Liu, Y.; Wang, S. Paraganglioma of the spermatic cord: Report of one case and review of literature. *Int. J. Clin. Exp. Pathol.* **2020**, *13*, 1779–1786.

10. Dorobisz, K.; Dorobisz, T.; Temporale, H.; Zatoński, T.; Kubacka, M.; Chabowski, M.; Dorobisz, A.; Krecicki, T.; Janczak, D. Diagnostic and Therapeutic Difficulties in Carotid Body Paragangliomas, Based on Clinical Experience and a Review of the Literature. *Adv. Clin. Exp. Med. Off. Organ Wroc. Med. Univ.* **2016**, *25*, 1173–1177. [CrossRef]

11. Else, T.; Greenberg, S.; Fishbein, L. Hereditary Paraganglioma-Pheochromocytoma Syndromes. In *GeneReviews®*; Adam, M.P., Ardinger, H.H., Pagon, R.A., Wallace, S.E., Bean, L.J., Stephens, K., Amemiya, A., Eds.; University of Washington: Seattle, WA, USA, 1993.

12. Boedeker, C.C.; Hensen, E.F.; Neumann, H.P.; Maier, W.; Van Nederveen, F.H.; Suárez, C.; Kunst, H.P.; Rodrigo, J.P.; Takes, R.P.; Pellitteri, P.K.; et al. Genetics of hereditary head and neck paragangliomas. *Head Neck* **2013**, *36*, 907–916. [CrossRef] [PubMed]

13. Dahia, P.L.M. Pheochromocytoma and paraganglioma pathogenesis: Learning from genetic heterogeneity. *Nat. Rev. Cancer* **2014**, *14*, 108–119. [CrossRef]

14. Katabathina, V.S.; Rajebi, H.; Chen, M.M.; Restrepo, C.S.; Salman, U.; Vikram, R.; Menias, C.O.; Prasad, S.R. Genetics and imaging of pheochromocytomas and paragangliomas: Current update. *Abdom. Radiol. N. Y.* **2020**, *45*, 928–944. [CrossRef] [PubMed]

15. Cavenagh, T.; Patel, J.; Nakhla, N.; Elstob, A.; Ingram, M.; Barber, B.; Snape, K.; Bano, G.; Vlahos, I. Succinate dehydrogenase mutations: Paraganglioma imaging and at-risk population screening. *Clin. Radiol.* **2019**, *74*, 169–177. [CrossRef] [PubMed]

16. Turchini, J.; Cheung, V.K.Y.; Tischler, A.S.; Krijger, R.R.D.; Gill, A.J. Pathology and genetics of phaeochromocytoma and paraganglioma. *Histopathology* **2017**, *72*, 97–105. [CrossRef] [PubMed]

17. Sen, I.; Young, W.F.; Kasperbauer, J.L.; Polonis, K.; Harmsen, W.S.; Colglazier, J.J.; DeMartino, R.R.; Oderich, G.S.; Kalra, M.; Bower, T.C. Tumor-specific prognosis of mutation-positive patients with head and neck paragangliomas. *J. Vasc. Surg.* **2020**, *71*, 1602–1612.e2. [CrossRef]

18. Chase, W.H. Familial and bilateral tumours of the carotid body. *J. Pathol. Bacteriol.* **1933**, *36*, 1–12. [CrossRef]

19. Martin, T.P.; Irving, R.M.; Maher, E.R. The genetics of paragangliomas: A review. *Clin. Otolaryngol.* **2007**, *32*, 7–11. [CrossRef]

20. Burnichon, N.; Abermil, N.; Buffet, A.; Favier, J.; Gimenez-Roqueplo, A.-P. The genetics of paragangliomas. *Eur. Ann. Otorhinolaryngol. Head Neck Dis.* **2012**, *129*, 315–318. [CrossRef]

21. Calsina, B.; Currás-Freixes, M.; Buffet, A.; Pons, T.; Contreras, L.; Letón, R.; Méndez, I.C.; Remacha, L.; Calatayud, M.; Obispo, B.; et al. Role of MDH2 pathogenic variant in pheochromocytoma and paraganglioma patients. *Genet. Med. Off. J. Am. Coll. Med. Genet.* **2018**, *20*, 1652–1662. [CrossRef]

22. Chen, H.; Zhu, W.; Li, X.; Xue, L.; Wang, Z.-Y.; Wu, H. Genetic and epigenetic patterns in patients with the head-and-neck paragangliomas associate with differential clinical characteristics. *J. Cancer Res. Clin. Oncol.* **2017**, *23*, 8812–8960. [CrossRef] [PubMed]

23. Fishbein, L.; Khare, S.; Wubbenhorst, B.; Desloover, D.; D'Andrea, K.; Merrill, S.; Cho, N.W.; Greenberg, R.A.; Else, T.; Montone, K.; et al. Whole-exome sequencing identifies somatic ATRX mutations in pheochromocytomas and paragangliomas. *Nat. Commun.* **2015**, *6*, 6140. [CrossRef] [PubMed]

24. Muth, A.; Crona, J.; Gimm, O.; Elmgren, A.; Filipsson, K.; Askmalm, M.S.; Sandstedt, J.; Tengvar, M.; Tham, E. Genetic testing and surveillance guidelines in hereditary pheochromocytoma and paraganglioma. *J. Intern. Med.* **2018**, *285*, 187–204. [CrossRef]

25. Niemann, S.; Becker-Follmann, J.; Nurnberg, G.; Rüschendorf, F.; Sieweke, N.; Hügens-Penzel, M.; Traupe, H.; Wienker, T.F.; Reis, A.; Muller, U. Assignment of PGL3 to chromosome 1 (q21-q23) in a family with autosomal dominant non-chromaffin paraganglioma. *Am. J. Med. Genet.* **2001**, *98*, 32–36. [CrossRef]

26. Mannelli, M.; Castellano, M.; Schiavi, F.; Filetti, S.; Giacchè, M.; Mori, L.; Pignataro, V.; Bernini, G.; Giachè, V.; Bacca, A.; et al. Clinically Guided Genetic Screening in a Large Cohort of Italian Patients with Pheochromocytomas and/or Functional or Nonfunctional Paragangliomas. *J. Clin. Endocrinol. Metab.* **2009**, *94*, 1541–1547. [CrossRef] [PubMed]

27. Bayley, J.-P.; Kunst, H.P.; Cascón, A.; Sampietro, M.L.; Gaal, J.; Korpershoek, E.; Hinojar-Gutiérrez, A.; Timmers, H.J.; Hoefsloot, L.H.; Hermsen, M.A.; et al. SDHAF2 mutations in familial and sporadic paraganglioma and phaeochromocytoma. *Lancet Oncol.* **2010**, *11*, 366–372. [CrossRef]

28. Kunst, H.P.M.; Rutten, M.H.; Hoefsloot, L.H.; Timmers, H.J.L.M.; Marres, H.A.; Jansen, J.C.; Kremer, H.; Bayley, J.-P.; Cremers, C.W.R.J.; De Mönnink, J.-P. SDHAF2 (PGL2-SDH5) and Hereditary Head and Neck Paraganglioma. *Clin. Cancer Res.* **2011**, *17*, 247–254. [CrossRef] [PubMed]

29. Casey, R.; Garrahy, A.; Tuthill, A.; O'Halloran, D.; Joyce, C.; Casey, M.B.; O'Shea, P.M.; Bell, M. Universal Genetic Screening Uncovers a Novel Presentation of an SDHAF2 Mutation. *J. Clin. Endocrinol. Metab.* **2014**, *99*, E1392–E1396. [CrossRef]

30. Luchetti, A.; Walsh, D.; Rodger, F.; Clark, G.; Martin, T.; Irving, R.; Sanna, M.; Yao, M.; Robledo, M.; Neumann, H.P.H.; et al. Profiling of Somatic Mutations in Phaeochromocytoma and Paraganglioma by Targeted Next Generation Sequencing Analysis. *Int. J. Endocrinol.* **2015**, *2015*, 138573. [CrossRef] [PubMed]

31. Fishbein, L.; Leshchiner, I.; Walter, V.; Danilova, L.; Robertson, A.G.; Johnson, A.R.; Lichtenberg, T.M.; Murray, B.A.; Ghayee, H.K.; Else, T.; et al. Comprehensive Molecular Characterization of Pheochromocytoma and Paraganglioma. *Cancer Cell* **2017**, *31*, 181–193. [CrossRef]

32. Bausch, B.; Schiavi, F.; Ni, Y.; Welander, J.; Patocs, A.; Ngeow, J.; Wellner, U.; Malinoc, A.; Taschin, E.; Barbon, G.; et al. Clinical Characterization of the Pheochromocytoma and Paraganglioma Susceptibility Genes SDHA, TMEM127, MAX, and SDHAF2 for Gene-Informed Prevention. *JAMA Oncol.* **2017**, *3*, 1204–1212. [CrossRef] [PubMed]

33. Ding, Y.; Feng, Y.; Wells, M.; Huang, Z.; Chen, X. SDHx gene detection and clinical Phenotypic analysis of multiple paraganglioma in the head and neck. *Laryngoscope* **2018**, *129*, E67–E71. [CrossRef] [PubMed]

34. Griffiths, A.J.; Miller, J.H.; Suzuki, D.T.; Lewontin, R.C.; Gelbart, W.M. Somatic Versus Germinal Mutation. In *An Introduction to Genetic Analysis*, 7th ed.; W.H. Freeman: New York, NY, USA, 2000.

35. Guha, A.; Musil, Z.; Vicha, A.; Zelinka, T.; Pacak, K.; Astl, J.; Chovanec, M. A systematic review on the genetic analysis of paragangliomas: Primarily focused on head and neck paragangliomas. *Neoplasma* **2019**, *66*, 671–680. [CrossRef]

36. Rattenberry, E.; Vialard, L.; Yeung, A.; Bair, H.; McKay, K.; Jafri, M.; Canham, N.; Cole, T.R.; Denes, J.; Hodgson, S.V.; et al. A Comprehensive Next Generation Sequencing–Based Genetic Testing Strategy to Improve Diagnosis of Inherited Pheochromocytoma and Paraganglioma. *J. Clin. Endocrinol. Metab.* **2013**, *98*, E1248–E1256. [CrossRef]

37. Crona, J.; Taïeb, D.; Pacak, K. New Perspectives on Pheochromocytoma and Paraganglioma: Toward a Molecular Classification. *Endocr. Rev.* **2017**, *38*, 489–515. [CrossRef]

38. Taïeb, D.; Pacak, K. New Insights into the Nuclear Imaging Phenotypes of Cluster 1 Pheochromocytoma and Paraganglioma. *Trends Endocrinol. Metab.* **2017**, *28*, 807–817. [CrossRef]

39. Jochmanova, I.; Pacak, K. Genomic Landscape of Pheochromocytoma and Paraganglioma. *Trends Cancer* **2018**, *4*, 6–9. [CrossRef]

40. Gimenez-Roqueplo, A.-P.; Dahia, P.L.; Robledo, M. An Update on the Genetics of Paraganglioma, Pheochromocytoma, and Associated Hereditary Syndromes. *Horm. Metab. Res.* **2012**, *44*, 328–333. [CrossRef] [PubMed]

41. Gunawardane, P.T.K.; Grossman, A.B. The clinical genetics of phaeochromocytoma and paraganglioma. *Arch. Endocrinol. Metab.* **2017**, *61*, 490–500. [CrossRef]

42. Gupta, N.; Strome, S.E.; Hatten, K.M. Is routine genetic testing warranted in head and neck paragangliomas? *Laryngoscope* **2018**, *129*, 1491–1493. [CrossRef]

43. Khatami, F.; Mohammadamoli, M.; Tavangar, S.M. Genetic and epigenetic differences of benign and malignant pheochromocytomas and paragangliomas (PPGLs). *Endocr. Regul.* **2018**, *52*, 41–54. [CrossRef]

44. Welander, J.; Söderkvist, P.; Gimm, O. Genetics and clinical characteristics of hereditary pheochromocytomas and paragangliomas. *Endocr. Relat. Cancer* **2011**, *18*, R253–R276. [CrossRef] [PubMed]

45. Dahia, P.L.M. Pheochromocytomas and Paragangliomas, Genetically Diverse and Minimalist, All at Once! *Cancer Cell* **2017**, *31*, 159–161. [CrossRef] [PubMed]

46. Liu, P.; Li, M.; Guan, X.; Yu, A.; Xiao, Q.; Wang, C.; Hu, Y.; Zhu, F.; Yin, H.; Yi, X.; et al. Clinical Syndromes and Genetic Screening Strategies of Pheochromocytoma and Paraganglioma. *J. Kidney Cancer VHL* **2018**, *5*, 14–22. [CrossRef] [PubMed]

47. Opocher, G.; Schiavi, F. Genetics of pheochromocytomas and paragangliomas. *Best Pract. Res. Clin. Endocrinol. Metab.* **2010**, *24*, 943–956. [CrossRef] [PubMed]

48. Kantorovich, V.; King, K.S.; Pacak, K. SDH-related pheochromocytoma and paraganglioma. *Best Pract. Res. Clin. Endocrinol. Metab.* **2010**, *24*, 415–424. [CrossRef]

49. Williams, M.D. Paragangliomas of the Head and Neck: An Overview from Diagnosis to Genetics. *Head Neck Pathol.* **2017**, *11*, 278–287. [CrossRef]

50. Ong, R.K.S.; Flores, S.K.; Reddick, R.L.; Dahia, P.L.M.; Shawa, H. A Unique Case of Metastatic, Functional, Hereditary Paraganglioma Associated with anSDHC Germline Mutation. *J. Clin. Endocrinol. Metab.* **2018**, *103*, 2802–2806. [CrossRef]

51. Weinhold, B. Epigenetics: The Science of Change. *Environ. Health Perspect.* **2006**, *114*, A160–A167. [CrossRef]

52. Rao, S.; Chiu, T.-P.; Kribelbauer, J.F.; Mann, R.S.; Bussemaker, H.J.; Rohs, R. Systematic prediction of DNA shape changes due to CpG methylation explains epigenetic effects on protein–DNA binding. *Epigenetics Chromatin* **2018**, *11*, 6. [CrossRef]

53. Ross, J.S.; Cronin, M. Whole Cancer Genome Sequencing by Next-Generation Methods. *Am. J. Clin. Pathol.* **2011**, *136*, 527–539. [CrossRef] [PubMed]

54. Currás-Freixes, M.; Inglada-Pérez, L.; Mancikova, V.; Montero-Conde, C.; Letón, R.; Comino-Méndez, I.; Apellániz-Ruiz, M.; Sánchez-Barroso, L.; Sánchez-Covisa, M.A.; Alcázar, V.; et al. Recommendations for somatic and germline genetic testing of single pheochromocytoma and paraganglioma based on findings from a series of 329 patients. *J. Med. Genet.* **2015**, *52*, 647–656. [CrossRef] [PubMed]

55. Roose, L.M.; Rupp, N.J.; Röösli, C.; Valcheva, N.; Weber, A.; Beuschlein, F.; Tschopp, O. Tinnitus with Unexpected Spanish Roots: Head and Neck Paragangliomas Caused by SDHAF2 Mutation. *J. Endocr. Soc.* **2020**, *4*. [CrossRef] [PubMed]

56. Brouwers, F.M.; Eisenhofer, G.; Tao, J.J.; Kant, J.A.; Adams, K.T.; Linehan, W.M.; Pacak, K. High Frequency ofSDHBGermline Mutations in Patients with Malignant Catecholamine-Producing Paragangliomas: Implications for Genetic Testing. *J. Clin. Endocrinol. Metab.* **2006**, *91*, 4505–4509. [CrossRef]

57. Favier, J.; Amar, L.; Gimenez-Roqueplo, A.-P. Paraganglioma and phaeochromocytoma: From genetics to personalized medicine. *Nat. Rev. Endocrinol.* **2015**, *11*, 101–111. [CrossRef]

Publisher's Note: MDPI stays neutral with regard to jurisdictional claims in published maps and institutional affiliations.

International Journal of
Molecular Sciences

Review

The PI3K-AKT-mTOR Pathway and Prostate Cancer: At the Crossroads of AR, MAPK, and WNT Signaling

Boris Y. Shorning, Manisha S. Dass, Matthew J. Smalley and Helen B. Pearson *

The European Cancer Stem Cell Research Institute, Cardiff University, Hadyn Ellis Building, Maindy Road, Cardiff CF24 4HQ, Wales, UK; ShorningB@cardiff.ac.uk (B.Y.S.); DassMS@cardiff.ac.uk (M.S.D.); SmalleyMJ@cardiff.ac.uk (M.J.S.)
* Correspondence: PearsonH2@cardiff.ac.uk

Received: 1 June 2020; Accepted: 22 June 2020; Published: 25 June 2020

Abstract: Oncogenic activation of the phosphatidylinositol-3-kinase (PI3K), protein kinase B (PKB/AKT), and mammalian target of rapamycin (mTOR) pathway is a frequent event in prostate cancer that facilitates tumor formation, disease progression and therapeutic resistance. Recent discoveries indicate that the complex crosstalk between the PI3K-AKT-mTOR pathway and multiple interacting cell signaling cascades can further promote prostate cancer progression and influence the sensitivity of prostate cancer cells to PI3K-AKT-mTOR-targeted therapies being explored in the clinic, as well as standard treatment approaches such as androgen-deprivation therapy (ADT). However, the full extent of the PI3K-AKT-mTOR signaling network during prostate tumorigenesis, invasive progression and disease recurrence remains to be determined. In this review, we outline the emerging diversity of the genetic alterations that lead to activated PI3K-AKT-mTOR signaling in prostate cancer, and discuss new mechanistic insights into the interplay between the PI3K-AKT-mTOR pathway and several key interacting oncogenic signaling cascades that can cooperate to facilitate prostate cancer growth and drug-resistance, specifically the androgen receptor (AR), mitogen-activated protein kinase (MAPK), and WNT signaling cascades. Ultimately, deepening our understanding of the broader PI3K-AKT-mTOR signaling network is crucial to aid patient stratification for PI3K-AKT-mTOR pathway-directed therapies, and to discover new therapeutic approaches for prostate cancer that improve patient outcome.

Keywords: AKT; AR; castration-resistant prostate cancer (CRPC); MAPK; mTOR; PI3K; prostate cancer; therapeutic resistance; WNT

---

1. Introduction

Prostate cancer is the second leading cause of cancer-related deaths in men worldwide, despite extensive efforts to raise awareness and significant advancements in detection, screening, and treatment approaches [1–3]. Although patients with localized prostate cancer generally have a good prognosis, the 5-year relative survival rate is significantly reduced for patients that present with metastatic prostate cancer at diagnosis [4]. ADT and/or radiotherapy remains the mainstay treatment for patients that relapse post-surgery. ADT involves blocking the production of androgen in the testes via the hypothalamus-pituitary-gonadal axis with luteinizing hormone releasing hormone (LHRH) agonists (e.g., Leuprolide) or antagonists (e.g., Degorelix). Although prostate tumors respond initially to ADT, the emergence of androgen-independent, castration-resistant prostate cancer (CRPC) invariably occurs and the outcome is poor [5–8]. Treatment options for CRPC and patients with metastatic disease at diagnosis include chemotherapy, radium-223, second generation anti-androgens (e.g., the Cytochrome P450 17A1 (CYP17A1) inhibitor abiraterone acetate that prevents androgen biosynthesis, or enzalutamide that targets AR directly), and clinical trials [5,6,8–10]. However, CRPC remains incurable and new biomarkers and treatments for prostate cancer and CRPC are in high demand.

PI3K-AKT-mTOR signaling is elevated in a high proportion of prostate cancer patients, and CRPC is associated with increased activation of the PI3K-AKT-mTOR pathway [11–13]. Accordingly, PI3K-AKT-mTOR pathway inhibitors are currently being explored as therapeutic agents against hormone-sensitive prostate cancer and CRPC [11–17]. PI3Ks are a large family of lipid kinase enzymes divided into three classes termed Class I (subdivided into Class IA and IB), Class II, and Class III, reflecting substrate specificity and subunit organization [18–20]. Class IA PI3Ks are heterodimers containing a catalytic subunit (p110α, p110β, or p110δ, encoded by *PIK3CA*, *PIK3CB* and *PIK3CD* respectively) and a regulatory subunit (p85α/p55α/p50α, p85β or p55γ, encoded by *PIK3R1*, *PIK3R2* and *PIK3R3* respectively) that controls protein localization, receptor binding, and activation [19–21]. Class IA isoforms are ubiquitously expressed, except for p110δ and p55γ that are primarily expressed in the hematopoietic/central nervous systems and testes [19–22]. Receptor tyrosine kinases (RTKs) can activate p110α, p110β, and p110δ catalytic isoforms, whereas the p110β isoform can be additionally activated by G protein-coupled receptors (GPCRs) [19–22] (Figure 1). The small GTPase RAS can also directly activate p110α and p110δ, while Rho-GTPases (e.g., RAC) are reported to activate p110β [20]. Once activated, Class IA PI3Ks initiate a wave of downstream signaling events by synthesizing the lipid secondary messenger phosphatidylinositol 3, 4, 5 trisphosphate (PIP3) from phosphatidylinositol 4,5 bisphosphate (PIP2) to mediate cell growth, proliferation, autophagy, and apoptosis [19,21]. The tumor suppressor, phosphatase and tensin homolog deleted on chromosome 10 (PTEN), negatively regulates PI3K-AKT-mTOR signaling by converting PIP3 back to PIP2 [23] (Figure 1).

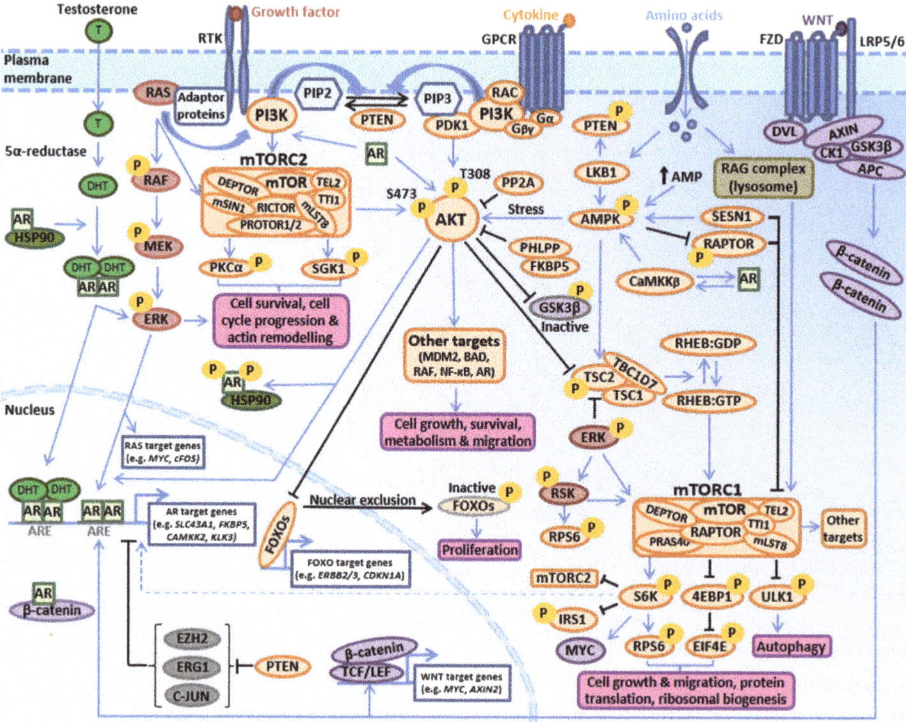

Figure 1. PI3K-AKT-mTOR signaling interaction with the AR, MAPK, and WNT pathways. Image displays a model of PI3K-AKT-mTOR signaling via Class IA PI3Ks, and crosstalk with AR, RAS/MAPK, and WNT signaling cascades. 4EBP1, eukaryotic initiation factor 4E binding protein 1; AMP, adenosine

monophosphate; AMPK, 5′ AMP-activated protein kinase; APC, adenomatous polyposis coli; ARE, androgen responsive element; AXIN, axis inhibition protein; BAD, Bcl-2-associated death promoter; c-JUN, transcription factor AP−1; CaMKKβ, Ca(2+)/calmodulin-dependent protein kinase kinase β; CK1, casein kinase 1; DEPTOR, DEP domain-containing mTOR-interacting protein; DHT, dihydrotestosterone; DVL, dishevelled; EIF4E, eukaryotic translation initiation factor 4E; *ERBB2/3*, Erb-B2 receptor tyrosine kinase 2, that encodes human epidermal growth factor 2/3 (HER2/3); ERG1, ETS-related gene 1; ERK, mitogen-activated protein kinase 1/3; EZH2, enhancer of zeste homolog 2; *FKBP5*, FK506 binding protein 5; FOXO, forkhead box protein O; FZD, frizzled family receptor; GDP, guanosine diphosphate; GPCR, G-protein coupled receptor; GSK3β, glycogen synthase kinase 3 beta; GTP, guanosine triphosphate; HSP90, heat-shock protein 90; IRS, insulin receptor substrate; *KLK3*, kallikrein related peptidase 3 (encoding prostate specific antigen, PSA); LKB1, liver kinase B1; LRP5/6, low-density lipoprotein receptor-related proteins 5 and 6; LEF, lymphoid enhancer binding factor 1; MAPK, mitogen-activated protein kinase; MDM2, mouse double minute 2 homolog; MEK, mitogen-activated protein kinase kinase; mLST8, mTOR associated protein LST8 homolog; mSIN1, mitogen-activated protein kinase associated protein 1 (MAPKAP1); mTOR, mammalian target of rapamycin; mTORC1/2, mTOR complex 1/2; NF-κB, nuclear factor kappa light chain enhancer of activated B cells; P, phosphorylation event; PDK1, phosphoinositide dependent kinase 1; PHLPP, PH domain leucine-rich repeat protein phosphatase; PKCα, protein kinase C alpha; PP2A, protein phosphatase 2A; PRAS40, proline-rich AKT substrate of 40 kDa; PROTOR1, protein observed with Rictor-1; PROTOR2, protein observed with Rictor-2; RAF, rapidly accelerated fibrosarcoma; RAG, recombination activating genes; RAPTOR, regulatory-associated protein of mTOR; RHEB, RAS homolog enriched in brain; RICTOR, rapamycin-insensitive companion of mTOR; RPS6, ribosomal protein S6; RSK, 90 kDa ribosomal S6 kinase; RTK, receptor tyrosine kinase; S6K, p70 ribosomal S6 kinase; SESN1, sestrin 1; SGK1, serum/glucocorticoid-regulated kinase 1; *SLC43A1*, solute carrier family 43 member 1 (encoding ʟ-type amino acid transporter 3, LAT3); T, testosterone; TBC1D7, Tre2-Bub2-Cdc16 domain family member 7; TCF, T cell factor; TEL2, telomere length regulation protein (or telomere maintenance 2, TELO2); TSC1, Tuberous sclerosis complex 1; TCS2, tuberous sclerosis complex 2; TTI1, TELO2 interacting protein 1; ULK1, Unc-51 like autophagy activating kinase 1; WNT, WNT ligand. Figure based on previous work [12,14,19–22,24,25].

Elevated PIP3 levels lead to the activation of multiple kinases, including PDK1, which phosphorylates downstream targets such as AKT at residue Thr308 [19,21,26–28]. Activated AKT phosphorylates numerous substrates to regulate vital cellular processes, including FOXOs, GSK3β, NF-κB, and TSC2 [19,21,26–28]. For instance, TSC2 phosphorylation by AKT inactivates RHEB, which potentiates mTORC1 signaling and results in the inhibition of autophagy and increases cell growth, protein translation and ribosomal biogenesis via the subsequent phosphorylation of mTORC1 substrates such as ULK1, S6K, and 4EBP1 [27,29]. Phospho-S6K can also phosphorylate RICTOR to regulate mTORC2 signaling [30]. mTORC2 phosphorylates multiple downstream targets to mediate cell survival, cell cycle progression, and actin remodeling. These include AKT at residue Ser473, which leads to AKT hyperactivation, serum/glucocorticoid-regulated kinase 1 (SGK1) and protein kinase Cα (PKCα) [31,32].

In addition to mediating PI3K-dependent signaling, AKT, PTEN and mTORC1/2 have also been shown to play a role in PI3K-independent signaling events (reviewed in [23,33–36]), and the PI3K-AKT-mTOR cascade interacts with multiple cooperative signal transduction cascades via a series of partially understood interactions and feedback loops to promote tumor growth (including MAPK, AR and WNT signaling, Figure 1). Hence, establishing the scope of this complex signaling program is fundamental for the identification of new and effective biomarkers and therapeutic approaches that will benefit patients with prostate cancer.

2. Genetic Aberrations in the PI3K-AKT-mTOR Pathway in Prostate Cancer Are Diverse

Augmented phosphorylation/activation of key PI3K-AKT-mTOR pathway components (e.g., p-AKT and p-mTOR) has been shown to correlate with prostate cancer progression in the clinic [37–41].

Furthermore, genomic and transcriptomic profiling has revealed that genetic alterations and deregulated gene expression of PI3K pathway components are common in patients with prostate cancer, occurring in as many as 42% of primary and 100% of metastatic prostate cancer samples [42–46]. Deregulation of the PI3K-AKT-mTOR pathway reflects a variety of genetic alterations, primarily PTEN loss-of-function [42–46]. To improve our understanding of the frequency and diversity of PI3K-AKT-mTOR pathway genetic aberrations in prostate cancer, we used the cBioPortal platform to survey three publicly available prostate cancer genomic datasets with primary and/or metastatic patient samples for a panel of 68 genes that encode key PI3K cascade components/effectors [47,48]. OncoPrints displaying the percentage frequency of each type of genetic aberration assessed within each dataset (i.e., gene mutation, amplification and deep deletion) highlight that PI3K-AKT-mTOR pathway genetic alterations are commonplace in primary and metastatic prostate cancer, and illustrate that the wide range of genetic events observed have a tendency to co-occur (Figures S1–S3 and Tables S1–S3, summarized in Table 1).

Table 1. Frequency of common genetic alterations in PI3K-AKT-mTOR pathway genes in prostate cancer.

Common Types of Genetic Alterations in PI3K-AKT-mTOR Pathway Genes	Frequency in Prostate Cancer [1]
PTEN deletion/mutation	16.4–32.0%
DEPTOR amplification	5.1–21.4%
SGK mutation/amplification	5.6–20.5% (*SGK3*)
	0.2–2.7% (*SGK1*)
FOXO deletion	0.0–15.2% (*FOXO1*)
	4.5–13.4% (*FOXO3*)
MAP3K7 deletion	5.9–14.8%
RRAGD deletion	6.5–14.4%
SESN1 mutation/deletion	5.4–13.6%
PIK3CA mutation/amplification	5.5–11.5%
PIK3C2B mutation/amplification	1.4–11.5%
PDPK1 amplification	0–8.1%

[1] Data sourced from the Memorial Sloan Kettering Cancer Centre/Dana-Farber Cancer Institute (MSKCC/DFCI) (n = 1013) [45] and The Cancer Genome Atlas (TCGA), Firehose Legacy (n = 492) prostate adenocarcinoma datasets, and the metastatic prostate adenocarcinoma Stand Up To Cancer & Prostate Cancer Foundation International Dream Team (SU2C-PCF IDT) dataset (n = 444) [46] using cBioPortal [47,48] (Tables S1–S3). Only samples with mutation and copy number alteration (CNA) data were analyzed. The percentage frequency range for each genetic alteration listed reflects the entire patient population across all the three datasets, irrespective of the disease stage or subtype.

Common genetic alterations within the three prostate cancer datasets analyzed were observed in *PTEN, DEPTOR, SGK3, FOXO1/3, MAP3K7, RRAGD, SESN1, PIK3CA, PIK3C2B,* and *PDPK1* (Table 1). In addition, a vast range of less frequent aberrations were also detected, including genes encoding AMPK subunits (e.g., amplification of *PRKAB1* and *PRKAB2*) and AMPK regulators (e.g., *CAMKK2* and *LKB1* deletion) (Figures S1–S3, Tables S1–S3), as described below.

2.1. PI3K Gain of Function

2.1.1. Class IA PI3Ks

Gain-of-function mutations in *PIK3CA* (encoding p110α) that activate the PI3K cascade are highly prevalent in a number of malignancies, including up to 40% of breast cancer patients [49] and as many as 53% of endometrial cancer patients [50]. In the prostate cancer datasets analyzed, *PIK3CA* mutation and high-level gene amplification occur in up to 4% and 9% of cases respectively (Tables S1–S3), although high-level amplification has been observed previously in as many as 29% of cases [13]. Our recent work identified that *PIK3CA* genetic alterations significantly correlate with poor prostate cancer prognosis, and that *Pik3ca* oncogenic mutation at a clinically relevant hotspot (H1047R) in mouse prostate epithelium can cause locally invasive prostate adenocarcinoma, demonstrating *Pik3ca* activation is a genetic driver of prostate cancer in vivo [13]. Although less common, *PIK3CB* mutation and amplification have also been detected in clinical prostate tumor specimens (0.6–1.8% and

1.8–3.1% respectively, Tables S1–S3), and activation of p110β (encoded by *PIK3CB*) predisposes prostate intra-epithelial neoplasia in mice [51]. Previous work has shown that p110α isoform-specific PI3K inhibitors can suppress *Pik3ca* mutant prostate cancer, whereas a p110β/δ inhibitor, or combined p110α/β blockade improves therapeutic outcome in *Pten*-deficient (p110β-dependent) prostate cancers [13,52,53]. Consequently, these findings have identified that selected p110 isoform-specific inhibitors may prove to hold efficacy against *PIK3CA* mutant and *PTEN*-deleted prostate cancer in the clinic.

Unlike the ubiquitous p110α and p110β PI3K catalytic isoforms, p110δ is predominantly expressed in cells of hematopoietic lineage and sensory neurons [54–56], and p110δ isoform-specific inhibitors are currently being explored in the clinic for B-cell malignancies and some autoimmune diseases [57]. However, several epithelial malignancies have also been shown to express p110δ [58], and 3% of patients with head and neck, germ cell, or colorectal cancer are reported to carry a *PIK3CD* mutation [59]. In patients with prostate cancer, *PIK3CD* mutation and amplification are infrequent events (≤1.1%, Tables S1–S3). However, a *PIK3CD* splice variant missing exon 20 (*PIK3CD-S*) has been identified in African American prostate cancer patients that can promote proliferation and AKT-mTOR signaling [60], and several CRPC cell lines have been shown to express p110δ at high levels, comparable to that detected in leukocytes [58]. In this study, inactivation of p110δ in p110δ-high CRPC cells suppressed PI3K-AKT signaling and inhibited cell proliferation, suggesting p110δ inhibitors may prove to hold therapeutic efficacy against p110δ-high prostate cancer [58].

The p85 regulatory subunit of PI3K is often present in a monomeric free form, and at a higher ratio relative to the p110 catalytic subunit, which suppresses p110 activity in the absence of stimuli [20,61,62]. Both free p85 monomers and p85–p110 heterodimers have been shown to bind to insulin receptor substrate (IRS), a cytoplasmic adaptor protein for the RTK insulin growth factor 1 (IGF-1) receptor, in addition to directly binding with activated RTKs [20,61]. Nonetheless PI3K-AKT-mTOR signaling activation is considered to require p85–p110 heterodimerization [20,61]. Interestingly, constitutive heterozygous deletion of *PIK3R1* (that encodes p85α and splice variants p55α/p50α) has been shown to lower blood glucose, enhance insulin sensitivity, potentiate insulin–stimulated glucose transport in skeletal muscle and adipocytes, and can stimulate insulin-dependent AKT phosphorylation in mouse liver [63,64]. Furthermore, liver-specific deletion of *Pik3r1* in mice is reported to not only enhance insulin and growth factor signaling, but causes development of aggressive hepatocellular carcinomas with pulmonary metastases associated with AKT activation and decreased PTEN expression [65]. *PIK3R1* shRNA-mediated knockdown in human breast cancer cell lines can also augment AKT signaling and anchorage-independent growth, illustrating a tumor suppressive role for p85α in breast cancer [66]. While p85α is generally viewed as a tumor suppressor, evidence in the literature also points toward an oncogenic role, similarly to p85β and p55γ [67–71].

In prostate cancer, *PIK3R1* is rarely mutated (0.4–1.6% of cases) yet deep deletions occur in 1–6% of patients (Tables S1–S3), which could potentially promote PI3K-AKT-mTOR signaling. Genetic alterations in *PIK3R2* (percentage incidence: mutation < 1.1%, amplification < 2.9%, deletion < 0.23%, Tables S1–S3) and *PIK3R3* (percentage incidence: mutation < 0.7%, amplification < 0.5%, deletion < 1%, Tables S1–S3) are also infrequent, yet the functional significance of these events remains unclear. Interestingly, down-regulation of *PIK3R1* in prostate cancer has been linked to reciprocal negative feedback between the AR and PI3K signaling cascades [72], and *PIK3R3* upregulation has been linked to prostate hyperplasia [73]. Furthermore, *PIK3R2* upregulation in prostate cancer specimens has recently been shown to inversely correlate with miR-126 expression [74]. Song and colleagues identified *PIK3R2* as a direct target of miR-126 in prostate cancer cell lines, and reported that enforced miR-126 expression in prostate cancer cell lines reduces *PIK3R2* mRNA expression and suppresses cell proliferation, migration, and invasion [74].

2.1.2. Class IB PI3Ks

The smaller Class IB PI3K family is comprised of the catalytic subunit p110γ and two regulatory subunits, p101 and p87 (also known as p84), which are encoded by *PIK3CG*, *PIK3R5*, and *PIK3R6* respectively. Like Class IA, Class IB PI3Ks generate PIP3 from PIP2 to stimulate downstream effectors [19]. Class IB PI3Ks transmit Gβγ-GPCR and RAS signals to coordinate immune, inflammatory and allergic responses, predominantly within hematopoietic cells [18–20,22]. However, Brazzatti and colleagues have shown that knockdown of p110γ or p101 in 4t1.2 and MDA-MB-231 triple-negative breast cancer cell lines reduces migration in vitro and metastatic potential in xenograft mouse models, whereas p87/p84 knockdown had the opposite effect [75]. *PIK3CG* mutation and amplification are frequent in multiple malignancies, including 9–11% of melanomas and uterine, stomach and squamous cell lung cancers, while genetic alterations in *PIK3R5* and *PIK3R6* are prevalent in uterine cancer and melanoma, occurring in 4–8% of cases [50,76–79]. In prostate cancer, Class IB PI3K genetic aberrations are less common, and include *PIK3CG* mutation and amplification (1.4–1.8% and 0.6–3.6% incidence respectively) as well as *PIK3R5* and *PIK3R6* deep deletions (0–3.3% incidence) that are indicative of a homozygous deletion (Tables S1–S3).

2.1.3. Class II PI3Ks

In comparison to Class I PI3Ks, the Class II family of PI3Ks (PI3KC2α, β and γ, encoded by *PIK3C2A*, *PIK3C2B*, and *PIK3C2G* respectively) is less well-characterized. Class II PI3Ks are generally considered to catalyze the production of lipid secondary messengers phosphatidylinositol 3-phosphate (PtdIns3*P* or PI(3)P) and phosphatidylinositol 3,4-bisphosphate (PI(3,4)P2) to mediate cell migration, channel regulation, endocytosis, and exocytosis [18,80]. The frequency of *PIK3C2A*, *PIK3C2B* and *PIK3C2G* mutation is generally low (0.2–1.4% incidence, Tables S1–S3), however *PIK3C2B* amplification has been observed in as many as 10% of cases (Table S3). Although the role of *PIK3C2B* amplification in prostate cancer is not clear, a recent study identified that PI3KC2β is highly expressed in PTEN-negative PC3 and LNCaP prostate cell lines compared to PTEN-positive DU145 prostate cancer cells (*PTEN*$^{+/-}$), and PNT2 immortalized "normal" prostate epithelial cells (*PTEN*$^{+/+}$) [81]. This study also reported that PI3KC2β regulates MAPK signaling to mediate prostate cancer cell invasion, thus the PI3KC2β-MEK-ERK signaling axis may present a novel therapeutic target for invasive prostate cancer [81].

2.1.4. Class III PI3Ks

The Class III PI3K subfamily is comprised of the catalytic subunit vacuolar protein sorting 34 (VPS34) encoded by *PIK3C3*, and the regulatory subunit vacuolar protein sorting 15 (VPS15, or p150) encoded by *PIK3R4*. VPS34 catalyzes the phosphorylation of phosphatidylinositol (PI) to produce PI(3)P, which plays a central role in the regulation of intracellular trafficking [82]. To regulate the fusion and maturation of endosomes, VPS34 binds to VPS15 and Beclin-1 to form either VPS34 Complex I or VPS34 Complex II that differ by binding to Autophagy Related 14 (ATG14) or UV radiation resistance associated protein (UVRAG) respectively [83]. The Class III PI3K family has also been shown to mediate autophagy, endosome–lysosome maturation, membrane trafficking, and AMPK-dependent insulin sensitivity [82,84–89].

PIK3C3 mutations are most frequently observed in uterine and gastric cancer patients (7% and 3.5% respectively), and *PIK3R4* gene mutation or amplification occur in up to 10% of squamous cell lung cancer and uterine cancer patients [50,76,77]. Although *PIK3C3/PIK3R4* mutation and *PIK3C3* gene amplification are infrequent events in prostate cancer (<1% of cases), *PIK3R4* high-level gene amplification is observed in up to 6.5% of cases and could potentially facilitate prostate cancer growth (Tables S1–S3).

Taken together, these data highlight the emerging diversity of genetic alterations within the PI3K family in prostate cancer, and emphasize the need for future work to gain further insight into the functional importance of these different genetic alterations during prostate cancer formation, progression, and recurrence. This is particularly important, as determining their non-redundant roles may present novel therapeutic targets and could aid patient stratification for future clinical trials.

2.2. Loss of Function of Phosphoinositide Phosphatases

Phosphoinositide phosphatases are a family of enzymes that dephosphorylate phosphoinositides to diminish phosphoinositide signals and regulate cellular functions [90]. The PI3K-AKT-mTOR pathway is regulated by multiple phosphoinositide phosphatases, including the tumor suppressor PTEN that dephosphorylates PIP3 into PIP2 to reduce PI3K-AKT-mTOR pathway activity (Figure 1). Genetic alterations in phosphoinositide phosphatases are strongly associated with human malignancies, and *PTEN* is one of the most frequently deleted genes in prostate cancer [91–95]. Here we review the frequency of genetic alterations in prostate cancer for genes encoding key phosphoinisitide phosphatases known to regulate the PI3K-AKT-mTOR cascade.

2.2.1. Loss or Inactivation of PTEN

PTEN is a lipid/protein phosphatase that has been shown to negatively regulate the PI3K-AKT-mTOR pathway by dephosphorylating phosphatidylinositol (3,4,5)-trisphosphates (PIP3) back to phosphatidylinositol 4,5-bisphosphates (PIP2) (Figure 1) [23,96,97]. *PTEN* genetic alterations, primarily homozygous deletion, are common in advanced prostate cancer and significantly correlate with poor outcome and elevated PI3K-AKT-mTOR signaling [13,14,37,39,98,99]. The functional consequence of *PTEN* loss has been studied in vivo using a number of genetically engineered mice, which have demonstrated *PTEN* loss is a genetic driver of invasive prostate cancer [24,100–104]. Homozygous *Pten* deletion within the murine prostate epithelium leads to aggressive, locally invasive prostate carcinoma that has an inherent ability to acquire castration-resistant disease [13,24,100–102,105]. However, metastatic disease is rare in these models, possibly owing to the primary tumor reaching ethical limits before disseminated cells can colonize distant sites, differences in genetic background, and/or *PTEN* loss-induced p21/p53-dependent senescence [102–104,106,107].

In primary prostate adenocarcinoma, *PTEN* mutation and deep deletion occur in 2% and 18% of cases respectively (Table S1), and the frequency appears to increase in metastatic disease (6% and 26% respectively, Table S3). Although the majority of *PTEN* mutations identified in prostate cancer are truncating mutations, missense mutations are also observed, which could differentially impact PTEN lipid and/or protein phosphatase function [108]. Thus, determining how each PTEN genetic alteration impacts PTEN function may inform clinical trial design.

PTEN heterozygous deletion and epigenetic silencing can also deplete PTEN expression/function [92,98,106,109]. Importantly, mono-allelic deletion of *PTEN* has been reported in up to 68% of prostate cancer surgical specimens and PTEN immunohistochemistry (IHC) and/or fluorescent *in situ* hybridization analysis has revealed PTEN loss may occur in as many as 60% of advanced/CRPC cases [92]. A subset of patients with prostate cancer have also been found to harbor intratumoral heterogeneous *PTEN* loss [92], which could have significant implications for therapeutic strategies.

2.2.2. Deregulation of Phosphoinositide Phosphatase Enzymes (other than PTEN)

In addition to PTEN, several other phosphatidylinositol phosphate phosphatase enzymes are also deregulated in human cancers that have the potential to facilitate malignant growth [90,91,110]. These phosphatases include; (a) proline-rich inositol polyphosphate 5-phosphatase (PIPP) encoded by polyphosphate-5-phosphatase J (*INPP5J*), (b) Src homology 2 (SH2) domain-containing inositol 5′-phosphatase 1 (SHIP1) encoded by inositol polyphosphate-5-phosphatase D (*INPP5D*), (c) Src homology 2 (SH2) domain-containing inositol 5′-phosphatase 2 (SHIP2) encoded by inositol polyphosphate phosphatase like 1 (*INPPL1*), and (d) inositol polyphosphate 4-phosphatase type II (INPP4B) encoded by *INPP4B*. While PTEN converts PIP3 to PIP2, PIPP and SHIP1/2 dephosphorylate PIP3 to phosphatidylinositol (3,4)-bisphosphate PI(3,4)P2, which is further hydrolyzed by INPP4B to form PI(3)P [90,111]. *INPP5D* deep deletion is observed in as many as 3.8% of patients with prostate cancer whereas *INPPL1* and *INPP4B* are amplified in up to 2.9% of cases (Tables S1–S3).

INPP5D/INPPL1/INPP5J/INPP4B mutation, *INPP5J* amplification and *INPPL1/INPP4B/INPP5J* deep deletion events are rare (≤1.2%, Tables S1–S3). Relative to PTEN, the frequency of genetic alterations in these phosphoinositide phosphatases is much lower, however they are gaining increasing attention in the literature [111]. Interestingly, *PIPP* deletion is reported to increase tumor growth in the mouse mammary tumor virus-polyoma middle tumor-antigen (MMTV-*PyMT*) breast cancer model, and is accompanied with elevated proliferation, plasma membrane PIP3 levels, and AKT activation [110]. However, *PIPP* deletion also significantly reduced the incidence of lung metastasis in this setting, suggesting PIPP mediates a critical metastatic process [110,112]. Furthermore, INPP4B can compensate for PTEN loss by acting as a "back-up" phosphatase, and is regarded as a tumor suppressor in several epithelial tissues including the prostate, breast, ovary, and thyroid [112–116]. Notably, *Inpp4B* loss and *Pten* heterozygous deletion can cooperate in mice to facilitate metastatic thyroid cancer by increasing PIP3 levels and AKT signaling relative to single mutants [115], and enforced INPP4B overexpression in PC3 ($PTEN^{-/-}$) and DU145 ($PTEN^{+/-}$) prostate cancer cells can suppress prostate cancer cell migration and invasion, both in vitro and in vivo [117]. Immunostaining to detect INPP4B in prostate carcinoma clinical samples has also identified INPP4B loss as an independent prognostic marker, correlating with reduced biochemical (PSA) relapse-free survival [118]. In contrast, SHIP2 is reported to play an oncogenic role. Unlike PTEN that catalyzes PIP3 into PIP2, SHIP2 converts PIP3 into PI(3,4)P2 to further potentiate AKT activity [119,120]. Moreover, increased SHIP2 expression directly correlates with poor survival in patients with colorectal cancer [120]. Consequently, genetic aberrations in phosphoinositide phosphatase enzymes could prove to differentially influence therapeutic responses to PI3K pathway-directed therapies.

2.3. AKT Gain of Function

AKT isoforms 1, 2, and 3 (encoded by *AKT1, AKT2,* and *AKT3* respectively) form a subfamily of serine/threonine protein kinases that possess both overlapping and distinct cellular functions to regulate a variety of cellular processes during normal tissue homeostasis and cell transformation [121,122]. PI3K activity elevates PIP3 levels to recruit AKT to the plasma membrane where is it activated (Figure 1). AKT is activated by multiple kinases, including PDK1 and mTORC2 that phosphorylate AKT at residues Thr308 and Ser473 respectively, triggering a wave of phosphorylation through multiple downstream targets that stimulate cell survival, proliferation, metabolism and differentiation to promote tumor growth [19,20,32,123,124]. AKT downstream targets include PRAS40 (a component of mTORC1), BAD, FOXOs, and MDM2 (reviewed in [31]). AKT signaling is negatively regulated by several protein phosphatases that dephosphorylate and inactivate AKT, including protein phosphatase 2 (PP2A), and PH domain and leucine-rich repeat protein phosphatase-1 and -2 (PHLPP1 and PHLPP2) [125,126]. Below, we outline the various genetic alterations within the *AKT* isoforms and their regulators that have been detected in prostate cancer, and discuss their potential to activate AKT signaling and promote prostate tumor growth.

2.3.1. AKT Mutation and Amplification

AKT genetic aberrations that increase AKT activity have been detected in multiple malignancies and are especially common in breast cancer, where *AKT3* amplification and *AKT1* E17K oncogenic mutation have been reported in up to 24% and 1–8% of cases respectively [127–129]. *AKT1, AKT2,* and *AKT3* activating mutations are rare in prostate cancer (≤0.9%, predominantly in *AKT1* at E17K), whereas *AKT1, AKT2,* and *AKT3* high-level gene amplification that can increase AKT activity is more common, particularly in advanced disease (up to 4.5%, 2%, and 4.7% incidence respectively, Tables S1–S3). Moreover, AKT activation in prostate cancer has been shown to positively correlate with Gleason score and invasive progression [37,130], and over-expression of myristoylated AKT (which causes constitutive AKT activation) causes prostate neoplasia in mice [131]. In support of an oncogenic role in prostate cancer and therapeutic resistance, conditional activation of AKT in either the LNCaP human prostate cancer cells or a transgenic mouse results in increased cell proliferation and inhibits cell death to promote tumor growth and castration-resistance in vivo [132]. Chen and colleagues have also demonstrated a requirement for AKT in *PTEN*-deficient prostate cancer, as *Akt1* haplodeficiency was found to suppress high-grade prostate intraepithelial neoplasia development within *Pten* heterozygous mice [133]. AKT inhibitors are being widely explored in the clinic to treat prostate cancer and have shown promise in *PTEN*-deficient patients [16,134].

2.3.2. Genetic Alteration of AKT Regulators

A number of genetic alterations in genes that encode AKT regulators have been linked to prostate cancer, including kinases (e.g., PDK1), binding proteins (e.g., FKBP5), and phosphatases (e.g., PHLPP1, PHLPP2, and PP2A) [42–46]. PDK1 (encoded by *PDPK1*) is recruited to the membrane by PIP3 to phosphorylate and activate multiple targets, including AKT at residue T308 (Figure 1). *PDPK1* amplification and PDK1 over-expression are observed in several human cancers, including breast cancer [135]. In prostate cancer, *PDPK1* mutations are rare (≤0.2%), yet *PDPK1* amplification occurs in up to 8.1% of patients (Tables S1–S3). Interestingly, PDK1 RNAi-mediated knockdown does not impair *Pten*-deleted prostate cancer growth in mice, possibly reflecting mTORC2-mediated activation of AKT, and/or compensatory augmentation of the MAPK cascade [136]. These findings suggest that PDK1 inhibitors are not likely to be efficacious against *PTEN*-deficient prostate cancer in the clinic as a single agent.

FKBP5 (also known as *FKBP51*) is an AR target gene that plays a key role in mediating the cellular distribution of steroid hormone receptors and has been shown to negatively regulate AKT signaling by stabilizing PHLPP1/2 (Figure 1) [11,24,137]. During androgen/AR-directed therapy, FKBP5-PHLPP1/2-AKT signaling forms a negative feedback loop between the AR and PI3K-AKT-mTOR pathways to facilitate ADT resistance [11,24,137], discussed in Section 3.2. Mutation and deep deletion of *FKBP5* are fairly infrequent in prostate cancer (≤1.22%, Tables S1–S3), however FKBP5 down-regulation has been linked to CRPC and increased AKT signaling [11].

PHLPP1 and PHLLP2 (encoded by *PHLPP1* and *PHLPP2)* are protein phosphatases that dephosphorylate and inactivate AKT. *PHLPP1* and *PHLPP2* deep deletion occurs in up to 3.9% and 6.5% of patients with prostate cancer respectively (Tables S1–S3), which could potentially sustain AKT-signaling. Interestingly, Chen and colleagues reported a strong tendency for *PTEN, PHLPP1, PHLPP2,* and *TP53* co-deletion in metastatic prostate cancer and that low *PHLPP1* expression correlates with reduced patient survival and relapse after surgery [138]. Additionally, the tumor suppressive function of PHLPP1 has been demonstrated in vivo, as *Phlpp1* loss causes prostate neoplasia in mice and promotes invasive carcinoma progression in *Pten*$^{+/-}$ transgenic mice [138]. In contrast, *Phlpp2* loss impairs *Pten/p53*-deleted prostate tumor growth in mice [139], indicating PHLPP1 and PHLPP2 mediate differential AKT-independent functions. Indeed, PHLPP2 can dephosphorylate MYC at residue Thr58 to prevent MYC degradation and promote tumor progression [139]. Consequently, in PHLPP2-positive MYC-driven advanced prostate cancer, it has been suggested that PHLPP2 may present a valuable therapeutic target [139].

In addition, genetic alterations in *PPP2CA* (protein phosphatase 2 catalytic subunit alpha) that encodes the negative AKT regulator PP2A have also been observed in prostate cancer [42–46], and PP2A loss has been linked to prostate cancer progression and metastatic potential in the clinic [140]. *PPP2CA* mutation and deep deletion events occur in 0.4–1.4% of patients with prostate cancer (Tables S1–S3), further highlighting the diversity of genetic aberrations in AKT regulators that could promote oncogenic PI3K signaling.

2.4. SGK Deregulation

The serum/glucocorticoid-regulated kinase isoforms SGK1, SGK2, and SGK3 belong to a subgroup of the AGC (cAMP-dependent, cGMP-dependent, and protein kinase C) family of protein kinases that play a role in multiple cellular processes including cell growth, proliferation, metabolism, intracellular trafficking and survival [141–143]. SGK1 and 3 are considered to be ubiquitously expressed, while

SGK2 expression is prominent in the liver, kidney, pancreas, and brain [144]. SGKs share structural similarities, upstream regulators, substrates and functions with the AKT isoforms (reviewed in [142]). For instance, all SGKs are phosphorylated and activated by PDK1, and SGK1 is a downstream target of mTORC2 [26,143,145–148] (Figure 1). SGKs are also activated by PI3K/PDK1-independent mechanisms, for example SGK1 is regulated by big mitogen-activated protein kinase-1 (BMK-1) and p38 mitogen-activated protein kinase in response to epidermal growth factor (EGF) and interleukin-6 (IL6) respectively [149,150]. Although the role of SGKs during prostate cancer is currently unclear, SGK1 over-expression has been shown to facilitate CRPC transition in a prostate cancer xenograft model, indicating that SGK1 can promote ADT-resistance [151]. Furthermore, the SGK1 inhibitor GSK650394 has been shown to induce autophagy and apoptosis in PC3, LNCaP, DU145, and CWR22RV1 prostate cancer cells in vitro [152]. Interestingly, SGK1 and SGK3 have also been linked to PI3K/AKT-targeted therapy resistance in breast cancer [145,148]. Gasser and colleagues have also shown that INPP4B over-expression leads to enhanced SGK3 activation in ZR-75-1 breast cancer cells, triggering a switch from AKT- to SGK-dependent signaling downstream of PDK1 [153].

Mutation of the SGK isoforms is a rare event in human cancers, however gene amplification is commonly detected [154]. In keeping with this, *SGK1, SGK2,* and *SGK3* are rarely mutated in prostate cancer (≤0.41%), whereas amplification occurs in up to 2.5%, 2.0%, and 20.3% of cases respectively (Tables S1–S3). Of note, the frequency of *SGK3* gene amplification is particularly high in the SUC2/PCF IDT metastatic prostate adenocarcinoma dataset (Table S3), underlining the need for future studies to establish how SGKs contribute to prostate cancer and metastatic progression.

2.5. Loss of FOXO Transcription Factors

The mammalian forkhead box O (FOXO) family consists of four transcription factors (FOXO1, 3, 4, and 6) that are highly similar in structure and function [155]. In response to insulin and growth factors, FOXOs modulate the transcription of several target genes to mediate key cellular processes including proliferation, apoptosis, autophagy, inflammation, metabolism and stress resistance, and they form an important regulatory circuit within the AKT and mTOR signaling cascades [156–158] (Figure 1). FOXOs are regulated by several kinases, including AKT and SGK isoforms, which phosphorylate and inactivate FOXO-mediated gene transcription by inhibiting FOXO DNA binding and triggering FOXO nuclear-to-cytoplasm translocation [157–159]. FOXOs are generally regarded as tumor suppressors, and are reported to inhibit mTORC1 via sestrins, however a number of oncogenic functions are emerging in the literature [156–158]. For instance, FOXO-mediated transcription of the mTORC2 component *RICTOR* in response to physiological stress is reported to promote mTORC2 signaling [156]. FOXOs also provide a reciprocal negative feedback loop between PI3K-AKT-mTOR pathway and AR signaling [12] (discussed in Section 3.2).

In prostate cancer, FOXO mutations are rare (<0.5% incidence), however *FOXO1* and *FOXO3* deep deletion is a frequent event, occurring in up to 15.2% and 13.4% of patients respectively (Tables S1–S3). *FOXO3* lies within the 6q21 locus that is frequently lost in prostate cancer [160], and reduced FOXO3 (also FOXO3a) activity via peptide driven inhibition is reported to accelerate prostate cancer progression in the transgenic adenocarcinoma mouse prostate (TRAMP) neuroendocrine prostate cancer model [161]. FOXO1 has also been shown to bind and inhibit the transcriptional activity of E26 transformation-specific (ETS) transcription factor ERG, which is over-expressed in 50% of prostate cancers owing to TMPRSS2-ERG (transmembrane protease, serine 2: ERG fusion) gene rearrangements [162]. Furthermore, *Foxo1* bi-allelic deletion and ERG overexpression can cooperate to cause prostate neoplasia in mice [162]. Together, these findings suggest FOXO1/3 act as tumor suppressors during prostate cancer.

FOXO4 gene amplification occurs in up to 8.8% of patients with metastatic prostate cancer (Table S3), however the functional importance of this genetic alteration remains to be clarified. Although *FOXO4* down-regulation is reported to correlate with reduced prostate cancer metastasis-free survival, conversely *FOXO4* knockdown in LNCaP cells can increase metastatic potential [163]. Thus, future work addressing the role of *FOXO4* during prostate cancer progression is warranted.

2.6. TSC1-TSC2-TBC1D7 Complex and RHEB Deregulation

To regulate mTORC1 signaling, TSC1, TSC2, and TBC1D7 form a complex to suppress RHEB GTPase, an upstream activator mTORC1 [164] (Figure 1). Activated AKT directly phosphorylates TSC2 at multiple residues to inhibit the TSC1:TSC2 complex, activate RHEB GTPase, and subsequently stimulate mTORC1 signaling [165,166]. TSC2 is also regulated by MAPK, WNT, and energy signals through coordinated phosphorylation by ERK, GSK3, and AMPK respectively, thus limiting mTORC1 activation and cell growth in response to poor growth conditions, and illustrating TSC2 as a central node for PI3K-AKT-mTOR crosstalk with multiple signaling cascades [164,166–168].

TSC1 and *TSC2* are frequently mutated/deleted in a variety of solid tumors, including lung (22%) and liver (16%) cancers, leading to deregulated PI3K-AKT-mTOR signaling [169,170]. In prostate cancer, the frequency of *TSC1* and *TBC1D7* mutation or deep deletion is low (≤0.8% incidence, Tables S1–S3), whereas *TSC2* mutation and deep deletion are more frequent (1–1.8% and up to 4.2% of cases respectively, Tables S1–S3). Interestingly an inactivating splice variant of *TSC2* unique to African American patients with prostate cancer has also recently been linked to aggressive prostate cancer and therapeutic resistance [60]. In mice, *Tsc1* conditional deletion in murine prostate epithelium is reported to cause prostate neoplasia associated with elevated mTORC1 signaling [171], and combined *Tsc2* and *Pten* heterozygosity has been shown to promote invasive prostate carcinoma relative to single mutants [172]. In lung cancer, TSC1 and TBC1D7 have been shown to function as oncoproteins [173], possibly reflecting mTORC1-independent functions such as TSC1-mediated activation of TGFβ-SMAD2/3 signaling [174]. Remarkably, up to 3%, 4%, and 7% of patients with prostate cancer also display *TBC1D7*, *TSC1*, and *TSC2* high-level amplification respectively (Tables S1–S3), yet the functional consequence is currently unclear.

RHEB GTPase has also been shown to act as a proto-oncogene in prostate cancer and up to 4% of patients with prostate cancer carry *RHEB* gene amplification, however *RHEB* oncogenic mutations are rare (≤0.1% incidence, Tables S1–S3). RHEB GTPase is over expressed in several prostate cancer cell lines and transgenic mice over-expressing *Rheb* specifically within the prostate epithelium develop low-grade prostatic intraepithelial neoplasia lesions by 10 months of age, accompanied with increased mTORC1 activity [175]. *Rheb* over-expression can also cooperate with *Pten* haploinsufficiency to promote prostate tumorigenesis [175], indicating *RHEB* amplification is likely to be a genetic driver of prostate tumorigenesis in the clinic.

2.7. Amplification of mTORC1 and mTORC2 Complex Components

The mTORC1 and mTORC2 protein complexes are functionally and structurally distinct, originally distinguished by their sensitivity to the mTOR inhibitor rapamycin [176–178]. Both mTORC1 and mTORC2 complexes contain mTOR, MLST8 (also known as G-protein beta-subunit like, GβL), TEL2, TTI1, and the negative regulator DEPTOR [179,180]. RAPTOR and PRAS40 (encoded by *AKT1S1*) are additional members of mTORC1 complex, whereas RICTOR, mSIN1, and PROTOR1/2 form the mTORC2 complex [180] (Figure 1). mTORC1 and mTORC2 are downstream effectors and regulators of PI3K/AKT signaling that mediate key cellular processes in response to growth factors and hormones [179,181–183]. mTORC1 is sensitive to rapamycin treatment and functions to regulate cell growth, autophagy, protein translation machinery, and cell-cycle progression by phosphorylating substrates such as ULK1, S6K and 4EBP1 [179,183–185]. The mTORC2 complex plays a critical role in PI3K/AKT signaling by increasing the activity of AKT, SGK1 and PKCα to regulate cell survival, metabolism and cytoskeletal dynamics [184] (Figure 1). mTORC2 is generally insensitive to rapamycin [179], however chronic exposure to the drug has been shown to impair mTORC2 assembly [185]. Crucially, mTORC1 and mTORC2 can also regulate each other via multiple mechanisms, including AKT regulation of PRAS40 to block suppression of mTORC1 activity and S6K regulation of mSIN1 to modulate mTORC2 activity [186].

In general, the frequency of genetic alterations in mTORC1 and mTORC2 components is low in prostate cancer. Genomic profiling data have shown that *mTOR* mutation occurs in 0.6–1.6% of cases, and the frequency of mutation or deep deletion in the other components of mTORC1/2 is ≤1% (Tables S1–S3). However, *DEPTOR* gene amplification is comparatively frequent, occurring in 5.1–21.4% of cases, with the highest incidence observed in the SUC2/PCF-IDT metastatic prostate adenocarcinoma dataset (Tables S1–S3). In addition, *DEPTOR* amplification directly correlates with worse disease/progression-free survival in the TCGA Firehose Legacy prostate adenocarcinoma dataset (Figure S4), indicating *DEPTOR* amplification may provide a valuable predictive biomarker in the clinic. DEPTOR is an endogenous suppressor of mTOR kinase activity, yet *DEPTOR* upregulation can reduce S6K1 activation, thus relieving feedback inhibition from mTORC1 to PI3K and mTORC2 signaling that results in increased AKT activation [187]. Nevertheless, *DEPTOR* knockdown in colorectal cancer cells reduced cell proliferation and induced differentiation [188], raising the possibility that *DEPTOR* can promote tumorigenesis in other epithelial cancers. DEPTOR has also been shown to exert mTORC1/2-independent functions in the nucleus as a transcriptional regulator in multiple myeloma cells [189] and is a transcriptional target of WNT/β-catenin/MYC signaling in colorectal cancer cells [188], adding further complexity to PI3K-AKT-mTOR and WNT pathway crosstalk.

In addition to *DEPTOR*, a number of other genes encoding mTOR components were also distinctly amplified in the SUC2/PCF-IDT metastatic prostate cancer dataset (*AKT1S1*, 2.7%; *MLST8*, 7.7%; *MAPKAP1*, 4.5%; *RPTOR*, 7%; *RICTOR*, 5%; *TELO2*, 6.5%; *TTI1*, 2.5%, Table S3), which could potentially facilitate tumor progression. However, none of these genetic alterations correlate with disease/progression-free survival (determined by cBioPortal analysis of the TCGA Firehose Legacy prostate adenocarcinoma dataset, *n* = 492, data not shown) [47,48]. Significantly, bi-allelic deletion of *Rictor* in mouse prostate epithelium has revealed RICTOR is not required for normal tissue homeostasis, yet RICTOR loss can suppress *Pten*-deleted prostate tumorigenesis in mice [190]. These findings indicate that mTORC2 signaling can contribute to *PTEN*-deleted prostate cancer growth, and that mTORC2 inhibition may be efficacious in the clinic against prostate cancers with *PTEN* loss [190].

Intracellular amino acids can also activate mTORC1 signaling by stimulating vacuolar H^+-ATPase (v-ATPase) to activate Ragulator, a guanine exchange factor that converts RAGA/B·GDP to RAGA/B·GTP, enabling formation of the active RAG complex where RAGA·GTP or RAGB·GTP form heterodimers with either RAGC·GDP or RAGD·GDP [36,191–195]. Similarly to RAGA/RAGB, RAGC/RAGD are functionally redundant and are 80–90% homologous [195]. When amino acids are sufficient, mTORC1 is recruited to the lysosome where it binds to the active RAG complex via RAPTOR, followed by its localization to RHEB that leads to mTORC1 activation [36,191,195] (Figure 1). Recent evidence also suggests that amino acids such as glutamine can activate mTORC1 in a RAG-complex independent manner, for example via the GTPase adenosine ribosylation factor 1 (ARF1) [196], highlighting the complex nature of mTORC1 regulation.

In prostate cancer, genetic alterations in *RRAGA* and *RRAGC* genes that encode RAGA and RAGC respectively are uncommon, however *RRAGB* (encoding RAGB) is amplified in up to 7.7% of cases and *RRAGD* (encoding RAGD) deep deletion occurs in 6.5–14.4% of cases (Tables S1–S3). Interestingly, *RRAGD* deep deletion in prostate adenocarcinoma strongly correlates with *FOXO3* deletion (one-sided Fisher's Exact test, *p*-value < 0.001; data sourced from the cBioPortal platform, TCGA Firehose Legacy prostate adenocarcinoma dataset, *n* = 492), however the functional consequence of *RRAGD/FOXO3* co-deletion and *RAGB* amplification during prostate cancer growth and therapeutic resistance is currently unknown and merits further investigation.

2.8. Aberrant AMPK Signaling

The metabolic sensor AMPK functions to maintain an adenosine triphosphate (ATP) equilibrium, influencing cell growth, lipid and glucose metabolism, autophagy and cell polarity [197]. AMPK is composed of a catalytic subunit (α1/α2, encoded by *PRKAA1/PRKAA2*), a β structural subunit (β1/β2, encoded by *PRKAB1/PRKAB2*) and a regulatory γ subunit (γ1/γ2/γ3, encoded by *PRKAG1/PRKAG2/PRKAG3*) [198]. AMPK activation plays a tumor suppressive role by inhibiting mTORC1 through the phosphorylation of TSC2 and RAPTOR in response to energy stress [199] (Figure 1), and by negatively regulating lipogenesis [200–202]. AMPK can also play an oncogenic role during stress (including hypoxia, oxidative stress, and glucose deprivation) to activate AKT, yet the molecular mechanisms involved remain to be fully elucidated [200]. Mutation and deep deletion of the AMPK subunits are uncommon in human malignancies [203], including prostate cancer (<1.2% incidence, Tables S1–S3). Instead, gene amplification of the AMPK subunits is more common [43,45,204]. In prostate cancer, high-level amplification of *PRKAB1, PRKAB2, PRKAG2,* and *PRKAG3* occurs in up to 6.3%, 6.8%, 4.1%, and 2% of cases respectively (Tables S1–S3). Whether AMPK amplification equates to increased activity remains to be determined, however AMPK phosphorylation/activation is reported to positively correlate with Gleason score and disease progression [205,206].

Interestingly, androgen-mediated activation of AMPK has been shown to increase the growth of prostate cancer cells, associated with elevated intracellular ATP levels and peroxisome proliferator-activated receptor gamma coactivator 1-alpha (PGC-1α)-mediated mitochondrial biogenesis [206]. Thus, AR-mediated AMPK activation could potentially function to avoid energy crisis and promote tumor growth. Upstream activators of AMPK include Ca^{2+}/calmodulin-dependent protein kinase kinase β (CAMKKβ), liver kinase B1 (LKB1), sestrins, and potentially mitogen-activated protein kinase kinase kinase 7 (MAP3K7) [207–209]. Below we explore several potential mechanisms underpinning deregulation of the AMPK-AKT/mTOR signaling axis in prostate cancer.

2.8.1. CAMKKβ Amplification

CAMKKβ is encoded by *CAMKK2* and phosphorylates AMPK in response to Ca^{2+} signaling. In prostate cancer, *CAMKK2* is amplified in up to 6.3% of patients (Tables S1–S3), however it is currently unknown if *CAMKK2* amplification promotes AMPK activity in the clinic. In a *Pten*-deleted prostate cancer mouse model, *Camkk2* deletion or CAMKKβ pharmacological inhibition has been shown to suppress prostate tumorigenesis and reduce *de novo* lipogenesis, whereas *Prkab1* (AMPK-β1) and *Pten* co-deletion accelerates tumor progression [210]. These findings indicate that CAMKKβ plays an oncogenic role in this setting and that CAMKKβ and AMPK-β1 play opposing roles in *Pten*-deficient prostate cancer, possibly reflecting their differential regulation of lipogenesis [210]. CAMKKβ has also been shown to activate AMPK in response to androgen signaling, and AMPK can subsequently inhibit AR function to form a negative feedback loop [210]. However, the impact on the PI3K-AKT-mTOR signaling cascade remains unclear. Interestingly, a recent report has shown that CAMKKβ can directly phosphorylate AKT at residue Thr308 in ovarian cancer cells [211], indicating CAMKKβ may regulate AKT/mTOR signaling both directly and indirectly via AMPK.

2.8.2. LKB1 Loss

LKB1 (encoded by serine/threonine kinase 11, *STK11*) is a multifaceted enzyme that plays a tumor suppressive role by phosphorylating multiple substrates (e.g., AMPK and PTEN) to regulate crucial cellular processes including cell metabolism, polarity, differentiation, and proliferation [212,213] (Figure 1). While *STK11* deletion or inactivating mutations are frequent in lung cancer (occurring in up to 50% of patients) [208], *STK11* mutations are rare in prostate cancer (0.2% incidence, Tables S1–S3) and the frequency of *STK11* deep deletion is also comparatively low (0–3.4% incidence, Tables S1–S3). We have previously shown that LKB1 exerts a tumor suppressive function in the prostate, as *Lkb1* homozygous deletion in murine prostate epithelial cells causes prostate intra-epithelial neoplasia (PIN), associated with elevated PI3K/AKT signaling [214]. The relatively mild effects of LKB1 loss are greatly enhanced when combined with *Pten* heterozygosity in the mouse prostate, which causes lethal metastatic prostate cancer [215]. Interestingly, the expression of either wild-type LKB1, or a kinase-dead form of LKB1 (LKB1^{K78I}) is sufficient to reduce tumor burden and impair metastatic potential of DU145 prostate cancer cells that lack LKB1, indicating LKB1 may also elicit a kinase-independent tumor suppressive function [215]. These in vivo findings indicate that deregulation of the LKB1-AMPK signaling axis is a potential mechanism whereby AKT/mTOR signaling is potentiated to facilitate prostate tumor formation and/or progression. Furthermore, a recent study has shown that LKB1 protein levels are reduced in immortalized prostate cancer cell lines relative to normal prostate epithelial cells, and siRNA-mediated *STK11* knockdown correlated with elevated hedgehog signaling and increased proliferation and invasion of prostate cancer cells in vitro, however PI3K-AKT-mTOR signaling was not assessed [216].

2.8.3. Sestrin Deletion

Sestrins are a family of stress inducible antioxidant proteins comprising of SESN1, SESN2, and SESN3, which play a key role in regulating autophagy, mitophagy, metabolic homeostasis, inflammation, hypoxia and oxidative stress [217–219]. SESN1 and SESN2 are p53 target genes that are induced upon DNA damage and oxidative stress [217]. SESN1 and SESN2 can directly bind to both the TSC1:TSC2 complex and AMPK, which leads to AMPK activation/autophosphorylation in a p53-dependent manner and stimulates AMPK-mediated phosphorylation of TSC2 to negatively regulate mTORC1 signaling [217]. In addition, sestrins are reported to negatively regulate mTORC1 signaling via GATOR2/RAG, indicating that sestrins can also mediate PI3K-AKT-mTOR signaling in response to energy stress (e.g., nutrient starvation) [220,221].

Genetic alterations in the genes encoding sestrins have been linked to non-small cell lung carcinoma (NSCLC) and colorectal cancer, and recent evidence in the literature has indicated sestrins play a tumor suppressive role [221,222]. Although sestrin mutations are rare (≤0.8%), *SESN1* deep deletion is a frequent event in prostate cancer occurring in 4.7–13.4% of cases (Tables S1–S3), potentially leading to increased mTORC1 signaling through alleviation of SESN1-mediated negative regulation of mTORC1. Interestingly, similarly to *FOXO3*, *SESN1* is located within the 6q21 locus that is commonly lost in prostate cancer [160]. *SESN1* is also reported to be transcriptionally repressed by AR [223], whereas p53 and FOXOs are known to mediate *SESN1* transcription [156,217]. Thus, future work exploring the functional significance and predictive value of *SESN1* depletion in prostate cancer could identify new therapeutic avenues or biomarkers to aid patient care.

2.8.4. MAP3K7 Deletion

MAP3K7 (also known as transforming growth factor (TGF) β-activated kinase 1, TAK1) is a serine/threonine protein kinase that mediates cell survival via NF-κB-dependent and NF-κB-independent signaling in response to TGFβ and cytokines [224]. Recent evidence in the literature has indicated that MAP3K7 may also mediate AMPK-AKT-mTOR signaling, as MAP3K7/TAK1 inactivation is associated with AMPK activation and reduced p-mTOR levels in skeletal muscle [209]. However, MAP3K7 is reported to mediate mTOR signaling independently of AMPK in hepatocellular carcinoma, possibly via p38 activation [225].

In prostate cancer, *MAP3K7* is a putative tumor suppressor gene and *MAP3K7* deletion has been shown to directly correlate with prostate cancer progression, lymph node metastasis, and biochemical recurrence [226,227]. *MAP3K7* deep deletion is a frequent event in prostate cancer, occurring in up to 14.8% of patients (Tables S1–S3). Furthermore, loss of *Map3k7* in mice has been shown to promote prostate tumorigenesis [227], suggesting MAP3K7 plays a tumor-suppressive function in the prostate. However, in an AML xenograft model MAP3K7 inhibition was found to attenuate leukemia development [228], indicating that MAP3K7 plays a dual role as a tumor suppressor and an oncogene depending on the malignancy.

3. The PI3K-AKT-mTOR Pathway Intersects with Multiple Oncogenic Signaling Cascades to Facilitate Prostate Cancer Growth

The PI3K-AKT-mTOR signaling cascade is one of the most frequently upregulated pathways in prostate cancer, which potentiates multiple downstream signaling events to mediate a plethora of cellular processes that promote tumor growth and therapeutic resistance to current treatment regimens. Targeting the PI3K-AKT-mTOR pathway using small molecules, such as pan-PI3K, PI3K-isoform specific, AKT, mTOR and dual PI3K/mTOR inhibitors has been challenging owing to their limited efficacy and poor tolerability (reviewed in [14–17,134,229,230]). Many clinical trials involving PI3K-AKT-mTOR-directed therapies have failed owing to incomplete inhibition of the pathway, reflecting the multiple modes of pathway redundancy and numerous positive/negative feedback loops that exist both within the PI3K-AKT-mTOR cascade and via crosstalk with other signaling pathways [15,231–236] (Figure 1). Here, we review PI3K-AKT-mTOR interactions with the RAS/MAPK, AR, and WNT signaling pathways, illustrating the need to improve our molecular understanding of the broader PI3K-AKT-mTOR signaling network. Delineating the complexity of the PI3K-AKT-mTOR pathway interactions with other signaling cascades during normal tissue homeostasis, tumorigenesis and therapeutic resistance is crucial for the discovery of new, efficacious personalized treatment approaches that overcome PI3K-AKT-mTOR inhibitor resistance.

3.1. PI3K-AKT-mTOR and RAS/MAPK Signaling Crosstalk

The RAS/MAPK cascade transduces extracellular growth signals via transmembrane receptors (e.g., RTKs and GPCRs) and a series of intracellular protein kinases to regulate gene expression in the nucleus, and to mediate a range of cellular functions including cell proliferation, migration, differentiation, senescence, and survival [25,237,238]. Growth factors bind to the extracellular surface of RTKs (e.g., epidermal growth factor receptor, EGFR, and fibroblast growth factor receptor, FGFR) leading to a conformational change that enables RTK dimerization and autophosphorylation of several tyrosine residues within the RTK cytoplasmic tail. This creates docking sites for adaptor proteins that stimulate downstream effector cascades, such as growth factor receptor-bound protein 2 (GRB2) that recruits Son of Sevenless (SOS) and the GTPase RAS to activate the MAPK cascade (RAF-MEK-ERK signaling) and drive transcription of RAS/MAPK target genes [237,238] (Figure 1).

The RAS/MAPK cascade is frequently deregulated in human cancers, including prostate cancer [238]. Activating genetic alterations (i.e., mutation/amplification) in *RAS (HRAS, NRAS, or KRAS)* and *BRAF* have been reported in primary and metastatic prostate cancer (1–8% incidence), and augmented MAPK signaling is reported to correlate with castration-resistance and metastatic progression [43,45,46,101,239]. The PI3K-AKT-mTOR and RAS/MAPK pathways are interconnected at multiple levels (Figure 1), predominantly owing to (a) shared upstream regulation mechanisms through RTKs/GPCRs and their associated adaptors, (b) the ability of respective cytosolic signaling components to interact and cross-regulate, and (c) the regulation of joint downstream targets (e.g., BAD and RPS6), reviewed in [25,240]. At the level of the receptor for example, the GRB2-SOS complex that is recruited to activated RTKs can bind to the scaffolding protein GAB1 (GRB2-associated binder-1), which interacts with RasGAP, SHP2, PI3K, and PIP3 to augment both RAS/MAPK and PI3K-AKT-mTOR signaling [25]. In addition, mTORC1 signaling can negatively regulate RTK signaling to reduce both PI3K-AKT-mTOR and RAS/MAPK activity, including mTORC1-S6K-mediated suppression of the insulin receptor substrate protein IRS1; a major IGF-1 receptor substrate and adaptor protein that can promote both PI3K and RAS activation by binding to p85 and GRB2 respectively [25,241]. S6K can also phosphorylate RICTOR to reduce mTORC2 signaling [25,242].

At the membrane, RAS-GTP can also bind to the RAS-binding domain (RBD) of p110α, p110δ, and p110γ to directly activate several Class I PI3K catalytic subunit isoforms [20,243]. Intracellular components of both cascades also interact to form multiple feedforward and feedback loops that enable PI3K-AKT-mTOR and RAS/MAPK pathway cross-regulation (Figure 1) [25,232,240]. For instance, RAS/MAPK activation has been shown to stimulate mTORC1 signaling through ERK, which can directly phosphorylate TSC2, RAPTOR and 90 kDa ribosomal S6 kinase (RSK) to inactivate/dissociate the TSC1:TSC2 complex and regulate the recruitment of mTORC1 substrates [244–246]. ERK-RSK signaling can also phosphorylate serum response factor (SRF), cAMP response element-binding protein (CREB) and RPS6, thus promoting cap-dependent translation independently of mTORC1-S6K signaling [247,248]. In addition, AKT is also reported to directly phosphorylate and negatively regulate RAF to suppress the MAPK cascade [249,250], and activated RAS has recently been shown to directly interact with mSIN1 to stimulate mTORC2 signaling in cancer cells (including prostate cancer cell lines) [251].

3.1.1. RAS/MAPK-PI3K-AKT-mTOR Interactions Promote Resistance to PI3K-AKT-mTOR Pathway-Directed Therapies

Clinical trials exploring the efficacy of inhibitors targeting the PI3K-AKT-mTOR pathway in prostate cancer have been extensively reviewed previously [14–17]. Despite promising results in early preclinical studies [252,253], allosteric mTORC1 inhibitors (e.g., rapamycin and rapamycin analogs/rapalogs such as Everolimus and Temsirolimus) have been ineffective in patients with prostate cancer, owing to their inability to suppress AKT activity and a number of adverse side effects [17,254]. Evidence in the literature has revealed several mechanisms of resistance, including activation of the RAS/MAPK pathway [235,255–257]. Both normal and transformed prostate epithelial cells have been shown to augment RAS/MAPK signaling in response to mTORC1 inhibition [235,255], and administration of Everolimus (RAD001) has been shown to induce MAPK signaling in a *Pten*-deleted mouse model of prostate cancer [235,255]. Although the mechanisms underpinning resistance to mTORC1 inhibitors are not completely understood, several signaling events that involve PI3K/AKT/PI3K and RAS/MAPK crosstalk have been identified. For instance, mTORC1 inhibition is reported to promote AKT and RAS/MAPK signaling by blocking mTORC1-S6K-mediated negative regulation of IRS1 and mTORC2 signaling [258,259] (Figure 2A). Inhibition of mTORC1 has also been shown to prevent mTORC1 stabilization of growth factor receptor bound protein 10 (GRB10), an RTK adaptor protein that negatively regulates RTK signaling [260].

Resistance to AKT inhibitors (e.g., capivasertib and ipatasertib) has also been linked to elevated RAS/MAPK signaling and mTORC2 activity [14,261]. AKT inhibition can lead to the nuclear accumulation of active FOXO1, resulting in increased transcription of FOXO1-regulated genes, such as *ERBB2/3* that encode human epidermal growth factor 2/3 (HER2/3) RTKs [25,237,262–264] (Figure 2B). PI3K inhibition with either pan-PI3K inhibitors (GDC0941 and XL-147) or a dual PI3K/mTOR inhibitor (BEZ235) has also been found to increase HER2/3 expression in breast cancer, resulting in increased RAS/MAPK signaling [265,266]. Furthermore, FOXO-dependent transcription is associated with p110α and PDK1 co-inhibition [145].

Additionally, the mTORC2 substrate SGK1 can replace AKT in response to PI3K/AKT inhibition, leading to the activation of shared AKT substrates that mediate oncogenic cellular processes such as cell growth, survival metabolism, and migration [145,267] (Figure 2B). In *PIK3CA* mutant breast cancer cells, PDK1-SGK1 signaling has been shown to sustain AKT-independent mTORC1 activation to promote resistance to the p110α-isoform-specific PI3K inhibitor BLY719, and PDK1 or SGK1 blockade can restore BYL719 sensitivity [145]. Furthermore, elevated SGK1 can predict for AKT inhibitor resistance in breast cancer cells [267]. Interestingly, Class I PI3K and AKT inhibition has also been shown to increase PI3K Class III hVsp34-SGK3 signaling in breast cancer cells, which can substitute for AKT by phosphorylating TSC2 to activate mTORC1 [148]. Whether SGK1/3 shares AKT's ability to phosphorylate and activate RAF is currently unknown.

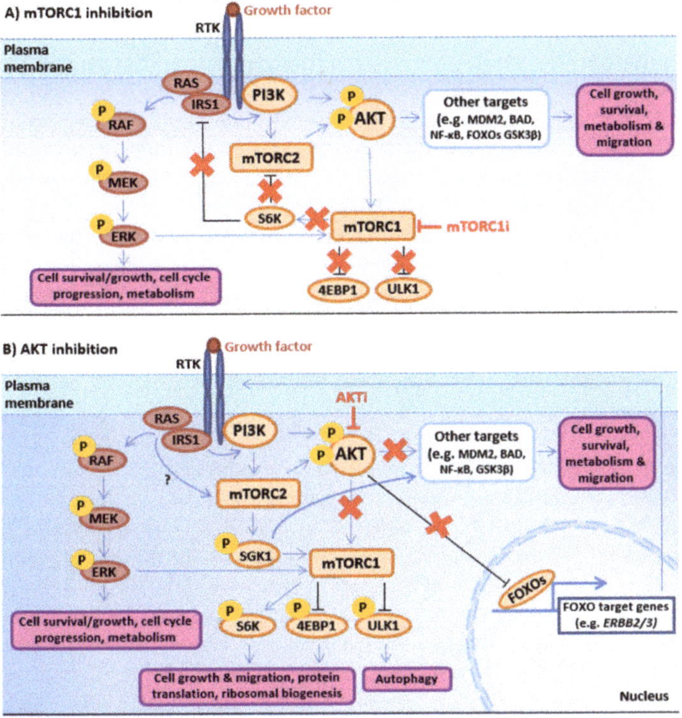

Figure 2. PI3K-AKT-mTOR and RAS/MAPK pathway crosstalk can contribute to mTORC1 and AKT inhibitor resistance. Model schematics illustrating reported mechanisms of therapeutic resistance to (**A**) mTORC1 inhibition and (**B**) AKT inhibition. mTORC1 and AKT blockade potentiates a series of feedback/feedforward loops between the PI3K-AKT-mTOR and RAS/MAPK signaling pathways, leading to augmented RAS/MAPK signaling and incomplete suppression of the PI3K-AKT-mTOR cascade that can promote drug-resistant tumor growth. AKTi, AKT inhibitor; mTORC1i, mTORC1 inhibitor.

3.1.2. Co-targeting RAS/MAPK and PI3K-AKT-mTOR Signaling in Prostate Cancer

Co-activation of the RAS/MAPK and PI3K-AKT-mTOR signaling pathways occurs frequently in human malignancies including prostate cancer, thus considerable research has been devoted to establishing how these two oncogenic cascades interact [101,256,257,268–270]. Nearly all metastatic prostate cancer patients are reported to show deregulation of both cascades [43]. To model this in vivo, genetically engineered mouse models of prostate cancer with prostate specific *Pten* homozygous deletion harboring either a *KRas*^G12D activating mutation or oncogenic *BRaf*^V600E with NK3 Homeobox 1 (*Nkx3.1*) depletion, promotes rapid tumor growth and metastatic progression relative to the single mutants [101,268,269]. To our knowledge, these tumor models were the first immunocompetent transgenic mouse models of prostate adenocarcinoma to display reproducible metastatic disease.

Taken together, these findings indicate that PI3K-AKT-mTOR and RAS/MAPK signaling synergize to promote prostate cancer growth and metastatic progression, and given the frequency of co-activation of these cascades in the clinic, this provides a clear justification for exploring the combination of PI3K-AKT-mTOR and RAS/MAPK pathway inhibitors in patients with advanced prostate cancer. This notion is further supported by the fact that MEK inhibition is associated with elevated PI3K-AKT-mTOR signaling in mammalian cancer cells, including prostate cancer cells [271,272]. Preclinical studies have also shown that co-inhibition of MEK and mTORC1 can significantly reduce tumor burden relative to monotherapy in a mouse model of prostate cancer driven by simultaneous heterozygous deletion of

Nkx3.1 and *Pten* [256], and can inhibit cell growth and increase cytotoxicity in the castration-resistant CWR22Rv1 human prostate cancer cell line [272]. However, MEK inhibition alone is reported to be sufficient to suppress the metastatic spread of *Pten*-deleted and *KRas* activated stem/progenitor murine prostate cancer cells orthotopically transplanted in vivo, similarly to combined mTORC1 and MEK inhibition [101]. This highlights the need to improve our molecular understanding of how these cascades interact during disease progression and in the presence of different genetic drivers to aid the stratification of patients that will benefit from (a) PI3K-AKT-mTOR inhibition, (b) MEK inhibition or (c) combined PI3K-AKT-mTOR and RAS/MAPK blockade.

Several prostate cancer clinical trials have been designed to investigate the therapeutic efficacy of targeting MEK (e.g.,MEK1/2 inhibitor trametinib, ClinicalTrials.gov identifiers: NCT02881242 and NCT01990196) or the PI3K-AKT-mTOR cascade (e.g., pan-AKT inhibitors including ipatasertib and capivasertib, ClinicalTrials.gov identifiers: NCT01485861/NCT03673787 and NCT02525068/ NCT02121639 respectively) [134,273]. Metformin (an oral type 2 anti-diabetic drug) is also currently being investigated in prostate cancer within the STAMPEDE trial [274]. Metformin targets the mitochondrial respiratory chain complex I, leading to reduced mitochondrial ATP production that causes cellular energy crisis with subsequent AMPK activation and mTORC1 inhibition [275]. Metformin has also been shown to inhibit MEK/ERK in response to growth factors, contrasting mTORC1 inhibitor treatment with rapamycin that increases MAPK signaling [276].

Although not currently specific to patients with prostate cancer, clinical trials exploring co-inhibition of the PI3K-AKT-mTOR and MAPK cascades to treat various advanced solid cancers have also been developed (e.g., ClinicalTrials.gov identifiers: NCT01390818, NCT01347866, and NCT02583542), although response rates appear to be low and are linked to *RAS* and *RAF* mutations [277]. For example, a recent Phase Ib study of combination therapy with the MEK1/2 inhibitor binimetinib (Mektovi) and the pan-PI3K inhibitor Buparlisib (BKM120) in advanced solid tumors reported promising efficacy in patients with advanced ovarian cancer with *RAS/RAF* genetic alterations, however continuous dosing resulted in intolerable toxicities and an intermittent schedule is suggested for future trials [278]. Additionally, the MATCH screening trial (targeted therapy directed by genetic testing in treating patients with advanced refractory solid tumors, lymphomas, or multiple myeloma, ClinicalTrials.gov identifier: NCT02465060) will investigate the efficacy of MEK and PI3K inhibitors as monotherapies in patients with progressive disease that carries a genetic alteration in either the RAS/MAPK or the PI3K-AKT-mTOR pathways respectively.

3.2. PI3K-AKT-mTOR and AR Signaling Crosstalk

AR signaling regulates cell growth, differentiation, migration and survival, and plays a critical role as a transcriptional regulator during prostate development, normal prostate tissue homeostasis, and prostate cancer [279–281]. AR is a steroid nuclear receptor that transmits androgen signals such as testosterone (T), or its more potent metabolite dihydrotestosterone (DHT), to regulate gene expression and coordinate cellular responses. T is derived from cholesterol through a cascade of biochemical reactions involving four enzymes: cytochrome P450 side-chain cleavage enzyme (P450scc), cytochrome P450 17α-hydroxylase/17,20-lyase (CYP17A1), 3β-hydroxysteroid dehydrogenase (3β-HSD) and 17β-hydroxysteroid dehydrogenase (17β-HSD) [282]. The conversion of cholesterol to pregnenolone is catalyzed by P450scc, and its subsequent conversion to progesterone is catalyzed by 3β-HSD. Pregnenolone and progesterone can be converted by CYP17A1 to 17-OH-pregnenolone and 17-OH-progesterone and subsequently to dehydroepiandrosterone (DHEA) and androstenedione (AD or A4). DHEA and AD may then be converted to androstenediol and T by 17β-HSD [282]. T synthesis and secretion predominantly occurs in the Leydig cells of the testes, and is stimulated by pituitary-derived luteinizing hormone (LH), which is secreted in response to hypothalamus-derived LHRH, (also known as gonadotrophin-releasing hormone, GnRH) [281,283]. In the prostate, 5-alpha-reductase converts T to DHT [281,283]. In addition, androgens can also be produced by the adrenal glands and in some instances by prostate tumor cells [284], which may contribute to prostate cancer growth post-orchiectomy [285].

In the absence of androgens, AR forms a cytoplasmic complex with chaperones (e.g., HSP90 and HSP70). Androgen binding displaces the chaperones and triggers a conformational change in AR, which enables AR homodimerization and nuclear translocation (Figure 1) [279,281,286]. Nuclear AR homodimers regulate the transcription of androgen-regulated genes (e.g., *SLC43A1*, *FKBP5, CAMKK2, NKX3.1* and *KLK3*) by directly binding to an androgen responsive element (ARE) in the promotor/enhancer region of target genes [279,286]. However, growth factors (e.g., EGF and IGF-1), cytokines (e.g., IL6) and intracellular signaling kinases (e.g., AKT and SRC) can also independently stimulate AR dependent transcriptional activity when androgen levels are low, which can facilitate therapeutic resistance to androgen/AR blockade [12,281,287]. In addition to regulating gene transcription, AR can also mediate a number of intracellular signaling pathways through direct protein–protein interactions within the cytoplasm (known as non-genomic AR signaling) [12,281]. For instance, AR is reported to activate SRC family kinases, PKC, RAS, ERK, PI3K, and AKT [12,288].

Aberrant AR signaling is a common feature of prostate cancer [289], with up to 56% of primary cases and 100% of metastatic cases reported to carry genetic alterations within key AR pathway components [43]. While the majority of men with prostate cancer initially respond to androgen/AR-directed therapy, they inevitably develop castration-resistant prostate cancer (CRPC), as malignant cells develop therapeutic resistance [5,290]. Several inherent and acquired resistance mechanisms have been identified, including AR genetic alterations (e.g., activating mutations, gene amplification, androgen-independent constitutively active splice variants, AR loss), augmented androgen biosynthesis, adrenal androgens, AR-bypass signaling (e.g., glucocorticoid receptor (GR) regulation of shared AR target genes), trans-differentiation to neuroendocrine prostate cancer and ligand-independent activation via crosstalk with another signaling cascade, such as the PI3K-AKT-mTOR pathway [12,283].

The PI3K/AKT/mTOR and AR pathways have been shown to cross-regulate through several reciprocal inhibitory loops [11,12,24] (Figure 3). Consequently, the PI3K-AKT-mTOR pathway can be inadvertently activated in response to androgen/AR-directed therapies, and vice versa PI3K-AKT-mTOR pathway inhibition can augment AR signaling, leading to therapeutic resistance. Human patient samples (both primary tumor and bone metastases), human prostate cancer cell lines and transgenic mouse models of prostate cancer have consistently demonstrated that AKT-mTOR signaling is increased in response to androgen/AR-directed blockade [11,13,24,291–293]. Mechanistically, it is reported that inhibiting AR signaling reduces expression of the AR target gene FK506-binding protein-5 (*FKBP5*), which leads to PHLPP destabilization and reduced PHLPP-mediated dephosphorylation of AKT at Ser473 to promote AKT signaling [11,24] (Figure 3A). Thus, compensatory activation of the PI3K-AKT-mTOR pathway in response to androgen/AR pathway inhibition can facilitate CRPC growth.

Conversely, PI3K-AKT-mTOR pathway inhibition is associated with augmented AR signaling that can contribute to drug resistance and promote prostate cancer progression [11,293–295]. Carver and colleagues showed that PI3K/mTOR inhibition activates AR signaling in human xenograft and transgenic mouse models of prostate cancer, and that co-treatment with the PI3K/mTOR inhibitor BEZ235 and the antiandrogen MDV3100 (enzalutamide) significantly reduced tumor burden relative to monotherapy [11]. In corroboration, resistance to the AKT inhibitor capivasertib (AZD5363) in LNCaP prostate cancer xenografts is also associated with elevated AR signaling, and combining AZD5363 treatment with the antiandrogen bicalutamide prolonged disease stabilization [295]. Furthermore, mTOR and EGFR co-inhibition with everolimus and gefitinib has shown limited sensitivity in patients owing to enhanced AR activity and PSA levels [293], providing further rationale for combining AR and PI3K-AKT-mTOR blockade to treat prostate cancer.

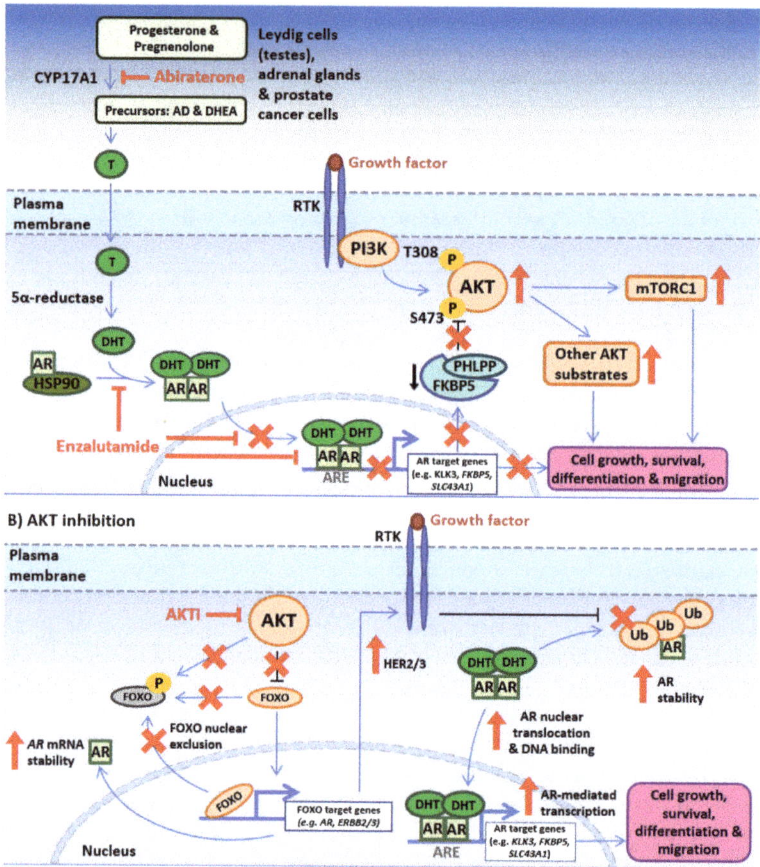

Figure 3. PI3K-AKT-mTOR and AR signaling crosstalk facilitates resistance to androgen/AR and AKT-directed monotherapy. Schematics depict reported model mechanisms for therapeutic resistance to (**A**) androgen/AR-directed therapy, which leads to increased AKT activation, and (**B**) AKT inhibition. AD, androstenedione; CYP17A1, cytochrome P450 17A1; DHEA, dehydroepiandrosterone; Ub, ubiquitination event.

Several distinct molecular mechanisms have been identified that underpin AR reactivation upon AKT inhibition. Notably, AKT inhibition can prevent AKT-mediated nuclear exclusion of FOXOs, which can lead to augmented transcription of FOXO-target genes such as RTKs (e.g., *ERBB2/3* encoding HER2/3) (Figure 3B) [11,262,296]. HER2/3 activity has been shown to promote AR signaling by protecting AR from ubiquitination and proteasomal degradation, and by enhancing AR binding to ARE target sequences and stimulating AR transcriptional activity [297–299] (Figure 3B). Nonetheless, the role of FOXO-dependent signaling in PI3K-AKT-mTOR and AR pathway crosstalk is complex. Although FOXO transcription factors upregulate the expression of RTKs [262] causing a subsequent increase in AR signaling [298], the ectopic expression of FOXO1 conversely dampens AR activity, which is further exacerbated when FOXO1 is co-transfected with the AR coregulator HDAC3 [300].

PTEN loss has also been shown to downregulate AR signaling via the upregulation of several factors that inhibit AR signaling through histone modification mechanisms such as early growth response 1 (EGR1), transcription factor AP-1 (c-JUN), and the catalytic subunit of polycomb repressive complex 2 enhance of zeste homolog 2 (EZH2) (Figure 1) [24]. PTEN protein-phosphatase activity has also been shown to protect the tumor suppressor NK3 homeobox 1 (NKX3.1) from degradation, which can derail the AR transcriptional network [301]. Of note, NKX3.1 and AR can cross-regulate [302], and enforced NKX3.1 expression can suppress *Pten*-deleted prostate tumorigenesis in transgenic mice [303].

Despite the antagonistic crosstalk between AR and AKT (Figure 3), AR signaling can boost mTORC1 activation through an AR-dependent increase in amino acid transport during tumorigenesis [304]. AR mediates expression of L-type amino acid transporters (e.g., LAT3 encoded by *SLC43A1*) to maintain sufficient levels of leucine needed for mTORC1 signaling and cell growth (Figures 1 and 3) [304]. Moreover, LAT1 and LAT3 transport inhibition is sufficient to decrease cell growth and mTORC1 signaling in prostate cancer cells in vitro [304]. Recent in vitro data have also revealed that mTOR can directly interact with AR in the nucleus of prostate cancer cells to promote metabolic rewiring, and high levels of nuclear mTOR correlate with poor prognosis in patients with prostate cancer [293]. Additionally, AKT has been shown to directly bind and phosphorylate AR when T levels are low, although the functional significance of this event remains to be determined [16,305,306].

The reciprocal feedback loop between AR and PI3K-AKT-mTOR signaling may also be perturbed by Speckle-type BTB/POZ protein (*SPOP*) loss of function mutations that lead to the stabilization of the SPOP substrate SRC3 (e.g., p.F133V), which consequentially increases PI3K activity [307]. SPOP is an adaptor protein of the Cullin 3 family E3 ligases that can target SRC3 for ubiquitin-mediated degradation and a known tumor suppressor [12,308–310]. In prostate cancer, *SPOP* is frequently mutated (9–11% incidence) [45,46]. Remarkably, *SPOP* mutation can also stabilize AR and potentiate AR signaling whilst the PI3K-AKT-mTOR signaling pathway is activated, allowing coordinated and cooperative signaling that drives tumorigenic growth [307]. Conversely, wildtype SPOP can trigger E3 ligase mediated degradation of AR via hinge domain binding when androgens levels are low [311]. Furthermore, AR has been shown to positively regulate the PI3K-AKT-mTOR pathway via a direct interaction with the SH2 domain of the Class IA PI3K regulatory subunit p85α, which has been shown to activate the PI3K-AKT-mTOR cascade [312], further highlighting the complexity of the interactions between these two cascades.

Taken together, these findings support the rationale for combining pharmacological inhibition of the AR and PI3K-AKT-mTOR cascades to treat prostate cancer in the clinic, and highlight the need for further work to delineate the molecular mechanisms underpinning crosstalk between these two oncogenic cascades. Importantly, clinical trials exploring co-targeting AR and PI3K-AKT-mTOR signaling are beginning to show promise. A randomized Phase Ib/II study combining the pan-AKT inhibitor ipatasertib with abiraterone in mCRPC patients has reported ipatasertib + abiraterone prolongs radiographic progression-free survival (rPFS), improves overall survival and extends time to PSA progression compared to abiraterone alone, particularly in patients with PTEN loss (ClinicalTrials.gov identifier: NCT01485861) [134]. This study also reports that the adverse effects common to PI3K-AKT-mTOR blockade (e.g., hyperglycemia) were generally clinically manageable [134]. A Phase I dose escalation study combining enzalutamide and capivasertib to treat mCRPC has also recently reported 3/16 patients responded (ClinicalTrials.gov identifier: NCT02525068) [313]. In this study, patients who met the response criteria had *PTEN* loss or *AKT* activating mutations, low/absent AR-V7 protein levels and elevated p-ERK [313]. Nevertheless, several additional clinical trials investigating the combination of AR and PI3K-AKT-mTOR blockade in men with mCRPC did not demonstrate a therapeutic benefit and were associated with poor tolerability (ClinicalTrials.gov identifiers: NCT01385293, NCT01634061 and NCT01717898) [314–316]. Interestingly, D'Abronzo and colleagues also recently showed that eIF4E phosphorylation at residue Ser209 in human CRPC cell lines pre-treated with the antiandrogen bicalutamide underpins resistance to subsequent combination therapy with bicalutamide + rapamycin treatment. Remarkably, suppression of eIF4E phosphorylation by MNK1/2 (MAP kinase interacting serine/threonine kinase1/2) or ERK1/2 inhibition was shown to sensitize bicalutamide pre-treated CRPC cells to combined anti-androgen and mTORC1 blockade [317], presenting a novel avenue for overcoming therapeutic resistance. Thus, despite some promising results, it is evident that further investigation into the molecular mechanisms underpinning AR and

PI3K-AKT-mTOR pathway crosstalk in prostate cancer is required to improve patient stratification and to discover new therapeutic approaches and predictive biomarkers that can inform future clinical trial design.

3.3. PI3K-AKT-mTOR and WNT Signaling Interactions

The WNT family is an evolutionarily conserved group of proteins essential for growth control, organ development, tissue homeostasis and stem cell renewal in multiple organs, and is crucial for normal prostate development [318,319]. WNT signaling is potentiated by secreted WNT ligands (a family of 19 lipoglycoproteins) that bind extracellularly to transmembrane frizzled receptors (FZD1-10) and their co-receptors, such as low-density lipoprotein receptors (e.g., LRP5 and LRP6), tyrosine protein-kinases (e.g., receptor tyrosine kinase–like orphan receptor-1 and -2, ROR1/2), and tyrosine kinase-related receptors (e.g., receptor-like tyrosine kinase, RYK, protein tyrosine kinase 7, PTK7, and muscle specific kinase, MuSK) [318–320]. The WNT signal is transduced intracellularly via dishevelled (DVL), which subsequently activates either β-catenin-dependent/canonical WNT signaling or β-catenin-independent/non-canonical WNT signaling [318–320]. In the absence of a canonical WNT ligand, cytosolic β-catenin levels are maintained at a low level via the β-catenin destruction complex that contains the scaffold protein AXIN, the tumor suppressor adenomatous polyposis coli (APC), GSK3β and casein kinase 1 (CK1). The β-catenin destruction complex phosphorylates β-catenin, leading to its ubiquitylation and proteasomal degradation [321]. Canonical WNT signals disrupt the β-catenin destruction complex, resulting in β-catenin stabilization and accumulation, nuclear translocation and interaction with TCF/LEF transcription factors to upregulate WNT target genes such as *MYC* and *AXIN2* [321] (Figure 1). Non-canonical WNT signaling involves WNT-mediated activation of RhoA/ROCK and RAC/JNK/NFAT signaling (planar cell polarity pathway), or phospholipase C (PLC) activation and the accumulation of intracellular Ca^{2+} that stimulates calmodulin-dependent kinase II (CamKII), calcineurin and protein kinase C (PKC) signaling (WNT/Ca^{2+} pathway) [320].

Activation of both the canonical and non-canonical WNT cascades has been reported in localized and advanced prostate cancer, and oncogenic deregulation of core WNT pathway components frequently occurs in primary and metastatic prostate cancer (up to 6% and 19% incidence respectively [45]), primarily via *APC* deep deletion/truncating mutations and *CTNNB1*/β-catenin activating mutations [45,46,320,322,323]. Furthermore, the WNT/β-catenin pathway is strongly linked to androgen/AR-directed therapy and chemotherapy resistance [324–327], thus WNT signaling presents an attractive therapeutic target for advanced prostate cancer. In addition, *AR* is in fact a WNT/β-catenin target gene, and AR and β-catenin can directly interact and co-localize in the nucleus to mediate transcriptional activity of AR-regulated genes [328–330] (Figure 1). WNT and AR signaling cascades have also been shown to reciprocally inhibit each other in murine prostate cancer [331].

Mouse models have been instrumental in determining the role of WNT signaling in prostate cancer, and we and others have shown that constitutive activation of β-catenin or *Apc* bi-allelic deletion predisposes to prostate adenocarcinoma in mice [269,331–333]. Moreover, β-catenin activation can cooperate with *Pten* heterozygous or homozygous deletion to promote prostate cancer progression, CRPC transition and metastatic potential [269,331–333], indicating a synergistic relationship exists between the PI3K-AKT-mTOR and WNT cascades.

Several molecular mechanisms that permit cross-regulation of the PI3K-AKT-mTOR and WNT signaling cascades have emerged in the literature, which could influence prostate cancer growth and resistance to anti-androgens and/or PI3K-AKT-mTOR pathway inhibitors [320,322,323,334,335] (Figure 4). While canonical WNT signaling mediates cellular β-catenin levels, the level of active β-catenin (unphosphorylated at residues Ser37 and Thr41) in melanoma, breast and prostate cancer cells is reported to be regulated by the PI3K-AKT-mTOR cascade, in a process that is dependent on PP2A activity [335]. PP2A is known to negatively regulate AKT, however it is currently speculated that this phosphatase may also directly dephosphorylate and activate β-catenin [335]. Additionally, the PI3K-AKT-mTOR pathway has also been shown to mediate β-catenin localization [335], and several transcription factors that directly interact with β-catenin are co-regulated by the PI3K-AKT-mTOR pathway, such as FOXO3a [336] and SOX4 [337]. In prostate cancer cells, FOXO3a has been shown to suppress β-catenin transcriptional activity and can be inhibited by AKT [336], whereas SOX4 is a positive regulator of canonical WNT and is stimulated by AKT [338] (Figure 4).

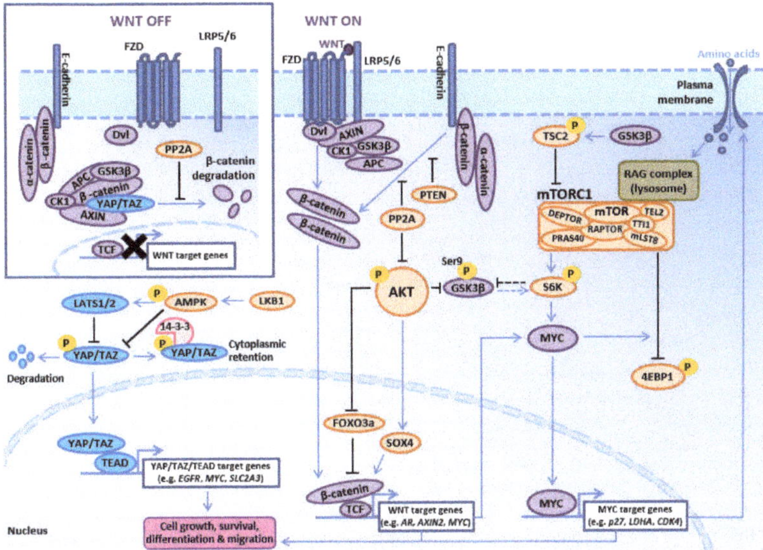

Figure 4. PI3K-AKT-mTOR and WNT signaling crosstalk. Upon ligand binding (WNT ON), the destruction complex is recruited to the plasma membrane leading to β-catenin accumulation in both the cytoplasm and nucleus, where it activates gene expression through TCF binding. Insert illustrates WNT signaling in the absence of WNT ligand (WNT OFF). The interplay between PI3K-AKT-mTOR and WNT may occur through shared pathway components (e.g., GSK3β, PTEN, PP2A) and/or the joint regulation of transcription factors such as MYC, FOXO3a, SOX4 or YAP/TAZ. CDK4, cyclin-dependent kinase 4; LATS1/2, ʟ-type amino acid transporter 1/2; LDHA, ʟ-lactate dehydrogenase A chain; SLC2A3, solute carrier family 2, facilitated glucose transporter member 3 (encoding GLUT3, glucose transporter 3); TAZ, transcriptional co-activator with PDZ-binding motif; TEAD, transcriptional enhanced associate domain transcription factor; YAP, Yes-associated protein.

In addition to transmitting canonical WNT signals, β-catenin also form adherens junctions with α-catenin and E-cadherin at cell–cell junctions to maintain tissue architecture and facilitate cell–cell signaling (Figure 4). PTEN has also been shown to modulate β-catenin nuclear localization and transcriptional activity through caveolin-1 (CAV1)-dependent dissociation of β-catenin from E-cadherin at the membrane independently of PI3K-AKT-GSK3β signaling, leading to increased tumor formation and metastatic progression in melanoma [339].

Crosstalk between the PI3K-AKT-mTOR and WNT/β-catenin signaling pathways is also mediated via GSK3β and the TSC1:TSC2 complex [167] (Figure 4). GSK3β play a critical role in both cascades, serving as a core member of the β-catenin destruction complex that helps to maintain low level of cytosolic/nuclear β-catenin in the absence of WNT signal [340], and as a direct substrate of AKT [167]. AKT inactivates GSK3β by phosphorylating residue Ser9 [167]. GSK3β can also phosphorylate and activate TSC2 resulting in inhibition of mTOR activity [167], and can restrict cellular growth by suppressing glucose uptake via TSC2 and mTOR [341]. Active WNT signaling inhibits GSK3β, abrogates the suppression of mTOR and stimulates phosphorylation of S6K, S6, and eukaryotic translation initiation factor 4E binding protein 1 (4EBP1) [167]. Interestingly, in the absence of TSC1 or TSC2, S6K has also been shown to inactivate GSK3β by directly phosphorylating residue Ser9 and active GSK3β has also been shown to phosphorylate/activate S6K, adding further complexity to GSK3β signaling between the WNT and PI3K-AKT-mTOR cascades [342]. However, previous work has also indicated that GSK3β does not mediate crosstalk between the PI3K-AKT-mTOR and WNT/β-catenin pathways [343,344], raising the possibility that GSK3β function is context/tissue dependent. Of note, WNT ligands can also activate mTOR through MYC-dependent suppression of TSC2 [345].

PI3K-AKT-mTOR and WNT signaling may also interact through Hippo signaling. Hippo signaling is tightly intertwined with cell size regulation and nutrient sensing through LKB1-AMPK, TSC1:TSC2 and mTOR [346,347], and the Hippo pathway signaling proteins yes-associated protein (YAP) and transcriptional co-activator with PDZ-binding motif (TAZ) are integral parts of both canonical and non-canonical WNT signaling [348,349]. YAP and TAZ are members of the β-catenin destruction complex, and in the presence of WNT signals, they dissociate from the complex and translocate to the nucleus to activate downstream targets [348] (Figure 4). AMPK activation has also been shown to negatively regulate YAP/TAZ activity [347].

Non-canonical WNT signaling has also been shown to activate the PI3K-AKT-mTOR pathway. For example, WNT/FZD7-dependent dissociation of Gβγ from Gαi enhances PI3K-AKT signaling and increases tumor cell invasive potential [350]. ROR1 can also activate PI3K-AKT signaling in response to trans-phosphorylation by tyrosine kinases, such as MET and SRC [351]. In addition, WNT receptor Frizzled2 (FZD2) can drive epithelial-to-mesenchymal transition (EMT) and cell migration through activation of Fyn [352], and activated Fyn kinase activity has been shown to suppress the AMPK-LKB1 signaling axis by blocking LKB1 redistribution into the cytoplasm [353].

Interestingly, the PI3K-AKT-mTOR, WNT, MAPK and AR signaling cascades all converge to regulate the transcription factor MYC, which is frequently amplified in prostate cancer/mCRPC [44,45,354]. While *MYC* is a WNT/β-catenin target gene [355], PI3K-AKT-mTOR signaling can mediate *MYC* mRNA stability, translation and protein stability [356–360]. AR signaling has also been shown to stimulate MYC in AR-driven prostate cancer, while in normal prostate tissue AR silences *MYC* to maintain normal homeostasis [361,362]. However, MYC is also reported to antagonize AR transcriptional activity in prostate cancer [363]. MYC upregulation is frequently observed in prostate cancer, and although targeting MYC remains a clinical challenge, preclinical studies have emphasized the potential efficacy of MYC blockade for patients with late-stage prostate cancer [364].

WNT inhibitors are beginning to enter clinical trials, including small-molecule inhibitors to the enzyme porcupine that block WNT ligand secretion, such as WNT974 (LGK974) [320,322]. Preliminary data from a WNT974 phase 1 clinical trial (NCT01351103) for a small range of human malignancies (excluding prostate cancer) report a manageable safety profile and suppression of canonical WNT/β-catenin target gene *AXIN2* [320], and recent preclinical studies have indicated that WNT974 treatment is efficacious against prostate cancer [331,365]. β-catenin has also been reported to facilitate resistance to PI3K and AKT inhibition in colon cancer [334]. Thus, further work exploring the therapeutic benefit of targeting the WNT pathway in prostate cancer is warranted.

4. Conclusions

In summary, the PI3K-AKT-mTOR cascade is frequently activated in prostate cancer, and genomic profiling has revealed that oncogenic genetic alterations occur within a diverse array of PI3K-AKT-mTOR pathway components. Significant research efforts have been devoted to delineating the mode of action of several of these aberrations (e.g., *PTEN* deletion and *PIK3CA* activating mutation), however our molecular understanding of how these events differentially mediate cell signaling programs is limited, and several genetic alterations remain to be studied functionally. Future work to gain novel insight into the functional consequence of these genetic alterations in prostate cancer is necessary to (a) identify passenger vs. driver alterations, (b) establish the ability of individual aberrations to synergize with additional oncogenic events, and (c) discover their mode of action during tumor growth, metastasis and therapeutic resistance. In conjunction with genomic and transcriptomic data, establishing the frequency and impact of post-transcriptional modifications and epigenetic events within core PI3K-AKT-mTOR pathway components during prostate tumorigenesis and disease progression/recurrence is also crucial, as these components are regulated at multiple levels and genomic/transcriptomic data do not consistently equate with protein activity.

Targeting the PI3K-AKT-mTOR pathway in prostate cancer remains a key clinical challenge. Therapeutic resistance emerges owing to various feedback/feedforward loops and redundancy mechanisms that prevent complete suppression of the pathway and cause compensatory augmentation of interacting signaling pathways, thus rationalizing the exploration of combination therapies. Encouragingly, clinical trials are beginning to report therapeutic efficacy when combining PI3K-AKT-mTOR and androgen/AR-directed therapies, particularly in patients with mCRPC that display PTEN loss. However, the mechanism of resistance to PI3K-AKT-mTOR pathway-targeted therapies is likely to vary dramatically between patients and within individual tumors owing to several factors. These include the activity status of the pathway components, the extent of intratumoral heterogeneity, the mode and concentration of upstream stimuli, the genetic alterations present and the composition of the tumor microenvironment. Furthermore, our ability to successfully translate preclinical findings to the clinic is currently hampered by the limited number of prostate cancer preclinical models available, which do not fully cover the broad range of prostate cancer subtypes or disease heterogeneity seen in the clinic. Accordingly, to discover newtherapeutic approaches that increase patient response rates and overall survival, further delineation of the complex signaling network that exists within the PI3K-AKT-mTOR pathway and the interacting MAPK, AR, and WNT pathways is needed, together with the development of a wider range of preclinical models that better recapitulate the clinic and a deeper understanding of the molecular biology underpinning prostate cancer disease subtypes and tissue heterogeneity.

Supplementary Materials: Supplementary Materials can be found at http://www.mdpi.com/1422-0067/21/12/4507/s1.

Author Contributions: Conceptualization, B.Y.S. and H.B.P.; writing—original draft preparation, B.Y.S., M.S.D. and H.B.P.; writing—review and editing, B.Y.S., M.S.D., M.J.S. and H.B.P.; project administration, H.B.P.; funding acquisition, M.J.S. and H.B.P. All authors have read and agreed to the published version of the manuscript.

Funding: This research was funded by the Prostate Cancer Research Centre, a charitable incorporated organization, with registered charity number 1156027 (project grant awarded to B.Y.S. and M.J.S.). M.S.D. is supported by a Knowledge Economy Skills Scholarship 2 (KESS2) Ph.D award, in partnership with Tenovus Cancer Care. H.B.P. is supported by a Cancer Research UK career development fellowship (#A27894).

Conflicts of Interest: The authors declare no conflict of interest.

Abbreviations

3β-HSD	3β hydroxysteroid dehydrogenase
4E-BP1	Eukaryotic translation initiation factor 4E binding protein 1
ADT	Androgen deprivation therapy
AD	Androstenedione
AGC	cAMP-dependent, cGMP-dependent and protein kinase C
AKTi	AKT inhibitor
AMP	Adenosine monophosphate
AMPK	5′ AMP-activated protein kinase
APC	Adenomatous polyposis coli
AR	Androgen receptor
ARE	Androgen responsive element
ARF1	Adenosine ribosylation factor 1
ATG14	Autophagy related 14 homolog
ATP	Adenosine triphosphate
AXIN	Axis inhibitor protein
BAD	Bcl-2-associated death promoter

BMK-1	Big mitogen-activated protein kinase-1
CaMKII	Calmodulin-dependent kinase 2
CAMKKβ	Ca(2+)/calmodulin-dependent protein kinase kinase β
CAV1	Caveolin-1
CDK4	Cyclin-dependent kinase 4
c-JUN	Transcription factor AP-1
CK1	Casein kinase 1
CAN	Copy number alteration
CREB	cAMP response element-binding protein
CRPC	Castrate resistant prostate cancer
CYP17A1	Cytochrome P450 17A1
DEPTOR	Dishevelled, EGL-10 and pleckstrin (DEP) domain-containing mTOR-interacting protein
DFCI	Dana-Farber Cancer Institute
DHEA	Dehydroepiandrosterone
DHTDVL	DihydrotestosteroneDisheveled
EIF4E	Eukaryotic translation initiation factor 4E
EGF	Epidermal growth factor
EGFR	Epidermal growth factor receptor
EGR1	Early growth response 1
EMT	Epithelial-to-mesenchymal transition
ERG1	ETS-related Gene 1
ERK1	Mitogen-activated protein kinase 3
ERK2	Mitogen-activated protein kinase 1
EZH2	Enhancer of zeste homolog 2
FKBP5	FK506 binding protein 5
FOXO	Forkhead box protein O
FGFR	Fibroblast growth factor receptor
FZD	Frizzled family receptor
GAB1	GRB2-associated binder-1
GDP	Guanosine diphosphate
GLUT3	Glucose transporter 3
GnRH	Gonadotrophin-releasing hormone
GPCR	G-protein coupled receptor
GR	Glucocorticoid receptor
GRB2	Growth factor receptor-bound protein 2
GRB10	Growth factor binding protein 10
GSK3β	Glycogen synthase kinase 3 beta
GTP	Guanosine triphosphate
HDAC3	Histone deacetylase 3
HER2/3	Human epidermal growth factor receptor 2/3
HSD3B/17B3	Hydroxysteroid dehydrogenase 3B/17B3
HSP70/90	Heat shock protein 70/90
IGF	Insulin growth factor
IGF-1	Insulin growth factor 1
IHC	Immunohistochemistry
IL6	Interleukin 6
INPP4B	Inositol polyphosphate 4-phosphatase type II
INPP5D	Inositol polyphosphate-5-phosphatase D
INPP5J	Inositol polyphosphate-5-phosphatase J
INPPL1	Inositol polyphosphate phosphatase like 1
IRS	Insulin receptor substrate
IRS1	Insulin receptor substrate protein 1
KLK3	Kallikrein related peptidase 3
LAT1/2/3	L-type amino acid transporter 1/2/3

LDHA	L-lactate dehydrogenase A chain
LEF	Lymphoid enhancer binding factor
LH	Luteinizing hormone
LHRH	Luteinizing hormone-releasing hormone
LKB1	Liver kinase B1
LRP5/6	Low-density lipoprotein receptor-related proteins 5 and 6
MAPK	Mitogen-activated protein kinase
MAP3K7	Mitogen-activated protein kinase kinase kinase 7
mCRPC	Metastatic castrate resistant prostate cancer
MDM2	Mouse double minute 2 homolog
MEK	Mitogen-activated protein kinase kinase
MSKCC	Memorial Sloan Kettering Cancer Centre
mLST8	MTOR associated protein LST8 homolog
MMTV-PyMT	Mouse mammary tumor virus-polyoma middle tumor-antigen
MNK1	MAP kinase interacting serine/threonine kinase 1
MNK2	MAP kinase interacting serine/threonine kinase 2
mSIN1	Mitogen-activated protein kinase associated protein 1 (MAPKAP1)
mTOR	Mammalian target of rapamycin
mTORC1	Mammalian target of rapamycin complex 1
mTORC1i	mTORC1 inhibitor
mTORC2	Mammalian target of rapamycin complex 2
MuSK	Muscle specific kinase
NF-κB	Nuclear factor kappa light chain enhancer of activated B cells
NKX3.1	NK3 Homeobox 1
NSCLC	Non-small-cell lung carcinoma
P	Phosphorylation event
PDK1	Phosphoinositide-dependent kinase 1
PGC-1α	Peroxisome proliferator-activated receptor gamma coactivator 1-alpha
PHLPP	PH domain leucine-rich repeat protein phosphatase
PHLPP1	PH domain and leucine rich repeat protein phosphatase 1
PHLPP2	PH domain and leucine rich repeat protein phosphatase 2
PI	Phosphatidylinositol
PI3K	Phosphoinositide 3-kinase
PI(3)P	Phosphatidylinositol 3-phosphate
PI(3,4)P2	Phosphatidylinositol 3,4-bisphosphate
PIN	Prostate intra-epithelial neoplasia
PIP2	Phosphatidylinositol 4,5-bisphosphate
PIP3	Phosphatidylinositol 3,4,5-trisphosphate
PIPP	Proline-rich inositol polyphosphate 5-phosphatase
PKB	Protein kinase B, AKT
PKC	Protein kinase
PKCα	Protein kinase C alpha
PLC	Phospholipase C
PP2A	Protein phosphatase 2A
PPP2CA	Protein phosphatase 2 catalytic subunit Alpha
PRAS40	Proline-rich AKT substrate of 40 kDa
PROTOR1	Protein observed with Rictor-1
PROTOR2	Protein observed with Rictor-2
PSA	Prostate specific antigen
PTEN	Phosphatase and tensin homologue deleted on chromosome 10
PTK7	Protein tyrosine kinase 7
RAF	Rapidly accelerated fibrosarcoma
RAG	Recombination activating genes
RAPTOR	Regulatory-associated protein of mTOR

RBD	Ras-binding domain
RHEB	Ras homolog enriched in brain
RICTOR	Rapamycin-insensitive companion of mTOR
ROR1/2	Receptor tyrosine kinase–like orphan receptor-1 and -2
RPS6	Ribosomal protein S6
rPFS	Radiographic progression-free survival
RSK	90 kDa Ribosomal S6 kinase
RTK	Tyrosine kinase receptor
RYK	Receptor-like tyrosine kinase
S6K	Ribosomal protein S6 kinase/p70 ribosomal S6 kinase
SESN1	Sestrin 1
SGK1	Serum and glucocorticoid regulated kinase 1
SGK2	Serum and glucocorticoid regulated kinase 2
SGK3	Serum and glucocorticoid regulated kinase 3
SH2	Src homology 2
SHIP1	SH2 domain-containing inositol 5′-phosphatase 1
SHIP2	SH2 domain-containing inositol 5′-Phosphatase 2
SLC2A3	Solute carrier family 2, facilitated glucose transporter member 3
SLC43A1	Solute carrier family 43 member 1
SOS	Son of Sevenless
SPOP	Speckle type BTB/POZ protein
SRC-3	Steroid receptor co-activator 3
SRF	Serum response factor
SU2C-PCF IDT	Stand Up To Cancer & Prostate Cancer Foundation International Dream Team
T	Testosterone
TAK1	TGFβ-activated kinase 1
TAZ	Transcriptional coactivator with PDZ binding motif
TBC1D7	TBC1 Domain Family Member 7
TCF	T cell Factor
TCGA	The Cancer Genome Atlas
TEAD	Transcriptional enhanced associate domain
TEL2	Telomere length regulation protein
TGF	Transforming growth factor
TMPRSS2-ERG	Transmembrane protease, serine 2: ETS Transcription Factor fusion
TRAMP	Transgenic adenocarcinoma mouse prostate
TSC1	Tuberous Sclerosis complex 1
TSC2	Tuberous Sclerosis complex 2
TTI1	TELO2 interacting protein 1
Ub	Ubiquitination event
ULK1	Unc-51 Like Autophagy Activating Kinase 1
UVRAG	UV radiation resistance-associated gene; v-ATPase, Vacuolar (H+)-ATPase
VPS15	Vacuolar protein sorting 15
V-ATPase	Vacuolar H+-ATPase
WNT	WNT ligand
YAP	Yes-associated protein

References

1. Ferlay, J.; Soerjomataram, I.; Dikshit, R.; Eser, S.; Mathers, C.; Rebelo, M.; Parkin, N.M.; Forman, D.; Bray, F. Cancer incidence and mortality worldwide: Sources, methods and major patterns in GLOBOCAN 2012. *Int. J. Cancer* **2014**, *136*, E359–E386. [CrossRef]
2. Jemal, A.; Fedewa, S.A.; Ma, J.; Siegel, R.; Lin, C.C.; Brawley, O.; Ward, E.M. Prostate Cancer Incidence and PSA Testing Patterns in Relation to USPSTF Screening Recommendations. *JAMA* **2015**, *314*, 2054–2061. [CrossRef]

3.   Steele, C.B.; Li, J.; Huang, B.; Weir, H.K. Prostate cancer survival in the United States by race and stage (2001–2009): Findings from the CONCORD-2 study. *Cancer* **2017**, *123*, 5160–5177. [CrossRef] [PubMed]

4.   Miller, K.D.; Siegel, R.L.; Lin, C.C.; Mariotto, A.B.; Kramer, J.L.; Rowland, J.H.; Stein, K.D.; Alteri, R.; Jemal, A. Cancer treatment and survivorship statistics, 2016. *CA Cancer J. Clin.* **2016**, *66*, 271–289. [CrossRef] [PubMed]

5.   Crawford, E.D.; Heidenreich, A.; Lawrentschuk, N.; Tombal, B.; Pompeo, A.C.L.; Mendoza-Valdes, A.; Miller, K.; Debruyne, F.M.J.; Klotz, L. Androgen-targeted therapy in men with prostate cancer: Evolving practice and future considerations. *Prostate Cancer Prostatic Dis.* **2018**, *22*, 24–38. [CrossRef] [PubMed]

6.   Culig, Z. Molecular Mechanisms of Enzalutamide Resistance in Prostate Cancer. *Curr. Mol. Biol. Rep.* **2017**, *3*, 230–235. [CrossRef] [PubMed]

7.   Giacinti, S.; Bassanelli, M.; Aschelter, A.M.; Milano, A.; Roberto, M.; Marchetti, P. Resistance to abiraterone in castration-resistant prostate cancer: A review of the literature. *Anticancer Res.* **2014**, *34*, 6265–6269. [PubMed]

8.   Rice, M.A.; Malhotra, S.V.; Stoyanova, T. Second-Generation Antiandrogens: From Discovery to Standard of Care in Castration Resistant Prostate Cancer. *Front. Oncol.* **2019**, *9*, 801. [CrossRef]

9.   Perlmutter, M.A.; Lepor, H. Androgen Deprivation Therapy in the Treatment of Advanced Prostate Cancer. *Rev. Urol.* **2007**, *9*, S3–S8.

10.   Mostaghel, E.A. Abiraterone in the treatment of metastatic castration-resistant prostate cancer. *Cancer Manag. Res.* **2014**, *6*, 39–51. [CrossRef]

11.   Carver, B.S.; Chapinski, C.; Wongvipat, J.; Hieronymus, H.; Chen, Y.; Chandarlapaty, S.; Arora, V.K.; Le, C.; Koutcher, J.; Scher, H.; et al. Reciprocal Feedback Regulation of PI3K and Androgen Receptor Signaling in PTEN-Deficient Prostate Cancer. *Cancer Cell* **2011**, *19*, 575–586. [CrossRef] [PubMed]

12.   Crumbaker, M.; Khoja, L.; Joshua, A.M. AR Signaling and the PI3K Pathway in Prostate Cancer. *Cancers* **2017**, *9*, 34. [CrossRef] [PubMed]

13.   Pearson, H.; Li, J.; Méniel, V.; Fennell, C.; Waring, P.; Montgomery, K.G.; Rebello, R.J.; MacPherson, A.A.; Koushyar, S.; Furic, L.; et al. Identification of Pik3ca Mutation as a Genetic Driver of Prostate Cancer That Cooperates with Pten Loss to Accelerate Progression and Castration-Resistant Growth. *Cancer Discov.* **2018**, *8*, 764–779. [CrossRef] [PubMed]

14.   Bitting, R.L.; Armstrong, A.J. Targeting the PI3K/Akt/mTOR pathway in castration-resistant prostate cancer. *Endocr. Relat. Cancer* **2013**, *20*, R83–R99. [CrossRef] [PubMed]

15.   Hsieh, A.C.; Edlind, M.P. PI3K-AKT-mTOR signaling in prostate cancer progression and androgen deprivation therapy resistance. *Asian J. Androl.* **2014**, *16*, 378–386. [CrossRef] [PubMed]

16.   Toren, P.; Zoubeidi, A. Targeting the PI3K/Akt pathway in prostate cancer: Challenges and opportunities (Review). *Int. J. Oncol.* **2014**, *45*, 1793–1801. [CrossRef]

17.   Yang, J.; Nie, J.; Ma, X.; Wei, Y.; Peng, Y.; Wei, X. Targeting PI3K in cancer: Mechanisms and advances in clinical trials. *Mol. Cancer* **2019**, *18*, 26. [CrossRef]

18.   Courtney, K.D.; Corcoran, R.B.; Engelman, J.A. The PI3K Pathway as Drug Target in Human Cancer. *J. Clin. Oncol.* **2010**, *28*, 1075–1083. [CrossRef]

19.   Vanhaesebroeck, B.; Guillermet-Guibert, J.; Graupera, M.; Bilanges, B. The emerging mechanisms of isoform-specific PI3K signalling. *Nat. Rev. Mol. Cell Biol.* **2010**, *11*, 329–341. [CrossRef]

20.   Thorpe, L.; Yuzugullu, H.; Zhao, J.J. PI3K in cancer: Divergent roles of isoforms, modes of activation and therapeutic targeting. *Nat. Rev. Cancer* **2014**, *15*, 7–24. [CrossRef]

21.   Liu, P.; Cheng, H.; Roberts, T.M.; Zhao, J.J. Targeting the phosphoinositide 3-kinase pathway in cancer. *Nat. Rev. Drug Discov.* **2009**, *8*, 627–644. [CrossRef] [PubMed]

22.   Guillermet-Guibert, J.; Bjorklof, K.; Salpekar, A.; Gonella, C.; Ramadani, F.; Bilancio, A.; Meek, S.; Smith, A.J.H.; Okkenhaug, K.; Vanhaesebroeck, B. The p110β isoform of phosphoinositide 3-kinase signals downstream of G protein-coupled receptors and is functionally redundant with p110γ. *Proc. Natl. Acad. Sci. USA* **2008**, *105*, 8292–8297. [CrossRef] [PubMed]

23.   Papa, A.; Pandolfi, P.P. The PTEN-PI3K Axis in Cancer. *Biomolecules* **2019**, *9*, 153. [CrossRef] [PubMed]

24.   Mulholland, D.J.; Tran, L.M.; Li, Y.; Cai, H.; Morim, A.; Wang, S.; Plaisier, S.; Garraway, I.P.; Huang, J.; Graeber, T.G.; et al. Cell Autonomous Role of PTEN in Regulating Castration-Resistant Prostate Cancer Growth. *Cancer Cell* **2011**, *19*, 792–804. [CrossRef]

25.   Mendoza, M.C.; Er, E.E.; Blenis, J. The Ras-ERK and PI3K-mTOR pathways: Cross-talk and compensation. *Trends Biochem. Sci.* **2011**, *36*, 320–328. [CrossRef]

26. Gagliardi, P.A.; Puliafito, A.; Primo, L. PDK1: At the crossroad of cancer signaling pathways. *Semin. Cancer Biol.* **2018**, *48*, 27–35. [CrossRef]

27. Manning, B.D.; Toker, A. AKT/PKB Signaling: Navigating the Network. *Cell* **2017**, *169*, 381–405. [CrossRef]

28. Berenjeno, I.M.; Piñeiro, R.; Castillo, S.D.; Pearce, W.; McGranahan, N.; Dewhurst, S.M.; Meniel, V.; Birkbak, N.J.; Lau, E.; Sansregret, L.; et al. Oncogenic PIK3CA induces centrosome amplification and tolerance to genome doubling. *Nat. Commun.* **2017**, *8*, 1773. [CrossRef]

29. Huang, J.; Manning, B.D. A complex interplay between Akt, TSC2 and the two mTOR complexes. *Biochem. Soc. Trans.* **2009**, *37*, 217–222. [CrossRef]

30. Julien, L.-A.; Carrière, A.; Moreau, J.; Roux, P.P. mTORC1-Activated S6K1 Phosphorylates Rictor on Threonine 1135 and Regulates mTORC2 Signaling. *Mol. Cell. Biol.* **2009**, *30*, 908–921. [CrossRef]

31. Cantley, L.C.; Wajant, H. The Phosphoinositide 3-Kinase Pathway. *Science* **2002**, *296*, 1655–1657. [CrossRef] [PubMed]

32. Sarbassov, D.D.; Guertin, D.A.; Ali, S.M.; Sabatini, D.M. Phosphorylation and Regulation of Akt/PKB by the Rictor-mTOR Complex. *Science* **2005**, *307*, 1098–1101. [CrossRef] [PubMed]

33. Mahajan, K.; Mahajan, N.P. PI3K-independent AKT activation in cancers: A treasure trove for novel therapeutics. *J. Cell. Physiol.* **2012**, *227*, 3178–3184. [CrossRef] [PubMed]

34. Faes, S.; Faes, O.D.S. PI3K and AKT: Unfaithful Partners in Cancer. *Int. J. Mol. Sci.* **2015**, *16*, 21138–21152. [CrossRef] [PubMed]

35. Lien, E.C.; Dibble, C.; Toker, A. PI3K signaling in cancer: Beyond AKT. *Curr. Opin. Cell Biol.* **2017**, *45*, 62–71. [CrossRef]

36. Jewell, J.L.; Russell, R.C.; Guan, K.-L. Amino acid signalling upstream of mTOR. *Nat. Rev. Mol. Cell Biol.* **2013**, *14*, 133–139. [CrossRef]

37. Malik, S.N.; Brattain, M.; Ghosh, P.M.; Troyer, D.A.; Prihoda, T.; Bedolla, R.; Kreisberg, J.I. Immunohistochemical demonstration of phospho-Akt in high Gleason grade prostate cancer. *Clin. Cancer Res.* **2002**, *8*, 1168–1171.

38. Kremer, C.L.; Klein, R.R.; Mendelson, J.; Browne, W.; Samadzedeh, L.K.; Vanpatten, K.; Highstrom, L.; Pestano, G.A.; Nagle, R. Expression of mTOR signaling pathway markers in prostate cancer progression. *Prostate* **2006**, *66*, 1203–1212. [CrossRef]

39. Sutherland, S.I.; Benito, R.P.; Henshall, S.M.; Horvath, L.; Kench, J.G. Expression of phosphorylated-mTOR during the development of prostate cancer. *Prostate* **2014**, *74*, 1231–1239. [CrossRef]

40. Liao, Y.; Grobholz, R.; Abel, U.; Trojan, L.; Michel, M.S.; Angel, P.; Mayer, D. Increase of AKT/PKB expression correlates with gleason pattern in human prostate cancer. *Int. J. Cancer* **2003**, *107*, 676–680. [CrossRef]

41. Evren, S.; Dermen, A.; Lockwood, G.; Fleshner, N.; Sweet, J. mTOR–RAPTOR and 14-3-3? immunohistochemical expression in high grade prostatic intraepithelial neoplasia and prostatic adenocarcinomas: A tissue microarray study. *J. Clin. Pathol.* **2011**, *64*, 683–688. [CrossRef]

42. Grasso, C.S.; Wu, Y.-M.; Robinson, D.R.; Cao, X.; Dhanasekaran, S.M.; Khan, A.P.; Quist, M.J.; Jing, X.; Lonigro, R.J.; Brenner, J.C.; et al. The mutational landscape of lethal castration-resistant prostate cancer. *Nature* **2012**, *487*, 239–243. [CrossRef] [PubMed]

43. Taylor, B.S.; Schultz, N.; Hieronymus, H.; Gopalan, A.; Xiao, Y.; Carver, B.S.; Arora, V.K.; Kaushik, P.; Cerami, E.; Reva, B.; et al. Integrative Genomic Profiling of Human Prostate Cancer. *Cancer Cell* **2010**, *18*, 11–22. [CrossRef] [PubMed]

44. Robinson, D.; Van Allen, E.M.; Wu, Y.-M.; Schultz, N.; Lonigro, R.J.; Mosquera, J.-M.; Montgomery, B.; Taplin, M.-E.; Pritchard, C.C.; Attard, G.; et al. Integrative clinical genomics of advanced prostate cancer. *Cell* **2015**, *161*, 1215–1228. [CrossRef] [PubMed]

45. Armenia, J.; Wankowicz, S.A.M.; Liu, D.; Gao, J.; Kundra, R.; Reznik, E.; Chatila, W.K.; Chakravarty, D.; Han, G.C.; Coleman, I.; et al. The long tail of oncogenic drivers in prostate cancer. *Nat. Genet.* **2018**, *50*, 645–651. [CrossRef] [PubMed]

46. Abida, W.; Cyrta, J.; Heller, G.; Prandi, D.; Armenia, J.; Coleman, I.; Cieslik, M.; Benelli, M.; Robinson, D.; Van Allen, E.M.; et al. Genomic correlates of clinical outcome in advanced prostate cancer. *Proc. Natl. Acad. Sci. USA* **2019**, *116*, 11428–11436. [CrossRef] [PubMed]

47. Cerami, E.; Gao, J.; Dogrusoz, U.; Gross, B.E.; Sumer, S.O.; Aksoy, B.A.; Skanderup, A.J.; Byrne, C.J.; Heuer, M.L.; Larsson, E.; et al. The cBio cancer genomics portal: An open platform for exploring multidimensional cancer genomics data. *Cancer Discov.* **2012**, *2*, 401–404. [CrossRef]

48. Gao, J.; Aksoy, B.A.; Dogrusoz, U.; Dresdner, G.; Gross, B.; Sumer, S.O.; Sun, Y.; Skanderup, A.J.; Sinha, R.; Larsson, E.; et al. Integrative Analysis of Complex Cancer Genomics and Clinical Profiles Using the cBioPortal. *Sci. Signal.* **2013**, *6*, pl1. [CrossRef]

49. The Cancer Genome Atlas Network. Comprehensive molecular portraits of human breast tumours. *Nature* **2012**, *490*, 61–70. [CrossRef]

50. Kandoth, C.; Schultz, N.; Cherniack, A.D.; Akbani, R.; Liu, Y.; Shen, H.; Robertson, A.G.; Pashtan, I.; Shen, R.; Benz, C.C.; et al. Integrated genomic characterization of endometrial carcinoma. *Nature* **2013**, *497*, 67–73. [CrossRef]

51. Lee, S.H.; Poulogiannis, G.; Pyne, S.; Jia, S.; Zou, L.; Signoretti, S.; Loda, M.; Cantley, L.C.; Roberts, T.M. A constitutively activated form of the p110β isoform of PI3-kinase induces prostatic intraepithelial neoplasia in mice. *Proc. Natl. Acad. Sci. USA* **2010**, *107*, 11002–11007. [CrossRef] [PubMed]

52. Schwartz, S.; Wongvipat, J.; Trigwell, C.B.; Hancox, U.; Carver, B.S.; Outmezguine, V.R.-; Will, M.; Yellen, P.; De Stanchina, E.; Baselga, J.; et al. Feedback suppression of PI3Kα signaling in PTEN-mutated tumors is relieved by selective inhibition of PI3Kβ. *Cancer Cell* **2014**, *27*, 109–122. [CrossRef] [PubMed]

53. Jia, S.; Gao, X.; Lee, S.H.; Maira, S.-M.; Wu, X.; Stack, E.C.; Signoretti, S.; Loda, M.; Zhao, J.J.; Roberts, T.M. Opposing effects of androgen deprivation and targeted therapy on prostate cancer prevention. *Cancer Discov.* **2012**, *3*, 44–51. [CrossRef] [PubMed]

54. Vanhaesebroeck, B.; Welham, M.J.; Kotani, K.; Stein, R.C.; Warne, P.H.; Zvelebil, M.; Higashi, K.; Volinia, S.; Downward, J.; Waterfield, M.D. p110, a novel phosphoinositide 3-kinase in leukocytes. *Proc. Natl. Acad. Sci. USA* **1997**, *94*, 4330–4335. [CrossRef]

55. Chantry, D.; Vojtek, A.; Kashishian, A.; Holtzman, D.A.; Wood, C.; Gray, P.W.; Cooper, J.A.; Hoekstra, M.F. p110δ, a Novel Phosphatidylinositol 3-Kinase Catalytic Subunit That Associates with p85 and Is Expressed Predominantly in Leukocytes. *J. Biol. Chem.* **1997**, *272*, 19236–19241. [CrossRef] [PubMed]

56. Eickholt, B.J.; Ahmed, A.; Davies, M.; Papakonstanti, E.; Pearce, W.; Starkey, M.L.; Bilancio, A.; Need, A.C.; Smith, A.J.H.; Hall, S.M.; et al. Control of Axonal Growth and Regeneration of Sensory Neurons by the p110δ PI 3-Kinase. *PLoS ONE* **2007**, *2*, e869. [CrossRef]

57. Tzenaki, N.; Papakonstanti, E. p110δ PI3 kinase pathway: Emerging roles in cancer. *Front. Oncol.* **2013**, *3*, 3. [CrossRef]

58. Tzenaki, N.; Andreou, M.; Stratigi, K.; Vergetaki, A.; Makrigiannakis, A.; Vanhaesebroeck, B.; Papakonstanti, E. High levels of p110δ PI3K expression in solid tumor cells suppress PTEN activity, generating cellular sensitivity to p110δ inhibitors through PTEN activation. *FASEB J.* **2012**, *26*, 2498–2508. [CrossRef]

59. Zehir, A.; Benayed, R.; Shah, R.; Syed, A.; Middha, S.; Kim, H.; Srinivasan, P.; Gao, J.; Chakravarty, D.; Devlin, S.M.; et al. Mutational landscape of metastatic cancer revealed from prospective clinical sequencing of 10,000 patients. *Nat. Med.* **2017**, *23*, 703–713. [CrossRef]

60. Wang, B.-D.; Ceniccola, K.; Hwang, S.; Andrawis, R.; Horvath, A.; Freedman, J.A.; Olender, J.; Knapp, S.; Ching, T.; Garmire, L.; et al. Alternative splicing promotes tumour aggressiveness and drug resistance in African American prostate cancer. *Nat. Commun.* **2017**, *8*, 15921. [CrossRef]

61. Ueki, K.; Yballe, C.M.; Brachmann, S.M.; Vicent, D.; Watt, J.M.; Kahn, C.R.; Cantley, L.C. Increased insulin sensitivity in mice lacking p85 subunit of phosphoinositide 3-kinase. *Proc. Natl. Acad. Sci. USA* **2001**, *99*, 419–424. [CrossRef] [PubMed]

62. Luo, J.; Field, S.J.; Lee, J.Y.; Engelman, J.A.; Cantley, L.C. The p85 regulatory subunit of phosphoinositide 3-kinase down-regulates IRS-1 signaling via the formation of a sequestration complex. *J. Cell Biol.* **2005**, *170*, 455–464. [CrossRef]

63. Terauchi, Y.; Tsuji, Y.; Satoh, S.; Minoura, H.; Murakami, K.; Okuno, A.; Inukai, K.; Asano, T.; Kaburagi, Y.; Ueki, K.; et al. Increased insulin sensitivity and hypoglycaemia in mice lacking the p85α subunit of phosphoinositide 3–kinase. *Nat. Genet.* **1999**, *21*, 230–235. [CrossRef] [PubMed]

64. Mauvais-Jarvis, F.; Ueki, K.; Fruman, D.A.; Hirshman, M.F.; Sakamoto, K.; Goodyear, L.J.; Iannacone, M.; Accili, M.; Cantley, L.C.; Kahn, C.R. Reduced expression of the murine p85α subunit of phosphoinositide 3-kinase improves insulin signaling and ameliorates diabetes. *J. Clin. Investig.* **2002**, *109*, 141–149. [CrossRef] [PubMed]

65. Taniguchi, C.M.; Winnay, J.; Kondo, T.; Bronson, R.T.; Guimaraes, A.R.; Aleman, J.O.; Luo, J.; Stephanopoulos, G.; Weissleder, R.; Cantley, L.C.; et al. The phosphoinositide 3-kinase regulatory subunit p85alpha can exert tumor suppressor properties through negative regulation of growth factor signaling. *Cancer Res.* **2010**, *70*, 5305–5315. [CrossRef] [PubMed]

66. Thorpe, L.M.; Spangle, J.M.; Ohlson, C.E.; Cheng, H.; Roberts, T.M.; Cantley, L.C.; Zhao, J.J. PI3K-p110α mediates the oncogenic activity induced by loss of the novel tumor suppressor PI3K-p85α. *Proc. Natl. Acad. Sci. USA* **2017**, *114*, 7095–7100. [CrossRef] [PubMed]

67. Philp, A.J.; Campbell, I.G.; Leet, C.; Vincan, E.; Rockman, S.P.; Whitehead, R.H.; Thomas, R.J.; Phillips, W.A. The phosphatidylinositol 3′-kinase p85alpha gene is an oncogene in human ovarian and colon tumors. *Cancer Res.* **2001**, *61*, 7426–7429. [PubMed]

68. Vallejo-Díaz, J.; Chagoyen, M.; Olazabal-Morán, M.; Gonzalez-Garcia, A.; Carrera, A. The Opposing Roles of PIK3R1/p85α and PIK3R2/p85β in Cancer. *Trends Cancer* **2019**, *5*, 233–244. [CrossRef]

69. Vallejo-Diaz, J.; Olazabal-Morán, M.; Cariaga-Martinez, A.E.; Pajares, M.J.; Flores, J.M.; Pio, R.; Montuenga, L.M.; Carrera, A. Targeted depletion of PIK3R2 induces regression of lung squamous cell carcinoma. *Oncotarget* **2016**, *7*, 85063–85078. [CrossRef]

70. Zhang, L.; Huang, J.; Yang, N.; Greshock, J.; Liang, S.; Hasegawa, K.; Giannakakis, A.; Poulos, N.; O'Brien-Jenkins, A.; Katsaros, D.; et al. Integrative genomic analysis of phosphatidylinositol 3′-kinase family identifies PIK3R3 as a potential therapeutic target in epithelial ovarian cancer. *Clin. Cancer Res.* **2007**, *13*, 5314–5321. [CrossRef]

71. Peng, Y.-P.; Zhu, Y.; Yin, L.-D.; Wei, J.-S.; Liu, X.-C.; Zhu, X.-L.; Miao, Y. PIK3R3 Promotes Metastasis of Pancreatic Cancer via ZEB1 Induced Epithelial-Mesenchymal Transition. *Cell. Physiol. Biochem.* **2018**, *46*, 1930–1938. [CrossRef] [PubMed]

72. Munkley, J.; Livermore, K.; McClurgg, U.L.; Kalna, G.; Knight, B.; Mccullagh, P.; McGrath, J.; Crundwell, M.; Leung, H.Y.; Robson, C.; et al. The PI3K regulatory subunit gene PIK3R1 is under direct control of androgens and repressed in prostate cancer cells. *Oncoscience* **2015**, *2*, 755–764. [CrossRef] [PubMed]

73. Han, R.; Zhang, L.; Gan, W.; Fu, K.; Jiang, K.; Ding, J.; Wu, J.; Han, X.; Li, D. piRNA-DQ722010 contributes to prostate hyperplasia of the male offspring mice after the maternal exposed to microcystin-leucine arginine. *Prostate* **2019**, *79*, 798–812. [CrossRef] [PubMed]

74. Song, L.; Xie, X.; Yu, S.; Peng, F.; Peng, L. MicroRNA-126 inhibits proliferation and metastasis by targeting pik3r2 in prostate cancer. *Mol. Med. Rep.* **2015**, *13*, 1204–1210. [CrossRef] [PubMed]

75. Brazzatti, J.; Klingler-Hoffmann, M.; Haylock-Jacobs, S.; Harata-Lee, Y.; Niu, M.; Higgins, M.D.; Kochetkova, M.; Hoffmann, P.; McColl, S.R. Differential roles for the p101 and p84 regulatory subunits of PI3Kγ in tumor growth and metastasis. *Oncogene* **2011**, *31*, 2350–2361. [CrossRef]

76. Rodgers, K.; Hammerman, P.S.; Lawrence, M.S.; Voet, D.; Jing, R.; Cibulskis, K.; Sivachenko, A.; Stojanov, P.; McKenna, A.; Lander, E.S.; et al. Comprehensive genomic characterization of squamous cell lung cancers. *Nature* **2012**, *489*, 519–525. [CrossRef]

77. Bass, A.J.; Thorsson, V.; Shmulevich, I.; Reynolds, S.M.; Miller, M.; Bernard, B.; Hinoue, T.; Laird, P.W.; Curtis, C.; Shen, H.; et al. Comprehensive molecular characterization of gastric adenocarcinoma. *Nature* **2014**, *513*, 202–209. [CrossRef]

78. Akbani, R.; Akdemir, K.C.; Aksoy, B.A.; Albert, M.; Ally, A.; Amin, S.; Arachchi, H.; Arora, A.; Auman, J.T.; Ayala, B.; et al. Genomic Classification of Cutaneous Melanoma. *Cell* **2015**, *161*, 1681–1696. [CrossRef]

79. Hoadley, K.A.; Yau, C.; Hinoue, T.; Wolf, D.M.; Lazar, A.J.F.; Drill, E.; Shen, R.; Taylor, A.M.; Cherniack, A.D.; Thorsson, V.; et al. Cell-of-Origin Patterns Dominate the Molecular Classification of 10,000 Tumors from 33 Types of Cancer. *Cell* **2018**, *173*, 291–304.e6. [CrossRef]

80. Falasca, M.; Maffucci, T. Regulation and cellular functions of class II phosphoinositide 3-kinases. *Biochem. J.* **2012**, *443*, 587–601. [CrossRef]

81. Mavrommati, I.; Cisse, O.; Falasca, M.; Maffucci, T. Novel roles for class II Phosphoinositide 3-Kinase C2β in signalling pathways involved in prostate cancer cell invasion. *Sci. Rep.* **2016**, *6*, 23277. [CrossRef] [PubMed]

82. Raiborg, C.; Schink, K.O.; Stenmark, H.A. Class III phosphatidylinositol 3-kinase and its catalytic product PtdIns3P in regulation of endocytic membrane traffic. *FEBS J.* **2013**, *280*, 2730–2742. [CrossRef] [PubMed]

83. Backer, J.M. The intricate regulation and complex functions of the Class III phosphoinositide 3-kinase Vps34. *Biochem. J.* **2016**, *473*, 2251–2271. [CrossRef] [PubMed]

84. Bilanges, B.; Alliouachene, S.; Pearce, W.; Morelli, D.; Szabadkai, G.; Chung, Y.-L.; Chicanne, G.; Valet, C.; Hill, J.M.; Voshol, P.J.; et al. Vps34 PI 3-kinase inactivation enhances insulin sensitivity through reprogramming of mitochondrial metabolism. *Nat. Commun.* **2017**, *8*, 1804. [CrossRef] [PubMed]

85. Simonsen, A.; Tooze, S.A. Coordination of membrane events during autophagy by multiple class III PI3-kinase complexes. *J. Cell Biol.* **2009**, *186*, 773–782. [CrossRef] [PubMed]

86. Nobukuni, T.; Joaquin, M.; Roccio, M.; Dann, S.G.; Kim, S.Y.; Gulati, P.; Byfield, M.P.; Backer, J.M.; Natt, F.; Bos, J.L.; et al. Amino acids mediate mTOR/raptor signaling through activation of class 3 phosphatidylinositol 3OH-kinase. *Proc. Natl. Acad. Sci. USA* **2005**, *102*, 14238–14243. [CrossRef]

87. Nezis, I.P.; Sagona, A.P.; Schink, K.O.; Stenmark, H. Divide and ProsPer: The emerging role of PtdIns3P in cytokinesis. *Trends Cell Biol.* **2010**, *20*, 642–649. [CrossRef]

88. Yoon, M.-S.; Son, K.; Arauz, E.; Han, J.M.; Kim, S.; Chen, J. Leucyl-tRNA Synthetase Activates Vps34 in Amino Acid-Sensing mTORC1 Signaling. *Cell Rep.* **2016**, *16*, 1510–1517. [CrossRef]

89. Backer, J.M. The regulation and function of Class III PI3Ks: Novel roles for Vps34. *Biochem. J.* **2008**, *410*, 1–17. [CrossRef]

90. Dyson, J.M.; Fedele, C.G.; Davies, E.M.; Becanovic, J.; Mitchell, C.A. Phosphoinositide Phosphatases: Just as Important as the Kinases. *Subcell. Biochem.* **2012**, *58*, 215–279. [CrossRef]

91. Rudge, S.A.; Wakelam, M.J. Phosphatidylinositolphosphate phosphatase activities and cancer. *J. Lipid Res.* **2015**, *57*, 176–192. [CrossRef] [PubMed]

92. Jamaspishvili, T.; Berman, D.M.; Ross, A.E.; Scher, H.I.; De Marzo, A.M.; Squire, J.A.; Lotan, T.L. Clinical implications of PTEN loss in prostate cancer. *Nat. Rev. Urol.* **2018**, *15*, 222–234. [CrossRef]

93. Cairns, P.; Okami, K.; Halachmi, S.; Halachmi, N.; Esteller, M.; Herman, J.G.; Jen, J.; Isaacs, W.B.; Bova, G.S.; Sidransky, D. Frequent inactivation of PTEN/MMAC1 in primary prostate cancer. *Cancer Res.* **1997**, *57*, 4997–5000. [PubMed]

94. Suzuki, H.; Freije, D.; Nusskern, D.R.; Okami, K.; Cairns, P.; Sidransky, D.; Isaacs, W.B.; Bova, G.S. Interfocal heterogeneity of PTEN/MMAC1 gene alterations in multiple metastatic prostate cancer tissues. *Cancer Res.* **1998**, *58*, 204–209. [PubMed]

95. Wang, S.I.; Parsons, R.; Ittmann, M. Homozygous deletion of the PTEN tumor suppressor gene in a subset of prostate adenocarcinomas. *Clin. Cancer Res.* **1998**, *4*, 811–815. [PubMed]

96. Maehama, T.; Dixon, J.E. The tumor suppressor, PTEN/MMAC1, dephosphorylates the lipid second messenger, phosphatidylinositol 3,4,5-trisphosphate. *J. Biol. Chem.* **1998**, *273*, 13375–13378. [CrossRef] [PubMed]

97. Myers, M.P.; Stolarov, J.P.; Eng, C.; Li, J.; Wang, S.I.; Wigler, M.; Parsons, R.; Tonks, N.K. P-TEN, the tumor suppressor from human chromosome 10q23, is a dual-specificity phosphatase. *Proc. Natl. Acad. Sci. USA* **1997**, *94*, 9052–9057. [CrossRef]

98. Geybels, M.S.; Fang, M.; Wright, J.L.; Qu, X.; Bibikova, M.; Klotzle, B.; Fan, J.-B.; Feng, Z.; Ostrander, E.A.; Nelson, P.S.; et al. PTEN loss is associated with prostate cancer recurrence and alterations in tumor DNA methylation profiles. *Oncotarget* **2017**, *8*, 84338–84348. [CrossRef]

99. McMenamin, M.E.; Soung, P.; Perera, S.; Kaplan, I.; Loda, M.; Sellers, W.R. Loss of PTEN expression in paraffin-embedded primary prostate cancer correlates with high Gleason score and advanced stage. *Cancer Res.* **1999**, *59*, 4291–4296.

100. Ratnacaram, C.K.; Telentin, M.; Jiang, M.; Meng, X.; Chambon, P.; Metzger, D. Temporally controlled ablation of PTEN in adult mouse prostate epithelium generates a model of invasive prostatic adenocarcinoma. *Proc. Natl. Acad. Sci. USA* **2008**, *105*, 2521–2526. [CrossRef]

101. Mulholland, D.J.; Kobayashi, N.; Ruscetti, M.; Zhi, A.; Tran, L.M.; Huang, J.; Gleave, M.; Wu, H. Pten loss and RAS/MAPK activation cooperate to promote EMT and metastasis initiated from prostate cancer stem/progenitor cells. *Cancer Res.* **2012**, *72*, 1878–1889. [CrossRef] [PubMed]

102. Wang, S.; Gao, J.; Lei, Q.-Y.; Rozengurt, N.; Pritchard, C.; Jiao, J.; Thomas, G.V.; Li, G.; Roy-Burman, P.; Nelson, P.S.; et al. Prostate-specific deletion of the murine Pten tumor suppressor gene leads to metastatic prostate cancer. *Cancer Cell* **2003**, *4*, 209–221. [CrossRef]

103. Chen, Z.; Trotman, L.C.; Shaffer, D.; Lin, H.-K.; Dotan, Z.A.; Niki, M.; Koutcher, J.A.; Scher, H.I.; Ludwig, T.; Gerald, W.; et al. Crucial role of p53-dependent cellular senescence in suppression of Pten-deficient tumorigenesis. *Nature* **2005**, *436*, 725–730. [CrossRef] [PubMed]

104. Ahmad, I.; Patel, R.; Singh, L.B.; Nixon, C.; Seywright, M.; Barnetson, R.J.; Brunton, V.G.; Muller, W.J.; Edwards, J.; Sansom, O.J.; et al. HER2 overcomes PTEN (loss)-induced senescence to cause aggressive prostate cancer. *Proc. Natl. Acad. Sci. USA* **2011**, *108*, 16392–16397. [CrossRef]

105. Kwak, M.K.; Johnson, D.T.; Zhu, C.; Lee, S.H.; Ye, D.-W.; Luong, R.; Sun, Z. Conditional Deletion of the Pten Gene in the Mouse Prostate Induces Prostatic Intraepithelial Neoplasms at Early Ages but a Slow Progression to Prostate Tumors. *PLoS ONE* **2013**, *8*, e53476. [CrossRef]

106. Trotman, L.C.; Niki, M.; Dotan, Z.A.; Koutcher, J.A.; Di Cristofano, A.; Xiao, A.; Khoo, A.S.; Roy-Burman, P.; Greenberg, N.M.; Van Dyke, T.; et al. Pten Dose Dictates Cancer Progression in the Prostate. *PLoS Biol.* **2003**, *1*, e59. [CrossRef]

107. Manda, K.R.; Tripathi, P.; Hsi, A.C.; Ning, J.; Ruzinova, M.B.; Liapis, H.; Bailey, M.; Zhang, H.; Maher, C.A.; Humphrey, P.A.; et al. NFATc1 promotes prostate tumorigenesis and overcomes PTEN loss-induced senescence. *Oncogene* **2015**, *35*, 3282–3292. [CrossRef]

108. Han, S.Y.; Kato, H.; Kato, S.; Suzuki, T.; Shibata, H.; Ishii, S.; Shiiba, K.; Matsuno, S.; Kanamaru, R.; Ishioka, C. Functional evaluation of PTEN missense mutations using in vitro phosphoinositide phosphatase assay. *Cancer Res.* **2000**, *60*, 3147–3151.

109. Correia, N.C.; Gírio, A.; Antunes, I.; Martins, L.R.; Barata, J. The multiple layers of non-genetic regulation of PTEN tumour suppressor activity. *Eur. J. Cancer* **2014**, *50*, 216–225. [CrossRef]

110. Ooms, L.M.; Binge, L.C.; Davies, E.M.; Rahman, P.; Conway, J.R.; Gurung, R.; Ferguson, D.T.; Papa, A.; Fedele, C.G.; Vieusseux, J.L.; et al. The Inositol Polyphosphate 5-Phosphatase PIPP Regulates AKT1-Dependent Breast Cancer Growth and Metastasis. *Cancer Cell* **2015**, *28*, 155–169. [CrossRef]

111. Rodgers, S.J.; Ferguson, D.T.; Mitchell, C.A.; Ooms, L.M. Regulation of PI3K effector signalling in cancer by the phosphoinositide phosphatases. *Biosci. Rep.* **2017**, *37*. [CrossRef] [PubMed]

112. Gewinner, C.; Wang, Z.C.; Richardson, A.; Teruya-Feldstein, J.; Etemadmoghadam, D.; Bowtell, D.D.; Barretina, J.; Lin, W.M.; Rameh, L.; Salmena, L.; et al. Evidence that Inositol Polyphosphate 4-Phosphatase Type II Is a Tumor Suppressor that Inhibits PI3K Signaling. *Cancer Cell* **2009**, *16*, 115–125. [CrossRef] [PubMed]

113. Hodgson, M.C.; Shao, L.-J.; Frolov, A.; Li, R.; Peterson, L.; Ayala, G.; Ittmann, M.M.; Weigel, N.L.; Agoulnik, I.U. Decreased expression and androgen regulation of the tumor suppressor gene INPP4B in prostate cancer. *Cancer Res.* **2011**, *71*, 572–582. [CrossRef]

114. Fedele, C.G.; Ooms, L.M.; Ho, M.; Vieusseux, J.; O'Toole, S.A.; Millar, E.; Knowles, E.L.; Sriratana, A.; Gurung, R.; Baglietto, L.; et al. Inositol polyphosphate 4-phosphatase II regulates PI3K/Akt signaling and is lost in human basal-like breast cancers. *Proc. Natl. Acad. Sci. USA* **2010**, *107*, 22231–22236. [CrossRef]

115. Kofuji, S.; Kimura, H.; Nakanishi, H.; Nanjo, H.; Takasuga, S.; Liu, H.; Eguchi, S.; Nakamura, R.; Itoh, R.; Ueno, N.; et al. INPP4B Is a PtdIns(3,4,5)P3 Phosphatase That Can Act as a Tumor Suppressor. *Cancer Discov.* **2015**, *5*, 730–739. [CrossRef] [PubMed]

116. Chew, C.L.; Lunardi, A.; Gulluni, F.; Ruan, D.T.; Chen, M.; Salmena, L.; Nishino, M.; Papa, A.; Ng, C.; Fung, J.; et al. In Vivo Role of INPP4B in Tumor and Metastasis Suppression through Regulation of PI3K-AKT Signaling at Endosomes. *Cancer Discov.* **2015**, *5*, 740–751. [CrossRef]

117. Chen, H.; Li, H.; Chen, Q. INPP4B overexpression suppresses migration, invasion and angiogenesis of human prostate cancer cells. *Clin. Exp. Pharmacol. Physiol.* **2017**, *44*, 700–708. [CrossRef]

118. Rynkiewicz, N.K.; Fedele, C.G.; Chiam, K.; Gupta, R.; Kench, J.G.; Ooms, L.M.; McLean, C.; Giles, G.G.; Horvath, L.; Mitchell, C.A. INPP4B is highly expressed in prostate intermediate cells and its loss of expression in prostate carcinoma predicts for recurrence and poor long term survival. *Prostate* **2014**, *75*, 92–102. [CrossRef]

119. Xie, J.; Erneux, C.; Pirson, I. How does SHIP1/2 balance PtdIns(3,4)P 2 and does it signal independently of its phosphatase activity? *BioEssays* **2013**, *35*, 733–743. [CrossRef]

120. Hoekstra, E.; Das, A.M.; Willemsen, M.; Swets, M.; Kuppen, P.J.; Van Der Woude, C.J.; Bruno, M.J.; Shah, J.P.; Hagen, T.L.T.; Chisholm, J.D.; et al. Lipid phosphatase SHIP2 functions as oncogene in colorectal cancer by regulating PKB activation. *Oncotarget* **2016**, *7*, 73525–73540. [CrossRef]

121. Chan, T.O.; Rittenhouse, S.E.; Tsichlis, P.N. AKT/PKB and Other D3 Phosphoinositide-Regulated Kinases: Kinase Activation by Phosphoinositide-Dependent Phosphorylation. *Annu. Rev. Biochem.* **1999**, *68*, 965–1014. [CrossRef] [PubMed]

122. Manning, B.D.; Cantley, L.C. AKT/PKB Signaling: Navigating Downstream. *Cell* **2007**, *129*, 1261–1274. [CrossRef] [PubMed]

123. Alessi, D.R.; Deak, M.; Casamayor, A.; Caudwell, F.B.; Morrice, N.; Norman, D.G.; Gaffney, P.; Reese, C.B.; MacDougall, C.N.; Harbison, D.; et al. 3-Phosphoinositide-dependent protein kinase-1 (PDK1): Structural and functional homology with the Drosophila DSTPK61 kinase. *Curr. Biol.* **1997**, *7*, 776–789. [CrossRef]

124. Burgering, B.; Coffer, P.J. Protein kinase B (c-Akt) in phosphatidylinositol-3-OH kinase signal transduction. *Nature* **1995**, *376*, 599–602. [CrossRef] [PubMed]

125. Andjelković, M.; Jakubowicz, T.; Cron, P.; Ming, X.F.; Han, J.W.; Hemmings, B.A. Activation and phosphorylation of a pleckstrin homology domain containing protein kinase (RAC-PK/PKB) promoted by serum and protein phosphatase inhibitors. *Proc. Natl. Acad. Sci. USA* **1996**, *93*, 5699–5704. [CrossRef]

126. Gao, T.; Furnari, F.; Newton, A.C. PHLPP: A Phosphatase that Directly Dephosphorylates Akt, Promotes Apoptosis, and Suppresses Tumor Growth. *Mol. Cell* **2005**, *18*, 13–24. [CrossRef]

127. Pereira, B.; Chin, S.-F.; Rueda, O.M.; Vollan, H.-K.M.; Provenzano, E.; Bardwell, H.A.; Pugh, M.; Jones, L.; Russell, R.; Sammut, S.-J.; et al. The somatic mutation profiles of 2,433 breast cancers refine their genomic and transcriptomic landscapes. *Nat. Commun.* **2016**, *7*, 11479. [CrossRef]

128. Bleeker, F.E.; Felicioni, L.; Buttitta, F.; Lamba, S.; Cardone, L.; Rodolfo, M.; Scarpa, A.; Leenstra, S.; Frattini, M.; Barbareschi, M.; et al. AKT1E17K in human solid tumours. *Oncogene* **2008**, *27*, 5648–5650. [CrossRef]

129. Troxell, M.L. PIK3CA/AKT1 Mutations in Breast Carcinoma: A Comprehensive Review of Experimental and Clinical Studies. *J. Clin. Exp. Pathol.* **2012**, S1-002. [CrossRef]

130. Shukla, S.; MacLennan, G.T.; Hartman, U.J.; Fu, P.; Resnick, M.I.; Gupta, S. Activation of PI3K-Akt signaling pathway promotes prostate cancer cell invasion. *Int. J. Cancer* **2007**, *121*, 1424–1432. [CrossRef]

131. Majumder, P.K.; Yeh, J.J.; George, D.J.; Febbo, P.G.; Kum, J.; Xue, Q.; Bikoff, R.; Ma, H.; Kantoff, P.W.; Golub, T.R.; et al. Prostate intraepithelial neoplasia induced by prostate restricted Akt activation: The MPAKT model. *Proc. Natl. Acad. Sci. USA* **2003**, *100*, 7841–7846. [CrossRef] [PubMed]

132. Li, B.; Sun, A.; Youn, H.; Hong, Y.; Terranova, P.F.; Thrasher, J.; Xu, P.; Spencer, D. Conditional Akt activation promotes androgen-independent progression of prostate cancer. *Carcinogenesis* **2006**, *28*, 572–583. [CrossRef]

133. Chen, M.-L.; Xu, P.-Z.; Peng, X.-D.; Chen, W.S.; Guzman, G.; Yang, X.; Di Cristofano, A.; Pandolfi, P.P.; Hay, N. The deficiency of Akt1 is sufficient to suppress tumor development in Pten± mice. *Genome Res.* **2006**, *20*, 1569–1574. [CrossRef] [PubMed]

134. De Bono, J.S.; De Giorgi, U.; Rodrigues, D.N.; Massard, C.; Bracarda, S.; Font, A.; Arija, J.A.A.; Shih, K.C.; Radavoi, G.D.; Xu, N.; et al. Randomized Phase II Study of Akt Blockade with or without Ipatasertib in Abiraterone-Treated Patients with Metastatic Prostate Cancer with and without PTEN Loss. *Clin. Cancer Res.* **2019**, *25*, 928–936. [CrossRef]

135. Maurer, M.; Su, T.; Saal, L.H.; Koujak, S.; Hopkins, B.D.; Barkley, C.R.; Wu, J.; Nandula, S.; Dutta, B.; Xie, Y.; et al. 3-Phosphoinositide-dependent kinase 1 potentiates upstream lesions on the phosphatidylinositol 3-kinase pathway in breast carcinoma. *Cancer Res.* **2009**, *69*, 6299–6306. [CrossRef] [PubMed]

136. Ellwood-Yen, K.; Keilhack, H.; Kunii, K.; Dolinski, B.; Connor, Y.; Hu, K.; Nagashima, K.; O'Hare, E.; Erkul, Y.; Di Bacco, A.; et al. PDK1 Attenuation Fails to Prevent Tumor Formation in PTEN-Deficient Transgenic Mouse Models. *Cancer Res.* **2011**, *71*, 3052–3065. [CrossRef]

137. Magee, J.A.; Chang, L.W.; Stormo, G.D.; Milbrandt, J. Direct, Androgen Receptor-Mediated Regulation of the FKBP5 Gene via a Distal Enhancer Element. *Endocrinology* **2006**, *147*, 590–598. [CrossRef]

138. Chen, M.; Pratt, C.; Zeeman, M.E.; Schultz, N.; Taylor, B.S.; O'Neill, A.; Castillo-Martin, M.; Nowak, D.G.; Naguib, A.; Grace, D.M.; et al. Identification of PHLPP1 as a Tumor Suppressor Reveals the Role of Feedback Activation in PTEN-Mutant Prostate Cancer Progression. *Cancer Cell* **2011**, *20*, 173–186. [CrossRef]

139. Nowak, D.G.; Katsenelson, K.C.; Watrud, K.E.; Chen, M.; Mathew, G.; D'Andrea, V.D.; Lee, M.F.; Swamynathan, M.M.; Casanova-Salas, I.; Jibilian, M.C.; et al. The PHLPP2 phosphatase is a druggable driver of prostate cancer progression. *J. Cell Biol.* **2019**, *218*, 1943–1957. [CrossRef]

140. Pandey, P.; Seshacharyulu, P.; Das, S.; Rachagani, S.; Ponnusamy, M.P.; Yan, Y.; Johansson, S.L.; Datta, K.; Lin, M.F.; Batra, S.K. Impaired expression of protein phosphatase 2A subunits enhances metastatic potential of human prostate cancer cells through activation of AKT pathway. *Br. J. Cancer* **2013**, *108*, 2590–2600. [CrossRef]

141. Malik, N.; Macartney, T.; Hornberger, A.; Anderson, K.E.; Tovell, H.; Prescott, A.R.; Alessi, D.R. Mechanism of activation of SGK3 by growth factors via the Class 1 and Class 3 PI3Ks. *Biochem. J.* **2018**, *475*, 117–135. [CrossRef] [PubMed]

142. Basnet, R.; Gong, G.; Li, C.; Wang, M. Serum and glucocorticoid inducible protein kinases (SGKs): A potential target for cancer intervention. *Acta Pharm. Sin. B* **2018**, *8*, 767–771. [CrossRef] [PubMed]

143. Kobayashi, T.; Deak, M.; Morrice, N.; Cohen, P. Characterization of the structure and regulation of two novel isoforms of serum- and glucocorticoid-induced protein kinase. *Biochem. J.* **1999**, *344*, 189–197. [CrossRef] [PubMed]

144. Tessier, M.; Woodgett, J.R. Serum and glucocorticoid-regulated protein kinases: Variations on a theme. *J. Cell. Biochem.* **2006**, *98*, 1391–1407. [CrossRef] [PubMed]

145. Castel, P.; Ellis, H.; Bago, R.; Toska, E.; Razavi, P.; Carmona, F.J.; Kannan, S.; Verma, C.S.; Dickler, M.; Chandarlapaty, S.; et al. PDK1-SGK1 Signaling Sustains AKT-Independent mTORC1 Activation and Confers Resistance to PI3Kα Inhibition. *Cancer Cell* **2016**, *30*, 229–242. [CrossRef]

146. Chou, M.M.; Hou, W.; Johnson, J.; Graham, L.K.; Lee, M.H.; Chen, C.S.; Newton, A.C.; Schaffhausen, B.S.; Toker, A. Regulation of protein kinase C zeta by PI 3-kinase and PDK-1. *Curr. Biol.* **1998**, *8*, 1069–1077. [CrossRef]

147. Mizuno, H.; Nishida, E. The ERK MAP kinase pathway mediates induction of SGK (serum- and glucocorticoid-inducible kinase) by growth factors. *Genes Cells* **2001**, *6*, 261–268. [CrossRef]

148. Bago, R.; Sommer, E.; Castel, P.; Crafter, C.; Bailey, F.P.; Shpiro, N.; Baselga, J.; Cross, D.; Eyers, P.A.; Alessi, D.R. The hVps34- SGK 3 pathway alleviates sustained PI3K/Akt inhibition by stimulating mTORC 1 and tumour growth. *EMBO J.* **2016**, *35*, 1902–1922. [CrossRef]

149. Hayashi, M.; Tapping, R.I.; Chao, T.-H.; Lo, J.-F.; King, C.C.; Yang, Y.; Lee, J.-D. BMK1 Mediates Growth Factor-induced Cell Proliferation through Direct Cellular Activation of Serum and Glucocorticoid-inducible Kinase. *J. Biol. Chem.* **2001**, *276*, 8631–8634. [CrossRef]

150. Meng, F.; Yamagiwa, Y.; Taffetani, S.; Han, J.; Patel, T. IL-6 activates serum and glucocorticoid kinase via p38α mitogen-activated protein kinase pathway. *Am. J. Physiol. Physiol.* **2005**, *289*, C971–C981. [CrossRef]

151. Isikbay, M.; Otto, K.; Kregel, S.; Kach, J.; Cai, Y.; Griend, D.V.; Conzen, S.D.; Szmulewitz, R. Glucocorticoid receptor activity contributes to resistance to androgen-targeted therapy in prostate cancer. *Horm. Cancer* **2014**, *5*, 72–89. [CrossRef] [PubMed]

152. Liu, W.; Wang, X.; Liu, Z.; Wang, Y.; Yin, B.; Yu, P.; Duan, X.; Liao, Z.; Chen, Y.; Liu, C.; et al. SGK1 inhibition induces autophagy-dependent apoptosis via the mTOR-Foxo3a pathway. *Br. J. Cancer* **2017**, *117*, 1139–1153. [CrossRef] [PubMed]

153. Gasser, J.A.; Inuzuka, H.; Lau, A.W.; Wei, W.; Beroukhim, R.; Toker, A. SGK3 mediates INPP4B-dependent PI3K signaling in breast cancer. *Mol. Cell* **2014**, *56*, 595–607. [CrossRef] [PubMed]

154. Bruhn, M.A.; Pearson, R.B.; Hannan, R.D.; Sheppard, K.E. Second AKT: The rise of SGK in cancer signalling. *Growth Factors* **2010**, *28*, 394–408. [CrossRef]

155. Zhang, Y.; Gan, B.; Liu, D.; Paik, J.-H. FoxO family members in cancer. *Cancer Biol. Ther.* **2011**, *12*, 253–259. [CrossRef]

156. Chen, C.-C.; Jeon, S.-M.; Bhaskar, P.T.; Nogueira, V.; Sundararajan, D.; Tonic, I.; Park, Y.; Hay, N. FoxOs Inhibit mTORC1 and Activate Akt by Inducing the Expression of Sestrin3 and Rictor. *Dev. Cell* **2010**, *18*, 592–604. [CrossRef]

157. Wang, Y.; Zhou, Y.; Graves, D. FOXO transcription factors: Their clinical significance and regulation. *Biomed. Res. Int.* **2014**, *2014*, 925350. [CrossRef]

158. Bach, D.-H.; Long, N.P.; Luu, T.-T.-T.; Anh, N.H.; Kwon, S.W.; Lee, S.K. The Dominant Role of Forkhead Box Proteins in Cancer. *Int. J. Mol. Sci.* **2018**, *19*, 3279. [CrossRef]

159. Van Der Heide, L.P.; Hoekman, M.F.M.; Smidt, M.P. The ins and outs of FoxO shuttling: Mechanisms of FoxO translocation and transcriptional regulation. *Biochem. J.* **2004**, *380*, 297–309. [CrossRef]

160. Hyytinen, E.-R.; Saadut, R.; Chen, C.; Paull, L.; Koivisto, P.A.; Vessella, R.L.; Frierson, H.F.; Dong, J.-T. Defining the region(s) of deletion at 6q16-q22 in human prostate cancer. *Genes Chromosom. Cancer* **2002**, *34*, 306–312. [CrossRef]

161. Shukla, S.; Bhaskaran, N.; MacLennan, G.T.; Gupta, S. Deregulation of FoxO3a accelerates prostate cancer progression in TRAMP mice. *Prostate* **2013**, *73*, 1507–1517. [CrossRef] [PubMed]

162. Yang, Y.; Blee, A.M.; Wang, D.; An, J.; Pan, Y.; Yan, Y.; Ma, T.; He, Y.; Dugdale, J.; Hou, X.; et al. Loss of FOXO1 Cooperates with TMPRSS2-ERG Overexpression to Promote Prostate Tumorigenesis and Cell Invasion. *Cancer Res.* **2017**, *77*, 6524–6537. [CrossRef] [PubMed]

163. Su, B.; Gao, L.; Baranowski, C.; Gillard, B.; Wang, J.; Ransom, R.; Ko, H.-K.; Gelman, I.H. A Genome-Wide RNAi Screen Identifies FOXO4 as a Metastasis-Suppressor through Counteracting PI3K/AKT Signal Pathway in Prostate Cancer. *PLoS ONE* **2014**, *9*, e101411. [CrossRef] [PubMed]

164. Dibble, C.; Elis, W.; Menon, S.; Qin, W.; Klekota, J.; Asara, J.M.; Finan, P.M.; Kwiatkowski, D.J.; Murphy, L.O.; Manning, B.D. TBC1D7 is a third subunit of the TSC1-TSC2 complex upstream of mTORC1. *Mol. Cell* **2012**, *47*, 535–546. [CrossRef] [PubMed]

165. Manning, B.D.; Tee, A.R.; Logsdon, M.N.; Blenis, J.; Cantley, L.C. Identification of the tuberous sclerosis complex-2 tumor suppressor gene product tuberin as a target of the phosphoinositide 3-kinase/akt pathway. *Mol. Cell* **2002**, *10*, 151–162. [CrossRef]

166. Inoki, K.; Zhu, T.; Guan, K.-L. TSC2 Mediates Cellular Energy Response to Control Cell Growth and Survival. *Cell* **2003**, *115*, 577–590. [CrossRef]

167. Inoki, K.; Ouyang, H.; Zhu, T.; Lindvall, C.; Wang, Y.; Zhang, X.; Yang, Q.; Bennett, C.; Harada, Y.; Stankunas, K.; et al. TSC2 Integrates Wnt and Energy Signals via a Coordinated Phosphorylation by AMPK and GSK3 to Regulate Cell Growth. *Cell* **2006**, *126*, 955–968. [CrossRef]

168. Shaw, R.J.; Bardeesy, N.; Manning, B.D.; Lopez, L.; Kosmatka, M.; DePinho, R.A.; Cantley, L.C. The LKB1 tumor suppressor negatively regulates mTOR signaling. *Cancer Cell* **2004**, *6*, 91–99. [CrossRef]

169. Liang, M.-C.; Ma, J.; Chen, L.; Kozlowski, P.; Qin, W.; Li, D.; Goto, J.; Shimamura, T.; Hayes, D.N.; Meyerson, M.; et al. TSC1 loss synergizes with KRAS activation in lung cancer development in the mouse and confers rapamycin sensitivity. *Oncogene* **2009**, *29*, 1588–1597. [CrossRef]

170. Ho, D.W.H.; Chan, L.K.; Chiu, Y.T.; Xu, I.M.J.; Poon, R.T.P.; Cheung, T.T.; Tang, C.N.; Tang, V.W.L.; Lo, I.L.; Lam, P.W.Y.; et al. TSC1/2mutations define a molecular subset of HCC with aggressive behaviour and treatment implication. *Gut* **2016**, *66*, 1496–1506. [CrossRef]

171. Kladney, R.D.; Cardiff, R.D.; Kwiatkowski, D.J.; Chiang, G.G.; Weber, J.D.; Arbeit, J.M.; Lu, Z.H. Tuberous sclerosis complex 1: An epithelial tumor suppressor essential to prevent spontaneous prostate cancer in aged mice. *Cancer Res.* **2010**, *70*, 8937–8947. [CrossRef] [PubMed]

172. Ma, L.; Teruya-Feldstein, J.; Behrendt, N.; Chen, Z.; Noda, T.; Hino, O.; Cordon-Cardo, C.; Pandolfi, P.P. Genetic analysis of Pten and Tsc2 functional interactions in the mouse reveals asymmetrical haploinsufficiency in tumor suppression. *Genes Dev.* **2005**, *19*, 1779–1786. [CrossRef] [PubMed]

173. Sato, N.; Koinuma, J.; Ito, T.; Tsuchiya, E.; Kondo, S.; Nakamura, Y.; Daigo, Y. Activation of an oncogenic TBC1D7 (TBC1 domain family, member 7) protein in pulmonary carcinogenesis. *Genes Chromosom. Cancer* **2010**, *49*, 353–367. [CrossRef]

174. Thien, A.; Prentzell, M.T.; Holzwarth, B.; Kläsener, K.; Kuper, I.; Boehlke, C.; Sonntag, A.G.; Ruf, S.; Maerz, L.; Nitschke, R.; et al. TSC1 Activates TGF-β-Smad2/3 Signaling in Growth Arrest and Epithelial-to-Mesenchymal Transition. *Dev. Cell* **2015**, *32*, 617–630. [CrossRef]

175. Nardella, C.; Chen, Z.; Salmena, L.; Carracedo, A.; Alimonti, A.; Egia, A.; Carver, B.; Gerald, W.; Cordon-Cardo, C.; Pandolfi, P.P. Aberrant Rheb-mediated mTORC1 activation and Pten haploinsufficiency are cooperative oncogenic events. *Genes Dev.* **2008**, *22*, 2172–2177. [CrossRef] [PubMed]

176. Loewith, R.; Jacinto, E.; Wullschleger, S.; Lorberg, A.; Crespo, J.L.; Bonenfant, D.; Oppliger, W.; Jenoe, P.; Hall, M.N. Two TOR complexes, only one of which is rapamycin sensitive, have distinct roles in cell growth control. *Mol. Cell* **2002**, *10*, 457–468. [CrossRef]

177. Sarbassov, D.D.; Ali, S.M.; Kim, D.-H.; Guertin, D.A.; Latek, R.R.; Erdjument-Bromage, H.; Tempst, P.; Sabatini, D.M. Rictor, a Novel Binding Partner of mTOR, Defines a Rapamycin-Insensitive and Raptor-Independent Pathway that Regulates the Cytoskeleton. *Curr. Biol.* **2004**, *14*, 1296–1302. [CrossRef]

178. Kim, L.C.; Cook, R.S.; Chen, J. mTORC1 and mTORC2 in cancer and the tumor microenvironment. *Oncogene* **2016**, *36*, 2191–2201. [CrossRef]

179. Laplante, M.; Sabatini, D.M. mTOR Signaling in Growth Control and Disease. *Cell* **2012**, *149*, 274–293. [CrossRef]

180. Conciatori, F.; Ciuffreda, L.; Bazzichetto, C.; Falcone, I.; Pilotto, S.; Bria, E.; Cognetti, F.; Milella, M. mTOR Cross-Talk in Cancer and Potential for Combination Therapy. *Cancers* **2018**, *10*, 23. [CrossRef]

181. Ma, X.M.; Blenis, J. Molecular mechanisms of mTOR-mediated translational control. *Nat. Rev. Mol. Cell Biol.* **2009**, *10*, 307–318. [CrossRef] [PubMed]

182. Fenton, T.R.; Gout, I. Functions and regulation of the 70kDa ribosomal S6 kinases. *Int. J. Biochem. Cell Biol.* **2011**, *43*, 47–59. [CrossRef] [PubMed]

183. Meyuhas, O.; Dreazen, A. Chapter 3 Ribosomal Protein S6 Kinase. *Prog. Mol. Biol. Transl. Sci.* **2009**, *90*, 109–153. [CrossRef] [PubMed]

184. García-Martínez, J.M.; Alessi, D.R. mTOR complex 2 (mTORC2) controls hydrophobic motif phosphorylation and activation of serum- and glucocorticoid-induced protein kinase 1 (SGK1). *Biochem. J.* **2008**, *416*, 375–385. [CrossRef]

185. Sarbassov, S.D.; Ali, S.M.; Sengupta, S.; Sheen, J.-H.; Hsu, P.P.; Bagley, A.F.; Markhard, A.L.; Sabatini, D.M. Prolonged Rapamycin Treatment Inhibits mTORC2 Assembly and Akt/PKB. *Mol. Cell* **2006**, *22*, 159–168. [CrossRef]

186. Jhanwar-Uniyal, M.; Wainwright, J.V.; Mohan, A.L.; Tobias, M.E.; Murali, R.; Gandhi, C.D.; Schmidt, M.H.; Jhanwar-Uniyal, M. Diverse signaling mechanisms of mTOR complexes: mTORC1 and mTORC2 in forming a formidable relationship. *Adv. Biol. Regul.* **2019**, *72*, 51–62. [CrossRef]

187. Peterson, T.R.; Laplante, M.; Thoreen, C.C.; Sancak, Y.; Kang, S.A.; Kuehl, W.M.; Gray, N.S.; Sabatini, D.M. DEPTOR Is an mTOR Inhibitor Frequently Overexpressed in Multiple Myeloma Cells and Required for Their Survival. *Cell* **2009**, *137*, 873–886. [CrossRef]

188. Wang, Q.; Zhou, Y.; Rychahou, P.; Harris, J.W.; Zaytseva, Y.Y.; Liu, J.; Wang, C.; Weiss, H.; Liu, C.; Lee, E.Y.; et al. Deptor is a novel target of Wnt/β-catenin/c-Myc and contributes to colorectal cancer cell growth. *Cancer Res.* **2018**, *78*, 3163–3175. [CrossRef]

189. Catena, V.; Bruno, T.; De Nicola, F.; Goeman, F.; Pallocca, M.; Iezzi, S.; Sorino, C.; Cigliana, G.; Floridi, A.; Blandino, G.; et al. Deptor transcriptionally regulates endoplasmic reticulum homeostasis in multiple myeloma cells. *Oncotarget* **2016**, *7*, 70546–70558. [CrossRef]

190. Guertin, D.A.; Stevens, D.M.; Saitoh, M.; Kinkel, S.; Crosby, K.; Sheen, J.-H.; Mullholland, D.J.; Magnuson, M.A.; Wu, H.; Sabatini, D.M. mTOR Complex 2 Is Required for the Development of Prostate Cancer Induced by Pten Loss in Mice. *Cancer Cell* **2009**, *15*, 148–159. [CrossRef]

191. Demetriades, C.; Plescher, M.; Teleman, A.A. Lysosomal recruitment of TSC2 is a universal response to cellular stress. *Nat. Commun.* **2016**, *7*, 10662. [CrossRef] [PubMed]

192. Saxton, R.A.; Sabatini, D.M. mTOR Signaling in Growth, Metabolism, and Disease. *Cell* **2017**, *168*, 960–976. [CrossRef] [PubMed]

193. Sabatini, D.M. Twenty-five years of mTOR: Uncovering the link from nutrients to growth. *Proc. Natl. Acad. Sci. USA* **2017**, *114*, 11818–11825. [CrossRef] [PubMed]

194. Demetriades, C.; Doumpas, N.; Teleman, A. Regulation of TORC1 in response to amino acid starvation via lysosomal recruitment of TSC2. *Cell* **2014**, *156*, 786–799. [CrossRef] [PubMed]

195. Nguyen, T.P.; Frank, A.R.; Jewell, J.L. Amino acid and small GTPase regulation of mTORC1. *Cell. Logist.* **2017**, *7*, e1378794. [CrossRef]

196. Jewell, J.L.; Kim, Y.C.; Russell, R.C.; Yu, F.-X.; Park, H.W.; Plouffe, S.W.; Tagliabracci, V.S.; Guan, K.-L. Differential regulation of mTORC1 by leucine and glutamine. *Science* **2015**, *347*, 194–198. [CrossRef]

197. Mihaylova, M.M.; Shaw, R.J. The AMPK signalling pathway coordinates cell growth, autophagy and metabolism. *Nature* **2011**, *13*, 1016–1023. [CrossRef]

198. Dasgupta, B.; Ju, J.-S.; Sasaki, Y.; Liu, X.; Jung, S.-R.; Higashida, K.; Lindquist, D.; Milbrandt, J. The AMPK β2 Subunit Is Required for Energy Homeostasis during Metabolic Stress. *Mol. Cell. Biol.* **2012**, *32*, 2837–2848. [CrossRef]

199. Gwinn, D.M.; Shackelford, D.B.; Egan, D.F.; Mihaylova, M.M.; Méry, A.; Vasquez, D.S.; Turk, B.E.; Shaw, R.J. AMPK Phosphorylation of Raptor Mediates a Metabolic Checkpoint. *Mol. Cell* **2008**, *30*, 214–226. [CrossRef]

200. Han, F.; Li, C.-F.; Cai, Z.; Zhang, X.; Jin, G.; Zhang, W.-N.; Xu, C.; Wang, C.-Y.; Morrow, J.; Zhang, S.; et al. The critical role of AMPK in driving Akt activation under stress, tumorigenesis and drug resistance. *Nat. Commun.* **2018**, *9*, 4728. [CrossRef]

201. Han, Y.; Hu, Z.; Cui, A.; Liu, Z.; Ma, F.; Xue, Y.; Liu, Y.; Zhang, F.; Zhao, Z.; Yu, Y.; et al. Post-translational regulation of lipogenesis via AMPK-dependent phosphorylation of insulin-induced gene. *Nat. Commun.* **2019**, *10*, 623. [CrossRef] [PubMed]

202. Zadra, G.; Photopoulos, C.; Tyekucheva, S.; Heidari, P.; Weng, Q.P.; Fedele, G.; Liu, H.; Scaglia, N.; Priolo, C.; Sicinska, E.; et al. A novel direct activator of AMPK inhibits prostate cancer growth by blocking lipogenesis. *EMBO Mol. Med.* **2014**, *6*, 519–538. [CrossRef] [PubMed]

203. Hardie, D.G. Molecular Pathways: Is AMPK a Friend or a Foe in Cancer? *Clin. Cancer Res.* **2015**, *21*, 3836–3840. [CrossRef] [PubMed]

204. Abeshouse, A.; Ahn, J.; Akbani, R.; Ally, A.; Amin, S.; Andry, C.D.; Annala, M.; Aprikian, A.; Armenia, J.; Arora, A.; et al. The Molecular Taxonomy of Primary Prostate Cancer. *Cell* **2015**, *163*, 1011–1025. [CrossRef] [PubMed]

205. Choudhury, Y.; Yang, P.Z.; Ahmad, I.; Nixon, C.; Salt, I.P.; Leung, H.Y. AMP-activated protein kinase (AMPK) as a potential therapeutic target independent of PI3K/Akt signaling in prostate cancer. *Oncoscience* **2014**, *1*, 446. [CrossRef] [PubMed]

206. Tennakoon, J.B.; Shi, Y.; Han, J.J.; Tsouko, E.; White, M.A.; Burns, A.R.; Zhang, A.; Xia, X.; Ilkayeva, O.R.; Xin, L.; et al. Androgens regulate prostate cancer cell growth via an AMPK-PGC-1α-mediated metabolic switch. *Oncogene* **2013**, *33*, 5251–5261. [CrossRef] [PubMed]

207. Hardie, D.G. AMP-activated protein kinase—An energy sensor that regulates all aspects of cell function. *Genes Dev.* **2011**, *25*, 1895–1908. [CrossRef]

208. Sanchez-Cespedes, M. The role of LKB1 in lung cancer. *Fam. Cancer* **2011**, *10*, 447–453. [CrossRef]

209. Hindi, S.M.; Sato, S.; Xiong, G.; Bohnert, K.R.; Gibb, A.A.; Gallot, Y.S.; McMillan, J.; Hill, B.G.; Uchida, S.; Kumar, A. TAK1 regulates skeletal muscle mass and mitochondrial function. *JCI Insight* **2018**, *3*. [CrossRef]

210. Penfold, L.; Woods, A.; Muckett, P.; Nikitin, A.Y.; Kent, T.R.; Zhang, S.; Graham, R.; Pollard, A.; Carling, D. CAMKK2 Promotes Prostate Cancer Independently of AMPK via Increased Lipogenesis. *Cancer Res.* **2018**, *78*, 6747–6761. [CrossRef]

211. Gocher, A.M.; Azabdaftari, G.; Euscher, L.M.; Dai, S.; Karacosta, L.G.; Franke, T.F.; Edelman, A.M. Akt activation by Ca^{2+}/calmodulin-dependent protein kinase kinase 2 (CaMKK2) in ovarian cancer cells. *J. Biol. Chem.* **2017**, *292*, 14188–14204. [CrossRef] [PubMed]

212. Mehenni, H.; Lin-Marq, N.; Buchet-Poyau, K.; Reymond, A.; Collart, M.A.; Picard, D.; Antonarakis, S.E. LKB1 interacts with and phosphorylates PTEN: A functional link between two proteins involved in cancer predisposing syndromes. *Hum. Mol. Genet.* **2005**, *14*, 2209–2219. [CrossRef] [PubMed]

213. Shorning, B.Y.; Clarke, A. Energy sensing and cancer: LKB1 function and lessons learnt from Peutz–Jeghers syndrome. *Semin. Cell Dev. Biol.* **2016**, *52*, 21–29. [CrossRef]

214. Pearson, H.; McCarthy, A.; Collins, C.M.; Ashworth, A.; Clarke, A. Lkb1 Deficiency Causes Prostate Neoplasia in the Mouse. *Cancer Res.* **2008**, *68*, 2223–2232. [CrossRef] [PubMed]

215. Hermanova, I.; Zúñiga-García, P.; Caro-Maldonado, A.; Fernandez-Ruiz, S.; Salvador, F.; Martín-Martín, N.; Zabala-Letona, A.; Nuñez-Olle, M.; Torrano, V.; Camacho, L.; et al. Genetic manipulation of LKB1 elicits lethal metastatic prostate cancer. *J. Exp. Med.* **2020**, *217*, e20191787. [CrossRef]

216. Xu, P.; Cai, F.; Liu, X.; Guo, L. LKB1 suppresses proliferation and invasion of prostate cancer through hedgehog signaling pathway. *Int. J. Clin. Exp. Pathol.* **2014**, *7*, 8480–8488.

217. Budanov, A.; Karin, M. p53 Target Genes Sestrin1 and Sestrin2 Connect Genotoxic Stress and mTOR Signaling. *Cell* **2008**, *134*, 451–460. [CrossRef]

218. Morrison, A.; Chen, L.; Wang, J.; Zhang, M.; Yang, H.; Ma, Y.; Budanov, A.; Lee, J.H.; Karin, M.; Li, J. Sestrin2 promotes LKB1-mediated AMPK activation in the ischemic heart. *FASEB J.* **2014**, *29*, 408–417. [CrossRef]

219. Wang, M.; Xu, Y.; Liu, J.; Ye, J.; Yuan, W.; Jiang, H.; Wang, Z.; Jiang, H.; Wan, J. Recent Insights into the Biological Functions of Sestrins in Health and Disease. *Cell. Physiol. Biochem.* **2017**, *43*, 1731–1741. [CrossRef]

220. Parmigiani, A.; Nourbakhsh, A.; Ding, B.; Wang, W.; Kim, Y.C.; Akopiants, K.; Guan, K.-L.; Karin, M.; Budanov, A. Sestrins inhibit mTORC1 kinase activation through the GATOR complex. *Cell Rep.* **2014**, *9*, 1281–1291. [CrossRef]

221. Cordani, M.; Sanchez-Alvarez, M.; Strippoli, R.; Bazhin, A.; Donadelli, M. Sestrins at the Interface of ROS Control and Autophagy Regulation in Health and Disease. *Oxidative Med. Cell. Longev.* **2019**, *2019*, 1283075. [CrossRef] [PubMed]

222. Pasha, M.; Eid, A.; Eid, A.A.; Gorin, Y.; Munusamy, S. Sestrin2 as a Novel Biomarker and Therapeutic Target for Various Diseases. *Oxidative Med. Cell. Longev.* **2017**, *2017*, 1–10. [CrossRef] [PubMed]

223. Wang, G.; Jones, S.J.M.; Marra, M.A.; Sadar, M.D. Identification of genes targeted by the androgen and PKA signaling pathways in prostate cancer cells. *Oncogene* **2006**, *25*, 7311–7323. [CrossRef] [PubMed]

224. Mihaly, S.R.; Ninomiya-Tsuji, J.; Morioka, S. TAK1 control of cell death. *Cell Death Differ.* **2014**, *21*, 1667–1676. [CrossRef] [PubMed]

225. Cheng, J.-S.; Tsai, W.-L.; Liu, P.-F.; Goan, Y.-G.; Lin, C.-W.; Tseng, H.-H.; Lee, C.-H.; Shu, C.-W. The MAP3K7-mTOR Axis Promotes the Proliferation and Malignancy of Hepatocellular Carcinoma Cells. *Front. Oncol.* **2019**, *9*, 474. [CrossRef]

226. Kluth, M.; Hesse, J.; Heinl, A.; Krohn, A.; Steurer, S.; Sirma, H.; Simon, R.; Mayer, P.-S.; Schumacher, U.; Grupp, K.; et al. Genomic deletion of MAP3K7 at 6q12-22 is associated with early PSA recurrence in prostate cancer and absence of TMPRSS2:ERG fusions. *Mod. Pathol.* **2013**, *26*, 975–983. [CrossRef]

227. Wu, M.; Shi, L.; Cimic, A.; Romero, L.; Sui, G.; Lees, C.J.; Cline, J.M.; Seals, D.F.; Sirintrapun, J.S.; McCoy, T.P.; et al. Suppression of Tak1 promotes prostate tumorigenesis. *Cancer Res.* **2012**, *72*, 2833–2843. [CrossRef]

228. Bosman, M.C.J.; Schepers, H.; Jaques, J.; Brouwers-Vos, A.Z.; Quax, W.J.; Schuringa, J.J.; Vellenga, E. The TAK1-NF-κB axis as therapeutic target for AML. *Blood* **2014**, *124*, 3130–3140. [CrossRef]

229. Janku, F.; Yap, T.A.; Meric-Bernstam, F. Targeting the PI3K pathway in cancer: Are we making headway? *Nat. Rev. Clin. Oncol.* **2018**, *15*, 273–291. [CrossRef]

230. Zou, Z.; Tao, T.; Li, H.; Zhu, X. mTOR signaling pathway and mTOR inhibitors in cancer: Progress and challenges. *Cell Biosci.* **2020**, *10*, 1–11. [CrossRef]

231. Sathe, A.; Chalaud, G.; Oppolzer, I.; Wong, K.Y.; Von Busch, M.; Schmid, S.C.; Tong, Z.; Retz, M.; Gschwend, J.E.; Schulz, W.A.; et al. Parallel PI3K, AKT and mTOR inhibition is required to control feedback loops that limit tumor therapy. *PLoS ONE* **2018**, *13*, e0190854. [CrossRef] [PubMed]

232. Rozengurt, E.; Soares, H.P.; Sinnet-Smith, J. Suppression of feedback loops mediated by PI3K/mTOR induces multiple overactivation of compensatory pathways: An unintended consequence leading to drug resistance. *Mol. Cancer Ther.* **2014**, *13*, 2477–2488. [CrossRef] [PubMed]

233. Zhang, H.H.; Lipovsky, A.I.; Dibble, C.; Sahin, M.; Manning, B.D. S6K1 Regulates GSK3 under Conditions of mTOR-Dependent Feedback Inhibition of Akt. *Mol. Cell* **2006**, *24*, 185–197. [CrossRef] [PubMed]

234. Rodrik-Outmezguine, V.S.; Chandarlapaty, S.; Pagano, N.C.; Poulikakos, P.I.; Scaltriti, M.; Moskatel, E.; Baselga, J.; Guichard, S.; Rosen, N. mTOR kinase inhibition causes feedback-dependent biphasic regulation of AKT signaling. *Cancer Discov.* **2011**, *1*, 248–259. [CrossRef] [PubMed]

235. O'Reilly, K.E.; Rojo, F.; She, Q.-B.; Solit, D.; Mills, G.B.; Smith, D.; Lane, H.; Hofmann, F.; Hicklin, D.J.; Ludwig, D.L.; et al. mTOR inhibition induces upstream receptor tyrosine kinase signaling and activates Akt. *Cancer Res.* **2006**, *66*, 1500–1508. [CrossRef] [PubMed]

236. Wan, X.; Harkavy, B.; Shen, N.; Grohar, P.; Helman, L.J. Rapamycin induces feedback activation of Akt signaling through an IGF-1R-dependent mechanism. *Oncogene* **2006**, *26*, 1932–1940. [CrossRef]

237. Margolis, B.; Skolnik, E.Y. Activation of Ras by receptor tyrosine kinases. *J. Am. Soc. Nephrol.* **1994**, *5*, 1288–1299.

238. Santarpia, L.; Lippman, S.M.; El-Naggar, A.K. Targeting the MAPK-RAS-RAF signaling pathway in cancer therapy. *Expert Opin. Ther. Targets* **2012**, *16*, 103–119. [CrossRef]

239. Fernández-Medarde, A.; Santos, E. Ras in Cancer and Developmental Diseases. *Genes Cancer* **2011**, *2*, 344–358. [CrossRef]

240. Steelman, L.S.; Chappell, W.H.; Abrams, S.L.; Kempf, C.R.; Long, J.; Laidler, P.; Mijatović, S.; Maksimovic-Ivanic, D.; Stivala, F.; Mazzarino, M.C.; et al. Roles of the Raf/MEK/ERK and PI3K/PTEN/Akt/mTOR pathways in controlling growth and sensitivity to therapy-implications for cancer and aging. *Aging* **2011**, *3*, 192–222. [CrossRef]

241. Copps, K.D.; White, M. Regulation of insulin sensitivity by serine/threonine phosphorylation of insulin receptor substrate proteins IRS1 and IRS2. *Diabetologia* **2012**, *55*, 2565–2582. [CrossRef] [PubMed]

242. Dibble, C.; Asara, J.M.; Manning, B.D. Characterization of Rictor Phosphorylation Sites Reveals Direct Regulation of mTOR Complex 2 by S6K1. *Mol. Cell. Biol.* **2009**, *29*, 5657–5670. [CrossRef] [PubMed]

243. Suire, S.; Hawkins, P.; Stephens, L. Activation of phosphoinositide 3-kinase gamma by Ras. *Curr. Biol.* **2002**, *12*, 1068–1075. [CrossRef]

244. Ma, L.; Chen, Z.; Erdjument-Bromage, H.; Tempst, P.; Pandolfi, P.P. Phosphorylation and Functional Inactivation of TSC2 by Erk. *Cell* **2005**, *121*, 179–193. [CrossRef]

245. Carrière, A.; Cargnello, M.; Julien, L.-A.; Gao, H.; Bonneil, É.; Thibault, P.; Roux, P.P. Oncogenic MAPK Signaling Stimulates mTORC1 Activity by Promoting RSK-Mediated Raptor Phosphorylation. *Curr. Biol.* **2008**, *18*, 1269–1277. [CrossRef]

246. Carriere, A.; Romeo, Y.; Acosta-Jaquez, H.A.; Moreau, J.; Bonneil, E.; Thibault, P.; Fingar, D.C.; Roux, P.P. ERK1/2 Phosphorylate Raptor to Promote Ras-dependent Activation of mTOR Complex 1 (mTORC1). *J. Biol. Chem.* **2011**, *286*, 567–577. [CrossRef]

247. Lara, R.; Seckl, M.J.; Pardo, O.E. The p90 RSK Family Members: Common Functions and Isoform Specificity. *Cancer Res.* **2013**, *73*, 5301–5308. [CrossRef]

248. Roux, P.P.; Shahbazian, D.; Vu, H.; Holz, M.K.; Cohen, M.S.; Taunton, J.; Sonenberg, N.; Blenis, J. RAS/ERK signaling promotes site-specific ribosomal protein S6 phosphorylation via RSK and stimulates cap-dependent translation. *J. Biol. Chem.* **2007**, *282*, 14056–14064. [CrossRef]

249. Zimmermann, S. Phosphorylation and Regulation of Raf by Akt (Protein Kinase B). *Science* **1999**, *286*, 1741–1744. [CrossRef]

250. Guan, K.-L. Negative regulation of the serine/threonine kinase B-Raf by Akt. *J. Biol. Chem.* **2000**, *275*, 275. [CrossRef]

251. Lone, M.-U.-D.; Miyan, J.; Asif, M.; Malik, S.A.; Dubey, P.; Singh, V.; Singh, K.; Mitra, K.; Pandey, D.; Haq, W.; et al. Direct physical interaction of active Ras with mSIN1 regulates mTORC2 signaling. *BMC Cancer* **2019**, *19*, 1–16. [CrossRef] [PubMed]

252. Majumder, P.K.; Febbo, P.G.; Bikoff, R.; Berger, R.; Xue, Q.; McMahon, L.M.; Manola, J.; Brugarolas, J.; McDonnell, T.J.; Golub, T.R.; et al. mTOR inhibition reverses Akt-dependent prostate intraepithelial neoplasia through regulation of apoptotic and HIF-1-dependent pathways. *Nat. Med.* **2004**, *10*, 594–601. [CrossRef] [PubMed]

253. Wu, L.; Birle, D.C.; Tannock, I.F. Effects of the Mammalian Target of Rapamycin Inhibitor CCI-779 Used Alone or with Chemotherapy on Human Prostate Cancer Cells and Xenografts. *Cancer Res.* **2005**, *65*, 2825–2831. [CrossRef] [PubMed]

254. Amato, R.J.; Jac, J.; Mohammad, T.; Saxena, S. Pilot Study of Rapamycin in Patients with Hormone-Refractory Prostate Cancer. *Clin. Genitourin. Cancer* **2008**, *6*, 97–102. [CrossRef]

255. Carracedo, A.; Ma, L.; Teruya-Feldstein, J.; Rojo, F.; Salmena, L.; Alimonti, A.; Egia, A.; Sasaki, A.T.; Thomas, G.; Kozma, S.C.; et al. Inhibition of mTORC1 leads to MAPK pathway activation through a PI3K-dependent feedback loop in human cancer. *J. Clin. Investig.* **2008**, *118*, 3065–3074. [CrossRef]

256. Kinkade, C.W.; Castillo-Martin, M.; Puzio-Kuter, A.; Yan, J.; Foster, T.H.; Gao, H.; Sun, Y.; Ouyang, X.; Gerald, W.L.; Cordon-Cardo, C.; et al. Targeting AKT/mTOR and ERK MAPK signaling inhibits hormone-refractory prostate cancer in a preclinical mouse model. *J. Clin. Investig.* **2008**, *118*, 3051–3064. [CrossRef]

257. Butler, D.E.; Marlein, C.; Walker, H.F.; Frame, F.M.; Mann, V.M.; Simms, M.S.; Davies, B.R.; Collins, A.T.; Maitland, N.J. Inhibition of the PI3K/AKT/mTOR pathway activates autophagy and compensatory Ras/Raf/MEK/ERK signalling in prostate cancer. *Oncotarget* **2017**, *8*, 56698–56713. [CrossRef]

258. Shi, Y.; Yan, H.; Frost, P.; Gera, J.; Lichtenstein, A. Mammalian target of rapamycin inhibitors activate the AKT kinase in multiple myeloma cells by up-regulating the insulin-like growth factor receptor/insulin receptor substrate-1/phosphatidylinositol 3-kinase cascade. *Mol. Cancer Ther.* **2005**, *4*, 1533–1540. [CrossRef]

259. Tremblay, F.; Gagnon, A.; Veilleux, A.; Sorisky, A.; Marette, A. Activation of the Mammalian Target of Rapamycin Pathway Acutely Inhibits Insulin Signaling to Akt and Glucose Transport in 3T3-L1 and Human Adipocytes. *Endocrinology* **2005**, *146*, 1328–1337. [CrossRef]

260. Yu, Y.; Yoon, S.-O.; Poulogiannis, G.; Yang, Q.; Ma, X.M.; Villén, J.; Kubica, N.; Hoffman, G.R.; Cantley, L.C.; Gygi, S.P.; et al. Phosphoproteomic Analysis Identifies Grb10 as an mTORC1 Substrate That Negatively Regulates Insulin Signaling. *Science* **2011**, *332*, 1322–1326. [CrossRef]

261. Porta, C.; Paglino, C.; Mosca, A. Targeting PI3K/Akt/mTOR Signaling in Cancer. *Front. Oncol.* **2014**, *4*, 64. [CrossRef]

262. Chandarlapaty, S.; Sawai, A.; Scaltriti, M.; Outmezguine, V.R.-; Grbovic-Huezo, O.; Serra, V.; Majumder, P.K.; Baselga, J.; Rosen, N. AKT Inhibition Relieves Feedback Suppression of Receptor Tyrosine Kinase Expression and Activity. *Cancer Cell* **2011**, *19*, 58–71. [CrossRef] [PubMed]

263. Biggs, W.H.; Meisenhelder, J.; Hunter, T.; Cavenee, W.K.; Arden, K.C. Protein kinase B/Akt-mediated phosphorylation promotes nuclear exclusion of the winged helix transcription factor FKHR1. *Proc. Natl. Acad. Sci. USA* **1999**, *96*, 7421–7426. [CrossRef] [PubMed]

264. Brunet, A.; Bonni, A.; Zigmond, M.J.; Lin, M.Z.; Juo, P.; Hu, L.S.; Anderson, M.J.; Arden, K.C.; Blenis, J.; Greenberg, M.E. Akt Promotes Cell Survival by Phosphorylating and Inhibiting a Forkhead Transcription Factor. *Cell* **1999**, *96*, 857–868. [CrossRef]

265. Chakrabarty, A.; Sánchez, V.; Kuba, M.G.; Rinehart, C.; Arteaga, C.L. Feedback upregulation of HER3 (ErbB3) expression and activity attenuates antitumor effect of PI3K inhibitors. *Proc. Natl. Acad. Sci. USA* **2011**, *109*, 2718–2723. [CrossRef]

266. Serra, V.; Scaltriti, M.; Prudkin, L.; Eichhorn, P.J.A.; Ibrahim, Y.H.; Chandarlapaty, S.; Markman, B.; Rodriguez, O.; Guzmán, M.; Rodriguez, S.; et al. PI3K inhibition results in enhanced HER signaling and acquired ERK dependency in HER2-overexpressing breast cancer. *Oncogene* **2011**, *30*, 2547–2557. [CrossRef]

267. Sommer, E.M.; Dry, H.; Cross, D.; Guichard, S.; Davies, B.R.; Alessi, D.R. Elevated SGK1 predicts resistance of breast cancer cells to Akt inhibitors. *Biochem. J.* **2013**, *452*, 499–508. [CrossRef]

268. Wang, J.; Kobayashi, T.; Floc'H, N.; Kinkade, C.W.; Aytes, A.; Dankort, D.; Lefebvre, C.; Mitrofanova, A.; Cardiff, R.D.; McMahon, M.; et al. B-Raf activation cooperates with PTEN loss to drive c-Myc expression in advanced prostate cancer. *Cancer Res.* **2012**, *72*, 4765–4776. [CrossRef]

269. Jefferies, M.T.; Cox, A.C.; Shorning, B.Y.; Meniel, V.; Griffiths, D.; Kynaston, H.G.; Smalley, M.; Clarke, A.R. PTEN loss and activation of K-RAS and β-catenin cooperate to accelerate prostate tumourigenesis. *J. Pathol.* **2017**, *243*, 442–456. [CrossRef]

270. Toren, P.; Kim, S.; Johnson, F.; Zoubeidi, A. Combined AKT and MEK Pathway Blockade in Pre-Clinical Models of Enzalutamide-Resistant Prostate Cancer. *PLoS ONE* **2016**, *11*, e0152861. [CrossRef]

271. Turke, A.B.; Song, Y.; Costa, C.; Cook, R.; Arteaga, C.L.; Asara, J.M.; Engelman, J.A. MEK inhibition leads to PI3K/AKT activation by relieving a negative feedback on ERBB receptors. *Cancer Res.* **2012**, *72*, 3228–3237. [CrossRef] [PubMed]

272. Gioeli, D.; Wunderlich, W.; Sebolt-Leopold, J.; Bekiranov, S.; Wulfkuhle, J.D.; Petricoin, E.F.; Conaway, M.; Weber, M.J. Compensatory pathways induced by MEK inhibition are effective drug targets for combination therapy against castration-resistant prostate cancer. *Mol. Cancer Ther.* **2011**, *10*, 1581–1590. [CrossRef] [PubMed]

273. Crabb, S.J.; Birtle, A.J.; Martin, K.; Downs, N.; Ratcliffe, I.; Maishman, T.; Ellis, M.; Griffiths, G.; Thompson, S.; Ksiazek, L.; et al. ProCAID: A phase I clinical trial to combine the AKT inhibitor AZD5363 with docetaxel and prednisolone chemotherapy for metastatic castration resistant prostate cancer. *Investig. New Drugs* **2017**, *35*, 599–607. [CrossRef]

274. Gillessen, S.; Gilson, C.; James, N.D.; Adler, A.; Sydes, M.R.; Clarke, N. Repurposing Metformin as Therapy for Prostate Cancer within the STAMPEDE Trial Platform. *Eur. Urol.* **2016**, *70*, 906–908. [CrossRef] [PubMed]

275. Viollet, B.; Guigas, B.; Garcia, N.S.; Leclerc, J.; Foretz, M.; Andreelli, F. Cellular and molecular mechanisms of metformin: An overview. *Clin. Sci.* **2011**, *122*, 253–270. [CrossRef] [PubMed]

276. Soares, H.P.; Ni, Y.; Kisfalvi, K.; Sinnett-Smith, J.; Rozengurt, E. Different Patterns of Akt and ERK Feedback Activation in Response to Rapamycin, Active-Site mTOR Inhibitors and Metformin in Pancreatic Cancer Cells. *PLoS ONE* **2013**, *8*, e57289. [CrossRef] [PubMed]

277. Jokinen, E.; Koivunen, J. MEK and PI3K inhibition in solid tumors: Rationale and evidence to date. *Ther. Adv. Med. Oncol.* **2015**, *7*, 170–180. [CrossRef] [PubMed]

278. Bardia, A.; Gounder, M.; Rodon, J.; Janku, F.; Lolkema, M.P.; Stephenson, J.J.; Bedard, P.L.; Schuler, M.; Sessa, C.; Lorusso, P.; et al. Phase Ib Study of Combination Therapy with MEK Inhibitor Binimetinib and Phosphatidylinositol 3-Kinase Inhibitor Buparlisib in Patients with Advanced Solid Tumors with RAS/RAF Alterations. *Oncologist* **2019**, *25*, e160–e169. [CrossRef]

279. Tindall, N.J.; Lonergan, P.E. Androgen receptor signaling in prostate cancer development and progression. *J. Carcinog.* **2011**, *10*, 20. [CrossRef]

280. Zhou, Y.; Bolton, E.C.; Jones, J.O. Androgens and androgen receptor signaling in prostate tumorigenesis. *J. Mol. Endocrinol.* **2014**, *54*, R15–R29. [CrossRef]

281. Tan, E.; Li, J.; Xu, H.E.; Melcher, K.; Yong, E.-L. Androgen receptor: Structure, role in prostate cancer and drug discovery. *Acta Pharmacol. Sin.* **2014**, *36*, 3–23. [CrossRef] [PubMed]

282. Miller, W.L.; Auchus, R.J. The molecular biology, biochemistry, and physiology of human steroidogenesis and its disorders. *Endocr. Rev.* **2010**, *32*, 81–151. [CrossRef] [PubMed]

283. Watson, P.A.; Arora, V.K.; Sawyers, C.L. Emerging mechanisms of resistance to androgen receptor inhibitors in prostate cancer. *Nat. Rev. Cancer* **2015**, *15*, 701–711. [CrossRef] [PubMed]

284. Cai, C.; Balk, S.P. Intratumoral androgen biosynthesis in prostate cancer pathogenesis and response to therapy. *Endocr. Relat. Cancer* **2011**, *18*, R175–R182. [CrossRef]

285. Dai, C.; Chung, Y.-M.; Kovac, E.; Zhu, Z.; Li, J.; Magi-Galluzzi, C.; Stephenson, A.J.; Klein, E.A.; Sharifi, N. Direct Metabolic Interrogation of Dihydrotestosterone Biosynthesis from Adrenal Precursors in Primary Prostatectomy Tissues. *Clin. Cancer Res.* **2017**, *23*, 6351–6362. [CrossRef]

286. Dehm, S.M.; Tindall, N.J. Androgen Receptor Structural and Functional Elements: Role and Regulation in Prostate Cancer. *Mol. Endocrinol.* **2007**, *21*, 2855–2863. [CrossRef]

287. Lamont, K.R.; Tindall, N.J. Minireview: Alternative activation pathways for the androgen receptor in prostate cancer. *Mol. Endocrinol.* **2011**, *25*, 897–907. [CrossRef]

288. Zamagni, A.; Cortesi, M.; Zanoni, M.; Tesei, A. Non-nuclear AR Signaling in Prostate Cancer. *Front. Chem.* **2019**, *7*, 651. [CrossRef]

289. Coutinho, I.; Day, T.K.; Tilley, W.D.; Selth, L.A. Androgen receptor signaling in castration-resistant prostate cancer: A lesson in persistence. *Endocr. Relat. Cancer* **2016**, *23*, T179–T197. [CrossRef]

290. Saranyutanon, S.; Srivastava, S.K.; Pai, S.; Singh, S.; Singh, A.P. Therapies Targeted to Androgen Receptor Signaling Axis in Prostate Cancer: Progress, Challenges, and Hope. *Cancers* **2019**, *12*, 51. [CrossRef]

291. Murillo, H.; Huang, H.; Schmidt, L.J.; Smith, D.I.; Tindall, N.J. Role of PI3K Signaling in Survival and Progression of LNCaP Prostate Cancer Cells to the Androgen Refractory State. *Endocrinology* **2001**, *142*, 4795–4805. [CrossRef]

292. Zoubeidi, A.; Gleave, M. Co-targeting driver pathways in prostate cancer: Two birds with one stone. *EMBO Mol. Med.* **2018**, *10*, e8928. [CrossRef] [PubMed]

293. Audet-Walsh, É.; Dufour, C.R.; Yee, T.; Zouanat, F.Z.; Yan, M.; Kalloghlian, G.; Vernier, M.; Caron, M.; Bourque, G.; Scarlata, E.; et al. Nuclear mTOR acts as a transcriptional integrator of the androgen signaling pathway in prostate cancer. *Genes Dev.* **2017**, *31*, 1228–1242. [CrossRef] [PubMed]

294. Qi, W.; Morales, C.; Cooke, L.S.; Johnson, B.; Somer, B.; Mahadevan, D. Reciprocal feedback inhibition of the androgen receptor and PI3K as a novel therapy for castrate-sensitive and -resistant prostate cancer. *Oncotarget* **2015**, *6*, 41976–41987. [CrossRef] [PubMed]

295. Thomas, C.; Lamoureux, F.; Crafter, C.; Davies, B.R.; Beraldi, E.; Fazli, L.; Kim, S.; Thaper, D.; Gleave, M.E.; Zoubeidi, A. Synergistic Targeting of PI3K/AKT Pathway and Androgen Receptor Axis Significantly Delays Castration-Resistant Prostate Cancer Progression In Vivo. *Mol. Cancer Ther.* **2013**, *12*, 2342–2355. [CrossRef] [PubMed]

296. Yao, E.; Zhou, W.; Lee-Hoeflich, S.T.; Truong, T.; Haverty, P.M.; Eastham-Anderson, J.; Lewin-Koh, N.; Günter, B.; Belvin, M.; Murray, L.J.; et al. Suppression of HER2/HER3-Mediated Growth of Breast Cancer Cells with Combinations of GDC-0941 PI3K Inhibitor, Trastuzumab, and Pertuzumab. *Clin. Cancer Res.* **2009**, *15*, 4147–4156. [CrossRef]

297. Mahajan, N.P.; Liu, Y.; Majumder, S.; Warren, M.R.; Parker, C.E.; Mohler, J.L.; Earp, H.S.; Whang, Y.E. Activated Cdc42-associated kinase Ack1 promotes prostate cancer progression via androgen receptor tyrosine phosphorylation. *Proc. Natl. Acad. Sci. USA* **2007**, *104*, 8438–8443. [CrossRef]

298. Mellinghoff, I.K.; Vivanco, I.; Kwon, A.; Tran, C.; Wongvipat, J.; Sawyers, C.L. HER2/neu kinase-dependent modulation of androgen receptor function through effects on DNA binding and stability. *Cancer Cell* **2004**, *6*, 517–527. [CrossRef]

299. Yeh, S.; Lin, H.-K.; Kang, H.-Y.; Thin, T.H.; Chang, C. From HER2/Neu signal cascade to androgen receptor and its coactivators: A novel pathway by induction of androgen target genes through MAP kinase in prostate cancer cells. *Proc. Natl. Acad. Sci. USA* **1999**, *96*, 5458–5463. [CrossRef]

300. Liu, P.; Li, S.; Gan, L.; Kao, T.P.; Huang, H. A Transcription-Independent Function of FOXO1 in Inhibition of Androgen-Independent Activation of the Androgen Receptor in Prostate Cancer Cells. *Cancer Res.* **2008**, *68*, 10290–10299. [CrossRef]

301. Bowen, C.; Ostrowski, M.C.; Leone, G.; Gelmann, E.P. Loss of PTEN Accelerates NKX3.1 Degradation to Promote Prostate Cancer Progression. *Cancer Res.* **2019**, *79*, 4124–4134. [CrossRef] [PubMed]

302. Tan, P.Y.; Chang, C.W.; Chng, K.R.; Wansa, K.D.S.A.; Sung, W.-K.; Cheung, E. Integration of Regulatory Networks by NKX3-1 Promotes Androgen-Dependent Prostate Cancer Survival. *Mol. Cell. Biol.* **2011**, *32*, 399–414. [CrossRef] [PubMed]

303. Lei, Q.-Y.; Jiao, J.; Xin, L.; Chang, C.-J.; Wang, S.; Gao, J.; Gleave, M.E.; Witte, O.N.; Liu, X.; Wu, H. NKX3.1 stabilizes p53, inhibits AKT activation, and blocks prostate cancer initiation caused by PTEN loss. *Cancer Cell* **2006**, *9*, 367–378. [CrossRef] [PubMed]

304. Wang, Q.; Bailey, C.; Ng, C.; Tiffen, J.; Thoeng, A.; Minhas, V.; Lehman, M.L.; Hendy, S.C.; Buchanan, G.; Nelson, C.C.; et al. Androgen Receptor and Nutrient Signaling Pathways Coordinate the Demand for Increased Amino Acid Transport during Prostate Cancer Progression. *Cancer Res.* **2011**, *71*, 7525–7536. [CrossRef] [PubMed]

305. Wen, Y.; Hu, M.C.; Makino, K.; Spohn, B.; Bartholomeusz, G.; Yan, D.H.; Hung, M.C. HER-2/neu promotes androgen-independent survival and growth of prostate cancer cells through the Akt pathway. *Cancer Res.* **2000**, *60*, 6841–6845. [PubMed]

306. Lin, H.-K.; Yeh, S.; Kang, H.-Y.; Chang, C. Akt suppresses androgen-induced apoptosis by phosphorylating and inhibiting androgen receptor. *Proc. Natl. Acad. Sci. USA* **2001**, *98*, 7200–7205. [CrossRef] [PubMed]

307. Blattner, M.; Liu, D.; Robinson, B.D.; Huang, D.; Poliakov, A.; Gao, D.; Nataraj, S.; Deonarine, L.D.; Augello, M.A.; Sailer, V.; et al. SPOP Mutation Drives Prostate Tumorigenesis In Vivo through Coordinate Regulation of PI3K/mTOR and AR Signaling. *Cancer Cell* **2017**, *31*, 436–451. [CrossRef]

308. Mani, R.S. The emerging role of speckle-type POZ protein (SPOP) in cancer development. *Drug Discov. Today* **2014**, *19*, 1498–1502. [CrossRef]

309. Agoulnik, I.U.; Weigel, N.L. Coactivator selective regulation of androgen receptor activity. *Steroids* **2009**, *74*, 669–674. [CrossRef]

310. Ferry, C.; Gaouar, S.; Fischer, B.; Boeglin, M.; Paul, N.; Samarut, E.; Piskunov, A.; Pankotai-Bodó, G.; Brino, L.; Rochette-Egly, C. Cullin 3 mediates SRC-3 ubiquitination and degradation to control the retinoic acid response. *Proc. Natl. Acad. Sci. USA* **2011**, *108*, 20603–20608. [CrossRef]

311. An, J.; Wang, C.; Deng, Y.; Yu, L.; Huang, H. Destruction of full-length androgen receptor by wild-type SPOP, but not prostate-cancer-associated mutants. *Cell Rep.* **2014**, *6*, 657–669. [CrossRef] [PubMed]

312. Baron, S.; Manin, M.; Beaudoin, C.; Leotoing, L.; Communal, Y.; Veyssiere, G.; Morel, L. Androgen Receptor Mediates Non-genomic Activation of Phosphatidylinositol 3-OH Kinase in Androgen-sensitive Epithelial Cells. *J. Biol. Chem.* **2003**, *279*, 14579–14586. [CrossRef] [PubMed]

313. Kolinsky, M.; Rescigno, P.; Bianchini, D.; Zafeiriou, Z.; Mehra, N.; Mateo, J.; Michalarea, V.; Riisnaes, R.; Crespo, M.; Figueiredo, I.; et al. A phase I dose-escalation study of enzalutamide in combination with the AKT inhibitor AZD5363 (capivasertib) in patients with metastatic castration-resistant prostate cancer. *Ann. Oncol.* **2020**, *31*, 619–625. [CrossRef] [PubMed]

314. Armstrong, A.J.; Halabi, S.; Healy, P.; Alumkal, J.J.; Winters, C.; Kephart, J.; Bitting, R.L.; Hobbs, C.; Soleau, C.F.; Beer, T.M.; et al. Phase II trial of the PI3 kinase inhibitor buparlisib (BKM-120) with or without enzalutamide in men with metastatic castration resistant prostate cancer. *Eur. J. Cancer* **2017**, *81*, 228–236. [CrossRef] [PubMed]

315. Massard, C.; Chi, K.N.; Castellano, D.; De Bono, J.; Gravis, G.; Dirix, L.; Machiels, J.-P.; Mita, A.; Mellado, B.; Turri, S.; et al. Phase Ib dose-finding study of abiraterone acetate plus buparlisib (BKM120) or dactolisib (BEZ235) in patients with castration-resistant prostate cancer. *Eur. J. Cancer* **2017**, *76*, 36–44. [CrossRef] [PubMed]

316. Wei, X.X.; Hsieh, A.C.; Kim, W.; Friedlander, T.; Lin, A.M.; Louttit, M.; Ryan, C.J. A Phase I Study of Abiraterone Acetate Combined with BEZ235, a Dual PI3K/mTOR Inhibitor, in Metastatic Castration Resistant Prostate Cancer. *Oncologist* **2017**, *22*, 503-e43. [CrossRef]

317. D'Abronzo, L.S.; Bose, S.; Crapuchettes, M.E.; Beggs, R.E.; Vinall, R.L.; Tepper, C.G.; Siddiqui, S.; Mudryj, M.; Melgoza, F.U.; Durbin-Johnson, B.P.; et al. The androgen receptor is a negative regulator of eIF4E phosphorylation at S209: Implications for the use of mTOR inhibitors in advanced prostate cancer. *Oncogene* **2017**, *36*, 6359–6373. [CrossRef]

318. Clevers, H.; Loh, K.M.; Nusse, R. An integral program for tissue renewal and regeneration: Wnt signaling and stem cell control. *Science* **2014**, *346*, 1248012. [CrossRef]

319. Nusse, R.; Clevers, H. Wnt/β-Catenin Signaling, Disease, and Emerging Therapeutic Modalities. *Cell* **2017**, *169*, 985–999. [CrossRef]

320. Murillo-Garzón, V.; Kypta, R. WNT signalling in prostate cancer. *Nat. Rev. Urol.* **2017**, *14*, 683–696. [CrossRef]

321. Stamos, J.L.; Weis, W.I. The β-Catenin Destruction Complex. *Cold Spring Harb. Perspect. Biol.* **2012**, *5*, a007898. [CrossRef] [PubMed]

322. Schneider, J.A.; Logan, S.K. Revisiting the role of Wnt/β-catenin signaling in prostate cancer. *Mol. Cell. Endocrinol.* **2017**, *462*, 3–8. [CrossRef] [PubMed]

323. Ahmad, I.; Sansom, O.J. Role of Wnt signalling in advanced prostate cancer. *J. Pathol.* **2018**, *245*, 3–5. [CrossRef] [PubMed]

324. Zhang, Z.; Cheng, L.; Li, J.; Farah, E.; Lanman, N.A.; Pascuzzi, P.; Gupta, S.; Liu, X. Inhibition of the Wnt/β-catenin pathway overcomes resistance to enzalutamide in castration-resistant prostate cancer. *Cancer Res.* **2018**, *78*, 3147–3162. [CrossRef] [PubMed]

325. Velho, P.I.; Fu, W.; Wang, H.; Mirkheshti, N.; Qazi, F.; Lima, F.A.; Shaukat, F.; Carducci, M.A.; Denmeade, S.R.; Paller, C.J.; et al. Wnt-pathway Activating Mutations Are Associated with Resistance to First-line Abiraterone and Enzalutamide in Castration-resistant Prostate Cancer. *Eur. Urol.* **2019**, *77*, 14–21. [CrossRef] [PubMed]

326. Johnsen, J.I.; Wickström, M.; Baryawno, N. Wingless/β-catenin signaling as a modulator of chemoresistance in cancer. *Mol. Cell. Oncol.* **2016**, *3*, e1131356. [CrossRef]

327. Rajan, P.; Sudbery, I.; Villasevil, M.E.M.; Mui, E.; Fleming, J.; Davis, M.; Ahmad, I.; Edwards, J.; Sansom, O.J.; Sims, D.; et al. Next-generation Sequencing of Advanced Prostate Cancer Treated with Androgen-deprivation Therapy. *Eur. Urol.* **2014**, *66*, 32–39. [CrossRef]

328. Song, L.-N.; Gelmann, E.P. Interaction of -Catenin and TIF2/GRIP1 in Transcriptional Activation by the Androgen Receptor. *J. Biol. Chem.* **2005**, *280*, 37853–37867. [CrossRef]

329. Yang, X.; Chen, M.-W.; Terry, S.; Vacherot, F.; Bemis, D.L.; Capodice, J.; Kitajewski, J.; De La Taille, A.; Benson, M.C.; Guo, Y.; et al. Complex regulation of human androgen receptor expression by Wnt signaling in prostate cancer cells. *Oncogene* **2006**, *25*, 3436–3444. [CrossRef]

330. Wang, G.; Wang, J.; Sadar, M.D. Crosstalk between the androgen receptor and beta-catenin in castrate-resistant prostate cancer. *Cancer Res.* **2008**, *68*, 9918–9927. [CrossRef]

331. Patel, R.; Brzezinska, E.A.; Repiscak, P.; Ahmad, I.; Mui, E.; Gao, M.; Blomme, A.; Harle, V.; Tan, E.H.; Malviya, G.; et al. Activation of β-Catenin Cooperates with Loss of Pten to Drive AR-Independent Castration-Resistant Prostate Cancer. *Cancer Res.* **2019**, *80*, 576–590. [CrossRef] [PubMed]

332. Pearson, H.; Phesse, T.J.; Clarke, A. K-ras and Wnt Signaling Synergize to Accelerate Prostate Tumorigenesis in the Mouse. *Cancer Res.* **2009**, *69*, 94–101. [CrossRef] [PubMed]

333. Bruxvoort, K.J.; Charbonneau, H.M.; Giambernardi, T.A.; Goolsby, J.C.; Qian, C.-N.; Zylstra, C.R.; Robinson, D.R.; Roy-Burman, P.; Shaw, A.; Buckner-Berghuis, B.D.; et al. Inactivation of Apc in the Mouse Prostate Causes Prostate Carcinoma. *Cancer Res.* **2007**, *67*, 2490–2496. [CrossRef] [PubMed]

334. Tenbaum, S.P.; Ordóñez-Morán, P.; Puig, I.; Chicote, I.; Arqués, O.; Landolfi, S.; Fernández, Y.; Camacho, J.R.H.; Gispert, J.D.; Mendizabal, L.; et al. β-catenin confers resistance to PI3K and AKT inhibitors and subverts FOXO3a to promote metastasis in colon cancer. *Nat. Med.* **2012**, *18*, 892–901. [CrossRef] [PubMed]

335. Persad, A.; Venkateswaran, G.; Hao, L.; Garcia, M.E.; Yoon, J.; Sidhu, J.; Persad, S. Active β-catenin is regulated by the PTEN/PI3 kinase pathway: A role for protein phosphatase PP2A. *Genes Cancer* **2017**, *7*, 368–382. [CrossRef]

336. Liu, H.; Yin, J.; Wang, H.; Jiang, G.; Deng, M.; Zhang, G.; Bu, X.; Cai, S.; Du, J.; He, Z. FOXO3a modulates WNT/β-catenin signaling and suppresses epithelial-to-mesenchymal transition in prostate cancer cells. *Cell. Signal.* **2015**, *27*, 510–518. [CrossRef]

337. Sinner, D.; Kordich, J.J.; Spence, J.R.; Opoka, R.; Rankin, S.; Lin, S.-C.J.; Jonatan, D.; Zorn, A.M.; Wells, J.M. Sox17 and Sox4 Differentially Regulate β-Catenin/T-Cell Factor Activity and Proliferation of Colon Carcinoma Cells. *Mol. Cell. Biol.* **2007**, *27*, 7802–7815. [CrossRef]

338. Bilir, B.; Osunkoya, A.O.; Wiles, W.G.; Sannigrahi, S.; Lefebvre, V.; Metzger, D.; Spyropoulos, D.D.; Martin, W.D.; Moreno, C.S. SOX4 Is Essential for Prostate Tumorigenesis Initiated by PTEN Ablation. *Cancer Res.* **2015**, *76*, 1112–1121. [CrossRef]

339. Conde-Perez, A.; Gros, G.; Longvert, C.; Pedersen, M.; Petit, V.; Aktary, Z.; Viros, A.; Gesbert, F.; Delmas, V.; Rambow, F.; et al. A caveolin-dependent and PI3K/AKT-independent role of PTEN in β-catenin transcriptional activity. *Nat. Commun.* **2015**, *6*, 8093. [CrossRef]

340. Wu, D.; Pan, W. GSK3: A multifaceted kinase in Wnt signaling. *Trends Biochem. Sci.* **2010**, *35*, 161–168. [CrossRef]

341. Buller, C.L.; Loberg, R.D.; Fan, M.-H.; Zhu, Q.; Park, J.L.; Vesely, E.; Inoki, K.; Guan, K.-L.; Brosius, F.C. A GSK-3/TSC2/mTOR pathway regulates glucose uptake and GLUT1 glucose transporter expression. *Am. J. Physiol. Physiol.* **2008**, *295*, C836–C843. [CrossRef]

342. Evangelisti, C.; Chiarini, F.; Paganelli, F.; Marmiroli, S.; Martelli, A.M. Crosstalks of GSK3 signaling with the mTOR network and effects on targeted therapy of cancer. *Biochim. Biophys. Acta Mol. Cell Res.* **2020**, *1867*, 118635. [CrossRef] [PubMed]

343. Ding, V.W.; Chen, R.-H.; McCormick, F. Differential Regulation of Glycogen Synthase Kinase 3β by Insulin and Wnt Signaling. *J. Biol. Chem.* **2000**, *275*, 32475–32481. [CrossRef]

344. Ng, S.S.; Mahmoudi, T.; Danenberg, E.; Bejaoui, I.; De Lau, W.; Korswagen, H.C.; Schutte, M.; Clevers, H. Phosphatidylinositol 3-Kinase Signaling Does Not Activate the Wnt Cascade. *J. Biol. Chem.* **2009**, *284*, 35308–35313. [CrossRef] [PubMed]

345. Ravitz, M.J.; Chen, L.; Lynch, M.; Schmidt, E.V. c-myc Repression of TSC2 contributes to control of translation initiation and Myc-induced transformation. *Cancer Res.* **2007**, *67*, 11209–11217. [CrossRef] [PubMed]

346. Csibi, A.; Blenis, J. Hippo–YAP and mTOR pathways collaborate to regulate organ size. *Nat. Cell Biol.* **2012**, *14*, 1244–1245. [CrossRef] [PubMed]

347. Santinon, G.; Pocaterra, A.; Dupont, S.; Information, P.E.K.F.C. Control of YAP/TAZ Activity by Metabolic and Nutrient-Sensing Pathways. *Trends Cell Biol.* **2016**, *26*, 289–299. [CrossRef]

348. Azzolin, L.; Panciera, T.; Soligo, S.; Enzo, E.; Bicciato, S.; Dupont, S.; Bresolin, S.; Frasson, C.; Basso, G.; Guzzardo, V.; et al. YAP/TAZ Incorporation in the β-Catenin Destruction Complex Orchestrates the Wnt Response. *Cell* **2014**, *158*, 157–170. [CrossRef]

349. Park, H.W.; Kim, Y.C.; Yu, B.; Moroishi, T.; Mo, J.-S.; Plouffe, S.W.; Meng, Z.; Lin, K.C.; Yu, F.-X.; Alexander, C.M.; et al. Alternative Wnt Signaling Activates YAP/TAZ. *Cell* **2015**, *162*, 780–794. [CrossRef]

350. Aznar, N.; Midde, K.K.; Dunkel, Y.; Lopez-Sanchez, I.; Pavlova, Y.; Marivin, A.; Barbazan, J.; Murray, F.; Nitsche, U.; Janssen, K.-P.; et al. Daple is a novel non-receptor GEF required for trimeric G protein activation in Wnt signaling. *eLife* **2015**, *4*, e07091. [CrossRef]

351. Hojjat-Farsangi, M.; Moshfegh, A.; Daneshmanesh, A.H.; Khan, S.; Mikaelsson, E.; Österborg, A.; Mellstedt, H. The receptor tyrosine kinase ROR1—An oncofetal antigen for targeted cancer therapy. *Semin. Cancer Biol.* **2014**, *29*, 21–31. [CrossRef] [PubMed]

352. Gujral, T.; Chan, M.; Peshkin, L.; Sorger, P.K.; Kirschner, M.W.; MacBeath, G. A noncanonical Frizzled2 pathway regulates epithelial-mesenchymal transition and metastasis. *Cell* **2014**, *159*, 844–856. [CrossRef] [PubMed]

353. Yamada, E.; Pessin, J.E.; Kurland, I.J.; Schwartz, G.J.; Bastie, C.C. Fyn-Dependent Regulation of Energy Expenditure and Body Weight Is Mediated by Tyrosine Phosphorylation of LKB1. *Cell Metab.* **2010**, *11*, 113–124. [CrossRef] [PubMed]

354. Koh, C.M.; Bieberich, C.J.; Dang, C.V.; Nelson, W.G.; Yegnasubramanian, S.; De Marzo, A.M. MYC and Prostate Cancer. *Genes Cancer* **2010**, *1*, 617–628. [CrossRef] [PubMed]

355. Sansom, O.J.; Meniel, V.S.; Muncan, V.; Phesse, T.; Wilkins, J.A.; Reed, K.; Vass, J.K.; Athineos, D.; Clevers, H.; Clarke, A. Myc deletion rescues Apc deficiency in the small intestine. *Nature* **2007**, *446*, 676–679. [CrossRef] [PubMed]

356. Marderosian, M.; Sharma, A.; Funk, A.P.; Vartanian, R.; Masri, J.; Jo, O.D.; Gera, J. Tristetraprolin regulates Cyclin D1 and c-Myc mRNA stability in response to rapamycin in an Akt-dependent manner via p38 MAPK signaling. *Oncogene* **2006**, *25*, 6277–6290. [CrossRef]

357. Gera, J.; Mellinghoff, I.K.; Shi, Y.; Rettig, M.B.; Tran, C.; Hsu, J.-H.; Sawyers, C.L.; Lichtenstein, A. AKT Activity Determines Sensitivity to Mammalian Target of Rapamycin (mTOR) Inhibitors by Regulating Cyclin D1 and c-mycExpression. *J. Biol. Chem.* **2003**, *279*, 2737–2746. [CrossRef]

358. Shi, Y.; Sharma, A.; Wu, H.; Lichtenstein, A.; Gera, J. Cyclin D1 and c-myc Internal Ribosome Entry Site (IRES)-dependent Translation Is Regulated by AKT Activity and Enhanced by Rapamycin through a p38 MAPK- and ERK-dependent Pathway. *J. Biol. Chem.* **2005**, *280*, 10964–10973. [CrossRef]

359. Wall, M.; Poortinga, G.; Hannan, K.M.; Pearson, R.B.; Hannan, R.D.; McArthur, G. Translational control of c-MYC by rapamycin promotes terminal myeloid differentiation. *Blood* **2008**, *112*, 2305–2317. [CrossRef]

360. Gregory, M.A.; Qi, Y.; Hann, S.R. Phosphorylation by Glycogen Synthase Kinase-3 Controls c-Myc Proteolysis and Subnuclear Localization. *J. Biol. Chem.* **2003**, *278*, 51606–51612. [CrossRef]

361. Griend, D.V.; Litvinov, I.V.; Isaacs, J.T. Conversion of Androgen Receptor Signaling From a Growth Suppressor in Normal Prostate Epithelial Cells to an Oncogene in Prostate Cancer Cells Involves a Gain of Function in c-Myc Regulation. *Int. J. Biol. Sci.* **2014**, *10*, 627–642. [CrossRef] [PubMed]

362. Antony, L.; Van Der Schoor, F.; Dalrymple, S.L.; Isaacs, J.T. Androgen receptor (AR) suppresses normal human prostate epithelial cell proliferation via AR/β-catenin/TCF-4 complex inhibition ofc-MYCtranscription. *Prostate* **2014**, *74*, 1118–1131. [CrossRef] [PubMed]

363. Barfeld, S.J.; Urbanucci, A.; Itkonen, H.M.; Fazli, L.; Hicks, J.L.; Thiede, B.; Rennie, P.S.; Yegnasubramanian, S.; DeMarzo, A.M.; Mills, I.G. c-Myc Antagonises the Transcriptional Activity of the Androgen Receptor in Prostate Cancer Affecting Key Gene Networks. *EBioMed.* **2017**, *18*, 83–93. [CrossRef] [PubMed]

364. Rebello, R.J.; Pearson, R.B.; Hannan, R.D.; Furic, L. Therapeutic Approaches Targeting MYC-Driven Prostate Cancer. *Genes* **2017**, *8*, 71. [CrossRef]

365. Ma, F.; Ye, H.; He, H.H.; Gerrin, S.J.; Chen, S.; Tanenbaum, B.A.; Cai, C.; Sowalsky, A.G.; He, L.; Wang, H.; et al. SOX9 drives WNT pathway activation in prostate cancer. *J. Clin. Investig.* **2016**, *126*, 1745–1758. [CrossRef]

 International Journal of
Molecular Sciences

Article

Integrated Analysis to Study the Relationship between Tumor-Associated Selenoproteins: Focus on Prostate Cancer

Francesca Capone [†], Andrea Polo [†], Angela Sorice [†], Alfredo Budillon [*,‡] and Susan Costantini [*,‡]

Unità di Farmacologia Sperimentale-Laboratori di Mercogliano, Istituto Nazionale Tumori "Fondazione G. Pascale"—IRCCS, 80131 Napoli, Italy; f.capone@istitutotumori.na.it (F.C.); a.polo@istitutotumori.na.it (A.P.); a.sorice@istitutotumori.na.it (A.S.)
* Correspondence: a.budillon@istitutotumori.na.it (A.B.); s.costantini@istitutotumori.na.it (S.C.); Tel.: +39-081-590-3292 (A.B.); +39-0825-191-1729 (S.C.)
† Capone F., Polo A. and Sorice A. contributed equally to this work.
‡ Budillon A. and Costantini S. are co-senior authors of this article.

Received: 31 July 2020; Accepted: 11 September 2020; Published: 13 September 2020

Abstract: Selenoproteins are proteins that contain selenium within selenocysteine residues. To date, twenty-five mammalian selenoproteins have been identified; however, the functions of nearly half of these selenoproteins are unknown. Although alterations in selenoprotein expression and function have been suggested to play a role in cancer development and progression, few detailed studies have been carried out in this field. Network analyses and data mining of publicly available datasets on gene expression levels in different cancers, and the correlations with patient outcome, represent important tools to study the correlation between selenoproteins and other proteins present in the human interactome, and to determine whether altered selenoprotein expression is cancer type-specific, and/or correlated with cancer patient prognosis. Therefore, in the present study, we used bioinformatics approaches to (i) build up the network of interactions between twenty-five selenoproteins and identify the most inter-correlated proteins/genes, which are named HUB nodes; and (ii) analyze the correlation between selenoprotein gene expression and patient outcome in ten solid tumors. Then, considering the need to confirm by experimental approaches the correlations suggested by the bioinformatics analyses, we decided to evaluate the gene expression levels of the twenty-five selenoproteins and six HUB nodes in androgen receptor-positive (22RV1 and LNCaP) and androgen receptor–negative (DU145 and PC3) cell lines, compared to human nontransformed, and differentiated, prostate epithelial cells (EPN) by RT-qPCR analysis. This analysis confirmed that the combined evaluation of some selenoproteins and HUB nodes could have prognostic value and may improve patient outcome predictions.

Keywords: selenoproteins; cancer; HUB nodes; prostate cancer

1. Introduction

Selenoproteins are a class of proteins that contain selenium atoms inside selenocysteine (Sec) residues. Sec has been identified as the 21st amino acid, and is an analog of cysteine in which a selenol group replaces the sulfur-containing thiol group. It is encoded by the UGA codon that directs the translational decoding of UGA codons, rather than being used as a translational terminator. The corresponding mRNA includes a SEC Insertion Sequence (SECIS), which is present in eukaryotes in the 3′-untranslated region (UTR) of RNA [1].

Today, twenty-five selenoproteins have been identified in humans and twenty-four have been identified in mice. Mammalian selenoproteins can be classified mainly into two groups according to the Sec location. One group of selenoproteins possesses Sec at a site very close to the protein C terminus, and consists of three thioredoxin reductases (TXNRDs) and six additional proteins (methionine-R-sulfoxidereductase 1 (MSRB1), SELENOI, SELENOK, SELENOO, SELENOP, and SELENOS). The other group has Sec in the N-terminal part, and includes five glutathione peroxidases (GPX1, 2, 3, 4 and 6), three iodothyronine deiodinases (DIOs), and eight additional proteins (SELENOF, SELENOH, SELENOM, SELENON, SELENOT, SELENOV, SELENOW, and SEPHS2). These proteins are located in different sub-cellular compartments (nucleus, mitochondria, cytoskeleton, cytoplasm, endoplasmic reticulum (ER), Golgi apparatus and endosome). Some selenoproteins are secreted in the blood, such as SELENOP and GPX3 [2]. In addition, thioredoxin reductase 1 (TXRND1), a cytoplasmic and nuclear selenoprotein, can be secreted, and its serum levels are associated with poor prognosis in non-small cell lung cancer [3]. The majority of selenoproteins are involved in anti-oxidative activities associated with defending cells in different compartments against oxidative stress [4]. Some selenoproteins are located in the ER, and implicated in protein degradation, ER stress, and redox metabolism regulation [5]. These proteins are also involved in additional physiological functions, such as thyroid hormone metabolism, selenium transportation and storage, selenocysteine synthesis, protein folding, cell maintenance, calcium homeostasis, immune responses, and senescence [6]. Altered expression levels of selenoproteins have been associated with different disorders, such as type 2 diabetes, neuronal degenerative and cardiovascular diseases, and cancer [7,8].

Altered redox homeostasis can be involved in cancer initiation and progression because the oxidative insult can lead to genomic instability, DNA mutation, and carcinogenesis. Selenoprotein alteration has been reported to induce cancer initiation when associated with a low intake of selenium, thus increasing redox alterations. Indeed, several studies reported an inverse relationship between selenium levels and cancer risk. Recently, Lubiński et al. (2018) highlighted that in laryngeal cancer patients, the selenium level at the time of diagnosis was associated with the outcome [9]; similarly selenium supplementation in women undergoing treatment for breast cancer can favorably influence the patients' outcomes [10]. Recently, two reviews highlighted the association between single selenoproteins and either colorectal or prostate cancer initiation and progression, as well as patient outcome, suggesting that these proteins could represent potential biomarkers or therapeutic targets [11,12]. Several additional reports also demonstrated the role of specific selenoproteins in other cancer types. High levels of SELENOM were associated with poor prognosis of renal cell carcinoma (RCC), and preclinical evidence demonstrated that SELENOM silencing was able to block cancer proliferation, invasion, migration, and tumorigenesis, via PI3K/AKt/mTOR pathway inhibition and reduced metalloproteinases (MMP) 2 and 9 expression [13]. SELENOK was found to have a crucial role in the proliferation and activation of immune cells and the promotion of calcium flux that induces melanoma progression [14]. SELENOS was found to be highly expressed in insulinoma cells, and its silencing is able to induce apoptosis by decreasing Bcl-xL in β-cells and to block the cell cycle by downregulating the transcription factor E2F1 and increasing cyclin dependent kinase inhibitor p27 expression [15].

Our group previously reported for the first time an increase of SELENOM and GPX4 expression in HCC liver tissues, and that their expression was associated with the malignancy grade [16,17]. Based on a transcriptomic and interactomic approach, a list of dysregulated selenoproteins was observed in two human HCC cell lines, HepG2 and Huh7, compared to normal human hepatocytes [18]. Moreover, we identified human miR-544a as able to modulate SELENOK expression in the two HepG2 and Huh7 cell lines [19], and also analyzed the selenoproteins' transcriptomes in MCF-7 and MDA-MB231 human breast cancer cell lines, compared to MCF10A normal epithelial breast cells [20]. Recently, we reported that elevated tumor tissue expression of the selenoprotein SEPHS2 in triple-negative breast cancer (TNBC) patients was correlated with the malignancy grade [21].

Despite all the evidence reported above, the mechanism and the specific role of single or associated selenoproteins in cancer initiation and progression is still not clear. In silico approaches, such as bioinformatics, have been used to investigate signaling pathways as well as protein and gene interactions, in order to obtain a better understanding of the molecular mechanisms of diseases [22,23].

In detail, the construction of a gene/protein interaction network, and the application of a scoring algorithm based on the calculation of the quality, and the quantification, of interactions, followed by cluster analysis, might allow: (i) the selection of a list of genes/proteins putatively important in a given disease/cancer, according to the confidence and number of interactions derived from available databanks from experimental data; and (ii) the identification of "leader" genes/proteins that can be assumed to play important functional roles, because they present the highest number of interactions with the other genes/proteins within the network, thus being considered as HUB nodes in the interaction map [24–27]. It is noteworthy that the interaction networks comprise physical or functional correlations between all the genes/proteins involved in specific diseases/cancer types, indicated as direct or indirect interactions, respectively, and can be used to suggest the functional significance of the experimental results and clinical data that should be further confirmed in new targeted "wet" experiments. In this context, it can be useful to use bioinformatics approaches to identify associations between selenoprotein genes and proteins, and different cancer types.

Therefore, in the present study, publicly available datasets and different bioinformatics tools were used to, (i) analyze the protein–protein interactome of the twenty-five mammalian selenoproteins, and determine the most inter-correlated proteins, defined as HUB nodes; and (ii) correlate the gene expression, of the selenoproteins and the identified HUB nodes, in ten solid tumors, with patient outcomes.

Moreover, experimental approaches were used to confirm some of the correlations suggested by the bioinformatics analyses, based on a study on human prostate cancer cell lines. Indeed, within the last year, our group has focused on searching for new markers to predict prostate cancer initiation/progression [28,29] and improve the early definition of prostate cancer patient outcomes [30]. In detail, we evaluated the gene expression level of the twenty-five selenoproteins and the identified HUB nodes in prostate cancer cells, compared to normal epithelial prostate cells, by RT-qPCR analysis.

2. Results

2.1. Network Analysis

To better understand the relationships between selenoproteins and cancer, we conducted an interaction network analysis to define whether these proteins were inter-correlated, or correlated with other key proteins involved in cancer.

Therefore, based on the human molecular interactome, we created an interaction network, in which the twenty-five known selenoproteins were the nodes, and the relationships between them were the edges. As shown in Figure 1, seventeen selenoproteins in the same network were correlated through different nodes, indicating functional redundancy and/or cooperation between different selenoproteins. To understand the positions and roles of the protein nodes, and identify the HUB nodes, which are the nodes with the strongest role in coordinating the network, the network was analyzed in terms of centrality and topological measures (node degree distribution, clustering coefficient, stress centrality, closeness centrality, betweenness centrality, clustering coefficient, network centralization, characteristic path length, average number of neighbors, network density, and network heterogeneity), as reported in the methods section. In detail, these statistical analyses showed that the obtained network had a network centralization value of 0.364, indicating that the network had good centralization, with the presence of nodes with a high degree. The portion of the potential connections into the network, expressed by the network density value, was equal to 0.023, whereas the higher network heterogeneity value was 1.825, suggesting a tendency of the network to contain HUB nodes. Moreover, Figure S1 shows the decreasing plots of the node degree distribution, clustering coefficient, stress centrality, closeness centrality, and betweenness centrality, which all together confirmed the scale-free property of the obtained network and its tendency to contain HUB nodes. These analyses identified the presence of the six following HUB nodes in the network: ABL1, EP300, FYN, MYC, PSMB2, and SRPK2.

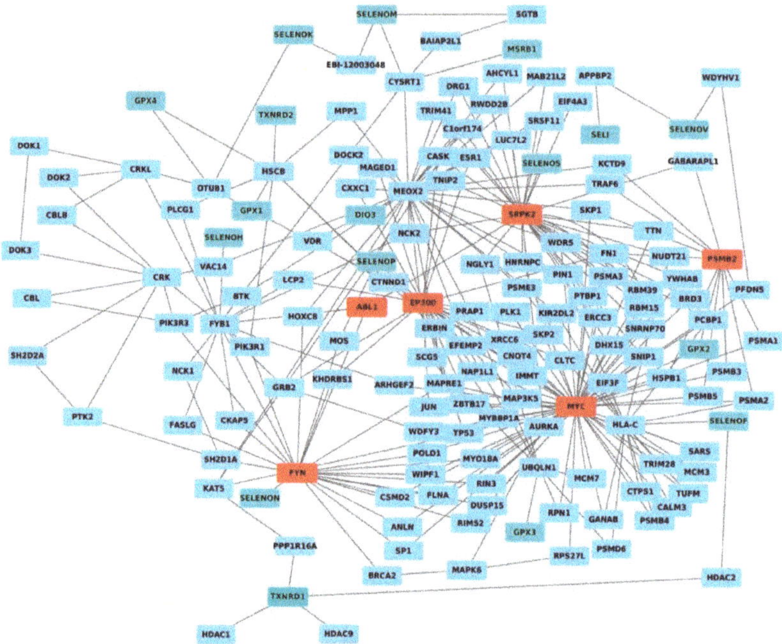

Figure 1. Network of the selenoproteins mapped on the human interactome. In detail, the selenoproteins are shown in cyan, HUB nodes are shown in red, whereas the other nodes are shown in blue.

The HUB nodes include the well-known oncogene MYC, which is amplified in various types of cancer, such as breast, colorectal, endometrial, ovarian, and prostate cancer, and controls the expression of genes and noncoding RNAs by regulating cell growth, cell cycle progression, apoptosis, and differentiation, as well as angiogenic switch, cellular metabolism, and drug resistance mechanisms [31,32]. Its expression has been correlated with prognostic/clinic-pathological outcomes in breast cancer [33], and occurs in the early stages of colorectal cancer [34].

EP300 has dual activities as a transcriptional factor and a histone acetyltransferase [35], and it is involved in different biological functions, such as proliferation, cell cycle regulation, apoptosis, differentiation, and DNA damage response [36]. EP300 gene has often been found to be mutated/truncated in lymphomas and different solid tumors, such as gastric, colorectal, breast, and pancreatic cancers, and over-expressed and correlated to poor prognosis in liver, nasopharyngeal, and small and non-small cell lung cancer [37].

FYN was reported to be correlated with cell motility and proliferation and over-expressed in chronic myeloid leukemia, breast cancer, squamous head and neck carcinoma, and melanoma [38]. In TNBC cells, FYN was shown to induce epithelial-mesenchymal transition (EMT), and its depletion reduced cell migration and invasion [39].

SRPK2 is part of a family of kinases that phosphorylate serine/arginine-rich (SR) proteins and regulate the conformation, subcellular localization, and/or interaction of SR enzymes, to regulate posttranscriptional mRNA processing [40]. In a colon cancer model, this protein was demonstrated to promote cell growth and migration, and control the expression of lipogenic enzymes [41].

ABL1 is a tyrosine kinase involved in different cellular signaling processes, controlling proliferation, survival, migration, and invasion of cells [42]. It was initially identified as a tumor suppressor but, since then, its oncogenic functions have been well-established [42]. However, data in solid tumors suggest that ABL1 can have tumor suppressive or oncogenic roles, depending on the cellular context.

PSMB2 is a component of the proteasome complex involved in the degradation of intracellular proteins, playing a role in maintaining protein homeostasis [43]. It was significantly associated with chronic leukemia [44], and its suppression was able to inhibit liver cancer proliferation [45]. Moreover, in our previous paper, PSMB2 was a HUB node in the bladder cancer network [46].

Overall, our network analysis can represent a starting point for planning targeted experiments, for verifying the correlations among selenoproteins (in terms of gene expression levels and/or functions), and with these HUB nodes, in different types of cancer via experimental approaches.

2.2. Correlation between Selenoproteins/HUB Nodes and Cancer Patient Overall Survival

To highlight eventual specific relationships between selenoproteins and cancers, we evaluated the correlation between the gene expression of the twenty-five selenoproteins and the overall survival (OS) of patients, for the ten most common solid tumor types, using different public datasets and the PROGgeneV2 online tool (http://genomics.jefferson.edu/proggene/index.php) [47].

This analysis demonstrated that (i) no correlation occurred between single selenoprotein expression and poor OS in the case of prostate cancer patients; (ii) no correlation occurred between ten selenoproteins (DIO1, DIO2, DIO3, MSRB1, SELENOF, SELENOH, SELENON, SELENOV, SELENOW, and TXNRD3) and poor OS in any of the ten analyzed cancer types; and iii) a statistically significant correlation occurred between the expression levels of the fifteen selenoproteins with poor OS in at least one cancer type (Figures 2 and 3).The largest number of selenoproteins with altered gene expression, either upregulation or downregulation, in tumor tissues compared with normal tissues, and correlations with poor OS, were found in kidney (eight), pancreas (six), and breast (five), cancers. The expression of some selenoproteins was altered in more than one cancer type; however, in the majority of cases, different patterns of selenoprotein gene expression were observed among the cancers examined. The only exceptions were (i) GPX2, whose high expression levels correlated with OS in breast and head and neck cancers; and (ii) TXNRD1, whose high expression levels correlated with poor OS in head and neck, kidney, liver, and pancreas cancers.

To analyze in more detail the correlation between the identified HUB nodes and the selenoproteins in ten examined cancers, we evaluated the correlation between HUB node expression and OS, in patients with these ten cancers. As shown in Figure 2 and Figure S2, all six HUB nodes correlated with at least one cancer type, and FYN and PSMB2 correlated with poor OS in patients with five different cancers. In particular, lower expression levels of FYN correlated with poor OS in patients with melanoma, head and neck, liver, lung, and pancreas cancers. On the other hand, higher expression levels of PSMB2 were correlated with poor OS in patients with kidney, liver, and prostate cancers, whereas lower levels of PSMB2 were correlated with poor OS in patients with bladder and lung cancers. Hence, in the case of prostate cancer, we did not observe a correlation between single selenoprotein expression and patient OS, and a significant correlation was only observed between the HUB node PSMB2 and poor OS.

	BLADDER	BREAST	COLON	HEAD&NECK	KIDNEY	LIVER	LUNG	MELANOMA	PANCREAS	PROSTATE
SELENOPROTEINS										
DIO1										
DIO2										
DIO3										
GPX1					0.0003					
GPX2		0.008		0.0035						
GPX3		0.0001			0.021				0.0015	
GPX4									0.043	
GPX6			0.0052							
MSRB1										
SELENOF										
SELENOH										
SELENOI		0.035								
SELENOK					0.044		0.038	0.011		
SELENOM		0.04			0.0002			0.01		
SELENON										
SELENOO	0.002				0.0014				0.0024	
SELENOP		0.0001			0.00001					
SELENOS				0.013				0.0049		
SELENOT	0.015							0.0021	0.026	
SELENOV										
SELENOW										
SEPHS2					0.0002					
TXRND1				0.036	0.0053	0.0041			0.025	
TXRND2									0.023	
TXRND3										
HUB NODES										
ABL1	0.035							0.037		
EP300		0.0079			0.0092					
FYN			0.03			0.043	0.019	0.025	0.016	
MYC					0.031					
PSMB2	0.047				0.00001	0.0013	0.005			0.019
SRPK2			0.0066				0.019			

Figure 2. Heat map related to the statistically significant correlations (with *p*-values < 0.05) between overall survival (OS) in solid tumor patients and selenoprotein/HUB node expression. In detail, we report significant *p*-values in red or in green, if the high or low expression of selenoproteins, was correlated with poor overall survival, respectively. White boxes indicate no significant correlations.

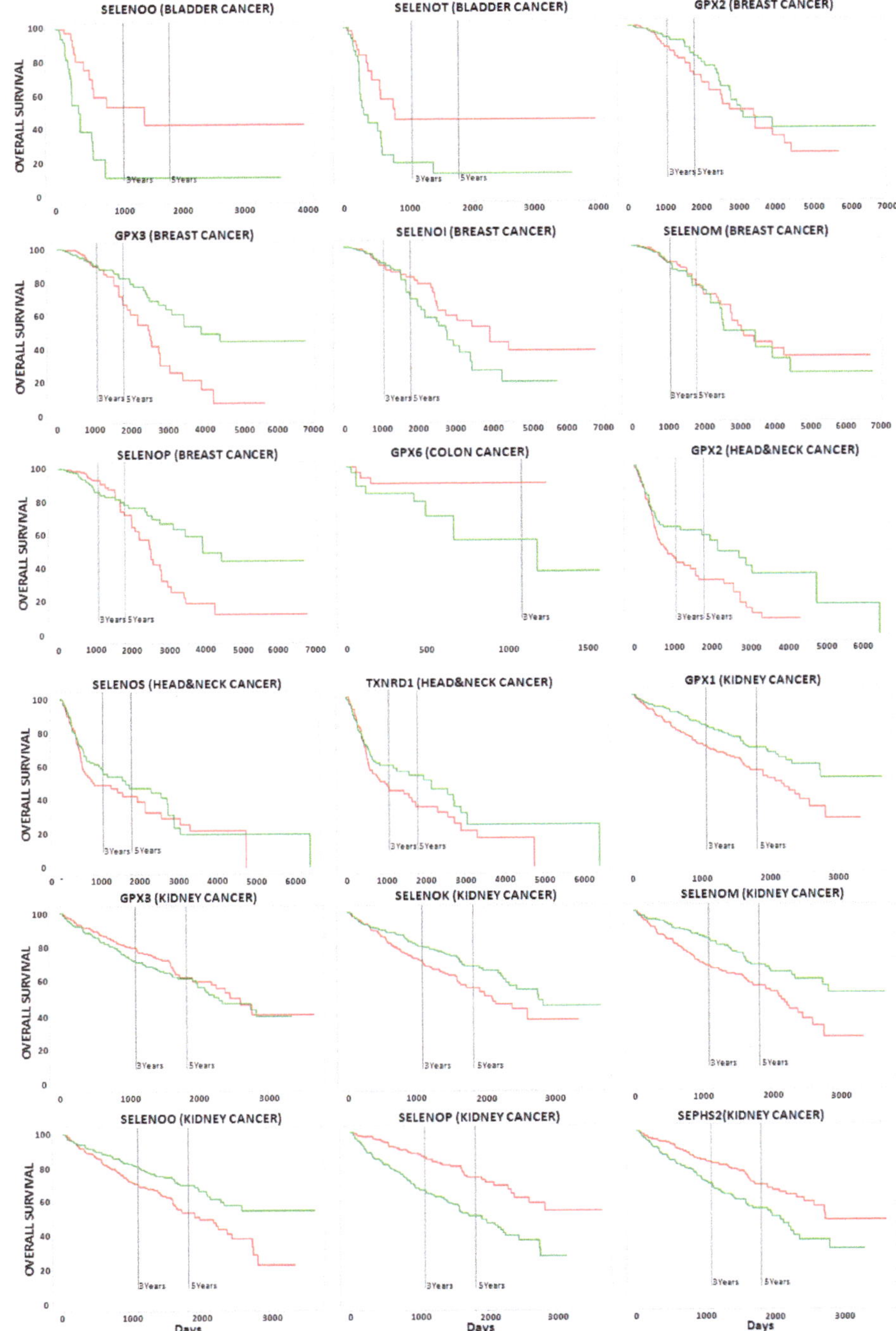

Figure 3. *Cont.*

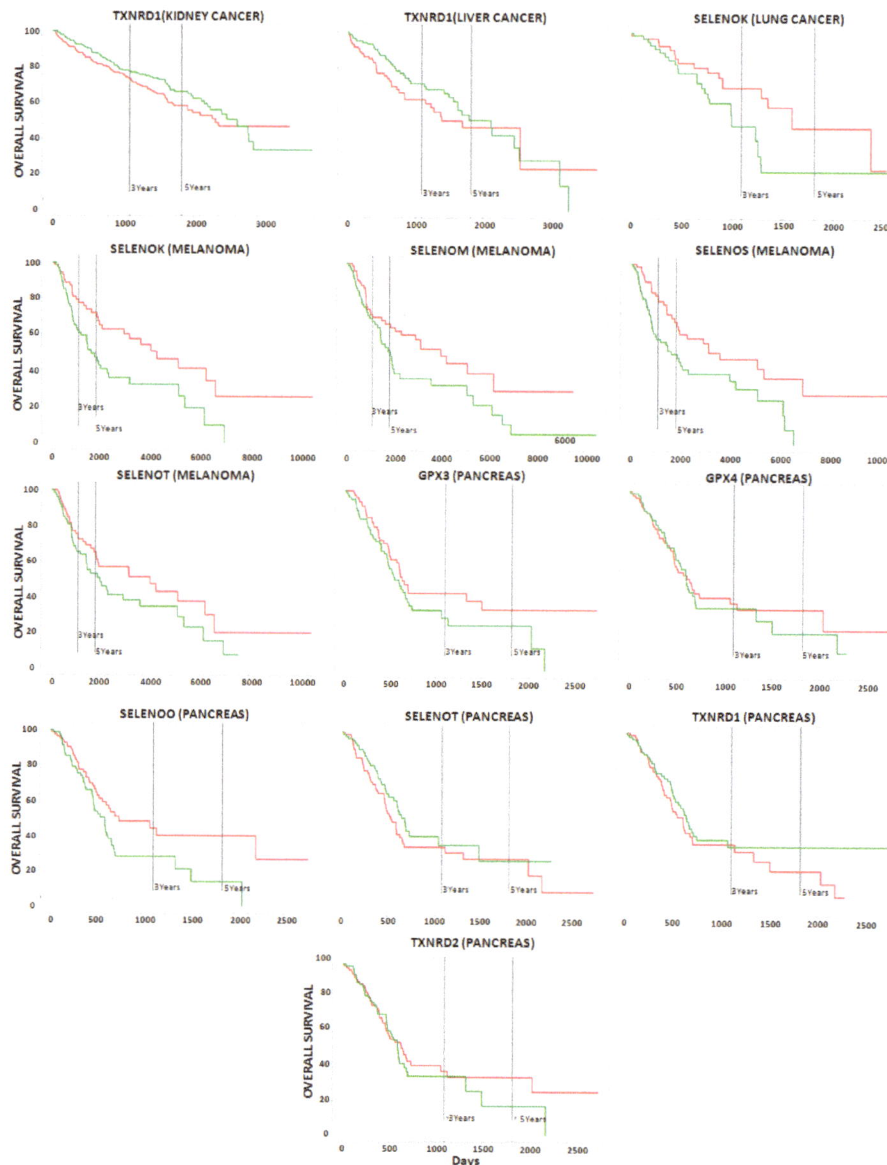

Figure 3. Kaplan-Mayer curves showing the overall survival (expressed in percentage) in solid cancer patients using the PROGgeneV2 online tool, in the case of high and low expression of selenoproteins, which are indicated by red and green curves, respectively.

2.3. Gene Expression Levels of Twenty-Five Selenoproteins in Prostate Cancer Cells Compared to Normal Prostate Cells

Next, to experimentally evaluate the data suggested by the bioinformatics analyses, and better define the selenoprotein alterations in prostate cancer as well as the correlation between them and the HUB nodes, we conducted a preliminary study on gene expression in five cell lines. In detail, the gene expression profiles for all twenty-five selenoproteins, in two androgen receptor-positive cell lines

(22RV1 and LNCaP), and two androgen receptor-negative cell lines (DU145 and PC3), were compared to human nontransformed, and differentiated, prostate epithelial cells (EPN), based on a RT-qPCR assay.

The results revealed, for the two androgen receptor-positive cell lines (Figure 4A), (i) a statistically significant increase (with $\log_2 2^{-\Delta\Delta Ct} > 1$) of the expression levels of DIO1, DIO2, GPX2, SELENOS, TXNRD1, and TXNRD2; and ii) a statistically significant decrease (with $\log_2 2^{-\Delta\Delta Ct} < 1$) of the expression levels of SELENOF and SELENOI. Compared to EPN cells, LNCaP cells showed higher DIO3 and SELENOT levels, but lower MSRB1, SELENOH, SELENON, and SELENOV levels. Compared to EPN cells, 22RV1 cells showed higher expression levels of GPX4, SELENOK, and SEPHS2, and lower levels of GPX1, GPX6, SELENOM, SELENOP, and SELENOT.

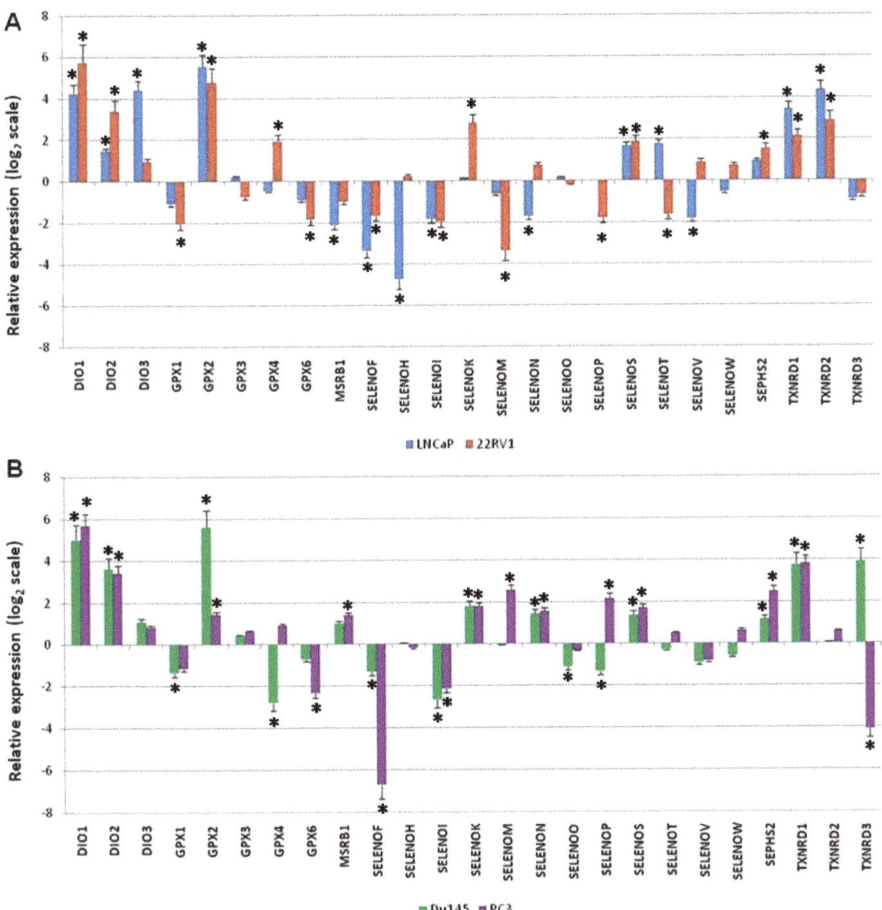

Figure 4. Fold change of gene expression level for each selenoprotein (indicated as the relative expression) in (**A**) two androgen receptor-positive (LNCaP and 22RV1) and (**B**) two androgen receptor-negative (DU145 and PC3) prostate cancer line cells, compared to the non-cancerous epithelial cells (EPN) cells, as evaluated by the $2^{-\Delta\Delta Cq}$ method, and reported on the $\log_2$ scale. We considered values higher and lower than +1 and −1 to be statistically significant, respectively. The statistically significant *p*-values at <0.05 are indicated by *.

As shown in Figure 4B, the two androgen receptor-negative cell lines showed: (i) a statistically significant increase (with $\log_2 2^{-\Delta\Delta Ct} > 1$) of the expression levels of DIO1, DIO2, GPX2, SELENOK, SELENON, SELENOS, SEPHS2, and TXNRD1; and (ii) a statistically significant decrease (with $\log_2 2^{-\Delta\Delta Ct} < 1$) of the expression levels of SELENOF and SELENOI. Moreover, compared to EPN cells, DU145 cells showed higher levels of TXNRD3 and lower levels of GPX1, GPX4, SELENOO, and SELENOP, whereas PC3 cells showed higher levels of MSRB1, SELENOM, and SELENOP, and lower levels of GPX6 and TXNRD3.

2.4. Gene Expression Levels of HUB Nodes in Prostate Cancer Cells, and Their Correlation with Selenoprotein Expression

Then, we evaluated the gene expression profiles for the six HUB nodes (ABL1, EP300, FYN, MYC, PSMB2, and SRPK2) in all the prostate cancer cells, compared to normal prostate EPN cells (Figure 5A).

These analyses showed that (i) ABL1 had lower statistically significant levels (with $\log_2 2^{-\Delta\Delta Ct} < -1$) in both androgen receptor-positive cell lines (22RV1 and LNCaP), but higher levels in the two androgen receptor-negative cell lines, although these values were not statistically significant; (ii) the EP300, FYN, MYC, PSMB2, and SRPK2 levels increased in all four prostate cancer cell lines, although the EP300 and FYN levels were not statistically significant. In both androgen receptor-negative cells, the MYC levels were not significant in the LNCaP cells, the PSMB2 levels were not significant in the DU145 cells, and the SRPK2 levels were not significant in the PC3 cells.

To determine whether a correlation occurred between the gene expression levels of all HUB nodes and the selenoproteins evaluated, we performed a Pearson correlation matrix analysis. Figure 5B shows that (i) MYC and SELENOK were very correlated with each other and with DIO1, DIO2, and SELENOS; (ii) ABL1 showed a slight correlation with MSRB1; and (iii) EP300, FYN, SRPK2, and PSMB2 were correlated with GPX2.

Therefore, these data confirmed the correlation between selenoproteins and HUB nodes in prostate cancer cells, and suggested the need to conduct more detailed studies to analyze the putative role of these proteins as combined markers of progression or response to therapeutic approaches in prostate cancer. In this regard, we preliminarily evaluated whether significant correlations occurred between the combined gene expression of selenoproteins and HUB nodes (based on clustering in the matrix analysis shown in Figure 5B), and the OS of prostate cancer patients; this analysis was performed using the PROGgeneV2 online tool (http://genomics.jefferson.edu/proggene/index.php) [47]. Furthermore, this analysis highlighted a statistically significant correlation (with *p*-value = 0.0091) between the combined gene expression of GPX2, EP300, and PSMB2 (Figure 6), and the OS of prostate cancer patients, thus confirming that the combined evaluation of some selenoproteins and HUB nodes, could represent a new strategy to predict patient outcomes.

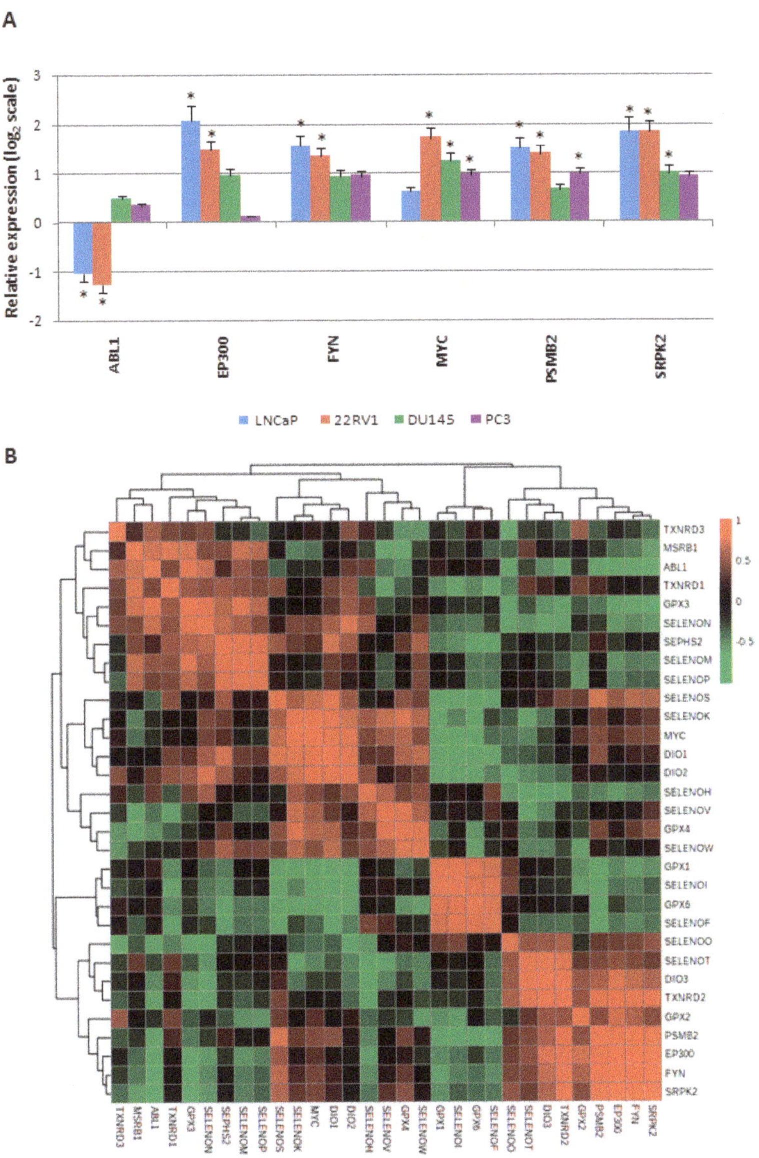

Figure 5. Evaluation of the gene expression levels of HUB nodes in five prostate cell lines and their correlation with the gene expression profiles of the selenoproteins. (**A**) Fold change of gene expression levels for each HUB node, in two androgen receptor-positive (LNCaP and 22RV1), and in two androgen receptor-negative (DU145 and PC3) prostate cancer line cells, compared to the non-cancerous EPN cells, as evaluated by the $2^{-\Delta\Delta Cq}$ method and reported as the log$_2$ scale. We considered values higher and lower than +1 and −1 to be statistically significant, respectively. The statistically significant *p*-values at <0.05 are indicated by *. (**B**) Pearson correlation matrix evaluated on the gene expression profiles of the selenoproteins and HUB nodes, evaluated for EPN and prostate cancer cells. Color scale, from red to green color, indicates from good to poor correlations between the gene expression levels of the analyzed proteins.

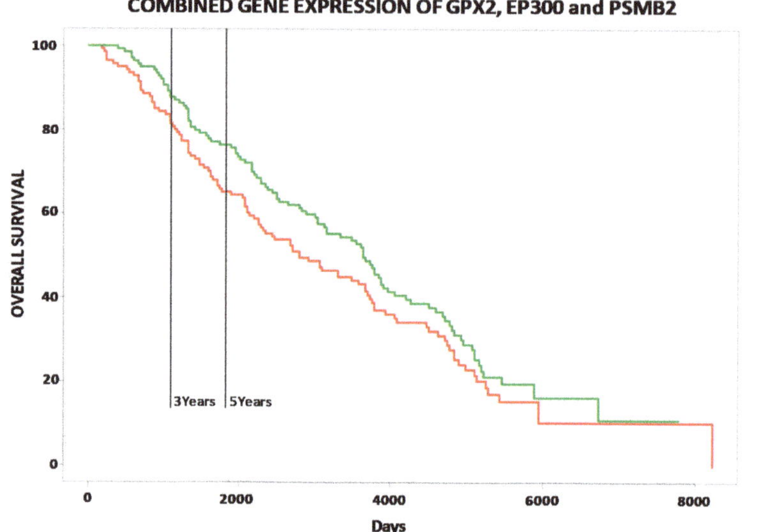

Figure 6. Kaplan-Mayer curves showing the correlations between overall survival (expressed in percentage) and the combined gene expression of GPX2, EP300, and PSMB2 in prostate cancer patients, which was based on the PROGgeneV2 online tool. High and low expression of selenoproteins are reported by the red and green curves, respectively.

3. Discussion

The selenoprotein family has a well-established function in regulating oxidative cell balance [48], and is involved in tumorigenesis and cancer progression [6]. Still, identifying tumor type-specific selenoprotein profiles, and determining whether these proteins can predict prognosis or could serve as therapeutic anticancer targets in cancer patients, represent critical challenges. In this study, we created an interaction network of the twenty-five known selenoproteins and highlighted the presence of six HUB nodes (ABL1, EP300, FYN, MYC, PSMB2, and SRPK2) that play the strongest role in coordinating the obtained network. Then, we evaluated the correlation between the gene expression of the twenty-five selenoproteins and HUB nodes, and the OS of patients with the ten most common solid tumor types, with the results demonstrating that (i) ten selenoproteins were not correlated with poor OS of cancer patients; (ii) more correlations between selenoprotein gene expression and patient OS were only observed for kidney, pancreas, and breast cancers; (iii) single selenoprotein expression and OS of prostate cancer patients was not observed; and (iv) all HUB nodes were correlated with poor OS for at least one cancer type.

To our knowledge, few selenoproteins have been studied in prostate cancer, and never altogether. For example, GPX1, SELENOF, and SELENOP levels were significantly reduced in prostate cancer models [49–51], whereas GPX2 was significantly increased [52]. Increased nuclear TXNRD1 levels were found in high-grade, versus low-grade, human prostate cancers [53], and correlated with prostate cancer progression and androgen-deprived castration-resistant prostate cancer (CRPC) cells, suggesting that CRPC possesses an enhanced dependency on TXNRD1 [54]. Hence, our study aimed to confirm, by experimental approaches, some of the correlations suggested by bioinformatics analyses, and, thus, represented the first systematic evaluation of the expression levels of all twenty-five selenoproteins in androgen receptor-positive and -negative prostate cancer cells. Through our analysis, we confirmed higher levels of GPX2 and TXNRD1, and lower levels of GPX1, SELENOF, and SELENOP in the cancer cell lines compared to normal epithelial cells, which has already been observed and reported in the literature. We also observed higher levels of DIO1, DIO2, and SELENOS, and lower levels of SELENOI in all the prostate cancer cell lines. The differences observed between androgen receptor-negative or androgen receptor-positive cell lines can also be correlated with their different molecular features [55–57]. Although the number of tested cell lines was small, the differences between the androgen receptor-positive and androgen receptor-negative cell lines were not highlighted in detail; rather, we only suggested that our analyses showed a different expression pattern for some selenoproteins in prostate cancer compared to normal prostate epithelium. This observation should be functionally evaluated and eventually expanded to a greater number of prostate cancer lines and prostate tumor tissues; however, this task is beyond the scope of the current study.

Still, an interesting point of discussion is represented by the finding of lower levels of SELENOI in prostate cancer cells vs. EPN. SELENOI, also known as ethanolphosphotransferase 1 (EPT1), is an enzyme responsible for the final step in the Kennedy pathway that transfers phosphoethanolamine from cytidine diphosphate ethanolamine to lipid acceptors, to produce ethanolamine glycerophospholipids, such as phosphatidylethanolamine (PE). Hence, it plays an important role in maintaining the normal homeostasis of ether-linked phospholipids in humans [58]. Interestingly, an analysis of the concentrations of plasma phospholipids in prostate cancer patients showed that the levels of phosphatidylethanolamine, and total phospholipids in these patients, were decreased, and correlated with an increased pathologic grade and Gleason score [59]. We can then speculate that the lower levels of SELENOI found in prostate cancer cells could be correlated to decreased levels of phospholipids, and could represent a new putative marker of prostate cancer progression that warrants further study.

Moreover, as reported above, in our study we evaluated the gene expression levels of HUB nodes and performed a correlation with the expression of the selenoproteins by correlation matrix analysis (Figure 5B). We observed that: (i) MYC and SELENOK expression was strongly correlated with each other, and with DIO1, DIO2, and SELENOS; (ii) ABL1 showed a slight correlation with MSRB1; and (iii) EP300, FYN, SRPK2, and PSMB2 were correlated with GPX2.

Considering the first correlation cluster, SELENOK and SELENOS are two ER selenoproteins that regulate ER stress and degradation [60]. Both of these proteins have a role in the ER-associated protein degradation (ERAD) pathway, and interact with the valosin containing protein (VCP/p97) for the retrotranslocation of misfolded proteins, from ER to the cytosol, via their polyubiquitination, and they also play a role in cell survival [61–63]. Moreover, SELENOK showed peroxidase activity capable of reducing harmful hydrophobic substrates, such as phospholipid hydroperoxides, and is involved in membrane repair [64]. SELENOK was also found to be over-expressed in gastric, glial, thyroid, testis, and cervix cancers, and its polymorphisms, in combination with selenium status, were related to prostate cancer progression [65,66]. Recently, we showed that SELENOK was over-expressed in two liver cancer cell lines, HepG2 and Huh7 [18]. Furthermore, SELENOK inhibition by miR-181 and miR-544a was able to suppress the proliferation of glioma and hepatocellular carcinoma cells, respectively [19,67]. Regarding SELENOS, its polymorphisms were associated with many tumors, such as colorectal and gastric cancer [68], and were able to affect the expression levels of inflammatory cytokines in plasma [69]. On the other hand, DIO1 and DIO2 are located on the plasma membrane and ER, respectively, and are part of the iodothyronine deiodinase family and involved in regulating the activity of thyroid hormones by deiodination reactions [61]. DIO2 was found to be highly expressed in mesothelioma cell lines [70], and its inhibition resulted in the suppressed expression of prostate specific antigen (PSA) in prostate cancer, thus representing a potential approach to overcoming castration resistance [71]. Moreover, in prostate cancer, MYC activity was correlated with dysregulation of the PI3K/AKT/mTOR pathway, which induced cellular survival. The therapeutic efficacy of targeting MYC activity by interfering with its transcriptional program was also evaluated [72]. Detailed data about the involvement of DIO1, SELENOK, and SELENOS in prostate cancer have not been available until now; thus, additional data on these selenoproteins must be obtained to confirm their putative role in prostate cancer and their possible functional correlation with MYC.

In the second correlation cluster, MSRB1 is a selenoprotein located in the cytosol and the nucleus. It is mainly known for its antioxidant and protein repair functions and its role as a switch for protein function, via reversible oxidation/reduction of specific methionine residues [73,74]. MSRB1 contributes to shaping cellular immune responses, and its silencing resulted in the induction of anti-inflammatory cytokines (IL-10 and IL-1ra) [75]. High levels of MSRB1 were previously found in hepatocellular carcinoma, and correlated with the MAPK pathway and epithelial-mesenchymal transition (EMT) [76], and in low metastatic MCF7 human breast cancer cells [77]. ABL1 has been previously reported to act as a switch between cellular, invasive and proliferative, states and can either promote invasion and prostate cancer aggressiveness, or inhibit its progression, depending on the signal [78]. Hence, as in the case of the first cluster, it will be interesting to plan further studies to analyze the correlation between MSRB1 and ABL1.

Finally, in the third correlation cluster, we found that GPX2 was correlated with four HUB nodes (EP300, FYN, SRPK2, and PSMB2). GPX2, a key molecule of the glutathione redox system that acts in concert to provide a coordinated network of protection against ROS accumulation and oxidative damage, has been suggested as a prognostic marker in CRPC, and its silencing is able to inhibit prostate cancer growth [52]. Recently, high levels of EP300 were correlated to both prostate cancer progression and chemotherapy resistance in metastatic CRPC patients, and it has emerged as a possible co-target in chemo-resistant prostate cancer treatment [79]. FYN over-expression was demonstrated in prostate cancer and suggested as an interesting therapeutic target [80]. Moreover, SRPK2 and PSMB2 over-expression was observed in prostate cancer, and correlated with a high Gleason score, advanced pathological stage, tumor metastasis [81], and significantly poor 10-year metastatic rate in prostate cancer, in younger men [82]. Overall, these data are consistent with the significant correlation found between the combined gene expression of GPX2, EP300, and PSMB2, and OS in prostate cancer patients (Figure 6). These observations confirm that the combined evaluation of some selenoproteins and HUB nodes could have a better prognostic value, and could improve the prediction of patient outcomes.

4. Materials and Methods

4.1. Network Analysis

Through the Cytoscape software platform, for the visualization of complex networks and their integration (http://www.cytoscape.org/), a network related to the interactions between the selenoproteins was constructed, and used as reference for the human molecular interactome (INTACT), where all interactions are derived from literature curation or direct user submissions. As already reported in our recent paper, some statistical analyses on the following centrality and topological measure parameters of networks were performed: node degree distribution, clustering coefficient, stress centrality, closeness centrality, betweenness centrality, clustering coefficient, network centralization, characteristic path length, avg. number of neighbors, network density, and network heterogeneity [46]. The obtained network has been deposited in the NDEX database (http://www.ndexbio.org/#/network/194d2011-7c7a-11e9-848d-0ac135e8bacf?accesskey=d37870a03d06010b5b49c17e5411056441c9e7bac99ddde7d7354c8f4c68fd11).

4.2. Survival Gene Analysis

Using the PROGgeneV2 online tool (http://genomics.jefferson.edu/proggene/index.php) [47], we evaluated the correlation between selenoproteins/HUB nodes gene expression and OS in the ten most common solid tumor types, using different public datasets. The following ten datasets were used to evaluate the correlation between selenoprotein gene expression and OS: TCGA_COAD (Colon Adenocarcinoma), TCGA_PAAD (Pancreatic Adenocarcinoma), TCGA_SKCM (Skin cutaneous melanoma), TCGA_LIHC (Liver hepatocellular carcinoma), TCGA_PRAD/GSE16560 (Prostate Cancer), TCGA_BRCA (Breast Cancer), TCGA_HNSC (Head and Neck squamous cell carcinoma), TCGA-LUAD (Lung adenocarcinoma), TCGA_KIRC (Kidney renal clear cell carcinoma), and TCGA_BLCA (Bladder Urothelial Carcinoma).

4.3. Prostate Cancer Cell Lines

The expression of all twenty-five selenoproteins and HUB nodes was investigated by RT-qPCR in an EPN line, two androgen receptor-positive cell lines (22RV1 and LNCaP), and two androgen receptor-negative cell lines (DU145 and PC3).

EPN cells are a novel epithelial cell line derived from human prostate tissue that does not form colonies in semisolid medium and does not form tumors once injected into nude mice. They express the functional androgen receptor cytokeratin (numbered 1, 5, 10, and 14) and have wild-type p53 [83]. EPN cells were grown in DMEM/Ham's F-12 50/50 supplemented with fetal bovine serum (5%) (Invitrogen, CA, USA), penicillin/streptomycin (100×) (Euroclone, Devon, UK), and Glutamax (100×) (Invitrogen, CA, USA).

22RV1, DU145, LNCaP, and PC3cells were grown in RPMI (Roswell Park Memorial Institute) supplemented with fetal bovine serum (Invitrogen, CA, USA) (10%), penicillin/streptomycin (100×) (Euroclone, Devon, UK), and Glutamax (100×) (Invitrogen, CA, USA).

4.4. RNA Preparation and Reverse Transcription-qPCR (RT-qPCR)

Total RNA was extracted from five prostate cell cultures (EPN, 22RV1, DU145, LNCaP, and PC3), using a RNAeasy Mini Kit (Qiagen Inc.) according to the manufacturer's instructions. The RNA concentration and purity were determined using a NanoDrop 2000 spectrophotometer (Thermo Scientific, Wilmington, DE, USA) at an optical density of 260/280 nm. Reverse transcription of RNA was performed with 2 µg of RNA using a SuperScript VILO cDNA Synthesis kit (Life Technologies) in a 20 µL reaction volume.

The mRNA sequence from the nucleotide data bank (NCBI, Bethesda, Maryland, USA) was used to design primer pairs for RT-qPCR with an amplicon <100 bp, according to the manufacturer's instructions. Oligonucleotides were obtained from Eurofins. The list of primers is reported in Table S1. RT-qPCR experiments were performed using a Step-One Real Time PCR System (Applied Biosystems). Each aliquot of cDNA (2 µL) was amplified in a mixture (25 µL) consisting of the reverse and forward primers (300 nM) and 2X SYBR Green PCR Master Mix (Applied Biosystems). The PCR conditions consisted of an initial denaturation step of 95 °C for 5 min followed by 44 cycles of a two-step program: (i) denaturation at 95 °C for 30 s; and (ii) annealing/extension at 60 °C for 1 min. Each assay included a no-template control for each primer pair, and to ensure that RNA samples were not contaminated with DNA, negative controls were obtained by performing the PCR assay on samples that were not reverse transcribed. Moreover, experiments were performed in triplicate to ensure reproducibility of the technique. β-Actin mRNA was used to normalize the data. All of the obtained data were analyzed statistically.

Sample ΔCq values were calculated as the difference between the mean Cq obtained for each selenoprotein transcript (seleno-mRNA) and the housekeeping gene. $2^{-\Delta\Delta Cq}$ values were determined to define the fold change of expression level for each seleno-mRNA in different prostate cancer cells compared to the non-cancerous EPN cells. We reported values on the $\log_2$ scale, and considered values higher and lower than +1 and −1 to be statistically significant, respectively. The statistical comparison between gene expression levels evaluated in the prostate cancer cell line vs EPN cells was calculated by the T-test. *P*-values lower than 0.05 were considered statistically significant.

Moreover, a Pearson correlation matrix analysis, comparing the gene expression profiles of the selenoproteins and HUB nodes evaluated for each prostate cancer cell line and the EPN cell line, was performed using the MetaboAnalyst tool (https://www.metaboanalyst.ca/).

5. Conclusions and Future Perspectives

In conclusion, our data add new insights into the cancer biology of selenoproteins, suggesting that further experimental studies are necessary for a deeper understanding of the role of these proteins, and the correlated hub nodes, as potential cancer prognostic markers and/or therapeutic targets. Certainly, because the members of the selenoprotein family are strongly interconnected, only a systematic approach should be used to obtain a global vision of the relationships between these proteins and cancer. We are aware that our approach has limitations, such as the potential risk of circular reasoning due to the lack of localization, and function annotations that are often incomplete or unavailable, since the interactomes of many species are unmapped. Hence, to improve the data quality in the interaction networks, it would be useful to analyze the properties of interacting gene/protein pairs, and to verify that they have similar process and functional annotations, common sub-cellular localizations, and shared interaction partners [84]. Moreover, experimental validation of the proteins/genes evidenced by in silico studies is always necessary, because these data must be considered as exploratory and used as a source of experimental and clinical hypotheses [23].

Indeed, using prostate cancer as a study model, our results suggest that some selenoproteins, such as MSRB1, SELENOI, SELENOK, SELENOS, and GPX2 (in particular) can represent interesting topics for further studies, being those proteins that are involved in inflammatory and metabolic/lipidic pathways, correlated with HUB nodes, and involved in prostate cancer. Moreover, whether these proteins are measurable in biological fluids, such blood, urine and saliva, should be verified, because they could represent ideal new biomarkers for the dynamic monitoring of prostate cancer progression and therapeutic approach responses.

Collectively, we confirmed that the role of all selenoproteins, rather than single members of the family, should be further evaluated in specific cancer types. To that end, our group is currently evaluating the expression of all twenty-five selenoproteins, at the mRNA and protein levels, in a group of selected breast cancer cell lines and tissues (manuscripts in preparation).

Supplementary Materials: Supplementary materials can be found at http://www.mdpi.com/1422-0067/21/18/6694/s1. Figure S1. Evaluation of topological properties of network. (A) node degree distribution, (B) average clustering coefficient, (C) stress centrality, (D) closeness cenytrality and (E) betweenness centrality measure. Figure S2. Kaplan-Mayer curves showing the overall survival (expressed in percentage) in solid cancer patients, using PROGgeneV2 online tool, in the case of high and low expression of HUB nodes reported by red and green curves, respectively. Table S1. List of primer sequences

Author Contributions: S.C. and A.B. conceived the study; A.P. and S.C. performed the bioinformatics analysis; F.C. and A.S. performed RT-qPCR analysis; and all authors contributed to writing the paper. All authors have read and agreed to the published version of the manuscript.

Funding: This research received no external funding.

Acknowledgments: A.P. was supported by a FIRC-AIRC fellowship for Italy. This study was supported by the Italian Ministry of Health to IRCCS Istituto Nazionale Tumori "Fondazione G. Pascale", Napoli (Italia) (Ricerca Corrente). We are grateful to Alessandra Trocino, Librarian at IRCCS "G. Pascale" of Naples, Italy, for the bibliographic assistance.

Conflicts of Interest: The authors declare no conflict of interest.

References

1. Diamond, A.M. The subcellular location of selenoproteins and the impact on their function. *Nutrients* **2015**, *7*, 3938–3948. [CrossRef]
2. Reszka, E.; Jablonska, E.; Gromadzinska, J.; Wasowicz, W. Relevance of selenoprotein transcripts for selenium status in humans. *Genes Nutr.* **2012**, *7*, 127–137. [CrossRef] [PubMed]
3. Chen, G.; Chen, Q.; Zeng, F.; Zeng, L.; Yang, H.; Xiong, Y.; Zhou, C.; Liu, L.; Jiang, W.; Yang, N.; et al. The serum activity of thioredoxin reductases 1 (TrxR1) is correlated with the poor prognosis in EGFR wild-type and ALK negative non-small cell lung cancer. *Oncotarget* **2017**, *8*, 115270–115279. [CrossRef] [PubMed]

4.	Zoidis, E.; Seremelis, I.; Kontopoulos, N.; Danezis, G.P. Selenium-Dependent Antioxidant Enzymes: Actions and Properties of Selenoproteins. *Antioxidants* **2018**, *7*, 66. [CrossRef] [PubMed]

5.	Varlamova, E.G. Participation of selenoproteins localized in the ER in the processes occurring in this organelle and in the regulation of carcinogenesis-associated processes. *J. Trace Elem. Med. Biol.* **2018**, *48*, 172–180. [CrossRef] [PubMed]

6.	Castets, P.; Lescure, A.; Guicheney, P.; Allamand, V. Selenoprotein N in skeletal muscle: From diseases to function. *J. Mol. Med.* **2012**, *90*, 1095–1107. [CrossRef]

7.	Short, S.P.; Williams, C.S. Selenoproteins in tumorigenesis and cancer progression. *Adv. Cancer Res.* **2017**, *136*, 49–83.

8.	Cheng, W.H.; Prabhu, K.S. Special Issue of "Optimal Selenium Status and Selenoproteins in Health". *Biol. Trace Elem. Res.* **2019**, *192*, 1–2. [CrossRef]

9.	Lubiński, J.; Marciniak, W.; Muszynska, M.; Jaworowska, E.; Sulikowski, M.; Jakubowska, A.; Kaczmarek, K.; Sukiennicki, G.; Falco, M.; Baszuk, P.; et al. Serum selenium levels and the risk of progression of laryngeal cancer. *PLoS ONE* **2018**, *13*, e0184873. [CrossRef]

10.	Lubinski, J.; Marciniak, W.; Muszynska, M.; Huzarski, T.; Gronwald, J.; Cybulski, C.; Jakubowska, A.; Debniak, T.; Falco, M.; Kladny, J.; et al. Serum selenium levels predict survival after breast cancer. *Breast Cancer Res. Treat.* **2018**, *167*, 591–598. [CrossRef]

11.	Peters, K.M.; Carlson, B.A.; Gladyshev, V.N.; Tsuji, P.A. Selenoproteins in colon cancer. *Free Radic. Biol. Med.* **2018**, *127*, 14–25. [CrossRef] [PubMed]

12.	Diamond, A.M. Selenoproteins of the Human Prostate: Unusual Properties and Role in Cancer Etiology. *Biol. Trace Elem. Res.* **2019**, *192*, 51–59. [CrossRef] [PubMed]

13.	Jiang, H.; Shi, Q.Q.; Ge, L.Y.; Zhuang, Q.F.; Xue, D.; Xu, H.Y.; He, X.Z. Selenoprotein M stimulates the proliferative and metastatic capacities of renal cell carcinoma through activating the PI3K/AKT/mTOR pathway. *Cancer Med.* **2019**, *8*, 4836–4844. [CrossRef] [PubMed]

14.	Marciel, M.P.; Hoffmann, P.R. Molecular Mechanisms by Which Selenoprotein K Regulates Immunity and Cancer. *Biol. Trace Elem. Res.* **2019**, *192*, 60–68. [CrossRef] [PubMed]

15.	Men, L.; Sun, J.; Ren, D. Deficiency of VCP-Interacting Membrane Selenoprotein (VIMP) Leads to G1 Cell Cycle Arrest and Cell Death in MIN6 Insulinoma Cells. *Cell. Physiol. Biochem.* **2018**, *51*, 2185–2197. [CrossRef] [PubMed]

16.	Guerriero, E.; Accardo, M.; Capone, F.; Colonna, G.; Castello, G.; Costantini, S. Assessment of the Selenoprotein M (SELM) over-expression on human hepatocellular carcinoma tissues by immunohistochemistry. *Eur. J. Histochem.* **2014**, *58*, 2433. [CrossRef] [PubMed]

17.	Guerriero, E.; Capone, F.; Accardo, M.; Sorice, A.; Costantini, M.; Colonna, G.; Castello, G.; Costantini, S. GPX4 and GPX7 over-expression in human hepatocellular carcinoma tissues. *Eur. J. Histochem.* **2015**, *59*, 2540. [CrossRef]

18.	Guariniello, S.; Di Bernardo, G.; Colonna, G.; Cammarota, M.; Castello, G.; Costantini, S. Evaluation of the selenotranscriptomeexpression in twohepatocellular carcinoma celllines. *Anal. Cell. Pathol.* **2015**, *2015*, 419561. [CrossRef]

19.	Potenza, N.; Castiello, F.; Panella, M.; Colonna, G.; Ciliberto, G.; Russo, A.; Costantini, S. Human MiR-544a Modulates SELK Expression in Hepatocarcinoma Cell Lines. *PLoS ONE* **2016**, *11*, e0156908. [CrossRef]

20.	Rusolo, F.; Capone, F.; Pasquale, R.; Angiolillo, A.; Colonna, G.; Castello, G.; Costantini, M.; Costantini, S. Comparison of the seleno-transcriptome expression between human non-cancerous mammary epithelial cells and two human breast cancer cell lines. *Oncol. Lett.* **2017**, *13*, 2411–2417. [CrossRef]

21.	Nunziata, C.; Polo, A.; Sorice, A.; Capone, F.; Accardo, M.; Guerriero, E.; Marino, F.Z.; Orditura, M.; Budillon, A.; Costantini, S. Structural analysis of human SEPHS2 protein, a selenocysteine machinery component, over-expressed in triple negative breast cancer. *Sci. Rep.* **2019**, *9*, 16131. [CrossRef] [PubMed]

22.	Sharma, A.; Costantini, S.; Colonna, G. The protein-protein interaction network of the human Sirtuin family. *Biochim. Biophys. Acta* **2013**, *1834*, 1998–2009. [CrossRef] [PubMed]

23.	Yu, T.; Acharya, A.; Mattheos, N.; Li, S.; Ziebolz, D.; Schmalz, G.; Haak, R.; Schmidt, J.; Sun, Y. Molecular mechanisms linking peri-implantitis and type 2 diabetes mellitus revealed by transcriptomic analysis. *Peer J.* **2019**, *7*, e7124. [CrossRef]

24. Covani, U.; Marconcini, S.; Derchi, G.; Barone, A.; Giacomelli, L. Relationship between human periodontitis and type 2 diabetes at a genomic level: A data-mining study. *J. Periodontol.* **2009**, *80*, 1265–1273. [CrossRef] [PubMed]

25. Giacomelli, L.; Nicolini, C. Gene expression of human T lymphocytes cell cycle: Experimental and bioinformatic analysis. *J. Cell. Biochem.* **2006**, *99*, 1326–1333. [CrossRef] [PubMed]

26. Jovanovic, V.; Giacomelli, L.; Sivozhelezov, V.; Degauque, N.; Lair, D.; Soulillou, J.-P.; Pechkova, E.; Nicolini, C.; Brouard, S. AKT1 leader gene and downstream targets are involved in a rat model of kidney allograft tolerance. *J. Cell. Biochem.* **2010**, *111*, 709–719. [CrossRef]

27. Barone, A.; Toti, P.; Giuca, M.R.; Derchi, G.; Covani, U. A gene network bioinformatics analysis for pemphigoid autoimmune blistering diseases. *Clin. Oral Investig.* **2015**, *19*, 1207–1222. [CrossRef]

28. Milone, M.R.; Pucci, B.; Bruzzese, F.; Carbone, C.; Piro, G.; Costantini, S.; Capone, F.; Leone, A.; Di Gennaro, E.; Caraglia, M.; et al. Acquired resistance to zoledronic acid and the parallel acquisition of an aggressive phenotype are mediated by p38-MAP kinase activation in prostate cancer cells. *Cell Death Dis.* **2013**, *4*, e641. [CrossRef]

29. Bizzarro, V.; Belvedere, R.; Milone, M.R.; Pucci, B.; Lombardi, R.; Bruzzese, F.; Popolo, A.; Parente, L.; Budillon, A.; Petrella, A. Annexin A1 is involved in the acquisition and maintenance of a stem cell-like/aggressive phenotype in prostate cancer cells with acquired resistance to zoledronic acid. *Oncotarget* **2015**, *6*, 25076. [CrossRef]

30. Ciardiello, C.; Leone, A.; Lanuti, P.; Roca, M.S.; Moccia, T.; Minciacchi, V.R.; Minopoli, M.; Gigantino, V.; De Cecio, R.; Rippa, M.; et al. Large oncosome soverexpressing integrin alpha-V promote prostate canceradhesion and invasion via AKT activation. *J. Exp. Clin. Cancer. Res.* **2019**, *38*, 317. [CrossRef]

31. Kalkat, M.; De Melo, J.; Hickman, K.A.; Lourenco, C.; Redel, C.; Resetca, D.; Tamachi, A.; Tu, W.B.; Penn, L.Z. MYC deregulation in primary human cancers. *Genes* **2017**, *8*, 151. [CrossRef] [PubMed]

32. Gabay, M.; Yulin, L.; Felsher, D.W. MYC Activation Is a Hallmark of Cancer Initiation and Maintenance. *Cold Spring Harb. Perspect. Med.* **2014**, *4*, a014241. [CrossRef] [PubMed]

33. Qu, J.; Zhao, X.; Wang, J.; Liu, X.; Yan, Y.; Liu, L.; Cai, H.; Qu, H.; Lu, N.; Sun, Y.; et al. MYC overexpression with its prognostic and clinicopathological significance in breast cancer. *Oncotarget* **2017**, *8*, 93998. [CrossRef] [PubMed]

34. Nikolaos, F.; Triggiani, E.; Teodossiu, G.; Renda, A. C-myc Overexpression in Preneoplastic Conditions and in Colorectal Cancer as Decisional Factor in Medical and Surgical Treatment. *Eur. J. Surg. Oncol.* **2011**, *37*, S4.

35. Attar, N.; Kurdistani, S.K. Exploitation of EP300 and CREBBP Lysine Acetyltransferases by Cancer. *Cold Spring Harb. Perspect. Med.* **2017**, *7*, a026534. [CrossRef]

36. Yan, G.; Eller, M.S.; Elm, C.; Larocca, C.A.; Ryu, B.; Panova, I.P.; Dancy, B.M.; Bowers, E.M.; Meyers, D.; Lareau, L.; et al. Selective inhibition of p300 HAT blocks cell cycle progression, induces cellular senescence, and inhibits the DNA damage response in melanoma cells. *J. Invest. Dermatol.* **2015**, *133*, 2444–2452. [CrossRef]

37. Kowalczyk, A.E.; Krazinski, B.E.; Godlewski, J.; Kiewisz, J.; Kwiatkowski, P.; Sliwinska-Jewsiewicka, A.; Kiezun, J.; Sulik, M.; Kmiec, Z. Expression of the EP300, TP53 and BAX genes in colorectal cancer: Correlations with clinicopathological parameters and survival. *Oncol. Rep.* **2017**, *38*, 201–210. [CrossRef]

38. Goel, R.K.; Lukong, E.K. Understanding the cellular roles of Fyn-related kinase (FRK): Implications in cancer biology. *Cancer Metastasis Rev.* **2016**, *35*, 179–199. [CrossRef]

39. Xie, Y.G.; Yu, Y.; Hou, L.K.; Wang, X.; Zhang, B.; Cao, X.C. FYN promotes breast cancer progression through epithelial-mesenchymal transition. *Oncol. Rep.* **2016**, *36*, 1000–1006. [CrossRef]

40. Wang, J.; Wu, H.F.; Shen, W.; Xu, D.Y.; Ruan, T.Y.; Tao, G.Q.; Lu, P.H. SRPK2 promotes the growth and migration of the colon cancer cells. *Gene* **2016**, *586*, 41–47. [CrossRef]

41. Lee, G.; Zheng, Y.; Cho, S.; Jang, C.; England, C.; Dempsey, J.M.; Yu, Y.; Liu, X.; He, L.; Cavaliere, P.M.; et al. Post-transcriptional Regulation of De Novo Lipogenesis by mTORC1-S6K1-SRPK2 Signaling. *Cell* **2017**, *171*, 1545–1558. [CrossRef] [PubMed]

42. Greuber, E.K.; Smith-Pearson, P.; Wang, J.; Pendergast, A.M. Role of ABL family kinases in cancer: From leukaemia to solid tumours. *Nat. Rev. Cancer* **2013**, *13*, 559. [CrossRef] [PubMed]

43. Yano, M.; Koumoto, Y.; Kanesaki, Y.; Wu, X.; Kido, H. 20S proteasome prevents aggregation of heat-denatured proteins without PA700 regulatory subcomplex like a molecular chaperone. *Biomacromolecules* **2004**, *5*, 1465. [CrossRef] [PubMed]

44. Bruzzoni-Giovanelli, H.; Gonzalez, J.R.; Sigaux, F.; Villoutreix, B.O.; Cayuela, J.M.; Guilhot, J.; Preudhomme, C.; Guilhot, F.; Poyet, J.L.; Rousselot, P. Genetic polymorphisms associated with increased risk of developing chronic myelogenous leukemia. *Oncotarget* **2015**, *6*, 36269. [CrossRef] [PubMed]

45. Tan, S.; Li, H.; Zhang, W.; Shao, Y.; Liu, Y.; Guan, H.; Wu, J.; Kang, Y.; Zhao, J.; Yu, Q.; et al. NUDT21 negatively regulates PSMB2 and CXXC5 by alternative polyadenylation and contributes to hepatocellular carcinoma suppression. *Oncogene* **2018**, *37*, 4887. [CrossRef] [PubMed]

46. Polo, A.; Crispo, A.; Cerino, P.; Falzone, L.; Candido, S.; Giudice, A.; De Petro, G.; Ciliberto, G.; Montella, M.; Budillon, A.; et al. Environment and bladdercancer: Molecularanalysis by interaction networks. *Oncotarget* **2017**, *8*, 65240. [CrossRef] [PubMed]

47. Goswami, C.P.; Nakshatri, H. PROGgeneV2: Enhancements on the existing database. *BMC Cancer* **2014**, *14*, 970. [CrossRef]

48. Rocca, C.; Pasqua, T.; Boukhzar, L.; Anouar, Y.; Angelone, T. Progress in the emerging role of selenoproteins in cardiovascular disease: Focus on endoplasmic reticulum-resident selenoproteins. *Cell Mol. Life Sci.* **2019**, *76*, 1–17. [CrossRef]

49. Rebsch, C.M.; Penna, F.J.; Copeland, P.R. Selenoprotein Expression Is Regulated at Multiple Levels in Prostate Cells. *Cell. Res.* **2006**, *16*, 940. [CrossRef]

50. Ekoue, D.N.; Ansong, E.; Liu, L.; Macias, V.; Deaton, R.; Lacher, C.; Picklo, M.; Nonn, L.; Gann, P.H.; Kajdacsy-Balla, A.; et al. Correlations of SELENOF and SELENOP Genotypes With Serum Selenium Levels and Prostate Cancer. *Prostate* **2018**, *78*, 279. [CrossRef]

51. Gonzalez-Moreno, O.; Boque, N.; Redrado, M.; Milagro, F.; Campion, J.; Endermann, T.; Takahashi, K.; Saito, Y.; Catena, R.; Schomburg, L.; et al. Selenoprotein-P Is Down-Regulated in Prostate Cancer, Which Results in Lack of Protection Against Oxidative Damage. *Prostate* **2011**, *71*, 824. [CrossRef] [PubMed]

52. Naiki, T.; Naiki-Ito, A.; Asamoto, M.; Kawai, N.; Tozawa, K.; Etani, T.; Sato, S.; Suzuki, S.; Shirai, T.; Kohri, K.; et al. GPX2 Overexpression Is Involved in Cell Proliferation and Prognosis of Castration-Resistant Prostate Cancer. *Carcinogenesis* **2014**, *35*, 1962. [CrossRef] [PubMed]

53. Shan, W.; Zhong, W.; Zhao, R.; Oberley, T.D. Thioredoxin 1 as a subcellular biomarker of redox imbalance in human prostate cancer progression. *Free Radic. Biol. Med.* **2010**, *49*, 2078. [CrossRef] [PubMed]

54. Samaranayake, G.J.; Troccoli, C.I.; Huynh, M.; Lyles, R.D.Z.; Kage, K.; Win, A.; Lakshmanan, V.; Kwon, D.; Ban, Y.; Chen, S.X.; et al. Thioredoxin-1 protects against androgen receptor-induced redox vulnerability in castration-resistant prostate cancer. *Nat. Commun.* **2017**, *8*, 1204. [CrossRef]

55. Cunningham, D.; You, Z. In Vitro and in Vivo Model Systems Used in Prostate Cancer Research. *J. Biol. Methods* **2015**, *2*, e17. [CrossRef]

56. Sramkoski, R.M.; Pretlow, T.G.; Giaconia, J.M.; Pretlow, T.P.; Schwartz, S.; Sy, M.S.; Marengo, S.R.; Rhim, J.S.; Zhang, D.; Jacobberger, J.W. A new human prostate carcinoma cell line, 22Rv1. *In Vitro Cell. Dev. Biol. Anim.* **1999**, *35*, 403. [CrossRef]

57. van Bokhoven, A.; Varella-Garcia, M.; Korch, C.; Johannes, W.U.; Smith, E.E.; Miller, H.L.; Nordeen, S.K.; Miller, G.J.; Lucia, M.S. Molecular characterization of human prostate carcinoma cell lines. *Prostate* **2003**, *57*, 205. [CrossRef]

58. Horibata, Y.; Elpeleg, O.; Eran, A.; Hirabayashi, Y.; Savitzki, D.; Tal, G.; Mandel, H.; Sugimoto, H. EPT1 (selenoprotein I) is critical for the neural development and maintenance of plasmalogen in humans. *J. Lipid. Res.* **2018**, *59*, 1015–1026. [CrossRef]

59. Cvetković, B.; Vučić, V.; Cvetković, Z.; Popović, T.; Glibetić, M. Systemic alterations in concentrations and distribution of plasma phospholipids in prostate cancer patients. *Med. Oncol.* **2012**, *29*, 809. [CrossRef]

60. Lee, J.H.; Jang, J.K.; Ko, K.Y.; Jin, Y.; Ham, M.; Kang, H.; Kim, I.Y. Degradation of selenoprotein S and selenoprotein K through PPARγ-mediated ubiquitination is required for adipocyte differentiation. *Cell Death Differ.* **2019**, *26*, 1007. [CrossRef]

61. Labunskyy, V.M.; Hatfield, D.L.; Gladyshev, V.N. Selenoproteins: Molecular pathways and physiological roles. *Physiol. Rev.* **2014**, *94*, 739. [CrossRef] [PubMed]

62. Lee, J.H.; Kwon, J.H.; Jeon, Y.H.; Ko, K.Y.; Lee, S.R.; Kim, I.Y. Pro178 and Pro183 of selenoprotein S are essential residues for interaction withp97(VCP) during endoplasmic reticulum-associated degradation. *J. Biol. Chem.* **2014**, *289*, 13758. [CrossRef] [PubMed]

63. Lee, J.H.; Park, K.J.; Jang, J.K.; Jeon, Y.H.; Ko, K.Y.; Kwon, J.H.; Lee, S.-R.; Kim, I.Y. Selenoprotein S-dependent Selenoprotein K binding top97(VCP) protein is essential for endoplasmic reticulum-associated degradation. *J. Biol. Chem.* **2015**, *290*, 29941. [CrossRef] [PubMed]

64. Liu, J.; Zhang, Z.; Rozovsky, S. Selenoprotein K form an intermolecular diselenide bond with unusually high redox potential. *FEBS Lett.* **2014**, *588*, 3311. [CrossRef]

65. Ben, S.B.; Peng, B.; Wang, G.C.; Li, C.; Gu, H.F.; Jiang, H.; Meng, X.L.; Lee, B.J.; Chen, C.L. Overexpression of SelenoproteinSelK in BGC-823 Cells Inhibits Cell Adhesion and Migration. *Biochemistry (Mosc)* **2015**, *80*, 1344–1353. [CrossRef] [PubMed]

66. Méplan, C.; Rohrmann, S.; Steinbrecher, A.; Schomburg, L.; Jansen, E.; Linseisen, J.; Hesketh, J. Polymorphisms in thioredoxinreductase and selenoprotein K genes and selenium status modulate risk of prostate cancer. *PLoS ONE* **2012**, *7*, e48709. [CrossRef]

67. Xu, C.H.; Xiao, L.M.; Zeng, E.M.; Chen, L.-K.; Zheng, S.-Y.; Li, D.-H.; Liu, Y. MicroRNA-181 inhibits the proliferation, drug sensitivity and invasion of human glioma cells by targeting Selenoprotein K (SELK). *Am. J. Transl. Res.* **2019**, *11*, 6632.

68. Sutherland, A.; Kim, D.H.; Relton, C.; Ahn, Y.O.; Hesketh, J. Polymorphisms in the selenoprotein S and 15-kDa selenoprotein genes are associated with altered susceptibility to colorectal cancer. *Genes Nutr.* **2010**, *5*, 215. [CrossRef]

69. Curran, J.E.; Jowett, J.B.; Elliott, K.S.; Gao, Y.; Gluschenko, K.; Wang, J.; AbelAzim, D.M.; Cai, G.; Mahaney, M.C.; Comuzzie, A.G.; et al. Genetic variation in selenoprotein S influences inflammatory response. *Nat Genet.* **2005**, *37*, 1234. [CrossRef]

70. Curcio, C.; Baqui, M.M.; Salvatore, D.; Rihn, B.H.; Mohr, S.; Harney, J.W.; Larsen, P.R.; Bianco, A.C. The human type 2 iodothyroninedeiodinase is a selenoprotein highly expressed in a mesothelioma cell line. *J. Biol. Chem.* **2001**, *276*, 30183. [CrossRef]

71. Hepburn, A.C.; Steele, R.E.; Veeratterapillay, R.; Wilson, L.; Kounatidou, E.E.; Barnard, A.; Berry, P.; Cassidy, J.R.; Moad, M.; El-Sherif, A.; et al. The induction of core pluripotency master regulators in cancers defines poor clinical outcomes and treatment resistance. *Oncogene* **2019**, *38*, 4412. [CrossRef] [PubMed]

72. Rebello, R.J.; Pearson, R.B.; Hannan, R.D.; Furic, L. Therapeutic Approaches Targeting MYC-Driven Prostate Cancer. *Genes* **2017**, *8*, 71. [CrossRef] [PubMed]

73. Drazic, A.; Miura, H.; Peschek, J.; Le, Y.; Bach, N.C.; Kriehuber, T.; Winter, J. Methionine oxidation activates a transcription factor in response to oxidative stress. *Proc. Natl. Acad. Sci. USA* **2013**, *110*, 9493. [CrossRef]

74. Lee, B.C.; Péterfi, Z.; Hoffmann, F.W.; Moore, R.E.; Kaya, A.; Avanesov, A.; Tarrago, L.; Zhou, Y.; Weerapana, E.; Fomenko, D.E.; et al. MsrB1 and MICALs regulate actin assembly and macrophage function via reversible stereoselective methionine oxidation. *Mol. Cell.* **2013**, *51*, 397. [CrossRef] [PubMed]

75. Lee, B.C.; Lee, S.G.; Choo, M.K.; Kim, J.H.; Lee, H.M.; Kim, S.; Fomenko, D.E.; Kim, H.-Y.; Park, J.M.; Gladyshev, V.N. Selenoprotein MsrB1 promotes anti-inflammatory cytokine gene expression in macrophages and controls immune response in vivo. *Sci. Rep.* **2017**, *7*, 5119. [CrossRef] [PubMed]

76. He, Q.; Li, H.; Meng, F.; Sun, X.; Feng, X.; Chen, J.; Li, L.; Liu, J. MethionineSulfoxideReductase B1 Regulates Hepatocellular Carcinoma Cell Proliferation and Invasion via the Mitogen-Activated Protein Kinase Pathway and Epithelial-Mesenchymal Transition. *Oxid. Med. Cell. Longev.* **2018**, *2018*, 5287971. [CrossRef]

77. De Luca, A.; Sacchetta, P.; Nieddu, M.; Di Ilio, C.; Favaloro, B. Important roles of multiple Sp1 binding sites and epigenetic modifications in the regulation of the methionine sulfoxidereductase B1 (MsrB1) promoter. *BMC Mol. Biol.* **2007**, *8*, 39. [CrossRef]

78. Tripathi, R.; Liu, Z.; Plattner, R. EnABLing Tumor Growth and Progression: Recent progress in unraveling the functions of ABL kinases in solid tumor cells. *Curr. Pharmacol. Rep.* **2018**, *4*, 367. [CrossRef]

79. Gruber, M.; Ferrone, L.; Puhr, M.; Santer, F.R.; Furlan, T.; Eder, I.E.; Sampson, N.; Schäfer, G.; Handle, F.; Culig, Z. p300 is up-regulated by docetaxel and is a target in chemoresistant prostate cancer. *Endocr. Relat. Cancer* **2020**, *27*, 187. [CrossRef]

80. Posadas, E.M.; Al-Ahmadie, H.; Robinson, V.L.; Jagadeeswaran, R.; Otto, K.; Kasza, K.E.; Tretiakov, M.; Siddiqui, J.; Pienta, K.J.; Stadler, W.M.; et al. FYN is overexpressed in human prostate cancer. *BJU Int.* **2009**, *103*, 171. [CrossRef]

81. Zhuo, Y.J.; Liu, Z.Z.; Wan, S.; Cai, Z.D.; Xie, J.J.; Cai, Z.D.; Song, S.D.; Wan, Y.P.; Hua, W.; Zhong, W.; et al. Enhanced expression of SRPK2 contributes to aggressive progression and metastasis in prostate cancer. *Biomed. Pharmacother.* **2018**, *102*, 531–538. [CrossRef] [PubMed]

82. Snider, J.; Kotlyar, M.; Saraon, P.; Yao, Z.; Jurisica, I.; Stagljar, I. Fundamentals of protein interaction network mapping. *Mol. Syst. Biol.* **2015**, *11*, 848. [CrossRef] [PubMed]

83. Sinisi, A.A.; Chieffi, P.; Pasquali, D.; Kisslinger, A.; Staibano, S.; Bellastella, A.; Tramontano, D. EPN: A novel epithelial cell line derived from human prostate tissue. *Cell. Dev. Biol. Anim.* **2002**, *38*, 165. [CrossRef]

84. Zhao, S.G.; Jackson, W.C.; Kothari, V.; Schipper, M.J.; Erho, N.; Evans, J.R.; Speers, C.; Hamstra, D.A.; Niknafs, Y.S.; Nguyen, P.L.; et al. High-throughput transcriptomic analysis nominates proteasomal genes as age-specific biomarkers and therapeutic targets in prostate cancer. *Prostate Cancer Prostatic Dis.* **2015**, *18*, 229. [CrossRef] [PubMed]

International Journal of
Molecular Sciences

Review

Biomarkers in Triple-Negative Breast Cancer: State-of-the-Art and Future Perspectives

Stefania Cocco, Michela Piezzo, Alessandra Calabrese, Daniela Cianniello, Roberta Caputo, Vincenzo Di Lauro, Giuseppina Fusco, Germira di Gioia, Marina Licenziato and Michelino de Laurentiis *

Istituto Nazionale Tumori IRCCS "Fondazione G. Pascale", Via Mariano Semmola, 53, 80131 Napoli NA, Italy; s.cocco@breastunit.org (S.C.); m.piezzo@breastunit.org (M.P.); a.calabrese@istitutotumori.na.it (A.C.); d.cianniello@breastunit.org (D.C.); r.caputo@breastunit.org (R.C.); dilaurovincenzo87@gmail.com (V.D.L.); g.fusco@breastunit.org (G.F.); germiradigioia@gmail.com (G.d.G.); m.licenziato@istitutotumori.na.it (M.L.)
* Correspondence: delauren@breastunit.org; Tel.: +39-081-5903-535

Received: 31 May 2020; Accepted: 25 June 2020; Published: 27 June 2020

Abstract: Triple-negative breast cancer (TNBC) is a heterogeneous group of tumors characterized by aggressive behavior, high risk of distant recurrence, and poor survival. Chemotherapy is still the main therapeutic approach for this subgroup of patients, therefore, progress in the treatment of TNBC remains an important challenge. Data derived from molecular technologies have identified TNBCs with different gene expression and mutation profiles that may help developing targeted therapies. So far, however, only a few of these have shown to improve the prognosis and outcomes of TNBC patients. Robust predictive biomarkers to accelerate clinical progress are needed. Herein, we review prognostic and predictive biomarkers in TNBC, discuss the current evidence supporting their use, and look at the future of this research field.

Keywords: TNBC; *BRCA1/2*; HRR; PDL1; TILs; *PI3KCA*; *PTEN*; CTCs; CSC

1. Introduction

Triple negative breast cancer (TNBC) is a subtype of breast cancer lacking expression of estrogen receptor (ER), progesterone receptor (PgR) and human epidermal growth factor receptor 2 (HER2) [1]. TNBC accounts for approximately 10–15% of all breast cancers and is characterized by aggressive behavior, with trend to early relapse, metastatic spread, and poor survival [2]. Tumor heterogeneity of TNBC has been addressed as reason for different clinical outcomes and response to therapies. Lehmann et al. proposed a division of TNBCs into seven molecular subtypes: immunomodulatory (IM), mesenchymal (M), mesenchymal stem-like (MSL), luminal androgen receptor (LAR), unstable (UNS) subtype, and two basal-like subtypes (BL1 and BL2) [3]. Then, a subclassification refinement was performed, to define only four groups BL1 (immune-activated), BL2 (immune-suppressed), M (including most of the MSL), and LAR [4]. These classifications can, in theory, be used as prognostic and predictive tool for better patient selection and personalized treatments. For instance, LAR tumors, characterized by the expression of the androgen receptor (AR), are a subtype of TNBC that shares with Luminal (ER+) tumors some biological and clinical features [5–7] and are potentially sensitive to endocrine manipulation with AR antagonists.

However, sound clinical applications of this molecular classification are yet to appear [8].

In the last years, a great effort has been spent to identify new biomarkers and relative therapies, but only few of these have proven useful in clinical trials. Beside poly ADP-ribose polymerase (PARP) inhibitors, that have successfully been incorporated in the clinical practice in the BRCA1/2 subgroup of patients [9,10], and checkpoint inhibitor Atezolizumab, that was recently approved as front-line therapy in the metastatic setting [11], traditional chemotherapy without biomarker guide still remains the main therapeutic options for a large part of TNBC patients [12]. In this context, the implementation of more refined "omics" assays, along with appropriately designed clinical trials, may lead to the identification of new biomarkers to select new molecularly targeted therapies in TNBC. Several papers have provided a critical overview of numerous biomarkers evaluated in the past years in TNBC. In this review, we discuss the biomarkers that contributed to the development of new approved therapies in TNBC. We also review the new biomarkers that are showing promising results in ongoing clinical trials.

2. BRCA1/2 and Other Genes Involved in DNA Repair

The relationship between the tumor-suppressive genes BReast CAncer type 1 and Type 2 (*BRCA1/2*) and hereditary breast and ovarian cancer syndrome (HBOC) revolutionized clinical cancer genetics. First, the identification of a germline *BRCA1/2* mutations impacted cancer screening and prevention practices in this subgroup of patients and their relatives. Later, the knowledge of the pathological mechanisms of these mutations, led to the development of new therapeutic approaches, such as poly (ADP-ribose) polymerase (PARP) inhibitors, selectively directed on *BRCA 1* or *BRCA 2* deficient cells [13].

BRCA1 and *BRCA2* are autosomal dominant and tumor suppressor genes involved in the preservation of genome integrity. Both genes play a crucial role in homologous recombination repair (HRR) of DNA. The *BRCA1* gene on chromosome 17q21 has a broader role than *BRCA2* in responding to DNA damage; it controls the signal transduction pathway involved in HHR, including recognition of genomic damage, checkpoint activation, recruitment of DNA repairing proteins, and decision of whether DNA double strand breaks (DSBs) needs to be resected; in addition, it is also involved in chromatin remodeling and transcription control [14,15]. The *BRCA2* gene on chromosome 13 plays the key role of recruiting the DNA recombinase RAD51 and localizing it to damaged DNA [16]. HRR is a conservative, error-free, mechanism of DNA repair due to its ability to restore the original DNA sequence. A small percentage of people (about one in 400, or 0.25% of the population) carry mutated *BRCA1* or *BRCA2* genes. However, compared to other subtypes of breast cancers, women with TNBC have a higher prevalence of germline BRCA mutations (gBRCAm), about 11–31% [17]. In addition to the well-known germline mutations, a smaller proportion of somatic mutations in *BRCA1/2* genes (sBRCAm) were also found in primary ovarian and breast carcinomas [18].

When *BRCA1/2* genes are defective, DNA damage is repaired by non-conservative mechanisms of DNA repair, such as non-homologous end joining (NHEJ), in order to maintain cell viability. This process of repairing DSBs is simpler than HHR and consists in joining the two broken DNA ends without the homologous DNA sequence to guide the repair: it is, therefore, prone to joining errors with mutation of the original sequence. In *BRCA1/2*-deficient cells, DNA DSBs repair is dependent on PARP1 protein [19,20]. PARP is an abundant, constitutively expressed nuclear enzyme that facilitates DNA repair, cellular proliferation, and signaling to other critical cell-cycle proteins and oncogenes. At sites of DNA damage, PARP activates intracellular signaling pathways that modulate DNA repair and cell survival [21]. Therefore, inhibition of PARP1 in *BRCA1/2*-deficient cells can lead to severe, highly selective toxicity in these cells [22]. This process has been called "synthetic lethality," to highlight the interaction that occurs between the two genes when the perturbation of either gene alone is viable but the simultaneous perturbation of both genes results in the loss of viability [23].

In the past years, PARP inhibitors (PARPi) have been extensively studied as targeted therapy for gBRCAm ovarian and breast cancer patients. In 2014, the US Food and Drug Administration (FDA) approved the first PARPi, Olaparib, as monotherapy for patients with deleterious or suspected deleterious gBRCAm advanced ovarian cancer [24]. Later, other PARPi, such as nirapararib and rucaparib were approved [25–28]. On January 2018, based on data from OlympiAD trial (NCT02000622), FDA granted regular approval to Olpararib for the treatment of patients with deleterious or suspected deleterious gBRCAm, HER2-negative metastatic breast cancer previously treated with chemotherapy either in the neoadjuvant, adjuvant, or metastatic setting [9]. FDA also approved the BRACAnalysis CDx® test (Myriad Genetic Laboratories, Inc., Salt Lake City, UT, USA), whose accuracy was established based on a retrospective/prospective analysis of the OlympiAD trial population. In the same year, FDA approved talazoparib for patients with deleterious or suspected deleterious gBRCAm, HER2 negative locally advanced or metastatic breast cancer, based on the phase III EMBRACA study results [10]. Based on promising results in the metastatic setting, PARPi are under investigation also in the early diseases. In particular, talazoparib is considered to be the most powerful PARPi candidate for single-agent treatment in neoadjuvant setting [29]. Promising results derived from a pilot trial of the MD Anderson Cancer Center (NCT03499353), in which gBRCA1/2m, HER2-negative, stage I-III BC patients, received neoadjuvant Talazoparib as single-agent during 4 to 6 months, without any chemotherapy, in order to evaluate the pathological complete response (pCR) rate and tolerance. The primary end point was residual cancer burden (RCB). Of 20 patients enrolled, 19 completed 6 months of treatment and 10 of them had pCR (RCB-0: 53%) while two additional patients had RCB-I. The rate of RCB 0-I was 63% overall, 57% in TNBC, and 80% in HR+, 53% in gBRCA1m, and 100% in gBRCA2m. Toxicities were managed by dose reduction and transfusions [30].

Other PARPis, such as Niraparib, Rucaparib, and Veliparib, are still under clinical evaluation in breast cancer both as monotherapy and in different combinations [31–33].

Several pre-clinical studies have shown that PARPis are able to inhibit cell growth and promote the death of breast cancer cells that are wild type for *BRCA1/2* [34]. A recent study assessed the efficacy of 13 different PARPis in the treatment of 12 different breast cancer cell lines that are either wild type or mutated for *BRCA1/2*. The TNBC cell lines MDA-MB-436 (*BRCA1*-deficient), MDA-MB231, and MDA-MB-468 resulted sensitive to Talazoparib, suggesting that the benefit of Talazoparib might extend to TNBC without *BRCA1/2* mutations [35]. This suggests that we should identify additional biomarkers for PARPis [36].

The term "BRCAness" has been used to describe a dysfunction in the BRCA-related DNA repair mechanism that is not due to mutations of the *BRCA1/2* genes.

Deficiencies in a number of tumor-suppressor genes involved in HRR, such as ATM and ATR, may share the same therapeutic vulnerabilities with BRCAm tumors and confer sensitivity to PARP inhibition. Therefore, tumors with mutations in other HRR genes may also respond to a PARP inhibitor treatment [37].

On 2015, Domagala et al. performed genetic testing for 36 common germline mutations in genes engaged in HRR, (i.e., *BRCA1, BRCA2, CHEK2, NBN, ATM, PALB2, BARD1,* and *RAD51D*), in 202 patients, including TNBC and hereditary non-TNBC patients. As a result, 22.2% of 158 patients in TNBC group carried mutations in genes involved in DNA repair by HR [38]. Confirming data have been reported by other group [39] and it has been demonstrated that homologous recombination deficiency (HRD) can occur in sporadic cancer through genetic and epigenetic inactivation of other components such as *PALB2, BARD1, BRIP1, RAD51, RAD51C, RAD51D, ATM, FAAP20, CHECK2, FAM1, FANCE, FANCM, POLQ* [5,34]. These findings confirm the hypothesis that HRD TNBC, shares similar characteristics with gBRCAm TNBC, identifying new possible biomarkers of response to PARPi [40]. Instead of quantifying the effect of genetic variation in the HR pathway, researchers have developed methods to score the competency of the HR pathway. Three scoring systems have emerged: HRD-loss of heterozygosity (HRD-LOD), HRD-large-scale transition (HRD-LST), and HRD-telomeric allelic imbalance (HRD-TAI) [41–43].

Based on these findings, some clinical trials are now testing the use of PARPis in patients with BRCAm or HRD to maximize the number of individuals who may benefit from PARP inhibition [44–46]. The Phase II study Violette (NCT03330847) aims to assess the efficacy and safety of olaparib monotherapy versus two combinations (olaparib in combination with AZD6738 and olaparib in combination AZD1775), in TNBC patients prospectively stratified by presence/absence of qualifying tumor mutations in 15 genes involved in HRR pathway (*BRCA1, BRCA2, ATM, BARD1, BRIP1, CDK12, CHEK1, CHEK2, FANCL, PALB2, PPP2R2A, RAD51B, RAD51C, RAD51D,* or *RAD54L*). AZD6738 is an ATP competitive, orally bioavailable inhibitor of the Serine/Threonine protein kinase Ataxia Telangiectasia and Rad3 related (ATR), while AZD1775 is a small molecule WEE1 inhibitor used in combination with DNA-damaging agents in several trials [47]. Both, WEE1 and ATR, are kinases involved in the regulation of cell-cycle and DNA repair. Preclinical in vitro and in vivo experiments demonstrate that AZD1775 has synergistic cytotoxic effects when administered in combination with PARPis, while cancer cells with defective DNA repair mechanisms or cell cycle checkpoints may be particularly sensitive to ATR inhibition; this has led to the development of early-phase trials on three ATR inhibitors (M6620, AZD6738, and BAY1895344) [48,49].

Somatic BRCA mutations (sBRCAm) were supposed to report a phenotype similar to tumors from patients with germline mutations and response to PARPi [50]. A recent metanalysis, including 236 patients, with sBRCAm and 1204 patients with gBRCAm treated with PARPi for different cancers, indicates similar response rates of PARPi therapy [51]. In ovarian cancer, where the rate of somatic mutation is higher compared to the other cancers, PARPi have shown efficacy in patients carrying mutations of *BRCA1/2*, either germline or somatic, but also in wild-type *BRCA1/2* [52]. In breast cancer, clinical trials evaluating the predictive role to PARPi of sBRCAm are ongoing (NCT03990896, NCT03286842, NCT04053322, NCT03344965, NCT03920488, NCT01434420, NCT03078036). To the last ASCO meeting were presented the results from TBCRC 048 trial (NCT03344965), a phase II study of olaparib monotherapy in 54 metastatic breast cancer patients, of which 40 patients ER+ HER2−, 3 HER2+, and 10 TNBC, divided in 2 cohorts based on germline mutations in non-*BRCA1/2* DDR-pathway genes (cohort 1) and on somatic mutations in these genes or *BRCA1/2* (cohort 2). Results showed an ORR of 29.6% in Cohort 1 and 38.5% in Cohort 2, with *sBRCA1/2* or *gPALB2* mutations predictor of response in this last one [53].

Since *BRCA1/2*-deficient tumors have defects of DNA repair, the use of agents causing DNA damage have been extensively investigated as candidate therapy to promote mechanisms of cell cycle arrest and apoptosis in these tumors. Platinum drugs, causes platination of genomic and mitochondrial DNA, forming intrastrand crosslink adducts which cause double-strand breaks that culminate in the activation of apoptosis, especially when DNA lesions cannot be repaired [54]. This suggests that BRCA mutational status is a promising biomarkers for platinum-based chemotherapy. In metastatic setting, the 4th ESMO guidelines recommend anthracycline-taxane chemotherapy as first line for treatment of advanced TNBC, while carboplatin may be considered for BRCA positive TNBC as second line treatment [55]. Regarding sensibility to platinum agents of BRCAness phenotype in this setting, the phase II TBCR009 trial has shown a correlation between high HRD scores and their predictive response to platinum-based chemotherapy, beyond *BRCA1/2* mutations, in advanced first and second line TNBC patients. In particular, ORR was 54.5% in patients with germline *BRCA1/2* mutations, while, in patients without *BRCA1/2* mutations, HRD-LOH/HRD-LST scores discriminated responding and nonresponding tumors (12.68 and 5.11, respectively). Five of the six long-term responders alive at a median of 4.5 years lacked germline *BRCA1/2* mutations, and two of them had increased tumor HRD-LOH/HRD-LST scores [56]. At the last ASCO 2020 meeting, were presented results of SWOGS1416, a phase II randomized trial of cisplatin +/− veliparib in metastatic TNBC and/or germline BRCA-associated breast cancer (NCT02595905). In this study 248 TNBC patients were classified into the three groups (1) 37 gBRCA+ (2) 101 BRCA-like (3) 110 non-BRCA-like, based on results of central gBRCA testing and of multi-pronged biomarker panel including myChoice HRD score, somatic *BRCA1/2* mutations, *BRCA1* methylation, and non-*BRCA1/2* HR germline mutations. Results showed that addition of Veliparib to cisplatin significantly improved PFS and numerically improved OS and ORR, of BRCA-like sub-group of patients, that included 76% of HRD ≥ 42, 17% of *BRCA1* promoter methylation, and 7% of HRR genes mutations [57].

In early settings, several neo-adjuvant clinical trials have evaluated the impact of adding platinum to standard chemotherapy, however, its use remains controversial and it is not routinely recommended in unselected TNBC or BRCA mutations carriers [58]. The neoadjuvant phase 2 trial of cisplatin in TNBC by Silver et al. showed a pCR rate of 22% among all TNBC patients (n = 28). Between these, germline *BRCA1*, low *BRCA1* mRNA expression or *BRCA1* promoter methylation patients achieved pCR [59]. The GeparSixto trial, which included stage II or III TNBC and HER2+ patients, demonstrated significant improvement in pCR with carboplatin added to an anthracycline/taxane-based neoadjuvant chemotherapy in TNBC patients [60]. This study tried also to prove a possible correlation between BRCAness phenotype and response to platinum. The exploratory endpoints of the trial investigated (1) whether the HRD assay predicts specifically for carboplatin response or independently of treatment and (2) whether there is an association of HRD with long-term outcome in patients with BRCA mutations detected in tumor tissue (tmBRCA) and in patients with high HRD without tmBRCA. Results showed that the HR deficiency (defined as HRD score 42 and/or presence of tmBRCA) and HRD score in non-tmBRCA were predictors of response, however, the HRD assay failed to identify a subset of patients most likely to derive benefit from the addition of carboplatin [61]. In this context, the PrECOG 0105 phase II study of gemcitabine and carboplatin plus iniparib as neoadjuvant therapy for TNBC and *BRCA1/2* mutation-associated breast cancer, revealed that combination of gemcitabine, carboplatin, and iniparib is active in the treatment of early-stage triple-negative and *BRCA1/2* mutation-associated breast cancer. In particular, responder patients with sporadic TNBC, lacking *BRCA1/2* mutation, had an elevated HRD-LOH score, suggesting that HRD assay could predict to platinum response [62]. In a recent metanalysis including seven studies, with a total of 808 TNBC patients, among which 159 BRCA mutated, was reported that the addition of platinum to chemotherapy regimens in the neoadjuvant setting increases pCR rate in BRCA mutated as compared to wild-type TNBC patients, however, this trend did not achieve statistical significance [63]. Recently, through a whole genomic sequencing technique, researches developed a new weighted model called HRDdetect with a sensibility of 98.7% to detect *BRCA1/2* deficient samples [64]. Elevated HRDetect was significantly associated with clinical improvement on platinum-based therapy in advanced breast cancer [65].

Preclinical data support a potential synergism between PARPi and platinum-based chemotherapy [66,67]. The addition of low-dose veliparib (VELI) to carboplatin–paclitaxel versus placebo was tested in HER2-negative, gBRCA1/2m, advanced BC (BROCADE 3 trial). Veliparib increased PFS while no impact on OS was reported [68]. In the neoadjuvant setting, Veliparib was assessed in the carboplatin-VELI arm of the randomized phase II ISP-Y 2 trial. In this trial, the carboplatin-VELI arm enrolled 72 patients with HER2-BC, who received Veliparib plus carboplatin during the paclitaxel sequence (VELI-CARBO) followed by doxorubicin and cyclophosphamide (AC). The benefit of VELI-CARBO seemed to be restricted to the TNBC patients (51% pCR for VELI-CARBO vs. 26% pCR in the control arm in TNBC) [69]. The randomized phase III trial (BrighTNess) was subsequently conducted in 634 TNBC patients receiving neoadjuvant chemotherapy evaluating VELI-CARBO vs. placebo-CARBO or double placebo, in combination with paclitaxel followed by four cycles of AC. The pCR rate was higher in the CARBO-containing regimen (53% in VELI-CARBO and 58% in the placebo-CARBO vs. 31% in the control arm). Within the patients with gBRCA1/2m, no significant differences were achieved, the pCR rate was 57% in the VELI-CARBO/paclitaxel arm, 50% in the placebo-CARBO/paclitaxel arm, and 41% in the control paclitaxel arm. In summary, the trial did not detect statistically significant differences between these subgroups [70].

Research is also focusing on biomarkers of resistance to PARPis. Acquired or innate resistance to single-agent PARPis has been frequently observed in both preclinical and clinical studies [71–74]. The main mechanisms described are the reversion of BRCA and HRR gene mutations to wild-type, the demethylation of promoter of HR genes, the mitigation of replication stress, the mutations in PARP itself and/or drug efflux pumps [75–77]. In preclinical patient-derived BRCAm-xenograft (PDX) models, the detection of RAD51 foci, a surrogate biomarker of HRR functionality, correlated with resistance to PARPis regardless of the underlying mechanism restoring HRR function [78–82]. By using in vitro and in vivo models of intrinsic resistance to PARPis, Yu-Yi Chu et al. described the role of proteins like RTK, c-MET in PARP inhibition resistance [83,84].

The identification of biomarkers of resistance can drive the research of pharmacological strategies to delay or prevent the development of the drug resistant phenotype. Emerging data suggest that olaparib-resistant cancer models can be re-sensitized to olaparib when combined with AZD1775 or AZD6738 [85–87], leading to early phase clinical trials combining ATR inhibitors and PARPis in different cancers (NCT02723864, NCT03462342, NCT03682289, NCT02576444).

3. Biomarkers of Immunotherapy in TNBC

It is well-known that the immune system and cancer have a complex interplay: it is a multi-step process, named cancer immunoediting, largely mediated by CD8+ cytotoxic T lymphocytes, and in which both immune-stimulatory and inhibitory factors are involved. During the first phase, called elimination phase, the innate and adaptive immune system recognize and reject tumor cells, then the surviving tumor subclones can progress into a state of dormancy, the equilibrium phase, in which tumor growth is limited and tumor cells are gradually selected through upregulation of pro-survival pathways, changes in expression of molecules involved in immune suppression or angiogenesis. These immune-edited tumor cells can then enter into the escape phase, in which the tumor growth is uncontrolled [88].

Breast tumors have been historically considered non-immunogenic diseases, with a relatively low mutation rate. However, among BC subtypes, TNBC is characterized by high mutation rate and greater tumor-infiltrating lymphocytes (TILs). TILs are present both intratumorally and in adjacent stromal tissues and are composed mainly of cytotoxic CD8+ lymphocytes, and, to a lesser extent, CD4+ T-helper cells, T-regulatory (Treg) cells, macrophages, mast cells, and plasma-cells.

The presence of intra-tumoral and stromal TILs has predictive and prognostic role; in TNBC increased TILs at diagnosis have been associated with pathologic complete responses with neoadjuvant chemotherapy and improved survival after adjuvant chemotherapy [89–91]. The association between high number of stromal TILs and more favorable survival outcomes, in terms of overall survival (OS) and disease-free survival (DFS) highlighted the prognostic value of immune antitumoral activity, while the association between high TILs and response to chemotherapy established the predictive value of TILs as marker of response to chemotherapy. This also suggests that the effect of chemotherapy may be partially mediated by the immune system, making the investigation of immunotherapy in TNBC particularly interesting [92–94].

Programmed cell death-1 (PD-1) is an immune checkpoint able to inhibit both adaptive and innate immune response, and is expressed on the surface of immune cells, such as T cells, B cells, natural killer (NK), macrophages, dendritic cells (DCs), and monocytes [95]. PD-1 controls induction of tolerance to antigens and termination of immune response, playing a key role, under physiological condition, in maintaining the immune tolerance and limiting the autoimmunity. In the tumor microenvironment PD-1 is involved in the development of tumor immuno-tolerance [96,97]. PD-1 ligand, programmed cell death-ligand 1 (PD-L1), is a transmembrane protein expressed both on tumor cells and immune cells (DCs, B cells, T cells, macrophages), and it represents an "adaptive immune mechanism" that cancer cells may use to escape anti-tumor immunity.

PD-1/PD-L1 ligation acts as pro-tumorigenic pathway, leading to deactivation of T-cell function and resulting in the escape from immune surveillance [98–102]. PD-1/PD-L1 expression can be regulated by various signals in cancer cells, such as (1) activation of PI3K/AKT pathway that promote the expression of PD-L1 through increased extrinsic signaling and by downregulation of PTEN [103]; (2) MAPK signaling pathway, involved in the conversion of extracellular signals into intracellular responses and associated to PD-1/PD-L1 axis [104,105]; (3) JAK-STAT signaling pathway that provides a key mechanism for extracellular signals to control gene expression, including the expression of PD-L1 [106]; (4) abnormal WNT signaling pathway, able to interfere with cancer immuno-monitoring and to promote immune escape by a crosstalk mechanism between WNT activity and PD-L1 expression [107,108]; (5) NF-κB signaling pathway that mediates INF-γ-induced PD-L1 expression [109,110]; (6) Hedgehog signaling pathway that promotes the expression of PD-1/PD-L1 axis and which inhibition may induce lymphocyte antitumor activity [111].

PD-L1 is commonly expressed in 20% of TNBC and has been related to distinctive characteristics of BC, such as younger age, large tumor size, high grade, high proliferation, ER-negative status, and HER2-positive status. PD-L1 is expressed on about 10% of tumor cells (TC), while its expression on tumor infiltrating immune cells (IC) is higher (40–65%). The predictive role of PD-L1 positivity on IC has been validated in several clinical trials. The expression of both PD-1 and PD-L1 is associated with a good outcome and it is correlated with better overall survival and higher sensitivity to chemotherapy, confirming that the cytotoxic effect of chemotherapy is partially mediated by the immune response against tumor [112–116]. The expression of PD-L1 on tumor infiltrating immune cells (IC) may also play a role as biomarker as it has been shown in several clinical trials [117–119].

The introduction of immune checkpoint inhibitors (PD-1 inhibitors and PD-L1 inhibitors) as strategy to wake up the immune cells and reduce the tumor growth, is playing a critical role in improving treatment of TNBC. Atezolizumab (anti-PD-L1) is the first in class receiving the FDA accelerated approval on March 2019, based on results from IMpassion 130 study that showed a significant OS improvement (25 months vs. 18 months) in patients with IC PD-L1 expression (PD-L1+ IC ≥ 1%), treated with Atezolizumab plus nab-paclitaxel versus nab-placlitaxel alone as first line therapy for metastatic TNBC [11,120]. FDA also approved the VENTANA PD-L1 (SP142) assay as a companion assay to determine PD-L1 expression on IC. The analytical power of the VENTANA assay was evaluated and compared with other two immunohistochemistry assays (22C3 and SP263) in a post-hoc exploratory analysis of IMpassion 130 study. Overall, the VENTANA assay was predictive of atezolizumab efficacy when performed either on the primary or on the metastatic tumor specimen. With the cutoff of ≥1% of PDL+ IC, it identified a smaller population of pts as compared with 22C3 and SP263 assays, but with a higher predictive performance [121].

In the same setting, results from Keynote-355 trial, evaluating the combination of Pembrolizumab (anti-PD-1) plus chemotherapy as first line treatment of metastatic TNBC, have lately been presented at ASCO meeting. This study confirms the importance of evaluating PDL1 expression as biomarker. PDL1 has here been evaluated with the 22C3 (DAKO PharmaDx) assay using the combined positive score (CPS) defined as which is the number of PD-L1 staining cells (tumor cells, lymphocytes, macrophages) divided by the total number of viable tumor cells, multiplied by 100. Indeed, pembrolizumab has shown to improve progression-free survival (PFS) only in patients with a CPS score ≥10 [122]. Of note, however, that the predictive value of the same assay had not previously confirmed in the Keynote-086 study, with Pembrolizumab used as monotherapy: in the cohort A of the study (enrolling previously treated metastatic TNBC patients), PFS was similar irrespective of PD-L1 expression status [123].

Finally, in neo-adjuvant setting, conflicting results from Keynote-522 and NeoTRIPaPDL1 studies, investigating the combination of standard chemotherapy with Pembrolizumab and Atezolizumab respectively, showed that the role of PD-L1 expression as predictive marker of response to immune checkpoint inhibitors is still controversial [124,125]. Particularly, matters of uncertainty are: What is the best assay; what is the best cutoff to define PDL1-positivity; what drugs (pembrolizumab vs. atezolizumab vs. others) these apply to; what disease setting (early vs. metastatic) these apply to.

Evaluating other predictive biomarkers, in addition to PD-L1 testing, may help to select patients who could benefit from ICI.

Microsatellite instability (MSI) is a hypermutable phenotype generally deriving from a deficit in the DNA mismatch repair mechanism (deficient mismatch repair; dMMR). MSI is evaluated by identifying mutations involving microsatellites located throughout much of the genome as short DNA sequences repeated in tandem. High levels of microsatellite instability (MSI-H), corresponding to dMMR, were found across several cancers, such as endometrial and gastrointestinal cancers (20%–30%). MSI-H is correlated to a high neoantigen burden and, therefore, to a high immunogenic potential and sensitivity to immuno-checkpoint inhibitors, irrespective of the tumor histologic type. MSI-H/dMMR was the first biomarker to grant a "site-agnostic" FDA approval to an anticancer drug: on October 2016, based on the results of the Keynote 158 trial, Pembrolizumab was approved for use as monotherapy on solid tumors with MSI-H or dMMR and without satisfactory therapeutic alternatives [126]. While the frequency of MSI-H/dMMR in BC is very low (0%–1.5%) and its use as prognostic or predictive biomarker is still under investigation [127–129], the FDA approval allows using Pembrolizumab for TNBC with MSI-H/dMMR.

Tumor mutational burden (TMB), calculated as the total number of mutations in a sample divided by the length of the genomic target region (mut/Mb), is a good marker of tumor antigenicity. Since the greater number of somatic mutations, it is probable that these mutations will yield to misfolded proteins (neoantigens) capable of being immunogenic and providing targets for T-cell response. In several tumors, such as lung cancer, melanoma, and colorectal cancers, TMB, easily evaluated by NGS techniques, represents a good predictive biomarker for ICI response, since high TMB is associated with high neoantigen burden, high T-cell infiltration and high response rate to ICI, regardless the PD-L1 status [130,131]. In BC, the predictive role of TMB is still controversial, recent data showed that overall 3.1%–5% of breast cancers are hypermutated, with high prevalence in TNBC and metastatic tumors; these tumors seem to be more likely sensitive to PD-1 inhibitors after a preliminary analysis of clinical and genomic data, also if no differences in terms of survival has been shown in patients with high TMB treated with ICI [132–134]. In neoadjuvant setting, the GeparNuevo trial, investigating the addition of Durvalumab (anti-PD-L1) to anthracycline/taxane-based chemotherapy, showed a significant trend for increased pCR rates for PD-L1-positive patients in both PD-L1-TC in Durvalumab arm and PD-L1-IC in placebo arm. In addition, they performed a predefined analysis of 149 samples assessing the predictive value of TMB alone or in combination with an immune gene expression profile (GEP) for pCR. Results from multivariate analysis showed an odds ratio for pCR per mut/Mb of 2.06 (95% CI 1.33–3.20, P = 0.001) among all patients, 1.77 (95% CI 1.00–3.13, P = 0.049) in the Durvalumab treatment arm, and 2.82 (95% CI 1.21–6.54, P = 0.016) in the placebo treatment arm, confirming that further analyses of TMB in combination with other immune parameters are still necessary, as well as the choice of a standardized assay and a cut off value to define high mutational load [135,136]. Very recently, FDA approved FoundationOne® CDx test to identify patients with solid tumors with TMB score who may benefit from immunotherapy treatment with Pembrolizumab monotherapy. FoundationOne CDx is a next-generation sequencing-based in vitro diagnostic device for detection of substitutions, insertion and deletion alterations (indels), and copy number alterations (CNAs) in 324 genes and select gene rearrangements, as well as genomic signatures including microsatellite instability (MSI) and tumor mutational burden (TMB) using DNA isolated from formalin-fixed paraffin-embedded (FFPE) tumor tissue specimens. The accelerated approval was based on data from a prospectively planned retrospective analysis of 10 cohorts of patients with various previously treated unresectable or metastatic solid tumors, who were enrolled in KEYNOTE-158 (NCT02628067), a multicenter, non-randomized, open-label trial evaluating Pembrolizumab (200 mg every three weeks). TMB status was assessed using the FoundationOne CDx assay and TMB-High (TMB-H) was defined as TMB ≥10 mut/Mb. The results showed that patients with TMB-H in solid tumors who were treated with Pembrolizumab had a higher overall response rate (29%) compared to patients with TMB [137].

Despite the promising results of PD-1/PD-L1 axis blockade, additional strategies to improve the response rate and to overcome resistance to immunotherapy are ongoing, such as testing novel immunomodulatory compounds, which might increase the activity of immunotherapy treatment and contribute to convert "cold tumors" into "hot tumors." This novel combinations are focusing on the dual immune checkpoint blockade, which includes the introduction of antibodies against co-inhibitory, such as anti-LAG-3 antibodies.

Lymphocyte-associated gene 3 (LAG3) is a transmembrane protein with structural homology to the CD4 co-receptor and mainly expressed in activated CD4+ T cells, T-regulator cell, Tr1 cells, activated CD8+ T cells, natural killer cells, dendritic cells, B cells, and exhausted effector T cells. LAG3 negatively regulates the proliferation, activation, and effector function of T cells [138]. The role of LAG3 as prognostic biomarker is still controversial and under investigation. Results from a recent meta-analysis investigating the role of LAG3 as prognostic biomarker in several solid tumors, including TNBC, showed that high expression of LAG3 can be associated with favorable outcome, particularly in early stage tumor, also if there was a borderline statistical significance in their results, suggesting that the role of LAG3 as prognostic biomarker should be evaluated together with the expression of other biomarkers reflecting an active host immunity, such as PD-L1 and CD8 [139].

The suggestion that other molecules can stimulate and increase the number of immune cells, enhancing the anti-tumor immune function, is leading to new combination approaches of immune checkpoint inhibitors and novel agents for the treatment of TNBC. The InCITe trial (NCT03971409) is evaluating the combination therapy of Avelumab (anti-PD-L1) and Binimetinib (MEK 1/2 inhibitor), Utomilumab (anti-4-1BB), or PF-04518600 (anti-OX40) in a multi-arm study for treatment of first or second line in metastatic TNBC patients.

4. PI3KCA and PTEN Mutations as Predictive Biomarkers

The phosphatidylinositol 3-kinase (PI3K) pathway is a key regulator of survival, growth, proliferation, angiogenesis, metabolism, and migration. It comprises a family of intracellular signal transducer enzymes with three key regulatory nodes, PI3K, AKT, and mammalian target of rapamycin (mTOR). PI3K activation phosphorylates and activates AKT, that regulates the functions of numerous cellular proteins, including the FoxO proteins, mTOR complex 1 (mTORC1), and S6 kinase [140,141]. In many cancers, this pathway is overactive due to *gain-of-function* mutations of phosphatidylinositol-4, 5-bisphosphate 3-kinase, catalytic subunit, alpha (*PIK3CA*), *loss-of-function* alterations of the tumor suppressor phosphatase and tensin homolog (*PTEN*), deregulation of receptor tyrosine kinase signaling, and amplification and mutations of receptor tyrosine kinases [142,143]. These alterations occur in approximately 35% of triple-negative and 45% of ER/PgR-positive, HER2-negative breast cancers [144]. PI3Ks are a family of lipid kinases that are divided into three classes based on their structures and substrate specificities. Class IA PI3Ks are heterodimers that contain a p110 catalytic subunit and a p85 regulatory subunit. The genes *PIK3CA*, *PIK3CB*, and *PIK3CD* encode three highly homologous class IA catalytic isoforms: p110α, p110β, and p110δ, respectively [145]. In TNBC, the majority of activating mutations occur in the p110a (alpha subunit encoded by *PIK3CA*), overall mutated in ~9% of primary TNBC. *PIK3CA* mutations result in activated alpha PI3K, leading to an activating downstream pathway. *PTEN* alterations are also frequent in TNBC, with genetic loss of function occurring in 15% [146,147]. The phosphatase PTEN exert its activity of tumor suppressor through dephosphorylating phosphatidylinositol (3,4,5)-trisphosphate (PIP3), resulting in an inhibition of AKT [148,149]. Inactivating mutations of *PTEN*, including truncating and frameshift mutations or homozygous deletion, cause loss of function with consequent hyperactivation of AKT activity. Single nucleotide variants hotspots mutations such as R130X, R233X, and R335X, allelic loss in loci of the10q23 region were also reported [147,150,151]. A single amino acid substitution E17K of AKT1 was described in several cancers, with the highest incidence 3.8%, in breast cancer. This mutation results in a pathologic association of AKT1 with the plasma membrane and its constitutive activation [152]. Finally, mutations in *mTOR* were found in 1.8% of breast cancer [153]. The high frequency of mutations of PI3K/AKT/mTOR pathway found in breast cancer provides the rationale to test new inhibitors in combination with standard therapies. Several PI3K and AKT inhibitors are currently under investigation in clinical trials, mostly in ER/PgR-positive HER2-negative subtypes. Despite intense research efforts, so far, only *PIK3CA* mutations have proven to have a predictive value for treatment with α-selective and β-sparing PI3K inhibitors, Alpelisib and Taselisib respectively, in the advanced setting [154–158]. Alpelisib, an oral α-specific PI3K inhibitor that selectively inhibits p110α, was recently approved, based on results of phase III SOLAR-1 trial (NCT02437318), for postmenopausal women, and men, with metastatic or advanced *PIK3CA*-altered, ER/PgR-positive, and HER2-negative breast cancer, indicating that the integration of genomic testing for *PIK3CA* mutation may be useful in the selection of therapy [154,155].

The predictive effect of *PIK3CA* mutations may also have future relevancy for TNBC. Recently, Yuan et al. showed that combination of CDK4/6 inhibitor Ribociclib and Alpelisib caused the reduction of p-RB and p-S6, of MCL-1, induction of apoptosis, and an enhanced reduction of tumor growth in a TNBC PDX model [159]. Other findings reported that same combination significantly increased tumor-infiltrating T-cell activation and cytotoxicity and decreased the frequency of immunosuppressive myeloid-derived suppressor cells in a syngeneic TNBC mouse model [160]. These studies support the development of new possible combinational approaches in TNBC.

LAR TNBC, in particular, are sensitive to endocrine manipulations with AR antagonists and are enriched (approximately 40%–50%) in activating *PIK3CA* mutations [161,162]: this could confer sensitivity to PI3K inhibitors and synergy with AR antagonists. Based on these findings, in the TBCRC 032 IB/II trial (NCT02457910) an oral antiandrogen, enzalutamide, has been evaluated with or without taselisib in patients with AR+ metastatic TNBC. By RNA seq analysis, the authors noticed that patients receiving the combination displayed decreased expression of genes involved in mTOR signaling and increased expression of genes related to adaptive immunity after treatment. Overall this study demonstrated that this combination can be given safely and appears to increase clinical benefit in TNBC patients with AR+ tumors [163].

Loss of PTEN tumor suppressor activity has been investigated as biomarker of response to AKT inhibitors, based on finding that *PTEN*-loss could enhance activation of AKT signaling. The randomized Phase II study (GO29227, LOTUS NCT02162719) compared activity of Ipatasertib [164–167], a potent, highly selective inhibitor of all three isoforms of Akt, plus Paclitaxel versus placebo plus Paclitaxel as first-line treatment for patients with inoperable locally advanced or metastatic TNBC. Patients were classified according to PTEN expression by immunohistochemistry (PTEN-low or PTEN-high) and *PIK3CA/AKT1/PTEN* genomic alterations (*PIK3CA*, *AKT1* or *PTEN* mutations) characterized by NGS. Results have shown that the increase in median PFS was quite modest in the Intention-to-treat population and PTEN-low subgroup but more pronounced in predefined analyses of the patient population with *PIK3CA/AKT1/PTEN*-altered tumors, suggesting that a complete assessment of PI3K pathway could have a predictive value rather than a single alteration [168].

At present, Phase III Ipatunity 130 trial (NCT03337724) is aiming to confirm data from LOTUS trial, evaluating Ipatasertib in combination with Paclitaxel in *PIK3CA/AKT1/PTEN*-Altered, locally advanced or metastatic, TNBC or ER/PgR-positive, HER2-negative patients. Genetic alterations will be evaluated in relevant genes in the PI3K/Akt pathway, both in tissues and in blood samples, by NGS assay.

In the neoadjuvant setting, the phase II FAIRLANE study (NCT02301988), evaluating Paclitaxel plus Ipatasertib or placebo in TNBC, partially supports the potential utility as biomarker for the *PIK3CA/AKT1/PTEN* alterations; there was, indeed, in the trial, a numerically but non-significant increase in pCR rates, with more pronounced results in patients with PTEN-low tumors (32% versus 6%) and *PIK3CA/AKT1/PTEN*-altered tumors (39% versus 9%) [169]. In contrast, in the phase Ib study NCT03800836, evaluating the efficacy and safety of the combination of Ipatasertib, Tecentriq (Atezolizumab), and chemotherapy (Paclitaxel or Nab-paclitaxel) as a first-line treatment option for people with advanced TNBC, the objective response rate (ORR) was 73% (95% CI 53–88%), irrespective of tumor biomarker status [170,171]. A confirmatory phase III study NCT04177108 investigating the combination of ipatasertib, atezolizumab and paclitaxel as first-line therapy for locally advanced/metastatic TNBC cancer is still ongoing.

The predictive potential of *PIK3CA/AKT1/PTEN* alteration may is also supported by other AKT inhibitors trials. In the PAKT trial, Capivasertib (AZD5363), a potent and selective oral inhibitor of all three isoforms of the serine/threonine kinase AKT [172] has been evaluated in TNBC, showing that addition of capivasertib to first-line paclitaxel therapy prolonged PFS and OS. Also in this case, benefits were more pronounced in patients with *PIK3CA/AKT1/PTEN*-altered tumors [173].

In summary, alteration of PIK3CA/AKT1/PTEN pathway have not yet satisfied criteria for the clinical use in TNBC, but diverse evidences support further research on this topic.

5. Promising Molecular Biomarkers

5.1. New Targets of Antibody–Drug Conjugates in Triple Negative Breast Cancer

Antibody drug conjugates (ADCs) are a new class of anticancer drugs that share the same general mechanism of action; they are designed as a monoclonal antibody that are conjugated with a potent cytotoxin (so called payload). The monoclonal antibody is directed against an antigen on the surface of the target cancer cell, and upon binding to the target antigen, they are internalized and release the payload inside the cell, leading to selective cytotoxicity. This allows selective intracellular delivery of very potent payload that, because of their inherent toxicity, could not be infused as free molecules to the patient. As an additional mechanism of action, ADCs can elicit a potent immune response by inducing dendritic cell maturation and CD8 and CD4+ T-cell infiltration [174,175].

For an ADCs to be effective, a critical factor is the target antigen, that has to be selectively expressed (or, overexpressed) on the intended cancer cell. Therefore, the presence (or the overexpression) of the target antigen can be tested as biomarker to identify potentially sensitive patients. Several molecules have been identified in TNBC cells that meet these characteristics. The most promising ones are: (1) the glycoprotein non-metastatic b (GPNMB); (2) trophoblast cell-surface antigen 2 (Trop-2); (3) LIV-1; (4) the mucin 1-attached sialoglycotope CA6.

GPNMB was found highly overexpressed in aggressive tumors like TNBC, or in advanced setting, where it is involved in processes like cell migration, invasion, angiogenesis, or epithelial-mesenchymal transition [176–178]; in addition, it represents a biomarker of poor prognosis in breast cancer [179]. It is the target of Glembatumumab vedotin (CDX-011) a potent ADC conjugated with the microtubule-disrupting agent monomethyl auristatin E (MMAE) [180]. The phase II EMERGE trial, designed to evaluate CDX-011 activity in advanced GPNMB-expressing breast cancer versus chemotherapy of the investigator's choice, showed that this drug is well tolerated and more effective in patients with TNBC and/or GPNMB-overexpressing breast cancers [181]. The following pivotal phase II trial METRIC, designed to evaluating CDX-011 versus capecitabine in TNBC GPNMB-over-expressing patients, confirmed results of safety from EMERGE trial, but the primary end point of PFS was not met [182,183]. Nonetheless GPNMB remain a potentially useful target for other ADCs agents.

Trop-2 is a type I transmembrane glycoprotein, with a relevant role in migration, cell proliferation, cell cycle progression, and metastasis [184–187]. Sacituzumab govitecan (IMMU-132) is the new promising antibody targeting Trop-2, linked to topoisomerase-I inhibitor SN-38, the active metabolite of irinotecan that induces DNA damage [188,189]. Results from a phase I/II IMMU-132-01 (NCT01631552) showed the efficacy of Sacituzumab Govitecan-hziy, with 33.3% ORR in heavily pretreated TNBC patients [190]. Based on these results, on 22 April 2020, the FDA granted accelerated approval to Sacituzumab Govitecan-hziy for adult patients with metastatic TNBC who received at least two prior therapies for metastatic disease [190]. Currently, the randomized Phase III ASCENT clinical trial, comparing Sacituzumab Govitecan-hziy versus treatment of physician's choice, in metastatic TNBC patients who progressed after at least two prior cytotoxic therapies, is ongoing (NCT0257445599).

LIV-1 is a zinc transporter protein downstream target of STAT3, implicated in cell adhesion and epithelial-to-mesenchymal transition [191–194]. Its target therapy, the monoclonal antibody against the extracellular domain of LIV-1, Ladiratuzumab vedotin (SGN-LIV1A), showed high efficacy in preclinical models [195], and is under evaluation in patients with metastatic breast cancers, with promising results in metastatic TNBC (NCT03310957, NCT01969643, NCT04032704, NCT03424005, NCT01042379) [196].

CA6 is selectively expressed on solid tumors and is, therefore, an ideal target for ADC therapy. SAR566658 is an ADC directed against CA6 which carries DM4, a maytansine-derived anti-microtubule agent as payload. Based on promising results from a phase I trial, a phase II study in CA6-positive TNBC (NCT02984683) is currently ongoing [197].

5.2. Circulating Tumor Cells as Prognostic and Predictive Biomarkers in TNBC

Detection of circulating tumor cells (CTCs) has a promising potential as minimally invasive "liquid biopsies" that can facilitate prognosis, target therapy, or monitoring therapeutic response to drugs in several cancers [198,199]. CTC counts in cancer patients have been used as a dynamic prognostic biomarker in both early and metastatic cancer [200,201], while isolation and analysis of CTCs have been shown to provide information on dynamic changes in tumor [202,203].

Over the past decade, several strategies were developed to capture CTCs based on biological properties of the cells, like the expression of cell surface proteins, or biophysical properties using filtration, microfluidics, and dielectrophoresis, or by applying high throughput imaging to unpurified blood cell preparations [204]. Despite this, there are still certain technological limitations related to sensitivity and specificity, and a lack of consensus regarding the isolation technique to be used, the type of sample, the conditions of collection or storage, or the candidate biomarker to be used [205]. At present, the only one approved for application in the clinical practice is the CellSearch technology (Menarini Silicon Biosystems, Huntingdon Valley, PA, USA), based on epithelial cell adhesion molecule (EpCAM)-based CTC isolation technology, which was FDA patented on 2004 [206,207].

A consistent number of prospective studies have demonstrated that CTC counts in cancer patients can be used as a dynamic prognostic biomarker in metastatic disease. More than a decade ago, Cristofanilli et al. in a study involving 177 metastatic breast cancer (MBC) patients, demonstrated that CTC count detected using the CellSearch was an independent prognostic factor for PFS and OS in metastatic Breast Cancer. The cut-off of 5 CTCs/7.5 mL was identified to classify patients with good or poor clinical outcome [208] and subsequent studies have confirmed the prognostic value of CTCs with the same cut-off [202,209–212]. Moreover, other studies indicated that CTCs dynamics seem to reflect treatment response as an indicator to monitor the effectiveness of treatments and guide subsequent therapies in breast cancer [213]. Other studies revealed that CTCs counts loose its prognostic value in MBC treated with targeted therapies in HER2 positive tumors [214,215]. A recent large, retrospective study, involving 1944 MBC patients showed as CTCs enumeration should be used for prognostic stratification of MBC in two defined group of patients identified as Stage IV indolent and Stage IV aggressive, where Stage IV aggressive could better benefit from novel therapies compared with Stage IV indolent [216]. In early breast cancer, CTCs count is also a prognostic biomarker, not correlated with the other usual prognostic factors. The presence and also the quantity of CTCs has proven to be associated with worse outcome, however, CTC detection methods with higher sensitivity could be necessary considering the low number of cells found in early setting [217–223].

Several groups reported that the use of CellSearch platform for CTCs counts has limited prognostic power in TNBC, because of the fact that in these tumors, cells tend to switch from epithelial to mesenchymal phenotype, loosing EpCAM expression and exhibiting more stem cell-like properties [224–226]. Moreover, data from clinical trials, evaluating the prognostic and predictive role of CTCs in TNBC, are controversial. Munzone et al. retrospectively analyzed the CTC enumeration by CellSearch in 203 patients with MBC, and found that baseline CTC counts were significantly associated with OS but not with PFS in TNBC patients who were receiving new courses of systemic therapy [227]. CBCSG004 trial showed that baseline CTCs count is a prognostic but not a predictive factor to anticancer therapies in TNBC [228]. The phase II trial (TBCRC019) has analyzed if CellSearch was effective in TNBC, and whether CTC apoptosis and CTC clusters enhances the prognostic role of CTC in metastatic TNBC patients treated with nab-paclitaxel with or without tigatuzumab. Results showed that patients with elevated CTC at baseline, on day 15 and 29, had significant worse PFS versus not elevated, while there was no apparent prognostic effect comparing CTC apoptosis versus non-apoptosis, or the presence of CTC clusters [229]. The large, prospective, randomized study SWOG S0500, showed that patients with TNBC and low CTC levels at baseline, and those who had CTC clearance after chemotherapy treatment, had a longer OS compared with those who had elevated CTC levels [230]. A prospective study in TNBC comparing CellSearch and immunomagnetic enrichment/flow cytometry methods revealed that CTC enumeration by two different assays was highly concordant. Both assays showed an association between baseline CTC levels and OS, and changes in CTC levels during chemotherapy were significantly associated with time to progression and OS [231].

CTC analysis of the tnAcity trial reported better outcomes among patients with CTC levels at baseline that were reduced or eliminated in subsequent cycles of chemotherapy, compared with CTC levels that persisted post-baseline, suggesting that CTC clearance may predict the chemosensitivity of metastatic TNBC tumors. However, as reported by the authors, this study has some limitations because of the use of only CellSerch platform that can exclude populations of low- or non-EpCAM-expressing cells, the possibility that during chemotherapy cells underwent an epithelial to mesenchymal transition change, and the low number of patient samples (N = 126) [232]. Finally, Liu et al. developed a combined CTC-NK enumeration strategy that allows to predict PFS in TNBC. They reported that baseline CTC counts can predict PFS only in first-line TNBC patients but not in other TNBC lines of therapy, while baseline CTC combined with NK enumeration (CTC-NK) can predict PFS of TNBC patients regardless of their lines of therapy [233]. The 2019 AACR annual meeting reported a significant correlation between high levels of co-expression of CCR5 and HER2 in CTCs in MBC. CCR5 has been associated with cancer stem cells and believed to drive metastatic process. The researchers suggest that identifying CCR5 expression in CTCs could be used as a potential new biomarker for MBC with potential therapeutic implications in patients with TNBC [234].

Single CTCs have been extensively studied in recent years because its detection could be particularly useful for certain types of cancers, however, also actively growing, aggressive tumors tend to release relatively low numbers of detectable CTCs into the circulation, and technical improvements in their isolation are needed [235]. Factors leading to the generation of CTCs from a primary tumor are unknown. The number of CTCs released in the bloodstream is enormously higher compared to the number of metastatic lesions in patients, indicating that the majority of CTCs die in the bloodstream, with only a minor fraction representing viable metastatic precursors. Mesenchymal transformation, stromal-derived factors, or persistent interepithelial cell junctions may provide survival signals that attenuate this apoptotic outcome [236,237]. Gene characterization of CTCs from TNBC revealed their attitude to epithelial—mesenchymal transition (EMT) associated with increased plasticity and aggressiveness, increase of resistance to cell death and chemotherapy, capability to metastasize and senescence [238–242]. Since a consistent group of TNBC patients are negative to CellSearch system, Abeu et al. used a method based on CellSearch system for enumeration and a combination of immunoisolation and gene expression profiling, to molecularly characterize the population. Gene profiling revealed the expression of hybrid EMT and stem cell markers associated with poor prognosis and high aggressiveness, such as *VIM, SNAIL1, TIMP1, CRIPTO1, CD49F, ALDH2, CD44*, and *BCL11A* [242]. Razmara et al. by using a PDOX models of triple-negative breast cancer (TNBC), showed that CTC clusters and CTCs expressing a mesenchymal marker (vimentin) were associated with metastatic burden in lung and liver [243]. While, Thangavel et al. reported that evaluation of EMT-specific signature did not show significant differences between CTC cluster+ and CTC cluster tumors, in a TNBC PDX model [244]. In particular, CTCs cluster, ranging from 2–50 cells, were detected in the circulation of patients with metastatic cancers [245,246]. Although rare compared with single CTC, CTC clusters are more efficient than individual CTCs in seeding metastatic colonies, they are more resistant to apoptosis and may have an advantage in physically lodging in the Lumia of vessels [235,238]. Several studies have detected CTC clusters in breast cancer [247–249] where the expression of mesenchymal markers was found, rather than in single migratory cells [250]. Moreover, most of the evidences of the clinical impact of CTC-clusters in breast cancer have been gathered from prospectively designed clinical studies on rather homogeneous and well-selected cohorts of metastatic patients [229,248,251–253].

In summary, CTCs, CTC clusters detection and their relative molecular characterization represent a significant source of biomarkers. However, in TNBC, detection methods need to be improved in order to not lose a consistent number of cells with mesenchymal phenotypes. Moreover, large trials are needed to confirm their prognostic and predictive role.

Cancer cells can also disseminate their contents as free DNA fragments and exosomes into the bloodstream via different mechanisms [254,255]. The quantitative analysis of DNA fragments in the blood called circulating tumor DNA (ctDNA) has emerged as potential prognostic biomarker to reveal the presence or absence of tumors or to predict relapse and metastasis. The identification of tumor-specific genetic alterations could be used to lead personalized therapies, while, in metastatic setting, liquid biopsy appears to be a good alternative to tissue biopsy [256]. Despite the potential benefits of the use of DNA sequencing assays, ctDNA is still far to be integrated in the clinical practice. The prognostic value of ctDNA is still under evaluation in TNBC tumors. In a retrospective cohort of 164 patients with metastatic TNBC, the presence of a cell-free DNA fraction greater than 10% was associated with worse outcomes, regardless of clinicopathological data [257]. A study from Parsons et al. of NGS analysis of plasma-derived and tissue biopsies DNA from 26 patients with metastatic TNBC, revealed a concordance of 70%, demonstrating the potential of liquid biopsies for mutational profiling and serial monitoring [258]. High rate of concordance, 75% and 100% for *PIK3CA* and *AKT1* respectively, was also found in the analysis of comparison between plasma-based and tissue-based DNA sequencing of LOTUS trial [259]. While, Vidula et al. revealed the presence of *BRCA* somatic mutations in ctDNA not detected in primary tumors materials [260]. In early setting, a prospective study on a cohort of 101 patients, showed that ctDNA levels at diagnosis was higher in TNBC compared to HER2+ and ER+/HER2− [261]. In a study on 46 TNBC patients, ctDNA levels detected by digital PCR, showed that no patient had detectable ctDNA after surgery, expect one patient who experienced tumor progression during neoadjuvant chemotherapy. Despite this, pCR rate was not correlated with ctDNA detection at any time point, while ctDNA positivity after one cycle of chemotherapy was correlated with shorter DFS and OS [262]. Furthermore, Cavallone et al. reported that ctDNA detection after neoadjuvant chemotherapy and before surgery was associated to DFS and OS [263]. A phase II clinical trial (NCT03145961) is recruiting patients to evaluate whether ctDNA detection can be used to detect residual disease after standard primary treatment in early-stage TNBC.

5.3. CSCs and Drug Resistance in TNBC

In the past years, numerous evidences highlighted the contribution of cancer stem cells (CSCs) in tumorigenic potential, high risk of metastasis, and drugs resistance of TNBC [264]. CSCs represent a small population of cancer cells with staminal phenotype; they expose CD44+/CD24− and high ALDH expression [265–267]. This signature is associated with high capability of self-renewal, of proliferation and mammalian spheroids forming [268–270]. It is unclear if CSCs arise from pathogenic mutations in resident stem cells, or if they are the result of mutations of quiescent cells [271], anyway these CD44+/CD24− cells show an EMT phenotype with a great tumorigenic ability invasion and metastasis [265,272,273]. In this context, TNBC seems to have a significant number of stem cells CD44+/CD24− [266], and high expression of ALDH1 [3]. Breast cancer tissues of TNBC patients have shown to express high stem cell markers [272,274,275] and the gene signature of TNBC cells seems remarkably similar to that of mammary stem cells TNBC [264]. In an analysis of 466 invasive breast carcinomas and eight breast cancer cell lines, basal-like breast cancer harbored the highest percentage of tumor cells with the CSC phenotype CD44+CD24−/low and ALDH1 positivity [276]. In clinical studies, CD44+CD24−/low expression was associated with worse chemotherapy response, lymph node metastasis, distant metastasis, recurrence, and worse DFS and OS [277,278]; while ALDH1-expression predicted poor prognosis in TNBC patients [279–281]. As reported above, EMT phenotype confers capacity to metastasize and drug resistance in TNBC, and several researches have highlighted that EMT transition and enrichment for CSCs in TNBC tumors were correlated to invasiveness and drug resistance [282–284].

In particular, it seems that over-expression of EMT pathways promote the generation of mammary CSCs and is responsible of capacity of CSCs to survive in hard metabolic conditions because of reduction of nutrients or hypoxia [282,285,286].

Several evidences have reported that self-renewal activity of CSCs could be ascribed to the alteration of different signal transduction pathways such as STAT signaling, SRC signaling, Wnt/β-catenin signaling. In particular, STAT3 signaling is involved in the mechanism for self-renewal regulation of CSCs, and conversion of non-CSCs into CSCs, through the regulation of IL-6-Jak1-STAT3-OCT3 [287] or in tumorigenic potential, mammosphere-forming efficiency, and ALDH activity of breast cancer cells through VEGFR-2/STAT3 signaling [288].

In TNBC patients, STAT3 activation is a biomarker of poor prognosis. The phosphorylated isoform is preferentially expressed in TNBC cell lines [289], and it seems associated to initiation, progression, metastasis, and chemotherapy resistance [290]. In this regard, STAT3 signaling inhibitors are in clinical trials evaluation also in TNBC patients [291–293].

6. Discussion

TNBC is an intrinsically heterogeneous group of breast cancer and there is a need for effective biomarkers that can help physicians in selecting the most appropriate treatment.

Several proposed biomarkers for TNBC have been studied in clinical trials demonstrating, so far, modest clinical benefits. Mutations of *BRCA1/2* genes turned out to be factors predicting the efficacy of PARPis and alterations of other genes involved in homologous recombination seem promising in this setting.

PD-L1 protein expression either in IC, tumor cells or both can be used as a predictive biomarker for response to immunocheckpoint inhibitors. There are different commercially available assays that use different antibodies as long as different scoring systems. However, there still is uncertainty as to what is the best assay in TNBC and if the results apply to all immunocheckpoint inhibitors. At present, however, data clearly support the use of the VENTANA SP142 assay to predict the efficacy of Atezolizumab in metastatic TNBC.

Table 1 summarizes the main discussed biomarkers and their prognostic and predictive significance.

Table 1. Summary of biomarkers in triple negative breast cancer.

Biomarker	Main Function	Assay	Prognostic/Predictive Significance	Target Therapy	Ref.
BRCA1 and *BRCA2* genes	DNA-double strand break repair	BRACAnalysis CDx test HRDetect assay HRD assay myChoice CDx	Poor prognostic factor. High response to platinum-based therapy and predictor for response to PARP inhibitors	PARP inhibitors	[9,10,30,55, 58–60,63–65]
HRR genes	Homologous recombination repair of DNA		Predictor of response to platinum therapy in neoadjuvant setting	ATR inhibitor * WEE1 inhibitor *	[41,47,56,57, 59–62]
Stromal TILs	Tumor infiltrating lymphocytes involved in immune response against the tumor	Tissue Immunohistochemistry	High TILs correlates with more favorable survival outcomes and are predictive for increased response to neoadjuvant CT	NA	[92–94]
PD-L1 protein	Tumor immune evasion process	VENTANA PD-L1 (SP142) Assay	High expression correlates with higher survival rates in trials with ICI	Immune checkpoint inhibitors	[11,117–125]
Microsatellite instability (MSI)	High Immunogenic activity	Histologically/cytologically confirmed MSI-H/dMMR	Predictor of response to Pembrolizumab	Pembrolizumab	[126]
PI3-kinase pathway	Cell proliferation	Tissue Immunohistochemistry of PI3KCA/PTEN or PI3k pathway genomic sequencing by NGS	Higher sensitivity to AKT inhibitors and to combination therapy of PI3K and androgen receptor inhibitors in LAR tumors	PI3K inhibitor * AKT inhibitor *	[163,168–171, 173]
GPNMB	Cell migration, invasion, angiogenesis, epithelial-mesenchymal transition	Tissue Immunohistochemistry	Poor prognostic factor	Glembatumumab vedotin (Antibody-drug conjugate) *	[179,181–183]
Trop-2	Cell cycle progression, migration, proliferation, metastasis	Tissue Immunohistochemistry	Poor prognostic factor	Sacituzumab Govitecan-hziy (Antibody-drug conjugate) *	[189,190]
LIV-1	Cell adhesion, epithelial-mesenchymal transition	Tissue Immunohistochemistry	Under investigation	Ladiratuzumab vedotin (Antibody-drug conjugate) *	[196]
CA6	Tumor cell survival and proliferation	Tissue Immunohistochemistry	Under investigation	SAR566658 (Antibody-drug conjugate) *	[197]

* Under clinical investigation; ICI: immune checkpoint inhibitors.

7. Conclusions

Despite many research efforts, only a few useful biomarkers have been identified so far in TNBC. Some of these are already in the clinical practice. As new therapeutic agents are developed, parallel preclinical and clinical research is needed to identify biomarkers for the responsive, or conversely, the resistant patient population.

Author Contributions: S.C., M.P., and A.C. conceived, designed the present study, and wrote the manuscript. M.d.L. reviewed and edited the manuscript and supervised the study. S.C., M.P., A.C., D.C., R.C., V.D.L., G.F., G.d.G., M.L., and M.d.L. have actively contributed to the interpretation and data curation of the current review. They read and approved the final manuscript and agree to be accountable for all aspects of the research in ensuring that the accuracy or integrity of any part of the work are appropriately investigated and resolved. All authors have read and agreed to the published version of the manuscript.

Funding: This research received no external funding.

Conflicts of Interest: The authors declare no conflict of interest.

Abbreviations

ADC	Antibody Drug Coniugated
BC	Breast Cancer
BL1/2	Basal-Like 1/2
BRCA1/2	BReast CAncer type 1 and Type 2
CSCs	Cancer Stem Cells
CTCs	Circulating Tumor Cells
dMMR	Mismatch Repair deficiency
DSBs	DNA Double Strand Breaks
DCs	Dendritic Cells
DFS	Disease Free Survival
EMT	Epithelial Mesenchymal Transition
ER	Estrogen Receptor
FDA	Food and Drug Administration
FFPE	Formalin-fixed paraffin embedded
GEP	Gene Expression Profile
gBRCAm	germline BRCA mutations
GPNMB	Glycoprotein non-metastatic b
HBOC	Hereditary Breast and Ovarian Cancer Syndrome
HRD	Homologous Recombination Deficiency
HRD-LOD	HRD-Loss of Heterozygosity
HRD-LST	HDR-Large Scale Transition
HRD-TAI	HDR-Telomeric Allelic Imbalance
HER2	Human Epidermal Growth Factor Receptor 2
HR	Homologous Recombination
HRR	Homologous Recombination Repair
IC	Immune Cells
ICI	Immune Checkpoint Inhibitor
IM	Immunomodulatory
LAG3	Lymphocyte-associated gene 3
LAR	Luminal Androgen Receptor
MAP	Mitogen-Activated Protein
MBC	Metastatic Breast Cancer
MHC	Major Histocompatibility Complex
MMAE	Microtubule-disrupting agent monomethyl auristatin E
MMR	Mismatch Repair
MSI	Microsatellite Instability
MSI-H	High Microsatellite Instability

MSL	Mesenchymal Stem-Like
mTOR	Mammalian Target Of Rapamycin
mTORC1	mTOR complex 1
NGS	Next Generation Sequencing
NHEJ	Non-Homologous End Joining
NK	Natural Killer
ORR	Objective Response Rate
OS	Overall Survival
OXPHOS	Oxidative Phosphorylation
pCR	Pathological Complete Response
PARP	Poly ADP-ribose polymerase
PARPi	Poly ADP-ribose polymerase Inhibitors
PD-1	Programmed Cell Death-1
PD-L1	Programmed Cell Death Ligand-1
PDX	Patient-Derived-Xenograft
PFS	Progression Free Survival
PI3K	Phosphatidylinositol 3-kinase
PIK3CA	Phosphatidylinositol-4, 5-bisphosphate 3-kinase, catalytic subunit, alpha
PIP3	Phosphatidylinositol (3,4,5)-trisphosphate
PTEN	Phosphatase and Tensin Homolog
PR	Progesterone Receptor
RCB	Residual Cancer Burden
sBRCAm	Somatic BRCA mutations
TC	Tumor Cells
TIL	Tumor Infiltrating Lymphocytes
TMB	Tumor Mutational Burden
TNBC	Triple Negative Breast Cancer
UNS	Unstable

References

1. Bastien, R.R.; Rodríguez-Lescure, Á.; Ebbert, M.T.; Prat, A.; Munárriz, B.; Rowe, L.; Miller, P.; Ruiz-Borrego, M.; Anderson, D.; Lyons, B.; et al. PAM50 breast cancer subtyping by RT-qPCR and concordance with standard clinical molecular markers. *Bmc Med. Genom.* **2012**, *5*, 44. [CrossRef] [PubMed]

2. Dent, R.; Trudeau, M.; Pritchard, K.I.; Hanna, W.M.; Kahn, H.K.; Sawka, C.A.; Lickley, L.A.; Rawlinson, E.; Sun, P.; Narod, S.A. Triple-negative breast cancer: Clinical features and patterns of recurrence. *Clin. Cancer Res.* **2007**, *13*, 4429–4434. [CrossRef] [PubMed]

3. Lehmann, B.D.; Bauer, J.A.; Chen, X.; Sanders, M.E.; Chakravarthy, A.B.; Shyr, Y.; Pietenpol, J.A. Identification of human triple-negative breast cancer subtypes and preclinical models for selection of targeted therapies. *J. Clin. Investig.* **2011**, *121*, 2750–2767. [CrossRef] [PubMed]

4. Lehmann, B.D.; Jovanović, B.; Chen, X.; Estrada, M.V.; Johnson, K.N.; Shyr, Y.; Moses, H.L.; Sanders, M.E.; Pietenpol, J.A. Refinement of Triple-Negative Breast Cancer Molecular Subtypes: Implications for Neoadjuvant Chemotherapy Selection. *PLoS ONE* **2016**, *11*, e0157368. [CrossRef] [PubMed]

5. Rahim, B.; O'Regan, R. AR Signaling in Breast Cancer. *Cancers* **2017**, *9*, 21. [CrossRef] [PubMed]

6. Wang, D.Y.; Jiang, Z.; Ben-David, Y.; Woodgett, J.R.; Zacksenhaus, E. Molecular stratification within triple-negative breast cancer subtypes. *Sci. Rep.* **2019**, *9*, 19107. [CrossRef] [PubMed]

7. Gerratana, L.; Basile, D.; Buono, G.; De Placido, S.; Giuliano, M.; Minichillo, S.; Coinu, A.; Martorana, F.; De Santo, I.; Del Mastro, L.; et al. Androgen receptor in triple negative breast cancer: A potential target for the targetless subtype. *Cancer Treat. Rev.* **2018**, *68*, 102–110. [CrossRef] [PubMed]

8. Da Silva, J.L.; Cardoso Nunes, N.C.; Izetti, P.; de Mesquita, G.G.; de Melo, A.C. Triple negative breast cancer: A thorough review of biomarkers. *Crit. Rev. Oncol. Hematol.* **2020**, *145*, 102855. [CrossRef]

9. Robson, M.; Ruddy, K.J.; Im, S.A.; Senkus, E.; Xu, B.; Domchek, S.M.; Masuda, N.; Li, W.; Tung, N.; Armstrong, A.; et al. Patient-reported outcomes in patients with a germline BRCA mutation and HER2-negative metastatic breast cancer receiving olaparib versus chemotherapy in the OlympiAD trial. *Eur. J. Cancer* **2019**, *120*, 20–30. [CrossRef]

10. Ettl, J.; Quek, R.G.W.; Lee, K.H.; Rugo, H.S.; Hurvitz, S.; Gonçalves, A.; Fehrenbacher, L.; Yerushalmi, R.; Mina, L.A.; Martin, M.; et al. Quality of life with talazoparib versus physician's choice of chemotherapy in patients with advanced breast cancer and germline BRCA1/2 mutation: Patient-reported outcomes from the EMBRACA phase III trial. *Ann. Oncol.* **2018**, *29*, 1939–1947. [CrossRef]

11. Schmid, P.; Adams, S.; Rugo, H.S.; Schneeweiss, A.; Barrios, C.H.; Iwata, H.; Diéras, V.; Hegg, R.; Im, S.A.; Shaw Wright, G.; et al. Atezolizumab and Nab-Paclitaxel in Advanced Triple-Negative Breast Cancer. *N. Engl. J. Med.* **2018**, *379*, 2108–2121. [CrossRef] [PubMed]

12. Wahba, H.A.; El-Hadaad, H.A. Current approaches in treatment of triple-negative breast cancer. *Cancer Biol. Med.* **2015**, *12*, 106–116. [PubMed]

13. Lord, C.J.; Tutt, A.N.J.; Ashworth, A. Synthetic Lethality and Cancer Therapy: Lessons Learned from the Development of PARP Inhibitors. *Annu. Rev. Med.* **2015**, *66*, 455–470. [CrossRef] [PubMed]

14. Filipponi, D.; Bulavin, D. Wip1 and ATM in tumorevolution: Role for BRCA1. *Oncotarget* **2013**, *4*, 2170. [CrossRef] [PubMed]

15. Savage, K.I.; Harkin, D.P. BRCA1, a 'complex' proteininvolved in the maintenance of genomicstability. *FEBS J.* **2015**, *282*, 630–646. [CrossRef]

16. D'Alessandro, G.; Whelan, D.R.; Howard, S.M.; Vitelli, V.; Renaudin, X.; Adamowicz, M.; Iannelli, F.; Jones-Weinert, C.W.; Lee, M.; Matti, V.; et al. BRCA2 controls DNA:RNA hybrid level at DSBs by mediating RNase H2 recruitment. *Nat. Commun.* **2018**, *9*, 5376. [CrossRef]

17. Sharma, P.; Klemp, J.R.; Kimler, B.F.; Mahnken, J.D.; Geier, L.J.; Khan, Q.J.; Elia, M.; Connor, C.S.; McGinness, M.K.; Mammen, J.M.; et al. Germline BRCA mutation evaluation in a prospective triple-negative breast cancer registry: Implications for hereditary breast and/or ovarian cancer syndrome testing. *Breast Cancer Res. Treat.* **2014**, *145*, 707–714. [CrossRef]

18. Futreal, P.A.; Liu, Q.; Shattuck-Eidens, D.; Cochran, C.; Harshman, K.; Tavtigian, S.; Bennett, L.M.; Haugen-Strano, A.; Swensen, J.; Miki, Y.; et al. BRCA1 mutations in primary breast and ovarian carcinomas. *Science* **1994**, *266*, 120–122. [CrossRef]

19. Prakash, R.; Zhang, Y.; Feng, W.; Jasin, M. Homologous recombination and human health: The roles of BRCA1, BRCA2, and associated proteins. *Cold Spring Harb. Perspect. Biol.* **2015**, *7*. [CrossRef]

20. Ronson, G.E.; Piberger, A.L.; Higgs, M.R.; Olsen, A.L.; Stewart, G.S.; McHugh, P.J.; Petermann, E.; Lakin, N.D. PARP1 and PARP2 stabilise replication forks at base excision repair intermediates through Fbh1-dependent Rad51 regulation. *Nat. Commun.* **2018**, *9*, 746. [CrossRef]

21. Audebert, M.; Salles, B.; Calsou, P. Involvement of poly(ADP-ribose) polymerase-1 and XRCC1/DNA ligase III in an alternative route for DNA double-strand breaks rejoining. *J. Biol. Chem.* **2004**, *279*, 55117–55126. [CrossRef] [PubMed]

22. Dhillon, K.K.; Swisher, E.M.; Taniguchi, T. Secondary mutations of BRCA1/2 and drug resistance. *Cancer Sci.* **2011**, *102*, 663–669. [CrossRef] [PubMed]

23. Pommier, Y.; O'Connor, M.J.; de Bono, J. Laying a trap to kill cancer cells: PARP inhibitors and their mechanisms of action. *Sci. Transl. Med.* **2016**, *8*, 362ps17. [CrossRef]

24. Kaufman, B.; Shapira-Frommer, R.; Schmutzler, R.K.; Audeh, M.W.; Friedlander, M.; Balmaña, J.; Mitchell, G.; Fried, G.; Stemmer, S.M.; Hubert, A.; et al. Olaparib monotherapy in patients with advanced cancer and a germline BRCA1/2 mutation. *J. Clin. Oncol.* **2015**, *33*, 244–250. [CrossRef] [PubMed]

25. Tarapchak, P. FDA Approves Maintenance Treatment for Recurrent Epithelial Ovarian, Fallopian Tube, or PrimaryPeritoneal Cancers. *Oncology Times* 2017. Available online: http://journals.lww.com/oncology-times/blog/fdaactionsandupdates/pages/post.aspx?PostID=231 (accessed on 24 June 2020).

26. Scott, L.J. Niraparib: First Global Approval. *Drugs* **2017**, *77*, 1029–1034. [CrossRef]

27. Dockery, L.E.; Gunderson, C.C.; Moore, K.N. Rucaparib: The past, present, and future of a newly approved PARP inhibitor for ovarian cancer. *Oncol. Targets* **2017**, *10*, 3029–3037. [CrossRef] [PubMed]

28. Kristeleit, R.S.; Oaknin, A.; Ray-Coquard, I.; Leary, A.; Balmaña, J.; Drew, Y.; Oza, A.M.; Shapira-Frommer, R.; Domchek, S.M.; Cameron, T.; et al. Antitumor activity of the poly(ADP-ribose) polymerase inhibitor rucaparib as monotherapy in patients with platinum-sensitive, relapsed, BRCA-mutated, high-grade ovarian cancer, and an update on safety. *Int. J. Gynecol. Cancer* **2019**, *29*, 1396–1404. [CrossRef] [PubMed]

29. Litton, J.K.; Scoggins, M.; Ramirez, D.L.; Murthy, R.K.; Whitman, G.J.; Hess, K.R.; Adrada, B.E.; Moulder, S.L.; Barcenas, C.H.; Valero, V.; et al. A feasibility study of neoadjuvant talazoparib for operable breast cancer patients with a germline BRCA mutation demonstrates marked activity. *Npj Breast Cancer* **2017**, *3*, 49. [CrossRef]

30. Litton, J.K.; Scoggins, M.E.; Hess, K.R.; Adrada, B.E.; Murthy, R.K.; Damodaran, S.; DeSnyder, S.M.; Brewster, A.M.; Barcenas, C.H.; Valero, V.; et al. Neoadjuvant talazoparib for patients with operable breast cancer with a germline BRCA pathogenic variant. *J. Clin. Oncol.* **2020**, *38*, 388–394. [CrossRef]

31. Vinayak, S.; Tolaney, S.M.; Schwartzberg, L.; Mita, M.; McCann, G.; Tan, A.R.; Wahner-Hendrickson, A.E.; Forero, A.; Anders, C.; Wulf, G.M.; et al. Open-Label Clinical Trial of Niraparib Combined With Pembrolizumab for Treatment of Advanced or Metastatic Triple-Negative Breast Cancer. *JAMA Oncol.* **2019**, *5*, 1132–1140. [CrossRef]

32. Patsouris, A.; Tredan, O.; Nenciu, D.; Tran-Dien, A.; Campion, L.; Goncalves, A.; Arnedos, M.; Sablin, M.-P.; Gouraud, W.; Jimenez, M.; et al. RUBY: A phase II study testing rucaparib in germline (g) BRCA wild-type patients presenting metastatic breast cancer (mBC) with homologous recombination deficiency (HRD). *J. Clin. Oncol.* **2019**, *37*, 1092. [CrossRef]

33. Nuthalapati, S.; Stodtmann, S.; Shepherd, S.P.; Ratajczak, C.K.; Mensing, S.; Menon, R.; Xiong, H. Exposure-response analysis to inform the optimal dose of veliparib in combination with carboplatin and paclitaxel in BRCA-mutated advanced breast cancer patients. *Cancer Chemother. Pharm.* **2019**, *84*, 977–986. [CrossRef] [PubMed]

34. Sun, K.; Mikule, K.; Wang, Z.; Poon, G.; Vaidyanathan, A.; Smith, G.; Zhang, Z.Y.; Hanke, J.; Ramaswamy, S.; Wang, J. A comparative pharmacokinetic study of PARP inhibitors demonstrates favorable properties for niraparib efficacy in preclinical tumor models. *Oncotarget* **2018**, *9*, 37080–37096. [CrossRef] [PubMed]

35. Keung, M.Y.; Wu, Y.; Badar, F.; Vadgama, J.V. Response of Breast Cancer Cells to PARP Inhibitors Is Independent of BRCA Status. *J. Clin. Med.* **2020**, *9*, 940. [CrossRef]

36. Pilié, P.G.; Gay, C.M.; Byers, L.A.; O'Connor, M.J.; Yap, T.A. PARP Inhibitors: Extending Benefit Beyond BRCA-Mutant Cancers. *Clin. Cancer Res.* **2019**, *25*, 3759–3771. [CrossRef]

37. Lord, C.J.; Ashworth, A. BRCAness revisited. *Nat. Rev. Cancer* **2016**, *16*, 110–120. [CrossRef]

38. Domagala, P.; Jakubowska, A.; Jaworska-Bieniek, K.; Kaczmarek, K.; Durda, K.; Kurlapska, A.; Cybulski, C.; Lubinski, J. Prevalence of Germline Mutations in Genes Engaged in DNA Damage Repair by Homologous Recombination in Patients with Triple-Negative and Hereditary Non-Triple-Negative Breast Cancers. *PLoS ONE* **2015**, *10*, e0130393. [CrossRef]

39. Lin, P.H.; Chen, M.; Tsai, L.W.; Lo, C.; Yen, T.C.; Huang, T.Y.; Chen, C.K.; Fan, S.C.; Kuo, S.H.; Huang, C.S. Using next-generation sequencing to redefine BRCAness in triple-negative breast cancer. *Cancer Sci.* **2020**, *111*, 1375–1384. [CrossRef]

40. Domagala, P.; Hybiak, J.; Cybulski, C.; Lubinski, J. BRCA1/2-negative hereditary triple-negative breast cancers exhibit BRCAness. *Int. J. Cancer* **2017**, *140*, 1545–1550. [CrossRef]

41. Timms, K.M.; Abkevich, V.; Hughes, E.; Neff, C.; Reid, J.; Morris, B.; Kalva, S.; Potter, J.; Tran, T.V.; Chen, J.; et al. Association of BRCA1/2 defects with genomic scores predictive of DNA damage repair deficiency among breast cancer subtypes. *Breast Cancer Res.* **2014**, *16*, 475. [CrossRef]

42. Watkins, J.A.; Irshad, S.; Grigoriadis, A.; Tutt, A.N. Genomic scars as biomarkers of homologous recombination deficiency and drug response in breast and ovarian cancers. *Breast Cancer Res.* **2014**, *16*, 211. [CrossRef] [PubMed]

43. Takaya, H.; Nakai, H.; Takamatsu, S.; Mandai, M.; Matsumura, N. Homologous recombination deficiency status-based classification of high-grade serous ovarian carcinoma. *Sci. Rep.* **2020**, *10*, 2757. [CrossRef] [PubMed]

44. O'Reilly, E.M.; Lee, J.W.; Lowery, M.A.; Capanu, M.; Stadler, Z.K.; Moore, M.J.; Dhani, N.; Kindler, H.L.; Estrella, H.; Maynard, H.; et al. Phase 1 trial evaluating cisplatin, gemcitabine, and veliparib in 2 patient cohorts: Germline BRCA mutation carriers and wild-type BRCA pancreatic ductal adenocarcinoma. *Cancer* **2018**, *124*, 1374–1382. [CrossRef] [PubMed]

45. Zhu, H.; Wei, M.; Xu, J.; Hua, J.; Liang, C.; Meng, Q.; Zhang, Y.; Liu, J.; Zhang, B.; Yu, X.; et al. PARP inhibitors in pancreatic cancer: Molecular mechanisms and clinical applications. *Mol. Cancer* **2020**, *19*, 49. [CrossRef] [PubMed]

46. Adashek, J.J.; Jain, R.K.; Zhang, J. Clinical Development of PARP Inhibitors in Treating Metastatic Castration-Resistant Prostate Cancer. *Cells* **2019**, *8*, 860. [CrossRef] [PubMed]

47. Webster, P.J.; Littlejohns, A.T.; Gaunt, H.J.; Prasad, K.R.; Beech, D.J.; Burke, D.A. AZD1775 induces toxicity through double-stranded DNA breaks independently of chemotherapeutic agents in p53-mutated colorectal cancer cells. *Cell Cycle* **2017**, *16*, 2176–2182. [CrossRef] [PubMed]

48. Meng, X.; Bi, J.; Li, Y.; Yang, S.; Zhang, Y.; Li, M.; Liu, H.; Li, Y.; McDonald, M.E.; Thiel, K.W.; et al. AZD1775 Increases Sensitivity to Olaparib and Gemcitabine in Cancer Cells with p53 Mutations. *Cancers* **2018**, *10*, 149. [CrossRef]

49. Bradbury, A.; Hall, S.; Curtin, N.; Drew, Y. Targeting ATR as Cancer Therapy: A new era for synthetic lethality and synergistic combinations? *Pharmacol. Ther.* **2020**, *207*, 107450. [CrossRef]

50. Lord, C.J.; Ashworth, A. PARP inhibitors: Synthetic lethality in the clinic. *Science* **2017**, *355*, 1152–1158. [CrossRef]

51. Mohyuddin, G.R.; Aziz, M.; Britt, A.; Wade, L.; Sun, W.; Baranda, J.; Al-Rajabi, R.; Saeed, A.; Kasi, A. Similar response rates and survival with PARP inhibitors for patients with solid tumors harboring somatic versus Germline BRCA mutations: A Meta-analysis and systematic review. *BMC Cancer* **2020**, *20*, 507. [CrossRef]

52. Murthy, P.; Muggia, F. PARP inhibitors: Clinical development, emerging differences, and the current therapeutic issues. *Cancer Drug Resistance* **2019**, *2*(3), 665–679. [CrossRef]

53. Tung, N.M.; Robson, M.E.; Ventz, S.; Santa-Maria, C.A.; Marcom, P.K.; Nanda, R.; Shah, P.D.; Ballinger, T.J.; Shih-Hsin Yang, E.; Melisko, M.E.; et al. TBCRC 048: A phase II study of olaparib monotherapy in metastatic breast cancer patients with germline or somatic mutations in DNA damage response (DDR) pathway genes (Olaparib Expanded). *J. Clin. Oncol.* **2020**, *1*, 38. [CrossRef]

54. Dasari, S.; Tchounwou, P.B. Cisplatin in cancer therapy: Molecular mechanisms of action. *Eur J. Pharm.* **2014**, *740*, 364–378. [CrossRef] [PubMed]

55. Cardoso, F.; Senkus, E.; Costa, A.; Papadopoulos, E.; Aapro, M.; André, F.; Harbeck, N.; Aguilar Lopez, B.; Barrios, C.H.; Bergh, J.; et al. 4th ESO-ESMO International Consensus Guidelines for Advanced Breast Cancer (ABC 4). *Ann. Oncol* **2018**, *29*, 1634–1657. [CrossRef] [PubMed]

56. Isakoff, S.J.; Mayer, E.L.; He, L.; Traina, T.A.; Carey, L.A.; Krag, K.J.; Rugo, H.S.; Liu, M.C.; Stearns, V.; Come, S.E.; et al. TBCRC009: A Multicenter Phase II Clinical Trial of Platinum Monotherapy With Biomarker Assessment in Metastatic Triple-Negative Breast Cancer. *J. Clin. Oncol* **2015**, *33*, 1902–1909. [CrossRef] [PubMed]

57. Sharma, P.; Rodler, E.; Barlow, E.W.; Gralow, J.; Puhalla, S.L.; Anders, C.K.; Goldstein, L.J.; Brown-Glaberman, U.A.; Huynh, T.T.; Szyarto, c.s.; et al. Results of a phase II randomized trial of cisplatin +/− veliparib in metastatic triple-negative breast cancer (TNBC) and/or germline BRCA-associated breast cancer (SWOG S1416). *J. Clin. Oncol.* **2020**, *38*, 1001. [CrossRef]

58. Tazzite, A.; Jouhadi, H.; Benider, A.; Nadifi, S. BRCA Mutational Status is a Promising Predictive Biomarker for Platinum-based Chemotherapy in Triple-Negative Breast Cancer. *Curr. Drug Targets* **2020**, 32013831. [CrossRef]

59. Silver, D.P.; Richardson, A.L.; Eklund, A.C.; Wang, Z.C.; Szallasi, Z.; Li, Q.; Juul, N.; Leong, C.-O.; Calogrias, D.; Buraimoh, A.; et al. Efficacy of neoadjuvant Cisplatin in triple-negative breast cancer. *J. Clin. Oncol. Off. J. Am. Soc. Clin. Oncol.* **2010**, *28*, 1145–1153. [CrossRef]

60. Von Minckwitz, G.; Schneeweiss, A.; Loibl, S.; Salat, C.; Denkert, C.; Rezai, M.; Blohmer, J.U.; Jackisch, C.; Paepke, S.; Gerber, B.; et al. Neoadjuvant carboplatin in patients with triple-negative and HER2-positive early breast cancer (GeparSixto; GBG 66): A randomised phase 2 trial. *Lancet Oncol.* **2014**, *15*, 747–756. [CrossRef]

61. Loibl, S.; Weber, K.E.; Timms, K.M.; Elkin, E.P.; Hahnen, E.; Fasching, P.A.; Lederer, B.; Denkert, C.; Schneeweiss, A.; Braun, S.; et al. Survival analysis of carboplatin added to an anthracycline/taxane-based neoadjuvant chemotherapy and HRD score as predictor of response-final results from GeparSixto. *Ann. Oncol.* **2018**, *29*, 2341–2347. [CrossRef]

62. Telli, M.L.; Jensen, K.C.; Vinayak, S.; Kurian, A.W.; Lipson, J.A.; Flaherty, P.J.; Timms, K.; Abkevich, V.; Schackmann, E.A.; Wapnir, I.L.; et al. Phase II Study of Gemcitabine, Carboplatin, and Iniparib As Neoadjuvant Therapy for Triple-Negative and BRCA1/2 Mutation-Associated Breast Cancer With Assessment of a Tumor-Based Measure of Genomic Instability: PrECOG 0105. *J. Clin. Oncol.* **2015**, *33*, 1895–1901. [CrossRef] [PubMed]

63. Telli, M.L.; Timms, K.M.; Reid, J.; Hennessy, B.; Mills, G.B.; Jensen, K.C.; Szallasi, Z.; Barry, W.T.; Winer, E.P.; Tung, N.M.; et al. Homologous Recombination Deficiency (HRD) Score Predicts Response to Platinum-Containing Neoadjuvant Chemotherapy in Patients with Triple-Negative Breast Cancer. *Clin. Cancer Res.* **2016**, *22*, 3764–3773. [CrossRef] [PubMed]

64. Davies, H.; Glodzik, D.; Morganella, S.; Yates, L.R.; Staaf, J.; Zou, X.; Ramakrishna, M.; Martin, S.; Boyault, S.; Sieuwerts, A.M.; et al. HRDetect is a predictor of BRCA1 and BRCA2 deficiency based on mutational signatures. *Nat. Med.* **2017**, *23*, 517–525. [CrossRef] [PubMed]

65. Zhao, E.Y.; Shen, Y.; Pleasance, E.; Kasaian, K.; Leelakumari, S.; Jones, M.; Bose, P.; Ch'ng, C.; Reisle, C.; Eirew, P.; et al. Homologous Recombination Deficiency and Platinum-Based Therapy Outcomes in Advanced Breast Cancer. *Clin. Cancer Res.* **2017**, *23*, 7521–7530. [CrossRef] [PubMed]

66. McAndrew, N.; DeMichele, A. Neoadjuvant Chemotherapy Considerations in Triple-Negative Breast Cancer. *J. Target. Cancer* **2018**, *7*, 52–69.

67. Donawho, C.K.; Luo, Y.; Luo, Y.; Penning, T.D.; Bauch, J.L.; Bouska, J.J.; Bontcheva-Diaz, V.D.; Cox, B.F.; DeWeese, T.L.; Dillehay, L.E.; et al. ABT-888, an orally active poly(ADP-ribose) polymerase inhibitor that potentiates DNA-damaging agents in preclinical tumor models. *Clin. Cancer Res.* **2007**, *13*, 2728–2737. [CrossRef]

68. Diéras, V.C.; Han, H.S.; Kaufman, B.; Wildiers, H.; Friedlander, M.; Ayoub, J.-P.; Puhalla, S.L.; Bondarenko, I.; Campone, M.; Jakobsen, E.H.; et al. LBA9—Phase III study of veliparib with carboplatin and paclitaxel in HER2-negative advanced/metastatic gBRCA-associated breast cancer. *Ann. Oncol.* **2019**, *30*, v857–v858. [CrossRef]

69. Barker, A.D.; Sigman, C.C.; KelloN, G.J.; Hylton, N.M.; Berry, D.A.; Esserman, L.J. I-SPY 2: An adaptive breast cancer trial design in the setting of neoadjuvant chemotherapy. *Clin. Pharmacol. Ther.* **2009**, *86*, 97–100. [CrossRef]

70. Loibl, S.; O'Shaughnessy, J.; Untch, M.; Sikov, W.M.; Rugo, H.S.; McKee, M.D.; Huober, J.; Golshan, M.; von Minckwitz, G.; Maag, D.; et al. Addition of the PARP inhibitor veliparib plus carboplatin or carboplatin alone to standard neoadjuvant chemotherapy in triple-negative breast cancer (BrighTNess): A randomised, phase 3 trial. *Lancet Oncol.* **2018**, *19*, 497–509. [CrossRef]

71. Noordermeer, S.M.; van Attikum, H. PARP Inhibitor Resistance: A Tug-of-War in BRCA-Mutated Cells. *Trends Cell Biol.* **2019**, *29*, 820–834. [CrossRef]

72. Rajawat, J.; Shukla, N.; Mishra, D.P. Therapeutic Targeting of Poly(ADP-Ribose) Polymerase-1 (PARP1) in Cancer: Current Developments, Therapeutic Strategies, and Future Opportunities. *Med. Res. Rev.* **2017**, *37*, 1461–1491. [CrossRef] [PubMed]

73. Pettitt, S.J.; Lord, C.J. Dissecting PARP inhibitor resistance with functional genomics. *Curr. Opin. Genet. Dev.* **2019**, *54*, 55–63. [CrossRef] [PubMed]

74. Zhang, Y.; Dang, C.V.; Zhang, L. BETting on combination to overcome PARPi resistance. *Oncotarget* **2017**, *8*, 84630–84631. [CrossRef] [PubMed]

75. Waks, A.G.; Cohen, O.; Kochupurakkal, B.; Kim, D.; Dunn, C.E.; Buendia Buendia, J.; Wander, S.; Helvie, K.; Lloyd, M.R.; Marini, L.; et al. Reversion and non-reversion mechanisms of resistance to PARP inhibitor or platinum chemotherapy in BRCA1/2-mutant metastatic breast cancer. *Ann. Oncol.* **2020**, *31*, 590–598. [CrossRef] [PubMed]

76. Rusan, M.; Andersen, R.F.; Jakobsen, A.; Steffensen, K.D. Circulating HOXA9-methylated tumour DNA: A novel biomarker of response to poly (ADP-ribose) polymerase inhibition in BRCA-mutated epithelial ovarian cancer. *Eur. J. Cancer* **2020**, *125*, 121–129. [CrossRef] [PubMed]

77. Lahiguera, Á.; Hyroššová, P.; Figueras, A.; Garzón, D.; Moreno, R.; Soto-Cerrato, V.; McNeish, I.; Serra, V.; Lazaro, C.; Barretina, P.; et al. Tumors defective in homologous recombination rely on oxidative metabolism: Relevance to treatments with PARP inhibitors. *Embo Mol. Med.* **2020**, *12*, e11217. [CrossRef] [PubMed]

78. Zhao, L.; Si, C.S.; Yu, Y.; Lu, J.W.; Zhuang, Y. Depletion of DNA damage binding protein 2 sensitizes triple-negative breast cancer cells to poly ADP-ribose polymerase inhibition by destabilizing Rad51. *Cancer Sci.* **2019**, *110*, 3543–3552. [CrossRef]

79. Cruz, C.; Castroviejo-Bermejo, M.; Gutiérrez-Enríquez, S.; Llop-Guevara, A.; Ibrahim, Y.H.; Gris-Oliver, A.; Bonache, S.; Morancho, B.; Bruna, A.; Rueda, O.M.; et al. RAD51 foci as a functional biomarker of homologous recombination repair and PARP inhibitor resistance in germline BRCA-mutated breast cancer. *Ann. Oncol.* **2018**, *29*, 1203–1210. [CrossRef]

80. Dev, H.; Chiang, T.W.; Lescale, C.; de Krijger, I.; Martin, A.G.; Pilger, D.; Coates, J.; Sczaniecka-Clift, M.; Wei, W.; Ostermaier, M.; et al. Shieldin complex promotes DNA end-joining and counters homologous recombination in BRCA1-null cells. *Nat. Cell Biol.* **2018**, *20*, 954–965. [CrossRef]

81. Castroviejo-Bermejo, M.; Cruz, C.; Llop-Guevara, A.; Gutiérrez-Enríquez, S.; Ducy, M.; Ibrahim, Y.H.; Gris-Oliver, A.; Pellegrino, B.; Bruna, A.; Guzmán, M.; et al. A RAD51 assay feasible in routine tumor samples calls PARP inhibitor response beyond BRCA mutation. *Embo Mol. Med.* **2018**, *10*, e9172. [CrossRef]

82. Marzio, A.; Puccini, J.; Kwon, Y.; Maverakis, N.K.; Arbini, A.; Sung, P.; Bar-Sagi, D.; Pagano, M. The F-Box Domain-Dependent Activity of EMI1 Regulates PARPi Sensitivity in Triple-Negative Breast Cancers. *Mol. Cell* **2019**, *73*, 224–237. [CrossRef] [PubMed]

83. Chu, Y.Y.; Yam, C.; Chen, M.K.; Chan, L.C.; Xiao, M.; Wei, Y.K.; Yamaguchi, H.; Lee, P.C.; Han, Y.; Nie, L.; et al. Blocking c-Met and EGFR reverses acquired resistance of PARP inhibitors in triple-negative breast cancer. *Am. J. Cancer Res.* **2020**, *10*, 648–661. [PubMed]

84. Mihailidou, C.; Karamouzis, M.V.; Schizas, D.; Papavassiliou, A.G. Co-targeting c-Met and DNA double-strand breaks (DSBs): Therapeutic strategies in BRCA-mutated gastric carcinomas. *Biochimie* **2017**, *142*, 135–143. [CrossRef]

85. Kim, H.; George, E.; Ragland, R.; Rafail, S.; Zhang, R.; Krepler, C.; Morgan, M.; Herlyn, M.; Brown, E.; Simpkins, F. Targeting the ATR/CHK1 Axis with PARP Inhibition Results in Tumor Regression in BRCA-Mutant Ovarian Cancer Models. *Clin. Cancer Res.* **2017**, *23*, 3097–3108. [CrossRef]

86. Lallo, A.; Frese, K.K.; Morrow, C.J.; Sloane, R.; Gulati, S.; Schenk, M.W.; Trapani, F.; Simms, N.; Galvin, M.; Brown, S.; et al. The Combination of the PARP Inhibitor Olaparib and the WEE1 Inhibitor AZD1775 as a New Therapeutic Option for Small Cell Lung Cancer. *Clin. Cancer Res.* **2018**, *24*, 5153–5164. [CrossRef] [PubMed]

87. Haynes, B.; Murai, J.; Lee, J.M. Restored replication fork stabilization, a mechanism of PARP inhibitor resistance, can be overcome by cell cycle checkpoint inhibition. *Cancer Treat Rev.* **2018**, *71*, 1–7. [CrossRef] [PubMed]

88. O'Donnell, J.S.; Teng, M.W.L.; Smyth, M.J. Cancer immunoediting and resistance to T cell-based immunotherapy. *Nat. Rev. Clin. Oncol.* **2019**, *16*, 151–167. [CrossRef]

89. García-Teijido, P.; Cabal, M.L.; Fernández, I.P.; Pérez, Y.F. Tumor-Infiltrating Lymphocytes in Triple Negative Breast Cancer: The Future of Immune Targeting. *Clin. Med. Insights Oncol.* **2016**, *10*, 31–39. [CrossRef]

90. Mao, Y.; Qu, Q.; Chen, X.; Huang, O.; Wu, J.; Shen, K. The Prognostic Value of Tumor-Infiltrating Lymphocytes in Breast Cancer: A Systematic Review and Meta-Analysis. *PLoS ONE* **2016**, *11*, e0152500. [CrossRef]

91. Loi, S.; Drubay, D.; Adams, S.; Pruneri, G.; Francis, P.A.; Lacroix-Triki, M.; Joensuu, H.; Dieci, M.V.; Badve, S.; Demaria, S.; et al. Tumor-Infiltrating Lymphocytes and Prognosis: A Pooled Individual Patient Analysis of Early-Stage Triple-Negative Breast Cancers. *J. Clin. Oncol.* **2019**, *37*, 559–569. [CrossRef]

92. Loi, S.; Adams, S.; Schmid, P.; Cortés, J.; Cescon, D.W.; Winer, E.P.; Toppmeyer, D.L.; Rugo, H.S.; De Laurentiis, M.; Nanda, R.; et al. LBA13Relationship between Tumor Infiltrating Lymphocyte (TIL) Levels and Response to Pembrolizumab (Pembro) in Metastatic Triple-Negative Breast Cancer (MTNBC): Results from KEYNOTE-086. *Ann. Oncol.* **2017**, *28*, v605–v649. [CrossRef]

93. Loi, S.; Schmid, P.; Aktan, G.; Karantza, V.; Salgado, R. Relationship between Tumor Infiltrating Lymphocytes (TILs) and Response to Pembrolizumab (Pembro)+chemotherapy (CT) as Neoadjuvant Treatment (NAT) for Triple-Negative Breast Cancer (TNBC): Phase Ib KEYNOTE-173 Trial. *Ann. Oncol.* **2019**, *30*, iii2. [CrossRef]

94. Borcherding, N.; Kolb, R.; Gullicksrud, J.; Vikas, P.; Zhu, Y.; Zhang, W. Keeping Tumors in Check: A Mechanistic Review of Clinical Response and Resistance to Immune Checkpoint Blockade in Cancer. *J. Mol. Biol.* **2018**, *430*, 2014–2029. [CrossRef] [PubMed]

95. Ahmadzadeh, M.; Johnson, L.A.; Heemskerk, B.; Wunderlich, J.R.; Dudley, M.E.; White, D.E.; Rosenberg, S.A. Tumor antigen-specific CD8 T cells infiltrating the tumor express high levels of PD-1 and are functionally impaired. *Blood* **2009**, *114*, 1537–1544. [CrossRef] [PubMed]

96. Salmaninejad, A.; Khoramshahi, V.; Azani, A.; Soltaninejad, E.; Aslani, S.; Zamani, M.R.; Zal, M.; Nesaei, A.; Hosseini, S.M. PD-1 and cancer: Molecular mechanisms and polymorphisms. *Immunogenetics* **2018**, *70*, 73–86. [CrossRef] [PubMed]

97. Bayraktar, S.; Batoo, S.; Okuno, S.; Glock, S. Immunotherapy in breast cancer. *J. Carcinog.* **2019**, *18*, 2. [CrossRef]

98. Boussiotis, V.A. Molecular and Biochemical Aspects of the PD-1 Checkpoint Pathway. *N. Engl. J. Med.* **2016**, *375*, 1767–1778. [CrossRef]

99. Chen, L.; Han, X. Anti-PD-1/PD-L1 therapy of human cancer: Past, present, and future. *J. Clin. Investig.* **2015**, *125*, 3384–3391. [CrossRef] [PubMed]

100. Ishida, Y.; Agata, Y.; Shibahara, K.; Honjo, T. Induced expression of PD-1, a novel member of the immunoglobulin gene superfamily, upon programmed cell death. *Embo J.* **1992**, *11*, 3887–3895. [CrossRef] [PubMed]

101. Ohaegbulam, K.C.; Assal, A.; Lazar-Molnar, E.; Yao, Y.; Zang, X. Human cancer immunotherapy with antibodies to the PD-1 and PD-L1 pathway. *Trends Mol. Med.* **2015**, *21*, 24–33. [CrossRef] [PubMed]

102. Wherry, E.J.; Kurachi, M. Molecular and cellular insights into T cell exhaustion. *Nat. Rev. Immunol.* **2015**, *15*, 486–499. [CrossRef] [PubMed]

103. Chen, J.; Jiang, C.C.; Jin, L.; Zhang, X.D. Regulation of PD-L1: A novel role of pro-survival signalling in cancer. *Ann. Oncol.* **2016**, *27*, 409–416. [CrossRef] [PubMed]

104. Stutvoet, T.S.; Kol, A.; de Vries, E.G.; de Bruyn, M.; Fehrmann, R.S.; Terwisscha van Scheltinga, A.G.; de Jong, S. MAPK pathway activity plays a key role in PD-L1 expression of lung adenocarcinoma cells. *J. Pathol.* **2019**, *249*, 52–64. [CrossRef] [PubMed]

105. Jalali, S.; Price-Troska, T.; Bothun, C.; Villasboas, J.; Kim, H.J.; Yang, Z.Z.; Novak, A.J.; Dong, H.; Ansell, S.M. Reverse signaling via PD-L1 supports malignant cell growth and survival in classical Hodgkin lymphoma. *Blood Cancer J.* **2019**, *9*, 22. [CrossRef] [PubMed]

106. Li, P.; Huang, T.; Zou, Q.; Liu, D.; Wang, Y.; Tan, X.; Wei, Y.; Qiu, H. FGFR2 Promotes Expression of PD-L1 in Colorectal Cancer via the JAK/STAT3 Signaling Pathway. *J. Immunol.* **2019**, *202*, 3065–3075. [CrossRef] [PubMed]

107. Galluzzi, L.; Spranger, S.; Fuchs, E.; López-Soto, A. WNT Signaling in Cancer Immunosurveillance. *Trends Cell Biol.* **2019**, *29*, 44–65. [CrossRef] [PubMed]

108. Castagnoli, L.; Cancila, V.; Cordoba-Romero, S.L.; Faraci, S.; Talarico, G.; Belmonte, B.; Iorio, M.V.; Milani, M.; Volpari, T.; Chiodoni, C.; et al. WNT signaling modulates PD-L1 expression in the stem cell compartment of triple-negative breast cancer. *Oncogene* **2019**, *38*, 4047–4060. [CrossRef]

109. Lim, W.; Jeong, M.; Bazer, F.W.; Song, G. Curcumin Suppresses Proliferation and Migration and Induces Apoptosis on Human Placental Choriocarcinoma Cells via ERK1/2 and SAPK/JNK MAPK Signaling Pathways. *Biol. Reprod.* **2016**, *95*, 83. [CrossRef]

110. Bi, X.W.; Wang, H.; Zhang, W.W.; Wang, J.H.; Liu, W.J.; Xia, Z.J.; Huang, H.Q.; Jiang, W.Q.; Zhang, Y.J.; Wang, L. PD-L1 is upregulated by EBV-driven LMP1 through NF-κB pathway and correlates with poor prognosis in natural killer/T-cell lymphoma. *J. Hematol. Oncol.* **2016**, *9*, 109. [CrossRef]

111. Wu, F.; Zhang, Y.; Sun, B.; McMahon, A.P.; Wang, Y. Hedgehog Signaling: From Basic Biology to Cancer Therapy. *Cell Chem. Biol.* **2017**, *24*, 252–280. [CrossRef]

112. Van Berckelaer, C.; Rypens, C.; van Dam, P.; Pouillon, L.; Parizel, M.; Schats, K.A.; Kockx, M.; Tjalma, W.A.A.; Vermeulen, P.; van Laere, S.; et al. Infiltrating stromal immune cells in inflammatory breast cancer are associated with an improved outcome and increased PD-L1 expression. *Breast Cancer Res.* **2019**, *21*, 28. [CrossRef] [PubMed]

113. Bertucci, F.; Gonçalves, A. Immunotherapy in Breast Cancer: The Emerging Role of PD-1 and PD-L1. *Curr. Oncol. Rep.* **2017**, *19*, 64. [CrossRef] [PubMed]

114. Sabatier, R.; Finetti, P.; Mamessier, E.; Adelaide, J.; Chaffanet, M.; Ali, H.R.; Viens, P.; Caldas, C.; Birnbaum, D.; Bertucci, F. Prognostic and predictive value of PDL1 expression in breast cancer. *Oncotarget* **2015**, *6*, 5449–5464. [CrossRef] [PubMed]

115. DeNardo, D.G.; Coussens, L.M. Inflammation and breast cancer. Balancing immune response: Crosstalk between adaptive and innate immune cells during breast cancer progression. *Breast Cancer Res.* **2007**, *9*, 212. [CrossRef] [PubMed]

116. Schmidt, M.; Böhm, D.; von Törne, C.; Steiner, E.; Puhl, A.; Pilch, H.; Lehr, H.A.; Hengstler, J.G.; Kölbl, H.; Gehrmann, M. The humoral immune system has a key prognostic impact in node-negative breast cancer. *Cancer Res.* **2008**, *68*, 5405–5413. [CrossRef] [PubMed]

117. Marra, A.; Viale, G.; Curigliano, G. Recent advances in triple negative breast cancer: The immunotherapy era. *BMC Med.* **2019**, *17*, 90. [CrossRef] [PubMed]

118. Cortés, J.; André, F.; Gonçalves, A.; Kümmel, S.; Martín, M.; Schmid, P.; Schuetz, F.; Swain, S.M.; Easton, V.; Pollex, E.; et al. IMpassion132 Phase III trial: Atezolizumab and chemotherapy in early relapsing metastatic triple-negative breast cancer. *Future Oncol.* **2019**, *15*, 1951–1961. [CrossRef]

119. Adams, S.; Loi, S.; Toppmeyer, D.; Cescon, D.W.; De Laurentiis, M.; Nanda, R.; Winer, E.P.; Mukai, H.; Tamura, K.; Armstrong, A.; et al. Pembrolizumab monotherapy for previously untreated, PD-L1-positive, metastatic triple-negative breast cancer: Cohort B of the phase II KEYNOTE-086 study. *Ann. Oncol.* **2019**, *30*, 405–411. [CrossRef]

120. Schmid, P.; Rugo, H.S.; Adams, S.; Schneeweiss, A.; Barrios, C.H.; Iwata, H.; Diéras, V.; Henschel, V.; Molinero, L.; Chui, S.Y.; et al. Atezolizumab plus nab-paclitaxel as first-line treatment for unresectable, locally advanced or metastatic triple-negative breast cancer (IMpassion130): Updated efficacy results from a randomised, double-blind, placebo-controlled, phase 3 trial. *Lancet Oncol.* **2020**, *21*, 44–59. [CrossRef]

121. Rugo, H.S.; Loi, S.; Adams, S.; Schmid, P.; Schneeweiss, A.; Barrios, C.H.; Iwata, H.; Dieras, V.C.; Winer, E.P.; Kockx, M.; et al. LBA20—Performance of PD-L1 immunohistochemistry (IHC) assays in unresectablelocallyadvanced or metastatic triple-negative breastcancer (mTNBC): Post-hoc analysis of IMpassion130. In Proceedings of the 44th ESMO Congress 27 September–1 October 2019, Barcelona, Spain. *Ann. Oncol.* **2019**, *30* (Suppl. 5), v858–v859. [CrossRef]

122. Cortes, J.; Cescon, D.W.; Rugo, S.; Nowecki, Z.; Im, S.A.; Yusof, M.; Gallardo, C.; Lipatov, O.; Barrios, C.H.; Holgado, E.; et al. KEYNOTE-355: Randomized, double-blind, phase III study of pembrolizumab + chemotherapy versus placebo + chemotherapy for previously untreated locally recurrent inoperable or metastatic triple-negative breast cancer. *Asco Meet. Libr.* **2020**, *38*, 1000.

123. Adams, S.; Schmid, P.; Rugo, H.S.; Winer, E.P.; Loirat, D.; Awada, A.; Cescon, D.W.; Iwata, H.; Campone, M.; Nanda, R.; et al. Pembrolizumab monotherapy for previously treated metastatic triple-negative breast cancer: Cohort A of the phase II KEYNOTE-086 study. *Ann. Oncol.* **2019**, *30*, 397–404. [CrossRef] [PubMed]

124. Schmid, P.; Cortés, J.; Dent, R.; Pusztai, L.; McArthur, H.L.; Kuemmel, S.; Bergh, J.; Denkert, C.; Park, Y.H.; Hui, R.; et al. KEYNOTE-522: Phase III Study of Pembrolizumab (Pembro) + Chemotherapy (Chemo) vs. Placebo (Pbo) + ChemoasNeoadjuvant Treatment, Followed by Pembro vs. PboasAdjuvant Treatment for Early Triple-Negative BreastCancer (TNBC). *Ann. Oncol.* **2019**, *30*, v853–v854. [CrossRef]

125. Gianni, L.; Huang, C.S.; Egle, D.; Bermejo, B. Pathologic Complete Response (pCR) to Neoadjuvant Treatment with or withoutAtezolizumab in Triple Negative, Early High-Risk and Locally Advanced BreastCancer. NeoTRIPaPDL1 Michelangelo randomizedstudy. In Proceedings of the 2019 San Antonio Breast Cancer Symposium, San Antonio, TX, USA, 10–14 December 2019. Abstract GS3-04.

126. Marabelle, A.; Le, D.T.; Ascierto, P.A.; Di Giacomo, A.M.; De Jesus-Acosta, A.; Delord, J.P.; Geva, R.; Gottfried, M.; Penel, N.; Hansen, A.R.; et al. Efficacy of Pembrolizumab in Patients With Noncolorectal High Microsatellite Instability/Mismatch Repair-Deficient Cancer: Results From the Phase II KEYNOTE-158 Study. *J. Clin. Oncol.* **2020**, *38*, 1–10. [CrossRef]

127. Le, D.T.; Uram, J.N.; Wang, H.; Bartlett, B.R.; Kemberling, H.; Eyring, A.D.; Skora, A.D.; Luber, B.S.; Azad, N.S.; Laheru, D.; et al. PD-1 Blockade in Tumors with Mismatch-Repair Deficiency. *N. Engl. J. Med.* **2015**, *372*, 2509–2520. [CrossRef]

128. Obeid, E.; Ellerbrock, A.; Handorf, E.; Goldstein, L.; Gatalica, Z.; Arguello, D.; Swain, S.; Isaacs, C.; Vacirca, J.; Tan, A.; et al. Abstract PD6-03: Distribution of microsatellite instability, tumor mutational load, and PD-L1 status in molecularly profiled invasive breast cancer. *Cancer Res.* **2018**, *78*, PD6-03.

129. Takano, K.; Ichikawa, Y.; Ueno, E.; Ohwada, M.; Suzuki, M.; Tsunoda, H.; Miwa, M.; Uchida, K.; Yoshikawa, H. Microsatellite instability and expression of mismatch repair genes in sporadic endometrial cancer coexisting with colorectal or breast cancer. *Oncol. Rep.* **2005**, *13*, 11–16. [CrossRef] [PubMed]

130. Snyder, A.; Makarov, V.; Merghoub, T.; Yuan, J.; Zaretsky, J.M.; Desrichard, A.; Walsh, L.A.; Postow, M.A.; Wong, P.; Ho, T.S.; et al. Genetic basis for clinical response to CTLA-4 blockade in melanoma. *N. Engl. J. Med.* **2014**, *371*, 2189–2199. [CrossRef] [PubMed]

131. Rizvi, N.A.; Hellmann, M.D.; Snyder, A.; Kvistborg, P.; Makarov, V.; Havel, J.J.; Lee, W.; Yuan, J.; Wong, P.; Ho, T.S.; et al. Cancer immunology. Mutational landscape determines sensitivity to PD-1 blockade in non-small cell lung cancer. *Science* **2015**, *348*, 124–128. [CrossRef] [PubMed]

132. Barroso-Sousa, R.; Jain, E.; Kim, D.; Partridge, A.H.; Cohen, O.; Wagle, N. Determinants of high tumor mutational burden (TMB) and mutational signatures in breast cancer. *J. Clin. Oncol.* **2018**, *36*, 1010. [CrossRef]

133. Thomas, A.; Routh, E.D.; Pullikuth, A.; Jin, G.; Su, J.; Chou, J.W.; Hoadley, K.A.; Print, C.; Knowlton, N.; Black, M.A.; et al. Tumor mutational burden is a determinant of immune-mediated survival in breast cancer. *Oncoimmunology* **2018**, *7*, e1490854. [CrossRef] [PubMed]

134. Barroso-Sousa, R.; Jain, E.; Cohen, O.; Kim, D.; Buendia-Buendia, J.; Winer, E.; Lin, N.; Tolaney, S.M.; Wagle, N. Prevalence and mutational determinants of high tumor mutation burden in breast cancer. *Ann. Oncol.* **2020**, *31*, 387–394. [CrossRef] [PubMed]

135. Loibl, S.; Sinn, B.V.; Karn, T.; Untch, M.; Sinn, H.-P.; Weber, K.E.; Hanusch, C.; Huober, J.B.; Staib, P.; Lorenz, R.; et al. Exome analysis of oncogenic pathways and tumor mutational burden (TMB) in triple-negative breast cancer (TNBC): Results of the translational biomarker program of the neoadjuvant double-blind placebo controlled GeparNuevo trial. *J. Clin. Oncol.* **2019**, *37*, 509. [CrossRef]

136. Karn, T.; Denkert, C.; Weber, K.E.; Holtrich, U.; Hanusch, C.; Sinn, B.V.; Higgs, B.W.; Jank, P.; Sinn, H.P.; Huober, J.; et al. Tumor mutational burden and immune infiltration as independent predictors of response to neoadjuvant immune checkpoint inhibition in early TNBC in GeparNuevo. *Ann. Oncol. Off. J. Eur. Soc. Med. Oncol.* **2020**. S0923-7534(20)39836-7. [CrossRef] [PubMed]

137. Marabelle, A.; Fakih, M.; Lopez, J.; Shah, M.; Shapira-Frommer, R.; Nakagawa, K.; Chung, H.C.; Kindler, H.L.; Lopez-Martin, J.A.; Miller, W.; et al. Association of Tumor Mutational Burden with Outcomes in Patients with Select Advanced Solid Tumors Treated with Pembrolizumab in KEYNOTE-158. *Ann. Oncol.* **2019**, *30* (Suppl. 5), v475–v532. [CrossRef]

138. Anderson, A.C.; Joller, N.; Kuchroo, V.K. Lag-3, Tim-3, and TIGIT: Co-inhibitory Receptors with Specialized Functions in Immune Regulation. *Immunity* **2016**, *44*, 989–1004. [CrossRef]

139. Saleh, R.R.; Peinado, P.; Fuentes-Antrás, J.; Pérez-Segura, P.; Pandiella, A.; Amir, E.; Ocaña, A. Prognostic Value of Lymphocyte-Activation Gene 3 (LAG3) in Cancer: A Meta-Analysis. *Front. Oncol.* **2019**, *9*, 1040. [CrossRef]

140. Fruman, D.A.; Chiu, H.; Hopkins, B.D.; Bagrodia, S.; Cantley, L.C.; Abraham, R.T. The PI3K Pathway in Human Disease. *Cell* **2017**, *170*, 605–635. [CrossRef]

141. Engelman, J.A. Targeting PI3K signalling in cancer: Opportunities, challenges and limitations. *Nat. Rev. Cancer* **2009**, *9*, 550–562. [CrossRef]

142. Fruman, D.A.; Rommel, C. PI3K and cancer: Lessons, challenges and opportunities. *Nat. Rev. Drug Discov.* **2014**, *13*, 140–156. [CrossRef]

143. Liu, P.; Cheng, H.; Roberts, T.M.; Zhao, J.J. Targeting the phosphoinositide 3-kinase pathway in cancer. *Nat. Rev. Drug Discov.* **2009**, *8*, 627–644. [CrossRef] [PubMed]

144. Pereira, B.; Chin, S.F.; Rueda, O.M.; Vollan, H.K.; Provenzano, E.; Bardwell, H.A.; Pugh, M.; Jones, L.; Russell, R.; Sammut, S.J.; et al. The somatic mutation profiles of 2, 433 breast cancers refines their genomic and transcriptomic landscapes. *Nat. Commun.* **2016**, *7*, 11479. [CrossRef] [PubMed]

145. Thorpe, L.M.; Yuzugullu, H.; Zhao, J.J. PI3K in cancer: Divergent roles of isoforms, modes of activation and therapeutic targeting. *Nat. Rev. Cancer* **2015**, *15*, 7–24. [CrossRef] [PubMed]

146. Cancer Genome Atlas Network. Comprehensive molecular portraits of human breast tumours. *Nature* **2012**, *490*, 61–70. [CrossRef]

147. Pascual, J.; Turner, N.C. Targeting the PI3-kinase pathway in triple-negative breast cancer. *Ann. Oncol.* **2019**, *30*, 1051–1060. [CrossRef]

148. Jia, S.; Liu, Z.; Zhang, S.; Liu, P.; Zhang, L.; Lee, S.H.; Zhang, J.; Signoretti, S.; Loda, M.; Roberts, T.M.; et al. Essential roles of PI(3)K-p110beta in cell growth, metabolism and tumorigenesis. *Nature* **2008**, *454*, 776–779. [CrossRef]

149. Wee, S.; Wiederschain, D.; Maira, S.M.; Loo, A.; Miller, C.; deBeaumont, R.; Stegmeier, F.; Yao, Y.M.; Lengauer, C. PTEN-deficient cancers depend on PIK3CB. *Proc. Natl. Acad. Sci USA* **2008**, *105*, 13057–13062. [CrossRef]

150. Song, M.S.; Salmena, L.; Pandolfi, P.P. The functions and regulation of the PTEN tumour suppressor. *Nat. Rev. Mol. Cell Biol* **2012**, *13*, 283–296. [CrossRef] [PubMed]

151. Pezzolesi, M.G.; Platzer, P.; Waite, K.A.; Eng, C. Differential expression of PTEN-targeting microRNAs miR-19a and miR-21 in Cowden syndrome. *Am. J. Hum. Genet.* **2008**, *82*, 1141–1149. [CrossRef]

152. Carpten, J.D.; Faber, A.L.; Horn, C.; Donoho, G.P.; Briggs, S.L.; Robbins, C.M.; Hostetter, G.; Boguslawski, S.; Moses, T.Y.; Savage, S.; et al. A transforming mutation in the pleckstrin homology domain of AKT1 in cancer. *Nature* **2007**, *448*, 439–444. [CrossRef]

153. Shaw, R.J.; Bardeesy, N.; Manning, B.D.; Lopez, L.; Kosmatka, M.; DePinho, R.A.; Cantley, L.C. The LKB1 tumor suppressor negatively regulates mTOR signaling. *Cancer Cell* **2004**, *6*, 91–99. [CrossRef] [PubMed]

154. US Food and Drug Administration. FDA Approves Alpelisib for Metastatic Breast Cancer. Available online: https://www.fda.gov/drugs/resources-information-approved-drugs/fda-approves-alpelisib-metastatic-breast-cancer (accessed on 25 September 2019).

155. André, F.; Ciruelos, E.; Rubovszky, G.; Campone, M.; Loibl, S.; Rugo, H.S.; Iwata, H.; Conte, P.; Mayer, I.A.; Kaufman, B.; et al. Alpelisib for PIK3CA-Mutated, Hormone Receptor-Positive Advanced Breast Cancer. *N. Engl. J. Med.* **2019**, *380*, 1929–1940. [CrossRef] [PubMed]

156. Zumsteg, Z.S.; Morse, N.; Krigsfeld, G.; Gupta, G.; Higginson, D.S.; Lee, N.Y.; Morris, L.; Ganly, I.; Shiao, S.L.; Powell, S.N.; et al. Taselisib (GDC-0032), a Potent β-Sparing Small Molecule Inhibitor of PI3K, Radiosensitizes Head and Neck Squamous Carcinomas Containing Activating PIK3CA Alterations. *Clin. Cancer Res.* **2016**, *22*, 2009–2019. [CrossRef] [PubMed]

157. Ndubaku, C.O.; Heffron, T.P.; Staben, S.T.; Baumgardner, M.; Blaquiere, N.; Bradley, E.; Bull, R.; Do, S.; Dotson, J.; Dudley, D.; et al. Discovery of 2-{3-[2-(1-isopropyl-3-methyl-1H-1, 2-4-triazol-5-yl)-5,6-dihydrobenzo[f]imidazo[1,2-d][1,4]oxazepin-9-yl]-1H-pyrazol-1-yl}-2-methylpropanamide (GDC-0032): A β-sparing phosphoinositide 3-kinase inhibitor with high unbound exposure and robust in vivo antitumor activity. *J. Med. Chem.* **2013**, *56*, 4597–4610. [PubMed]

158. Baselga, J.; Dent, S.F.; Cortés, J.; Im, Y.-H.; Diéras, V.; Harbeck, N.; Krop, I.E.; Verma, S.; Wilson, T.R.; Jin, H.; et al. Phase III study of taselisib (GDC-0032) + fulvestrant (FULV) v FULV in patients (pts) with estrogen receptor (ER)-positive, PIK3CA-mutant (MUT), locally advanced or metastatic breast cancer (MBC): Primary analysis from SANDPIPER. *J. Clin. Oncol.* **2018**, *36* (Suppl. 18), LBA1006. [CrossRef]

159. Yuan, Y.; Wen, W.; Yost, S.E.; Xing, Q.; Yan, J.; Han, E.S.; Mortimer, J.; Yim, J.H. Combination therapy with BYL719 and LEE011 is synergistic and causes a greater suppression of p-S6 in triple negative breast cancer. *Sci. Rep.* **2019**, *9*, 7509. [CrossRef] [PubMed]

160. Teo, Z.L.; Versaci, S.; Dushyanthen, S.; Caramia, F.; Savas, P.; Mintoff, C.P.; Zethoven, M.; Virassamy, B.; Luen, S.J.; McArthur, G.A.; et al. Combined CDK4/6 and PI3Kα Inhibition Is Synergistic and Immunogenic in Triple-Negative Breast Cancer. *Cancer Res.* **2017**, *77*, 6340–6352. [CrossRef]

161. Lehmann, B.D.; Bauer, J.A.; Schafer, J.M.; Pendleton, C.S.; Tang, L.; Johnson, K.C.; Chen, X.; Balko, J.M.; Gómez, H.; Arteaga, C.L.; et al. PIK3CA mutations in androgen receptor-positive triple negative breast cancer confer sensitivity to the combination of PI3K and androgen receptor inhibitors. *Breast Cancer Res.* **2014**, *16*, 406. [CrossRef] [PubMed]

162. Coussy, F.; Lavigne, M.; de Koning, L.; Botty, R.E.; Nemati, F.; Naguez, A.; Bataillon, G.; Ouine, B.; Dahmani, A.; Montaudon, E.; et al. Response to mTOR and PI3K inhibitors in enzalutamide-resistant luminal androgen receptor triple-negative breast cancer patient-derived xenografts. *Theranostics* **2020**, *10*, 1531–1543. [CrossRef]

163. Lehmann, B.D.; Abramson, V.G.; Sanders, M.E.; Mayer, E.L.; Haddad, T.C.; Nanda, R.; Van Poznak, C.; Storniolo, A.M.; Nangia, J.R.; Gonzalez-Ericsson, P.I.; et al. TBCRC 032 IB/II Multicenter Study: Molecular Insights to AR Antagonist and PI3K Inhibitor Efficacy in Patients with AR(+) Metastatic Triple-Negative Breast Cancer. *Clin. Cancer Res.* **2020**, *26*, 2111–2123. [CrossRef] [PubMed]

164. Morgillo, F.; Della Corte, C.M.; Diana, A.; Mauro, C.D.; Ciaramella, V.; Barra, G.; Belli, V.; Franzese, E.; Bianco, R.; Maiello, E.; et al. Phosphatidylinositol 3-kinase (PI3Kα)/AKT axis blockade with taselisib or ipatasertib enhances the efficacy of anti-microtubule drugs in human breast cancer cells. *Oncotarget* **2017**, *8*, 76479–76491. [CrossRef] [PubMed]

165. Yan, Y.; Serra, V.; Prudkin, L.; Scaltriti, M.; Murli, S.; Rodríguez, O.; Guzman, M.; Sampath, D.; Nannini, M.; Xiao, Y.; et al. Evaluation and clinical analyses of downstream targets of the Akt inhibitor GDC-0068. *Clin. Cancer Res.* **2013**, *19*, 6976–6986. [CrossRef] [PubMed]

166. Lin, J.; Sampath, D.; Nannini, M.A.; Lee, B.B.; Degtyarev, M.; Oeh, J.; Savage, H.; Guan, Z.; Hong, R.; Kassees, R.; et al. Targeting activated Akt with GDC-0068, a novel selective Akt inhibitor that is efficacious in multiple tumor models. *Clin. Cancer Res.* **2013**, *19*, 1760–1772. [CrossRef] [PubMed]

167. Blake, J.F.; Xu, R.; Bencsik, J.R.; Xiao, D.; Kallan, N.C.; Schlachter, S.; Mitchell, I.S.; Spencer, K.L.; Banka, A.L.; Wallace, E.M.; et al. Discovery and preclinical pharmacology of a selective ATP-competitive Akt inhibitor (GDC-0068) for the treatment of human tumors. *J. Med. Chem.* **2012**, *55*, 8110–8127. [CrossRef] [PubMed]

168. Kim, S.-B.; Dent, R.; Im, S.-A.; Espié, M.; Blau, S.; Tan, A.R.; Isakoff, S.J.; Oliveira, M.; Saura, C.; Wongchenko, M.J.; et al. Ipatasertib plus paclitaxel versus placebo plus paclitaxel as first-line therapy for metastatic triple-negative breast cancer (LOTUS): A multicentre, randomised, double-blind, placebo-controlled, phase 2 trial. *Lancet. Oncol.* **2017**, *18*, 1360–1372. [CrossRef]

169. Oliveira, M.; Saura, C.; Nuciforo, P.; Calvo, I.; Andersen, J.; Passos-Coelho, J.L.; Gil Gil, M.; Bermejo, B.; Patt, D.A.; Ciruelos, E.; et al. FAIRLANE, a double-blind placebo-controlled randomized phase II trial of neoadjuvant ipatasertib plus paclitaxel for early triple-negative breast cancer. *Ann. Oncol.* **2019**, *30*, 1289–1297. [CrossRef] [PubMed]

170. Schmid, P.; Loirat, D.; Savas, P.; Espinosa, E.; Boni, V.; Italiano, A.; White, S.; Singel, S.M.; Withana, N.; Mani, A.; et al. Abstract CT049: Phase Ib study evaluating a triplet combination of ipatasertib (IPAT), atezolizumab (atezo), and paclitaxel (PAC) or nab-PAC as first-line (1L) therapy for locally advanced/metastatic triple-negative breast cancer (TNBC). *Cancer Res.* **2019**, *79* (Suppl. 13), CT049.

171. Jabbarzadeh Kaboli, P.; Salimian, F.; Aghapour, S.; Xiang, S.; Zhao, Q.; Li, M.; Wu, X.; Du, F.; Zhao, Y.; Shen, J.; et al. Akt-targeted therapy as a promising strategy to overcome drug resistance in breast cancer—A comprehensive review from chemotherapy to immunotherapy. *Pharm. Res.* **2020**, *156*, 104806. [CrossRef]

172. Capivasertib Active against AKT1-Mutated Cancers. *Cancer Discov.* **2019**, *9*, Of7. [CrossRef]

173. Schmid, P.; Abraham, J.; Chan, S.; Wheatley, D.; Brunt, A.M.; Nemsadze, G.; Baird, R.D.; Park, Y.H.; Hall, P.S.; Perren, T.; et al. Capivasertib Plus Paclitaxel Versus Placebo Plus Paclitaxel As First-Line Therapy for Metastatic Triple-Negative Breast Cancer: The PAKT Trial. *J. Clin. Oncol.* **2020**, *38*, 423–433. [CrossRef]

174. Thomas, A.; Teicher, B.A.; Hassan, R. Antibody-drug conjugates for cancer therapy. *Lancet Oncol.* **2016**, *17*, e254–e262. [CrossRef]

175. Müller, P.; Martin, K.; Theurich, S.; Schreiner, J.; Savic, S.; Terszowski, G.; Lardinois, D.; Heinzelmann-Schwarz, V.A.; Schlaak, M.; Kvasnicka, H.M.; et al. Microtubule-depolymerizing agents used in antibody-drug conjugates induce antitumor immunity by stimulation of dendritic cells. *Cancer Immunol. Res.* **2014**, *2*, 741–755. [CrossRef] [PubMed]

176. Rose, A.A.; Grosset, A.A.; Dong, Z.; Russo, C.; Macdonald, P.A.; Bertos, N.R.; St-Pierre, Y.; Simantov, R.; Hallett, M.; Park, M.; et al. Glycoprotein nonmetastatic B is an independent prognostic indicator of recurrence and a novel therapeutic target in breast cancer. *Clin. Cancer Res.* **2010**, *16*, 2147–2156. [CrossRef] [PubMed]

177. Rose, A.A.; Pepin, F.; Russo, C.; Abou Khalil, J.E.; Hallett, M.; Siegel, P.M. Osteoactivin promotes breast cancer metastasis to bone. *Mol. Cancer Res.* **2007**, *5*, 1001–1014. [CrossRef] [PubMed]

178. Maric, G.; Annis, M.G.; Dong, Z.; Rose, A.A.; Ng, S.; Perkins, D.; MacDonald, P.A.; Ouellet, V.; Russo, C.; Siegel, P.M. GPNMB cooperates with neuropilin-1 to promote mammary tumor growth and engages integrin $\alpha 5\beta 1$ for efficient breast cancer metastasis. *Oncogene* **2015**, *34*, 5494–5504. [CrossRef] [PubMed]

179. Maric, G.; Rose, A.A.; Annis, M.G.; Siegel, P.M. Glycoprotein non-metastatic b (GPNMB): A metastatic mediator and emerging therapeutic target in cancer. *Onco Targets* **2013**, *6*, 839–852.

180. Wolska-Washer, A.; Robak, T. Safety and Tolerability of Antibody-Drug Conjugates in Cancer. *Drug Saf.* **2019**, *42*, 295–314. [CrossRef]

181. Yardley, D.A.; Weaver, R.; Melisko, M.E.; Saleh, M.N.; Arena, F.P.; Forero, A.; Cigler, T.; Stopeck, A.; Citrin, D.; Oliff, I.; et al. EMERGE: A Randomized Phase II Study of the Antibody-Drug Conjugate Glembatumumab Vedotin in Advanced Glycoprotein NMB-Expressing Breast Cancer. *J. Clin. Oncol.* **2015**, *33*, 1609–1619. [CrossRef]

182. Yardley, D.A.; Melisko, M.E.; Forero, A.; Daniel, B.R.; Montero, A.J.; Guthrie, T.H.; Canfield, V.A.; Oakman, C.A.; Chew, H.K.; Ferrario, C.; et al. METRIC: A randomized international study of the antibody-drug conjugate glembatumumab vedotin (GV or CDX-011) in patients (pts) with metastatic gpNMB-overexpressing triple-negative breast cancer (TNBC). *J. Clin. Oncol.* **2015**, *33* (Suppl. 15), TPS1110. [CrossRef]

183. Vahdat, L.; Forero-Torres, A.; Schmid, P.; Blackwell, K.; Telli, M.; Melisko, M.; Holgado, E.; Moebus, V.; Cortes, J.; Fehrenbacher, L.; et al. Abstract P6-20-01: METRIC: A randomized international phase 2b study of the antibody-drug conjugate (ADC) glembatumumab vedotin (GV) in gpNMB-overexpressing, metastatic, triple-negative breast cancer (mTNBC). *Cancer Res.* **2019**, *79* (Suppl. 4), P6-20-01.

184. Wang, J.; Day, R.; Dong, Y.; Weintraub, S.J.; Michel, L. Identification of Trop-2 as an oncogene and an attractive therapeutic target in colon cancers. *Mol. Cancer* **2008**, *7*, 280–285. [CrossRef] [PubMed]

185. Cubas, R.; Zhang, S.; Li, M.; Chen, C.; Yao, Q. Trop2 expression contributes to tumor pathogenesis by activating the ERK MAPK pathway. *Mol. Cancer* **2010**, *9*, 253. [CrossRef] [PubMed]

186. Lin, H.; Huang, J.F.; Qiu, J.R.; Zhang, H.L.; Tang, X.J.; Li, H.; Wang, C.J.; Wang, Z.C.; Feng, Z.Q.; Zhu, J. Significantly upregulated TACSTD2 and Cyclin D1 correlate with poor prognosis of invasive ductal breast cancer. *Exp. Mol. Pathol* **2013**, *94*, 73–78. [CrossRef] [PubMed]

187. Ambrogi, F.; Fornili, M.; Boracchi, P.; Trerotola, M.; Relli, V.; Simeone, P.; La Sorda, R.; Lattanzio, R.; Querzoli, P.; Pedriali, M.; et al. Trop-2 is a determinant of breast cancer survival. *PLoS ONE* **2014**, *9*, e96993. [CrossRef] [PubMed]

188. Goldenberg, D.M.; Sharkey, R.M. Antibody-drug conjugates targeting TROP-2 and incorporating SN-38: A case study of anti-TROP-2 sacituzumab govitecan. *MAbs* **2019**, *11*, 987–995. [CrossRef]

189. Bardia, A.; Mayer, I.A.; Diamond, J.R.; Moroose, R.L.; Isakoff, S.J.; Starodub, A.N.; Shah, N.C.; O'Shaughnessy, J.; Kalinsky, K.; Guarino, M.; et al. Efficacy and Safety of Anti-Trop-2 Antibody Drug Conjugate Sacituzumab Govitecan (IMMU-132) in Heavily Pretreated Patients with Metastatic Triple-Negative Breast Cancer. *J. Clin. Oncol.* **2017**, *35*, 2141–2148. [CrossRef] [PubMed]

190. Bardia, A.; Mayer, I.A.; Vahdat, L.T.; Tolaney, S.M.; Isakoff, S.J.; Diamond, J.R.; O'Shaughnessy, J.; Moroose, R.L.; Santin, A.D.; Abramson, V.G.; et al. Sacituzumab Govitecan-hziy in Refractory Metastatic Triple-Negative Breast Cancer. *N. Engl J. Med.* **2019**, *380*, 741–751. [CrossRef]

191. Tray, N.; Adams, S.; Esteva, F.J. Antibody-drug conjugates in triple negative breast cancer. *Future Oncol.* **2018**, *14*, 2651–2661. [CrossRef]

192. Lue, H.W.; Yang, X.; Wang, R.; Qian, W.; Xu, R.Z.; Lyles, R.; Osunkoya, A.O.; Zhou, B.P.; Vessella, R.L.; Zayzafoon, M.; et al. LIV-1 promotes prostate cancer epithelial-to-mesenchymal transition and metastasis through HB-EGF shedding and EGFR-mediated ERK signaling. *PLoS ONE* **2011**, *6*, e27720. [CrossRef]

193. Zhao, L.; Chen, W.; Taylor, K.M.; Cai, B.; Li, X. LIV-1 suppression inhibits HeLa cell invasion by targeting ERK1/2-Snail/Slug pathway. *Biochem. Biophys. Res. Commun.* **2007**, *363*, 82–88. [CrossRef]

194. Huber, M.A.; Kraut, N.; Beug, H. Molecular requirements for epithelial-mesenchymal transition during tumor progression. *Curr. Opin. Cell Biol.* **2005**, *17*, 548–558. [CrossRef] [PubMed]

195. Sussman, D.; Smith, L.M.; Anderson, M.E.; Duniho, S.; Hunter, J.H.; Kostner, H.; Miyamoto, J.B.; Nesterova, A.; Westendorf, L.; Van Epps, H.A.; et al. SGN-LIV1A: A novel antibody-drug conjugate targeting LIV-1 for the treatment of metastatic breast cancer. *Mol. Cancer* **2014**, *13*, 2991–3000. [CrossRef] [PubMed]

196. Nejadmoghaddam, M.-R.; Minai-Tehrani, A.; Ghahremanzadeh, R.; Mahmoudi, M.; Dinarvand, R.; Zarnani, A.-H. Antibody-Drug Conjugates: Possibilities and Challenges. *Avicenna J. Med. Biotechnol.* **2019**, *11*, 3–23. [PubMed]

197. Gomez-Roca, C.A.; Boni, V.; Moreno, V.; Morris, J.C.; Delord, J.-P.; Calvo, E.; Papadopoulos, K.P.; Rixe, O.; Cohen, P.; Tellier, A.; et al. A phase I study of SAR566658, an anti CA6-antibody drug conjugate (ADC), in patients (Pts) with CA6-positive advanced solid tumors (STs)(NCT01156870). *J. Clin. Oncol.* **2016**, (Suppl. 15). [CrossRef]

198. Krebs, M.G.; Hou, J.M.; Ward, T.H.; Blackhall, F.H.; Dive, C. Circulating tumour cells: Their utility in cancer management and predicting outcomes. *Adv. Med. Oncol.* **2010**, *2*, 351–365. [CrossRef]

199. Toss, A.; Mu, Z.; Fernandez, S.; Cristofanilli, M. CTC enumeration and characterization: Moving toward personalized medicine. *Ann. Transl. Med.* **2014**, *2*, 108. [PubMed]

200. Pukazhendhi, G.; Glück, S. Circulating tumor cells in breast cancer. *J. Carcinog.* **2014**, *13*, 8.

201. Sparano, J.; O'Neill, A.; Alpaugh, K.; Wolff, A.C.; Northfelt, D.W.; Dang, C.T.; Sledge, G.W.; Miller, K.D. Association of Circulating Tumor Cells With Late Recurrence of Estrogen Receptor-Positive Breast Cancer: A Secondary Analysis of a Randomized Clinical Trial. *JAMA Oncol.* **2018**, *4*, 1700–1706. [CrossRef]

202. Bidard, F.C.; Peeters, D.J.; Fehm, T.; Nolé, F.; Gisbert-Criado, R.; Mavroudis, D.; Grisanti, S.; Generali, D.; Garcia-Saenz, J.A.; Stebbing, J.; et al. Clinical validity of circulating tumour cells in patients with metastatic breast cancer: A pooled analysis of individual patient data. *Lancet Oncol.* **2014**, *15*, 406–414. [CrossRef]

203. Yu, T.; Di, G. Role of tumor microenvironment in triple-negative breast cancer and its prognostic significance. *Chin. J. Cancer Res.* **2017**, *29*, 237–252. [CrossRef]

204. Yu, M.; Stott, S.; Toner, M.; Maheswaran, S.; Haber, D.A. Circulating tumor cells: Approaches to isolation and characterization. *J. Cell Biol.* **2011**, *192*, 373–382. [CrossRef] [PubMed]

205. Piñeiro, R.; Martínez-Pena, I.; López-López, R. Relevance of CTC Clusters in Breast Cancer Metastasis. In *Circulating Tumor Cells in Breast Cancer Metastatic Disease. Advances in Experimental Medicine and Biology*; Piñeiro, R., Ed.; Springer: Cham, Germany, 2020; Volume 1220.

206. Cristofanilli, M.; Hayes, D.F.; Budd, G.T.; Ellis, M.J.; Stopeck, A.; Reuben, J.M.; Doyle, G.V.; Matera, J.; Allard, W.J.; Miller, M.C.; et al. Circulating tumor cells: A novel prognostic factor for newly diagnosed metastatic breast cancer. *J. Clin. Oncol.* **2005**, *23*, 1420–1430. [CrossRef] [PubMed]

207. De Bono, J.S.; Scher, H.I.; Montgomery, R.B.; Parker, C.; Miller, M.C.; Tissing, H.; Doyle, G.V.; Terstappen, L.W.; Pienta, K.J.; Raghavan, D. Circulating tumor cells predict survival benefit from treatment in metastatic castration-resistant prostate cancer. *Clin. Cancer Res.* **2008**, *14*, 6302–6309. [CrossRef] [PubMed]

208. Cristofanilli, M.; Budd, G.T.; Ellis, M.J.; Stopeck, A.; Matera, J.; Miller, M.C.; Reuben, J.M.; Doyle, G.V.; Allard, W.J.; Terstappen, L.W.; et al. Circulating tumor cells, disease progression, and survival in metastatic breast cancer. *N. Engl. J. Med.* **2004**, *351*, 781–791. [CrossRef] [PubMed]

209. Hayes, D.F.; Cristofanilli, M.; Budd, G.T.; Ellis, M.J.; Stopeck, A.; Miller, M.C.; Matera, J.; Allard, W.J.; Doyle, G.V.; Terstappen, L.W. Circulating tumor cells at each follow-up time point during therapy of metastatic breast cancer patients predict progression-free and overall survival. *Clin. Cancer Res.* **2006**, *12 Pt 1*, 4218–4224. [CrossRef]

210. Nolé, F.; Munzone, E.; Zorzino, L.; Minchella, I.; Salvatici, M.; Botteri, E.; Medici, M.; Verri, E.; Adamoli, L.; Rotmensz, N.; et al. Variation of circulating tumor cell levels during treatment of metastatic breast cancer: Prognostic and therapeutic implications. *Ann. Oncol.* **2008**, *19*, 891–897. [CrossRef] [PubMed]

211. Madic, J.; Kiialainen, A.; Bidard, F.C.; Birzele, F.; Ramey, G.; Leroy, Q.; Rio Frio, T.; Vaucher, I.; Raynal, V.; Bernard, V.; et al. Circulating tumor DNA and circulating tumor cells in metastatic triple negative breast cancer patients. *Int. J. Cancer* **2015**, *136*, 2158–2165. [CrossRef] [PubMed]

212. Rossi, G.; Mu, Z.; Rademaker, A.W.; Austin, L.K.; Strickland, K.S.; Costa, R.L.B.; Nagy, R.J.; Zagonel, V.; Taxter, T.J.; Behdad, A.; et al. Cell-Free DNA and Circulating Tumor Cells: Comprehensive Liquid Biopsy Analysis in Advanced Breast Cancer. *Clin. Cancer Res.* **2018**, *24*, 560–568. [CrossRef]

213. Bidard, F.C.; Mathiot, C.; Degeorges, A.; Etienne-Grimaldi, M.C.; Delva, R.; Pivot, X.; Veyret, C.; Bergougnoux, L.; de Cremoux, P.; Milano, G.; et al. Clinical value of circulating endothelial cells and circulating tumor cells in metastatic breast cancer patients treated first line with bevacizumab and chemotherapy. *Ann. Oncol.* **2010**, *21*, 1765–1771. [CrossRef]

214. Giordano, A.; Giuliano, M.; De Laurentiis, M.; Arpino, G.; Jackson, S.; Handy, B.C.; Ueno, N.T.; Andreopoulou, E.; Alvarez, R.H.; Valero, V.; et al. Circulating tumor cells in immunohistochemical subtypes of metastatic breast cancer: Lack of prediction in HER2-positive disease treated with targeted therapy. *Ann. Oncol.* **2012**, *23*, 1144–1150. [CrossRef]

215. Giuliano, M.; Giordano, A.; Jackson, S.; Hess, K.R.; De Giorgi, U.; Mego, M.; Handy, B.C.; Ueno, N.T.; Alvarez, R.H.; De Laurentiis, M.; et al. Circulating tumor cells as prognostic and predictive markers in metastatic breast cancer patients receiving first-line systemic treatment. *Breast Cancer Res.* **2011**, *13*, R67. [CrossRef] [PubMed]

216. Cristofanilli, M.; Pierga, J.Y.; Reuben, J.; Rademaker, A.; Davis, A.A.; Peeters, D.J.; Fehm, T.; Nolé, F.; Gisbert-Criado, R.; Mavroudis, D.; et al. The clinical use of circulating tumor cells (CTCs) enumeration for staging of metastatic breast cancer (MBC): International expert consensus paper. *Crit Rev. Oncol. Hematol.* **2019**, *134*, 39–45. [CrossRef] [PubMed]

217. Lucci, A.; Hall, C.S.; Lodhi, A.K.; Bhattacharyya, A.; Anderson, A.E.; Xiao, L.; Bedrosian, I.; Kuerer, H.M.; Krishnamurthy, S. Circulating tumour cells in non-metastatic breast cancer: A prospective study. *Lancet Oncol.* **2012**, *13*, 688–695. [CrossRef]

218. Janni, W.J.; Rack, B.; Terstappen, L.W.; Pierga, J.Y.; Taran, F.A.; Fehm, T.; Hall, C.; de Groot, M.R.; Bidard, F.C.; Friedl, T.W.; et al. Pooled Analysis of the Prognostic Relevance of Circulating Tumor Cells in Primary Breast Cancer. *Clin. Cancer Res.* **2016**, *22*, 2583–2593. [CrossRef] [PubMed]

219. Rack, B.; Andergassen, U.; Janni, W.; Neugebauer, J. CTCs in primary breast cancer (I). *Recent Results Cancer Res.* **2012**, *195*, 179–185. [PubMed]

220. Mikulová, V.; Cabiňaková, M.; Janatková, I.; Mestek, O.; Zima, T.; Tesařová, P. Detection of circulating tumor cells during follow-up of patients with early breast cancer: Clinical utility for monitoring of therapy efficacy. *Scand. J. Clin. Lab. Investig.* **2014**, *74*, 132–142. [CrossRef] [PubMed]

221. Maltoni, R.; Gallerani, G.; Fici, P.; Rocca, A.; Fabbri, F. CTCs in early breast cancer: A path worth taking. *Cancer Lett* **2016**, *376*, 205–210. [CrossRef] [PubMed]

222. Rack, B.; Schindlbeck, C.; Jückstock, J.; Andergassen, U.; Hepp, P.; Zwingers, T.; Friedl, T.W.; Lorenz, R.; Tesch, H.; Fasching, P.A.; et al. Circulating tumor cells predict survival in early average-to-high risk breast cancer patients. *J. Natl. Cancer Inst.* **2014**, *106*, dju066. [CrossRef] [PubMed]

223. Riethdorf, S.; Müller, V.; Zhang, L.; Rau, T.; Loibl, S.; Komor, M.; Roller, M.; Huober, J.; Fehm, T.; Schrader, I.; et al. Detection and HER2 expression of circulating tumor cells: Prospective monitoring in breast cancer patients treated in the neoadjuvant GeparQuattro trial. *Clin. Cancer Res.* **2010**, *16*, 2634–2645. [CrossRef]

224. Yang, M.-H.; Imrali, A.; Heeschen, C. Circulating cancer stem cells: The importance to select. *Chin. J. Cancer Res.* **2015**, *27*, 437–449.

225. Bulfoni, M.; Gerratana, L.; Del Ben, F.; Marzinotto, S.; Sorrentino, M.; Turetta, M.; Scoles, G.; Toffoletto, B.; Isola, M.; Beltrami, C.A.; et al. In patients with metastatic breast cancer the identification of circulating tumor cells in epithelial-to-mesenchymal transition is associated with a poor prognosis. *Breast Cancer Res.* **2016**, *18*, 30. [CrossRef] [PubMed]

226. Yu, M.; Bardia, A.; Wittner, B.S.; Stott, S.L.; Smas, M.E.; Ting, D.T.; Isakoff, S.J.; Ciciliano, J.C.; Wells, M.N.; Shah, A.M.; et al. Circulating breast tumor cells exhibit dynamic changes in epithelial and mesenchymal composition. *Science* **2013**, *339*, 580–584. [CrossRef] [PubMed]

227. Munzone, E.; Botteri, E.; Sandri, M.T.; Esposito, A.; Adamoli, L.; Zorzino, L.; Sciandivasci, A.; Cassatella, M.C.; Rotmensz, N.; Aurilio, G.; et al. Prognostic value of circulating tumor cells according to immunohistochemically defined molecular subtypes in advanced breast cancer. *Clin. Breast Cancer* **2012**, *12*, 340–346. [CrossRef] [PubMed]

228. Jiang, Z.F.; Cristofanilli, M.; Shao, Z.M.; Tong, Z.S.; Song, E.W.; Wang, X.J.; Liao, N.; Hu, X.C.; Liu, Y.; Wang, Y.; et al. Circulating tumor cells predict progression-free and overall survival in Chinese patients with metastatic breast cancer, HER2-positive or triple-negative (CBCSG004): A multicenter, double-blind, prospective trial. *Ann. Oncol.* **2013**, *24*, 2766–2772. [CrossRef] [PubMed]

229. Paoletti, C.; Li, Y.; Muñiz, M.C.; Kidwell, K.M.; Aung, K.; Thomas, D.G.; Brown, M.E.; Abramson, V.G.; Irvin, W.J., Jr.; Lin, N.U.; et al. Significance of Circulating Tumor Cells in Metastatic Triple-Negative Breast Cancer Patients within a Randomized, Phase II Trial: TBCRC 019. *Clin. Cancer Res.* **2015**, *21*, 2771–2779. [CrossRef] [PubMed]

230. Smerage, J.B.; Barlow, W.E.; Hortobagyi, G.N.; Winer, E.P.; Leyland-Jones, B.; Srkalovic, G.; Tejwani, S.; Schott, A.F.; O'Rourke, M.A.; Lew, D.L.; et al. Circulating tumor cells and response to chemotherapy in metastatic breast cancer: SWOG S0500. *J. Clin. Oncol.* **2014**, *32*, 3483–3489. [CrossRef]

231. Magbanua, M.J.M.; Carey, L.A.; DeLuca, A.; Hwang, J.; Scott, J.H.; Rimawi, M.F.; Mayer, E.L.; Marcom, P.K.; Liu, M.C.; Esteva, F.J.; et al. Circulating Tumor Cell Analysis in Metastatic Triple-Negative Breast Cancers. *Clin. Cancer Res.* **2015**, *21*, 1098–1105. [CrossRef]

232. Liu, M.C.; Janni, W.; Georgoulias, V.; Yardley, D.A.; Harbeck, N.; Wei, X.; McGovern, D.; Beck, R. First-Line Doublet Chemotherapy for Metastatic Triple-Negative Breast Cancer: Circulating Tumor Cell Analysis of the tnAcity Trial. *Cancer Manag Res.* **2019**, *11*, 10427–10433. [CrossRef]

233. Liu, X.; Ran, R.; Shao, B.; Rugo, H.S.; Yang, Y.; Hu, Z.; Wei, Z.; Wan, F.; Kong, W.; Song, G.; et al. Combined peripheral natural killer cell and circulating tumor cell enumeration enhance prognostic efficiency in patients with metastatic triple-negative breast cancer. *Chin. J. Cancer Res.* **2018**, *30*, 315–326. [CrossRef]

234. Zhang, Q.; Gerratana, L.; Shah, A.N.; Davis, A.A.; Flaum, L.; Zhang, Y.; Pestell, R.G.; Wehbe, F.; Behdad, A.; Platanias, L.; et al. Abstract 408: Expression of CCR5 associated with HER2 in circulating tumor cells (CTCs) is a novel biomarker for patients with metastatic breast cancer (MBC). *Cancer Res.* **2019**, *79* (Suppl. 13), 408.

235. Aceto, N.; Bardia, A.; Miyamoto, D.T.; Donaldson, M.C.; Wittner, B.S.; Spencer, J.A.; Yu, M.; Pely, A.; Engstrom, A.; Zhu, H.; et al. Circulating tumor cell clusters are oligoclonal precursors of breast cancer metastasis. *Cell* **2014**, *158*, 1110–1122. [CrossRef] [PubMed]

236. Duda, D.G.; Duyverman, A.M.; Kohno, M.; Snuderl, M.; Steller, E.J.; Fukumura, D.; Jain, R.K. Malignant cells facilitate lung metastasis by bringing their own soil. *Proc. Natl. Acad. Sci. USA* **2010**, *107*, 21677–21682. [CrossRef] [PubMed]

237. Yu, M.; Ting, D.T.; Stott, S.L.; Wittner, B.S.; Ozsolak, F.; Paul, S.; Ciciliano, J.C.; Smas, M.E.; Winokur, D.; Gilman, A.J.; et al. RNA sequencing of pancreatic circulating tumour cells implicates WNT signalling in metastasis. *Nature* **2012**, *487*, 510–513. [CrossRef] [PubMed]

238. Lambert, A.W.; Pattabiraman, D.R.; Weinberg, R.A. Emerging Biological Principles of Metastasis. *Cell* **2017**, *168*, 670–691. [CrossRef] [PubMed]

239. Santamaría, P.G.; Moreno-Bueno, G.; Cano, A. Contribution of Epithelial Plasticity to Therapy Resistance. *J. Clin. Med.* **2019**, *8*, 676. [CrossRef] [PubMed]

240. Aktas, B.; Tewes, M.; Fehm, T.; Hauch, S.; Kimmig, R.; Kasimir-Bauer, S. Stem cell and epithelial-mesenchymal transition markers are frequently overexpressed in circulating tumor cells of metastatic breast cancer patients. *Breast Cancer Res.* **2009**, *11*, R46. [CrossRef]

241. Kasimir-Bauer, S.; Bittner, A.K.; König, L.; Reiter, K.; Keller, T.; Kimmig, R.; Hoffmann, O. Does primary neoadjuvant systemic therapy eradicate minimal residual disease? Analysis of disseminated and circulating tumor cells before and after therapy. *Breast Cancer Res.* **2016**, *18*, 20. [CrossRef]

242. Abreu, M.; Cabezas-Sainz, P.; Pereira-Veiga, T.; Falo, C.; Abalo, A.; Morilla, I.; Curiel, T.; Cueva, J.; Rodríguez, C.; Varela-Pose, V.; et al. Looking for a Better Characterization of Triple-Negative Breast Cancer by Means of Circulating Tumor Cells. *J. Clin. Med.* **2020**, *9*, 353. [CrossRef]

243. Razmara, A.M.; Sollier, E.; Kisirkoi, G.N.; Baker, S.W.; Bellon, M.B.; McMillan, A.; Lemaire, C.A.; Ramani, V.C.; Jeffrey, S.S.; Casey, K.M. Tumor shedding and metastatic progression after tumor excision in patient-derived orthotopic xenograft models of triple-negative breast cancer. *Clin. Exp. Metastasis* **2020**, *37*, 413–424. [CrossRef]

244. Thangavel, H.; De Angelis, C.; Vasaikar, S.; Bhat, R.; Jolly, M.K.; Nagi, C.; Creighton, C.J.; Chen, F.; Dobrolecki, L.E.; George, J.T.; et al. A CTC-Cluster-Specific Signature Derived from OMICS Analysis of Patient-Derived Xenograft Tumors Predicts Outcomes in Basal-Like Breast Cancer. *J. Clin. Med.* **2019**, *8*, 1772. [CrossRef]

245. Cho, E.H.; Wendel, M.; Luttgen, M.; Yoshioka, C.; Marrinucci, D.; Lazar, D.; Schram, E.; Nieva, J.; Bazhenova, L.; Morgan, A.; et al. Characterization of circulating tumor cell aggregates identified in patients with epithelial tumors. *Phys. Biol.* **2012**, *9*, 016001. [CrossRef] [PubMed]

246. Gkountela, S.; Castro-Giner, F.; Szczerba, B.M.; Vetter, M.; Landin, J.; Scherrer, R.; Krol, I.; Scheidmann, M.C.; Beisel, C.; Stirnimann, C.U.; et al. Circulating Tumor Cell Clustering Shapes DNA Methylation to Enable Metastasis Seeding. *Cell* **2019**, *176*, 98–112. [CrossRef] [PubMed]

247. Aceto, N.; Toner, M.; Maheswaran, S.; Haber, D.A. En Route to Metastasis: Circulating Tumor Cell Clusters and Epithelial-to-Mesenchymal Transition. *Trends Cancer* **2015**, *1*, 44–52. [CrossRef] [PubMed]

248. Mu, Z.; Benali-Furet, N.; Uzan, G.; Znaty, A.; Ye, Z.; Paolillo, C.; Wang, C.; Austin, L.; Rossi, G.; Fortina, P.; et al. Cristofanilli, M. Detection and Characterization of Circulating Tumor Associated Cells in Metastatic Breast Cancer. *Int. J. Mol. Sci.* **2016**, *17*.

249. Wei, R.R.; Sun, D.N.; Yang, H.; Yan, J.; Zhang, X.; Zheng, X.L.; Fu, X.H.; Geng, M.Y.; Huang, X.; Ding, J. CTC clusters induced by heparanase enhance breast cancer metastasis. *Acta Pharm. Sin.* **2018**, *39*, 1326–1337. [CrossRef] [PubMed]

250. Yu, M. Metastasis Stemming from Circulating Tumor Cell Clusters. *Trends Cell Biol.* **2019**, *29*, 275–276. [CrossRef] [PubMed]

251. Larsson, A.M.; Jansson, S.; Bendahl, P.O.; Levin Tykjaer Jörgensen, C.; Loman, N.; Graffman, C.; Lundgren, L.; Aaltonen, K.; Rydén, L. Longitudinal enumeration and cluster evaluation of circulating tumor cells improve prognostication for patients with newly diagnosed metastatic breast cancer in a prospective observational trial. *Breast Cancer Res.* **2018**, *20*, 48. [CrossRef]

252. Paoletti, C.; Miao, J.; Dolce, E.M.; Darga, E.P.; Repollet, M.I.; Doyle, G.V.; Gralow, J.R.; Hortobagyi, G.N.; Smerage, J.B.; Barlow, W.E.; et al. Circulating Tumor Cell Clusters in Patients with Metastatic Breast Cancer: A SWOG S0500 Translational Medicine Study. *Clin. Cancer Res.* **2019**, *25*, 6089–6097. [CrossRef]

253. Costa, C.; Muinelo-Romay, L.; Cebey-López, V.; Pereira-Veiga, T.; Martínez-Pena, I.; Abreu, M.; Abalo, A.; Lago-Lestón, R.M.; Abuín, C.; Palacios, P.; et al. Analysis of a Real-World Cohort of Metastatic Breast Cancer Patients Shows Circulating Tumor Cell Clusters (CTC-clusters) as Predictors of Patient Outcomes. *Cancers* **2020**, *12*, 1111. [CrossRef]

254. Pantel, K.; Speicher, M.R. The biology of circulating tumor cells. *Oncogene* **2016**, *35*, 1216–1224. [CrossRef]

255. Alimirzaie, S.; Bagherzadeh, M.; Akbari, M.R. Liquid biopsy in breast cancer: A comprehensive review. *Clin. Genet.* **2019**, *95*, 643–660. [CrossRef] [PubMed]

256. Shang, M.; Chang, C.; Pei, Y.; Guan, Y.; Chang, J.; Li, H. Potential Management of Circulating Tumor DNA as a Biomarker in Triple-Negative Breast Cancer. *J. Cancer* **2018**, *9*, 4627–4634. [CrossRef] [PubMed]

257. Stover, D.G.; Parsons, H.A.; Ha, G.; Freeman, S.S.; Barry, W.T.; Guo, H.; Choudhury, A.D.; Gydush, G.; Reed, S.C.; Rhoades, J.; et al. Association of Cell-Free DNA Tumor Fraction and Somatic Copy Number Alterations With Survival in Metastatic Triple-Negative Breast Cancer. *J. Clin. Oncol.* **2018**, *36*, 543–553. [CrossRef] [PubMed]

258. Parsons, H.A.; Beaver, J.A.; Cimino-Mathews, A.; Ali, S.M.; Axilbund, J.; Chu, D.; Connolly, R.M.; Cochran, R.L.; Croessmann, S.; Clark, T.A.; et al. Individualized Molecular Analyses Guide Efforts (IMAGE): A Prospective Study of Molecular Profiling of Tissue and Blood in Metastatic Triple-Negative Breast Cancer. *Clin. Cancer Res.* **2017**, *23*, 379. [CrossRef] [PubMed]

259. Kim, S.B.; Dent, R.; Wongchenko, M.J.; Singel, S.M.; Baselga, J. Concordance between plasma-based and tissue-based next-generation sequencing in LOTUS. *Lancet Oncol.* **2017**, *18*, e638. [CrossRef]

260. Vidula, N.; Isakoff, S.J.; Niemierko, A.; Malvarosa, G.; Park, H.; Abraham, E.; Spring, L.; Peppercorn, J.; Moy, B.; Ellisen, L.W.; et al. Abstract PD1-13: Somatic BRCA mutation detection by circulating tumor DNA analysis in patients with metastatic breast cancer: Incidence and association with tumor genotyping results, germline BRCA mutation status, and clinical outcomes. *Cancer Res.* **2018**, *78* (Suppl. 4), PD1-13.

261. Garcia-Murillas, I.; Chopra, N.; Comino-Méndez, I.; Beaney, M.; Tovey, H.; Cutts, R.J.; Swift, C.; Kriplani, D.; Afentakis, M.; Hrebien, S.; et al. Assessment of Molecular Relapse Detection in Early-Stage Breast Cancer. *JAMA Oncol.* **2019**, *5*, 1473–1478. [CrossRef] [PubMed]

262. Riva, F.; Bidard, F.C.; Houy, A.; Saliou, A.; Madic, J.; Rampanou, A.; Hego, C.; Milder, M.; Cottu, P.; Sablin, M.P.; et al. Patient-Specific Circulating Tumor DNA Detection during Neoadjuvant Chemotherapy in Triple-Negative Breast Cancer. *Clin. Chem.* **2017**, *63*, 691–699. [CrossRef] [PubMed]

263. Cavallone, L.; Aguilar, A.; Aldamry, M.; Lafleur, J.; Brousse, S.; Lan, C.; Alirezaie, N.; Bareke, E.; Majewski, J.; Pelmus, M.; et al. Circulating tumor DNA (ctDNA) during and after neoadjuvant chemotherapy and prior to surgery is a powerful prognostic factor in triple-negative breast cancer (TNBC). *J. Clin. Oncol.* **2019**, *37* (Suppl. 15), 594. [CrossRef]

264. Park, S.-Y.; Choi, J.-H.; Nam, J.-S. Targeting Cancer Stem Cells in Triple-Negative Breast Cancer. *Cancers* **2019**, *11*, 965. [CrossRef]

265. Al-Hajj, M.; Wicha, M.S.; Benito-Hernandez, A.; Morrison, S.J.; Clarke, M.F. Prospective identification of tumorigenic breast cancer cells. *Proc. Natl. Acad. Sci. USA* **2003**, *100*, 3983–3988. [CrossRef] [PubMed]

266. Ginestier, C.; Hur, M.H.; Charafe-Jauffret, E.; Monville, F.; Dutcher, J.; Brown, M.; Jacquemier, J.; Viens, P.; Kleer, C.G.; Liu, S.; et al. ALDH1 is a marker of normal and malignant human mammary stem cells and a predictor of poor clinical outcome. *Cell Stem Cell* **2007**, *1*, 555–567. [CrossRef] [PubMed]

267. Li, W.; Ma, H.; Zhang, J.; Zhu, L.; Wang, C.; Yang, Y. Unraveling the roles of CD44/CD24 and ALDH1 as cancer stem cell markers in tumorigenesis and metastasis. *Sci. Rep.* **2017**, *7*, 13856. [CrossRef] [PubMed]

268. Van Phuc, P.; Nhan, P.L.; Nhung, T.H.; Tam, N.T.; Hoang, N.M.; Tue, V.G.; Thuy, D.T.; Ngoc, P.K. Downregulation of CD44 reduces doxorubicin resistance of CD44CD24 breast cancer cells. *Oncol. Targets* **2011**, *4*, 71–78. [CrossRef] [PubMed]

269. Bartucci, M.; Dattilo, R.; Moriconi, C.; Pagliuca, A.; Mottolese, M.; Federici, G.; Benedetto, A.D.; Todaro, M.; Stassi, G.; Sperati, F.; et al. TAZ is required for metastatic activity and chemoresistance of breast cancer stem cells. *Oncogene* **2015**, *34*, 681–690. [CrossRef] [PubMed]

270. Palomeras, S.; Ruiz-Martínez, S.; Puig, T. Targeting Breast Cancer Stem Cells to Overcome Treatment Resistance. *Molecules* **2018**, *23*, 2193. [CrossRef]

271. Talukdar, S.; Bhoopathi, P.; Emdad, L.; Das, S.; Sarkar, D.; Fisher, P.B. Dormancy and cancer stem cells: An enigma for cancer therapeutic targeting. *Adv. Cancer Res.* **2019**, *141*, 43–84. [PubMed]

272. Liu, S.; Cong, Y.; Wang, D.; Sun, Y.; Deng, L.; Liu, Y.; Martin-Trevino, R.; Shang, L.; McDermott, S.P.; Landis, M.D.; et al. Breast cancer stem cells transition between epithelial and mesenchymal states reflective of their normal counterparts. *Stem Cell Rep.* **2014**, *2*, 78–91. [CrossRef]

273. Liu, H.; Patel, M.R.; Prescher, J.A.; Patsialou, A.; Qian, D.; Lin, J.; Wen, S.; Chang, Y.F.; Bachmann, M.H.; Shimono, Y.; et al. Cancer stem cells from human breast tumors are involved in spontaneous metastases in orthotopic mouse models. *Proc. Natl. Acad. Sci. USA* **2010**, *107*, 18115–18120. [CrossRef]

274. Yehiely, F.; Moyano, J.V.; Evans, J.R.; Nielsen, T.O.; Cryns, V.L. Deconstructing the molecular portrait of basal-like breast cancer. *Trends Mol. Med.* **2006**, *12*, 537–544. [CrossRef]

275. Brugnoli, F.; Grassilli, S.; Al-Qassab, Y.; Capitani, S.; Bertagnolo, V. CD133 in Breast Cancer Cells: More than a Stem Cell Marker. *J. Oncol.* **2019**, *2019*, 7512632. [CrossRef] [PubMed]

276. Collina, F.; Di Bonito, M.; Li Bergolis, V.; De Laurentiis, M.; Vitagliano, C.; Cerrone, M.; Nuzzo, F.; Cantile, M.; Botti, G. Prognostic Value of Cancer Stem Cells Markers in Triple-Negative Breast Cancer. *Biomed. Res. Int.* **2015**, *2015*, 158682. [CrossRef]

277. Ricardo, S.; Vieira, A.F.; Gerhard, R.; Leitão, D.; Pinto, R.; Cameselle-Teijeiro, J.F.; Milanezi, F.; Schmitt, F.; Paredes, J. Breast cancer stem cell markers CD44, CD24 and ALDH1: Expression distribution within intrinsic molecular subtype. *J. Clin. Pathol* **2011**, *64*, 937–946. [CrossRef] [PubMed]

278. Chen, Y.; Song, J.; Jiang, Y.; Yu, C.; Ma, Z. Predictive value of CD44 and CD24 for prognosis and chemotherapy response in invasive breast ductal carcinoma. *Int. J. Clin. Exp. Pathol.* **2015**, *8*, 11287–11295. [PubMed]

279. Lin, Y.; Zhong, Y.; Guan, H.; Zhang, X.; Sun, Q. CD44+/CD24- phenotype contributes to malignant relapse following surgical resection and chemotherapy in patients with invasive ductal carcinoma. *J. Exp. Clin. Cancer Res.* **2012**, *31*, 59. [CrossRef] [PubMed]

280. Li, H.; Ma, F.; Wang, H.; Lin, C.; Fan, Y.; Zhang, X.; Qian, H.; Xu, B. Stem cell marker aldehyde dehydrogenase 1 (ALDH1)-expressing cells are enriched in triple-negative breast cancer. *Int J. Biol Mark.* **2013**, *28*, e357–e364. [CrossRef] [PubMed]

281. Ma, F.; Li, H.; Li, Y.; Ding, X.; Wang, H.; Fan, Y.; Lin, C.; Qian, H.; Xu, B. Aldehyde dehydrogenase 1 (ALDH1) expression is an independent prognostic factor in triple negative breast cancer (TNBC). *Medicine (Baltim.)* **2017**, *96*, e6561. [CrossRef]

282. Ohi, Y.; Umekita, Y.; Yoshioka, T.; Souda, M.; Rai, Y.; Sagara, Y.; Sagara, Y.; Sagara, Y.; Tanimoto, A. Aldehyde dehydrogenase 1 expression predicts poor prognosis in triple-negative breast cancer. *Histopathology* **2011**, *59*, 776–780. [CrossRef]

283. Doherty, M.R.; Cheon, H.; Junk, D.J.; Vinayak, S.; Varadan, V.; Telli, M.L.; Ford, J.M.; Stark, G.R.; Jackson, M.W. Interferon-beta represses cancer stem cell properties in triple-negative breast cancer. *Proc. Natl. Acad. Sci. USA* **2017**, *114*, 13792–13797. [CrossRef]

284. Opyrchal, M.; Salisbury, J.L.; Iankov, I.; Goetz, M.P.; McCubrey, J.; Gambino, M.W.; Malatino, L.; Puccia, G.; Ingle, J.N.; Galanis, E.; et al. Inhibition of Cdk2 kinase activity selectively targets the CD44$^+$/CD24$^-$/Low stem-like subpopulation and restores chemosensitivity of SUM149PT triple-negative breast cancer cells. *Int. J. Oncol.* **2014**, *45*, 1193–1199. [CrossRef] [PubMed]

285. Wu, Q.; Wang, J.; Liu, Y.; Gong, X. Epithelial cell adhesion molecule and epithelial-mesenchymal transition are associated with vasculogenic mimicry, poor prognosis, and metastasis of triple negative breast cancer. *Int. J. Clin. Exp. Pathol.* **2019**, *12*, 1678–1689. [PubMed]

286. Mani, S.A.; Guo, W.; Liao, M.J.; Eaton, E.N.; Ayyanan, A.; Zhou, A.Y.; Brooks, M.; Reinhard, F.; Zhang, C.C.; Shipitsin, M.; et al. The epithelial-mesenchymal transition generates cells with properties of stem cells. *Cell* **2008**, *133*, 704–715. [CrossRef] [PubMed]

287. Chaffer, C.L.; Marjanovic, N.D.; Lee, T.; Bell, G.; Kleer, C.G.; Reinhardt, F.; D'Alessio, A.C.; Young, R.A.; Weinberg, R.A. Poised chromatin at the ZEB1 promoter enables breast cancer cell plasticity and enhances tumorigenicity. *Cell* **2013**, *154*, 61–74. [CrossRef] [PubMed]

288. Kim, S.Y.; Kang, J.W.; Song, X.; Kim, B.K.; Yoo, Y.D.; Kwon, Y.T.; Lee, Y.J. Role of the IL-6-JAK1-STAT3-Oct-4 pathway in the conversion of non-stem cancer cells into cancer stem-like cells. *Cell Signal.* **2013**, *25*, 961–969. [CrossRef] [PubMed]

289. Zhao, D.; Pan, C.; Sun, J.; Gilbert, C.; Drews-Elger, K.; Azzam, D.J.; Picon-Ruiz, M.; Kim, M.; Ullmer, W.; El-Ashry, D.; et al. VEGF drives cancer-initiating stem cells through VEGFR-2/Stat3 signaling to upregulate Myc and Sox2. *Oncogene* **2015**, *34*, 3107–3119. [CrossRef] [PubMed]

290. Marotta, L.L.; Almendro, V.; Marusyk, A.; Shipitsin, M.; Schemme, J.; Walker, S.R.; Bloushtain-Qimron, N.; Kim, J.J.; Choudhury, S.A.; Maruyama, R.; et al. The JAK2/STAT3 signaling pathway is required for growth of CD44$^+$CD24$^-$ stem cell-like breast cancer cells in human tumors. *J. Clin. Investig.* **2011**, *121*, 2723–2735. [CrossRef]

291. Tian, J.; Raffa, F.A.; Dai, M.; Moamer, A.; Khadang, B.; Hachim, I.Y.; Bakdounes, K.; Ali, S.; Jean-Claude, B.; Lebrun, J.J. Dasatinib sensitises triple negative breast cancer cells to chemotherapy by targeting breast cancer stem cells. *Br. J. Cancer* **2018**, *119*, 1495–1507. [CrossRef]

292. Stover, D.G.; Gil Del Alcazar, C.R.; Brock, J.; Guo, H.; Overmoyer, B.; Balko, J.; Xu, Q.; Bardia, A.; Tolaney, S.M.; Gelman, R.; et al. Phase II study of ruxolitinib, a selective JAK1/2 inhibitor, in patients with metastatic triple-negative breast cancer. *NPJ Breast Cancer* **2018**, *4*, 10. [CrossRef]
293. O'Shaughnessy, J.; DeMichele, A.; Ma, C.X.; Richards, P.; Yardley, D.A.; Wright, G.S.; Kalinsky, K.; Steis, R.; Diab, S.; Kennealey, G.; et al. A randomized, double-blind, phase 2 study of ruxolitinib or placebo in combination with capecitabine in patients with advanced HER2-negative breast cancer and elevated C-reactive protein, a marker of systemic inflammation. *Breast Cancer Res. Treat.* **2018**, *170*, 547–557. [CrossRef]

International Journal of
Molecular Sciences

Article

N-Cadherin mRNA Levels in Peripheral Blood Could Be a Potential Indicator of New Metastases in Breast Cancer: A Pilot Study

Takaaki Masuda [1,†], Hiroki Ueo [2,†], Yuichiro Kai [3], Miwa Noda [1], Qingjiang Hu [1], Kuniaki Sato [1], Atsushi Fujii [1], Naoki Hayashi [1], Yusuke Tsuruda [1], Hajime Otsu [1], Yosuke Kuroda [1], Hidetoshi Eguchi [1], Shinji Ohno [4], Koshi Mimori [1,*] and Hiroaki Ueo [3,*]

[1] Department of Surgery, Kyushu University Beppu Hospital, 4546 Tsurumihara, Beppu 874-0838, Japan; takaaki.masuda.280@m.kyushu-u.ac.jp (T.M.); nmiwa-1126@oita-u.ac.jp (M.N.); guitar5158@gmail.com (Q.H.); basement.kuni13@gmail.com (K.S.); afujii2003@yahoo.co.jp (A.F.); isaya_hikoan@yahoo.co.jp (N.H.); yxmms497@ybb.ne.jp (Y.T.); ootsu@surg2.med.kyushu-u.ac.jp (H.O.); kuro731976@yahoo.co.jp (Y.K.); heguchi@beppu.kyushu-u.ac.jp (H.E.)

[2] Department of Surgery, Saiseikai Karatsu General Hospital, 817 Motohata, Karatshu 847-0852, Japan; ueohiro@med.kyushu-u.ac.jp

[3] Ueo Breast Surgical Hospital, 188-2 haya, Oita 870-0854, Japan; kai@oita-mamma.jp

[4] Breast Oncology Center, The Cancer Institute Hospital Ariake of Japanese Foundation for Cancer Research, 3-8-31, Ariake, Koto, Tokyo 135-8550, Japan; shinji.ohno@jfcr.or.jp

* Correspondence: kmimori@beppu.kyushu-u.ac.jp (K.M.); ueo@oita-mamma.jp (H.U.); Tel.: +81-977-27-1645 (K.M.); +81-97-514-0025 (H.U.); Fax: +81-977-27-1651 (K.M.); +81-97-514-1155 (H.U.)

† These authors contributed equally to this work.

Received: 4 January 2020; Accepted: 12 January 2020; Published: 14 January 2020

Abstract: Background: There is growing evidence that patients with metastatic breast cancer whose disease progresses from a new metastasis (NM) have a worse prognosis than that of patients whose disease progresses from a pre-existing metastasis. The aim of this pilot study is to identify a blood biomarker predicting NM in breast cancer. Methods: The expression of epithelial (cytokeratin 18/19) or mesenchymal (plastin-3, vimentin, and N-cadherin) markers in the peripheral blood (PB) of recurrent breast cancer patients undergoing chemotherapy with eribulin or S-1 was measured over the course of treatment by RT-qPCR. The clinical significance of preoperative N-cadherin expression in the PB or tumor tissues of breast cancer patients undergoing curative surgery was assessed by RT-qPCR or using public datasets. Finally, N-cadherin expression in specific PB cell types was assessed by RT-qPCR. Results: The expression levels of the mesenchymal markers N-cadherin and vimentin were high in the NM cases, whereas that of the epithelial marker cytokeratin 18 was high in the pre-existing metastasis cases. High preoperative N-cadherin expression in PB or tumor tissues was significantly associated with poor recurrence-free survival. N-cadherin was expressed mainly in polymorphonuclear leukocytes in PB. Conclusion: N-cadherin mRNA levels in blood may serve as a novel prognostic biomarker predicting NM, including recurrence, in breast cancer patients.

Keywords: N-cadherin; EMT; breast cancer; new metastasis; eribulin; blood; biomarker

---

1. Introduction

Distant metastasis is the leading cause of mortality in patients with cancer, including breast cancer, which is the most common malignancy in women worldwide despite the advent of new treatments [1]. Thus, controlling distant metastasis is important for prolonging survival.

Recently, the importance of the type of distant metastasis progression, namely new metastasis (NM) versus growth of a pre-existing metastasis (PEM), has been highlighted, because the progression type can affect the prognosis of patients with metastatic cancer [2–4] (Figure S1). Interestingly, patients with recurrent breast cancer who develop NM have a worse prognosis than that of those with PEM [2,3,5,6]. However, the differences between these two progression types has not affected treatment determination, because both NM and PEM are classified clinically as a "progressive disease" (PD) according to the diagnostic criteria of the Response Evaluation Criteria in Solid Tumors [7]. There have also been cases of NM vs. PEM reported in colorectal and lung cancers [2] and liposarcoma [4]. This concept of distant metastasis progression stems from phase 3 clinical trials of eribulin treatment of metastatic breast cancer [8,9].

Eribulin mesylate (eribulin; Eisai Co., Tokyo, Japan) is a non-taxane microtubule inhibitor with a novel mechanism of action involving irreversible blockade of mitosis at the G2/M phase, followed by apoptosis [10]. Eribulin is currently approved for the treatment of certain patients with advanced breast cancer in many countries worldwide, including Japan [8]. Interestingly, two different phase 3 clinical trials of patients with metastatic breast cancer showed that eribulin has more pronounced effects on overall survival compared with progression-free survival [8,9]; one possible explanation is that eribulin suppresses the incidence of NM, thus providing an increased survival benefit to patients. The preclinical studies described herein were designed to assess whether eribulin has such an anti-metastatic property, via reversion of epithelial–mesenchymal transition (EMT) [11,12]. Yoshida et al. and Terashima et al. showed experimentally that eribulin suppresses metastasis of breast cancer cells by inducing conversion of EMT to mesenchymal–epithelial transition (MET) [11,13]. EMT is the process by which epithelial cells lose their cell-cell junctions and acquisition of front-rear polarization, resulting in the formation of mesenchymal cells with migratory properties, and it is characterized by loss of the epithelial markers E-cadherin and cytokeratins (CKs), together with increased expression of mesenchymal markers such as N-cadherin and vimentin. EMT is observed during cancer progression; it promotes invasion and metastasis by facilitating the motility of tumor cells [14,15]. These data suggest that EMT may induce NM.

Thus, we hypothesized that monitoring the EMT status in real-time can predict NM and is important for guiding the treatment of patients with recurrent breast cancer. We focused on liquid biopsy, which is used to identify biomarkers in body fluids, mainly blood, and provides non-invasive real-time information regarding tumor characteristics [16,17]. In this study, we monitored the expression of EMT (mesenchymal) and MET (epithelial) markers in the peripheral blood (PB) of patients with recurrent breast cancer undergoing chemotherapy with eribulin or S-1 (oral 5-fluorouracil derivative) and identified N-cadherin as a useful marker predicting NM. Furthermore, we assessed the clinical significance of preoperative N-cadherin expression in the PB of breast cancer patients undergoing curative surgery.

2. Results

2.1. Type of Metastasis Progression in Patients with Recurrent Breast Cancer who underwent Chemotherapy with Eribulin or S-1

The subjects comprised 56 and 19 patients who underwent chemotherapy with eribulin and S-1, respectively. Of the 56 patients treated with eribulin, 35 received S-1 prior to eribulin. The patient characteristics are shown in Table 1.

As shown in Table 2, the patients treated with eribulin had a significantly lower incidence of NM than patients treated with S-1, although the time to treatment failure (TTF) was shorter in the former ((a) in Table 2, $p = 0.043$). Moreover, the patients treated with S-1 followed by eribulin also had a significantly lower incidence of NM under eribulin than under S-1 ((b) in Table 2, $p = 0.025$). These results support previous clinical and experimental findings that eribulin suppresses NM via conversion of EMT to MET in tumor cells [11–13]. Furthermore, these findings led us to hypothesize that markers of EMT may be predictive of NM, which we tested using samples from the patients treated with eribulin or S-1.

Table 1. Characteristics of the 75 study patients.

(a) Total ($n = 75$)		
Factor	Eribulin ($n = 56$)	S-1 ($n = 19$)
Age, years Median (range)	57 (40–72)	59 (33–83)
Number of prior chemotherapy lines Median (range)	3 (0–8)	0 (0–3)
Luminal [a] n (%)	29 (51.8)	8 (42.1)
Luminal/HER2 n (%)	3 (5.4)	5 (26.3)
HER2-enriched [b] n (%)	6 (10.7)	2 (10.5)
TN[c] n (%)	18 (32.1)	4 (21.1)

(b) Patients treated with both Eribulin and S-1		
Factor	S-1 Followed by Eribulin ($n = 35$)	Eribulin followed by S-1 ($n = 21$)
Age, years Median (range)	59 (43–72)	55 (40–71)
Number of prior chemotherapy lines Median (range)	3 (1–8)	0 (0–5)
Luminal [a] n (%)	16 (45.7)	13 (61.9)
Luminal/HER2 n (%)	2 (5.7)	1 (4.8)
HER2-enriched [b] n (%)	6 (5.8)	0 (0.0)
TN[c] n (%)	11 (31.4)	7 (33.3)

[a] Luminal, ER, or PgR-positive [b] HER2-enriched, only HER2-positive [c] TN, triple negative (ER/PgR/HER2-negative).

Table 2. Type of distant metastasis progression in patients undergoing S-1 or eribulin treatment.

(a) Total ($n = 75$)					
Agent	Disease Progression	TTF [a], Median (Range)	NM (+) [b]	NM (−) [c]	p
S-1 ($n = 19$)	16	8 (2–56)	8 (50.0%)	8 (50.0%)	0.043
Eribulin ($n = 56$)	38	6 (1–43)	7 (22.6%)	31 (77.4%)	

(b) S-1 followed by eribulin ($n = 35$)					
Agent	Disease Progression	TTF, Median (Range)	NM (+)	NM (−)	p
S-1	29	10 (3–59)	14 (48.3)	15 (51.7)	0.025
Eribulin	27	6 (1–22)	5 (18.5)	22 (81.5)	

(c) Eribulin followed by S-1 ($n = 21$)					
Agent	Disease Progression	TTF, Median (Range)	NM (+)	NM (−)	p
S-1	4	7 (1–11)	1 (25.0)	3 (75.0)	1.000
Eribulin	11	5 (1–16)	2 (18.2)	9 (81.8)	

[a] TTF, time to treatment failure; [b] NM (new metastasis) (+), NM or NM + PEM (pre-existing metastasis); [c] NM (−), PEM only.

2.2. Expression of Epithelial and Mesenchymal Markers in Breast Cancer and Non-Epithelial Cell Lines

First, we examined if our selected markers reflect the EMT status by evaluating their expression in various cell lines. The quantitated gene expression levels of the different markers are shown in Figure 1a. The epithelial markers (CK18 and CK19) were expressed in all breast cancer cell lines, whereas the mesenchymal markers (PLS3, vimentin, and N-cadherin) were expressed mainly in the non-epithelial cell lines. PLS3, which is reportedly expressed in both epithelial and mesenchymal cells [18], was expressed in most of the cell lines albeit at varying levels among them.

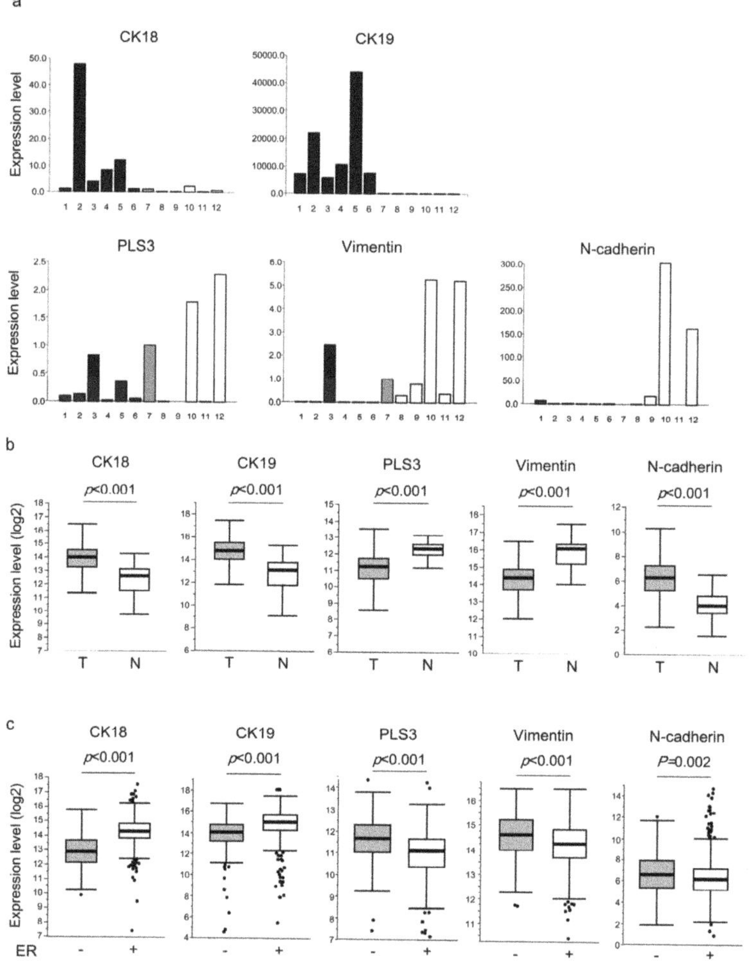

Figure 1. Expression of epithelial and mesenchymal markers in breast cancer cell lines and tissues. (**a**) Expression in invasive ductal carcinoma cell lines (#1–6: CRL1500, MCF-7, MDA-MB231, Mrknu1, SKBR3, and YMB1 cells, respectively), a normal mammary epithelial cell line (#7: HMECs), and non-epithelial cell lines (#8–12: Raji B lymphocytes, Jurkat T lymphocytes, HT1080 fibrosarcoma cells, THP-1 monocytes, KMST-6 fibroblasts, respectively). The expression levels are expressed relative to the level in HMECs (1.0). (**b**) Expression in normal and tumor tissues of breast cancer patients obtained from The Center Genome Atlas (TCGA) dataset. T: tumor tissues; N: normal mammary tissues. (**c**) Expression in tissues of breast cancer patients from TCGA dataset according to estrogen receptor (ER) status. +: ER-positive; −: ER-negative.

2.3. Expression of Epithelial and Mesenchymal Markers in Breast Cancer Tissues

We also assessed the expression levels of the epithelial and mesenchymal markers in breast cancer tissues using The Center Genome Atlas (TCGA) datasets (Figure 1b,c). As shown in Figure 1b, CK18, CK19, and N-cadherin levels were higher in tumor tissues ($n = 1093$) than in normal tissues ($n = 112$) of breast cancer patients ($p < 0.001$). Unexpectedly, PLS3 and vimentin expression levels were lower in tumor tissues than in normal tissues ($p < 0.001$).

Next, we compared marker expression between ER-positive ($n = 823$) and ER-negative ($n = 219$) cases (Figure 1c). As expected, the expression levels of the mesenchymal markers were higher in the ER-negative than ER-positive cases (PLS3, vimentin, and N-cadherin: $p < 0.001$, $p < 0.001$, and $p = 0.002$, respectively), whereas the expression levels of the epithelial markers were higher in the ER-positive than ER-negative cases (CK18 and CK19: both $p < 0.001$). These results suggest that mesenchymal markers are expressed in high-grade cancers with metastatic potential, because ER-negative tumors tend to be associated with earlier relapse and worse prognosis compared with ER-positive tumors [19–22].

2.4. Expression of Epithelial and Mesenchymal Markers in the PB of Breast Cancer Patients

Next, we assessed the mRNA expression levels of the epithelial and mesenchymal markers in the PB of 16 patients with recurrent breast cancer and 10 healthy volunteers (HVs) using Ueo and Beppu cohorts (Figure 2a). CK18, vimentin, and N-cadherin expression in PB was statistically higher in the patients with recurrent breast cancer than in HVs ($p = 0.031$, $p = 0.004$, and $p = 0.031$, respectively). Other markers also had a tendency to be higher in PB from patients compared with HVs. These findings indicate that these markers are expressed in circulating tumor cells (CTCs) or host cells in the PB of breast cancer patients.

2.5. Expression of Epithelial and Mesenchymal Markers in the PB of Patients with Recurrent Breast Cancer with NM or PEM

We compared the expression levels of epithelial and mesenchymal markers in the PB of patients with recurrent breast cancer with NM (the total number of samples; $n = 9$ from 4 patients) versus PEM (the total number of samples; $n = 16$ from 3 patients) using Ueo cohort. The changes in the expression levels of the markers in a representative case of NM or PEM are shown in Figure 2b. Interestingly, the expression level of N-cadherin increased consistently with time in the NM case, although the level of the tumor markers CEA and CA15-3 were not elevated. In the PEM case, the expression of N-cadherin was low, although CA15-3 was elevated. Furthermore, the expression levels of the mesenchymal markers tended to be higher in the NM, whereas epithelial marker expression was higher in the PEM cases (Figure 2c). Statistical differences in N-cadherin, vimentin, and CK18 levels between the NM and PEM cases were found ($p = 0.002$, $p = 0.027$, and $p = 0.011$, respectively). These data suggest that N-cadherin expression in PB is predictive of NM, and N-cadherin may reflect the real-time metastatic potential of tumor cells.

2.6. Expression of Epithelial and Mesenchymal Markers in the PB of Patients with Recurrent Breast Cancer undergoing Eribulin or S-1 Rreatment

Next, we compared the levels of epithelial and mesenchymal markers in the PB of patients with recurrent breast cancer undergoing eribulin (the total number of samples; $n = 19$ from 4 patients) versus S-1 (the total number of samples; $n = 17$ from 8 patients) treatment (Figure 2d). Interestingly, the expression of N-cadherin in PB was statistically lower in the patients treated with eribulin than in those treated with S-1 ($p < 0.001$).

Moreover, the expression level of N-cadherin was decreased by eribulin treatment compared with before treatment. The changes in N-cadherin expression in the breast cancer patients over the course of eribulin or S-1 treatment are shown in Figure 2e. Among the eight patients treated with S-1, six showed over a twofold increase in the N-cadherin expression compared with baseline, whereas only one of the four patients treated with eribulin showed over a twofold increase in N-cadherin expression compared with baseline. These data further support that eribulin can induce the conversion of EMT to MET, as reported previously [11–13].

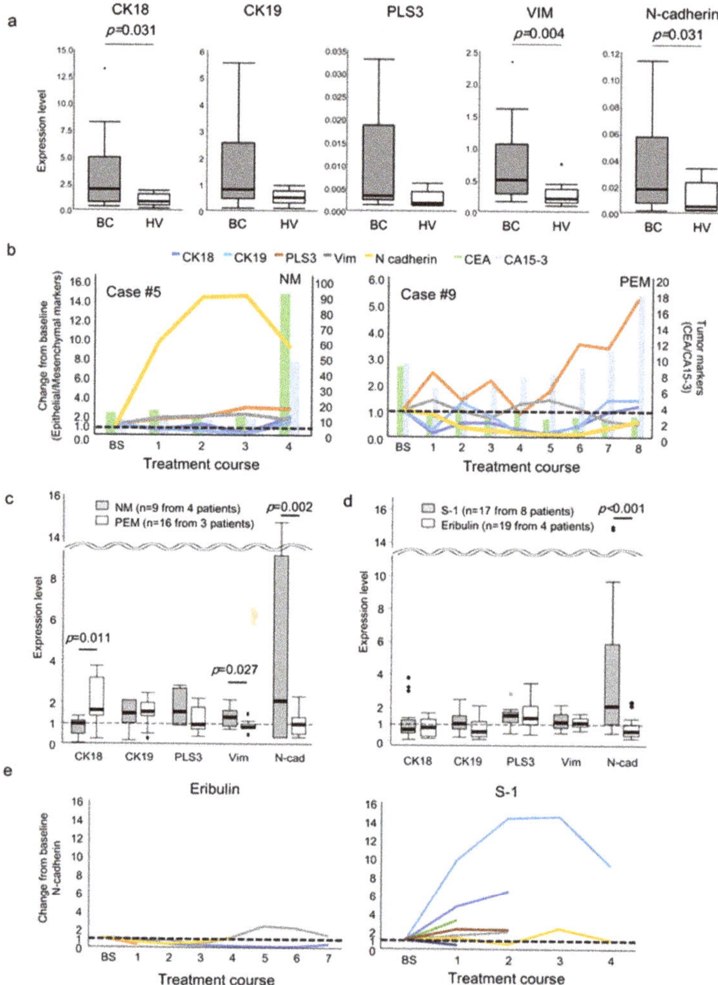

Figure 2. Expression of epithelial and mesenchymal markers in the peripheral blood (PB) of breast cancer patients undergoing chemotherapy. (**a**) A comparison of expression levels of the markers in PB between breast cancer patients just before eribulin or S-1 chemotherapy and healthy volunteers (HVs). (**b**) Changes in the expression levels of the markers in representative cases of NM and PEM over the treatment course of eribulin or S-1 treatment. Left: case #5 treated with S-1; right: case #9 treated with eribulin. The expression levels are expressed relative to the level at pre-treatment (baseline; 1.0). BS; baseline. (**c**) A comparison of expression levels of the markers in PB of breast cancer patients with NM or PEM. The expression levels are expressed relative to the level at pre-treatment (baseline; 1.0). BS; baseline. n; the total number of samples from patients. (**d**) A comparison of expression levels of the markers in PB of breast cancer patients undergoing eribulin or S-1 treatment. The expression levels are expressed relative to the level at pre-treatment (baseline; 1.0). BS; baseline. n; the total number of samples from patients. (**e**) Changes in N-cadherin expression in breast cancer patients over the course of eribulin or S-1 treatment. The expression levels are expressed relative to the level at pre-treatment (baseline; 1.0). BS; baseline.

2.7. Clinicopathological Significance of Preoperative N-Cadherin mRNA Expression in the PB of Breast Cancer Patients undergoing Curative Surgery

The results of the abovementioned expression analyses motivated us to investigate the value of preoperative N-cadherin expression in PB for predicting breast cancer recurrence in patients after curative surgery, because recurrence is considered a type of NM.

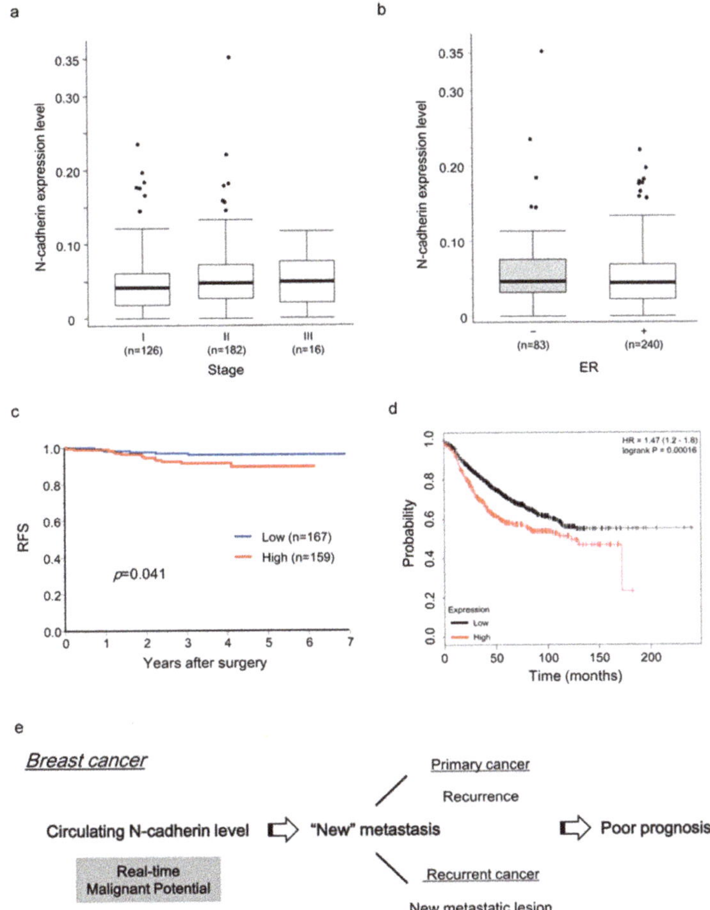

Figure 3. Preoperative expression levels of N-cadherin mRNA in the PB of breast cancer patients undergoing curative surgery. (**a**) N-cadherin expression in the PB of patients with breast cancer according to TNM stage. (**b**) N-cadherin expression in the PB of patients with breast cancer according to ER status. +: ER-positive; −: ER-negative. (**c**) The RFS of 326 patients with breast cancer after curative surgery according to N-cadherin expression in preoperative PB. (**d**) The RFS of 3951 patients with invasive ductal carcinoma from the Kaplan–Meier plotter dataset according to N-cadherin expression in breast cancer tissues. (**e**) A proposed model showing the clinical significance of the circulating N-cadherin level for predicting NM in breast cancer.

First, we assessed N-cadherin expression in the PB of the 326 patients with breast cancer using Kyushu cohort. The N-cadherin levels ranged from 5.805×10^{-5} to 0.352 (median, 0.045). The median expression level of N-cadherin in patients with TNM stage I, II, or III (126, 182, or 16 cases, respectively) was 0.042, 0.048, or 0.051, respectively (Figure 3a). The median expression levels of N-cadherin in the

ER-positive ($n = 240$) and ER-negative ($n = 83$) cases were 0.045 and 0.047, respectively. There was no significant difference in the N-cadherin level in PB according to TNM stage or ER status (Figure 3a,b).

Next, the relationships between clinicopathological factors and N-cadherin expression in blood were examined in the 326 patients with breast cancer. The patients were divided into two groups (high and low N-cadherin expression) as described in Materials and Methods. The cutoff level of N-cadherin expression was 0.046. As shown in Table 3, there were no significant differences between the high and low N-cadherin expression groups in terms of clinicopathological factors including age, tumor size, nuclear grade, venous involvement, lymphatic involvement, lymph node metastasis, ER/PgR/HER2 status, and subtype.

Table 3. Relationships between clinicopathological factors and N-cadherin expression in PB of breast cancer patients.

Variable	Low Expression ($n = 167$) n (%)	High Expression ($n = 159$) n (%)	p
Age (years)			0.092
<65	119 (71.3)	126 (79.2)	
≥65	42 (25.1)	28 (17.6)	
unknown	6 (3.6)	5 (3.2)	
Pathological tumor size			0.634
1	90 (53.9)	82 (51.6)	
2, 3	75 (44.9)	76 (47.8)	
unknown	2 (1.2)	1 (0.6)	
Nuclear grade			0.956
1, 2	108 (64.7)	105 (66.0)	
3	49 (29.3)	47 (29.6)	
unknown	10 (6.0)	7 (4.4)	
Venous involvement			0.402
(−)	155 (92.8)	143 (89.9)	
(+)	8 (4.8)	11 (6.9)	
unknown	4 (2.4)	5 (3.1)	
Lymphatic involvement			0.070
(−)	111 (66.5)	91 (57.2)	
(+)	52 (31.1)	65 (40.9)	
unknown	4 (2.4)	3 (1.9)	
Lymph node metastasis			0.194
(−)	112 (67.1)	95 (59.7)	
(+)	55 (32.9)	63 (39.6)	
unknown	0 (0.0)	1 (0.6)	
ER			0.721
(−)	41 (24.6)	42 (26.4)	
(+)	124 (74.3)	116 (73.0)	
unknown	2 (1.2)	1 (0.6)	
PgR			0.159
(−)	62 (37.1)	72 (45.3)	
(+)	103 (61.7)	87 (54.7)	
unknown	2 (1.2)	0 (0.0)	
HER2			0.235
− (0, 1)	104 (62.3)	97 (61.0)	
+ (2, 3)	44 (26.3)	55 (34.6)	
unknown	19 (11.4)	7 (4.4)	
Subtype			0.076
HR [a]+/HER2−	92 (55.1)	76 (47.8)	
HR±/HER2+	44 (26.3)	55 (34.6)	
TN [b]	12 (7.2)	21 (13.2)	
unknown	19 (11.4)	7 (4.4)	

Correlations were analyzed by Fisher's exact test. [a] HR, hormone receptor; [b] TN, triple negative.

2.8. Prognostic Significance of Preoperative N-Cadherin mRNA Expression in the PB of Breast Cancer Patients undergoing Curative Surgery

Next, we assessed the prognostic significance, in terms of RFS, of N-cadherin expression in PB using Kyushu cohort. The high N-cadherin expression group (n = 159) had a significantly worse recurrence-free survival (RFS) (p = 0.041) than the low N-cadherin expression group (n = 167) (Figure 3c). Next, univariate and multivariate regression analyses of predictive factors for RFS were performed (Table 4). Univariate analyses showed that lymphatic involvement, lymph node metastasis, PgR-negative, and high N-cadherin expression were statistically significant prognostic factors for RFS (p = 0.002, p = 0.002, p = 0.026, and p = 0.040, respectively). N-cadherin expression and clinicopathological factors such as lymph node metastasis, and PgR status were included in the multivariate analysis. N-cadherin expression in PB was not a significant independent prognostic factor for RFS in patients with breast cancer according to the multivariate analysis (HR: 2.215, p = 0.092).

Table 4. Univariate and multivariate analyses of prognostic factors for recurrence-free survival (RFS) in breast cancer patients.

Variables	Univariate		Multivariate	
	HR (95% CI [a])	p	HR (95% CI)	p
Age (years) (≥65/<65)	0.410 (0.065–1.432)	0.181		
Pathological tumor size (2 or 3/1)	2.064 (0.845–5.494)	0.113		
Nuclear grade (3/1 or 2)	1.673 (0.648–4.135)	0.277		
Venous involvement (+/−)	1.862×10^{-9} (0.609–1.609)	0.119		
Lymphatic involvement (+/−)	4.127 (1.655–11.653)	0.002		
Lymph node metastasis (+/−)	4.199 (1.685–11.860)	0.002	4.023 (1.610–11.380)	0.003
ER (+/−)	0.497 (0.205–1.268)	0.138		
PgR (+/−)	0.363 (0.136–0.887)	0.026	0.380 (0.142–0.931)	0.034
HER2 (2 or 3/0 or 1)	2.13 (0.875–5.210)	0.094		
N-cadherin expression in PB (high/low)	2.611 (1.046–7.380)	0.040	2.215 (0.882–6.295)	0.092

[a] CI, confidence interval.

2.9. Prognostic Significance of N-Cadherin mRNA Expression in the Tumor Tissues of Breast Cancer Patients undergoing Curative Surgery

We investigated the prognostic significance of N-cadherin in the tumor tissues of breast cancer patients treated with curative surgery using the Kaplan–Meier plotter dataset. "Systemically untreated breast cancer patients" with any intrinsic subtype were selected for the analysis. These patients were divided into two groups based on N-cadherin expression by selecting the "Auto Select best cutoff" feature of the Kaplan–Meier plotter. As expected, the RFS was shorter in the high (n = 1173) than low N-cadherin expression group (n = 2778) among all cases (p < 0.001, 1% FDR, Figure 3d).

These clinical results suggest that N-cadherin expression in PB reflects the metastatic potential of tumor cells and is a potential biomarker predicting NM in breast cancer (Figure 3e).

2.10. Assessment of N-Cadherin mRNA Expression in PB Cell Fractions

Finally, we assessed the distribution of N-cadherin expression in PB cells of breast cancer patients using Beppu cohort. The expression levels of N-cadherin and CKs in polymorphonuclear leukocytes (PMNs), mononuclear cells (MCs), and red blood cells (RBCs) from 24 patients with breast cancer were measured by RT-qPCR (Figure 4). The average ratios of N-cadherin relative expression in RBCs, MCs and PMNs to that in whole blood were 0.021, 0.417, and 14.375, respectively. N-cadherin expression was much higher in PMNs than in MCs or RBCs (p < 0.001), indicating that N-cadherin is expressed predominantly in PMNs within PB (Figure 4a).

a

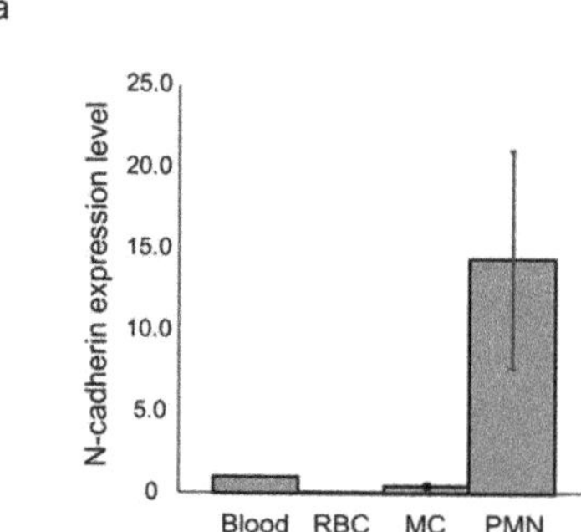

b

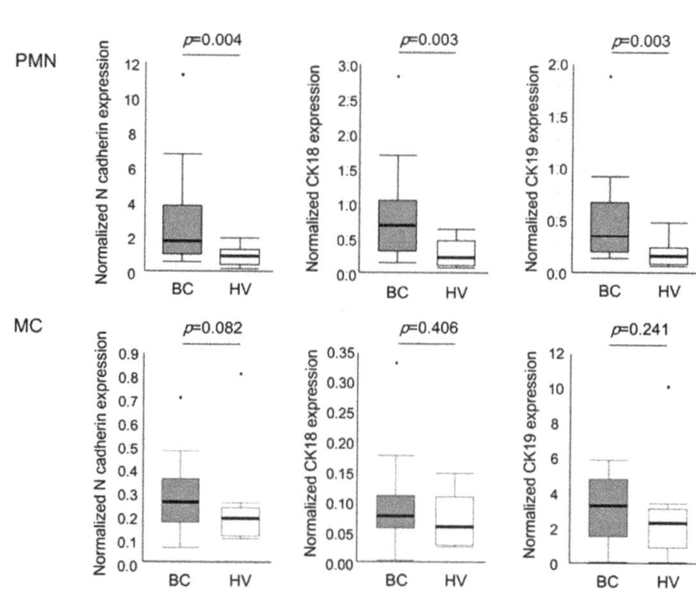

Figure 4. N-cadherin mRNA expression in different PB cell fractions. (**a**) N-cadherin expression in three PB cell types in breast cancer patients. RBC: red blood cells; MC: mononuclear cells; PMN: polymorphonuclear leukocytes. (**b**) N-cadherin, CK18, and CK19 expression in the PMNs and MCs of breast cancer patients and healthy volunteers.

Next, we compared N-cadherin expression in PMNs and MCs between breast cancer patients (n = 24) and HVs (*n* = 10) (Figure 4b). In PMNs, the median levels of N-cadherin, CK18, and CK19 expression were 1.698, 0.706, and 0.350 in breast cancer patients and 0.851, 0.249, and 0.167 in HVs, respectively. In MCs, the median levels of N-cadherin, CK18, and CK19 expression were 0.264, 0.079, and 3.214 in breast cancer patients and 0.195, 0.062, and 2.228 in HVs, respectively. The expression levels of N-cadherin, CK18, and CK19 in PMNs were higher in breast cancer patients than in HVs ($p = 0.004$, $p = 0.003$, and $p = 0.003$, respectively); however, in MCs, there was no statistical difference in expression levels between breast cancer patients and HVs. These findings suggest the possibility that N-cadherin-expressed cells are circulating tumor cells (CTCs) in PMN fraction of breast cancer patients because CKs-expressed cells in PB are mainly cancer cells in patients with various epithelial malignancies [23].

3. Discussion

N-cadherin is a transmembrane glycoprotein that mediates calcium-dependent cell-cell adhesion, mediated by post-translational modifications such as phosphorylation of the N-cadherin catenin complex [24]. This protein is multifunctional and can interact with many different proteins and be involved in many different functional events [25]. N-cadherin possesses seven glycation sites on its ectodomain with ectodomains. Therefore, in order to obtain a real picture about its function it is necessary to consider that: i) its adhesive function can be certainly regulated by gradual post-translational modifications; ii) for example, its structural organization can strongly depend on its phosphorylation status; and iii) its N-glycosylation sites can regulate N-cadherin-dependent cell adhesion. These points are very interesting and need further investigations to elucidate which form of the N-cadherin complex in blood can play a role in NM development.

In this pilot study, we demonstrated that N-cadherin expression in PB is a potentially useful biomarker for predicting NM, which is associated with a poor prognosis in breast cancer, by sequential monitoring of N-cadherin expression levels in the PB of patients undergoing chemotherapy. To our knowledge, this is the first study to identify a biomarker predictive of NM. Furthermore, we observed that breast cancer patients treated with eribulin, compared with S-1, have a low incidence of NM and low expression of the mesenchymal marker N-cadherin in PB. This adds to the growing evidence that eribulin suppresses NM by conversion of EMT to MET in tumor cells.

Cadherins are single-chain transmembrane glycoproteins that mediate calcium-dependent homophilic cell-cell adhesion and play a critical role in regulating signaling pathways that maintain essential gene transcription [26]. N-cadherin, one of most well-studied cadherins, is expressed mainly in neural and mesenchymal tissues. In various malignancies including breast cancer, cells acquire motility and invasiveness by upregulating N-cadherin during EMT [27–29]. We also reported that N-cadherin is associated with tumor aggressiveness in esophageal carcinoma [30], and others demonstrated the value of N-cadherin as a marker of invasive, malignant tumors [29,31], supporting our finding that N-cadherin expression in PB may be a predictive biomarker of NM. The expression level of N-cadherin may represent the real-time metastatic potential of tumor cells (Figure 3e).

Our prognostic analysis of N-cadherin expression in PB and tumor tissues showed that high preoperative levels are associated with early recurrence in breast cancer patients undergoing curative surgery. These observations provide new insight that recurrence in primary breast cancer after curative surgery is a type of NM. Furthermore, this insight suggests that sequential monitoring of N-cadherin expression in PB during postoperative follow-up may help determine the best treatment for patients with breast cancer.

We also evaluated the specific cell type in PB that expresses the highest level of N-cadherin in breast cancer patients. PMNs expressed the highest level of N-cadherin compared with MCs and RBCs. Furthermore, expression of N-cadherin as well as the epithelial markers CK18 and CK19 in PMNs was higher in breast cancer patients than in HVs, implying that N-cadherin is expressed mainly in CTCs, similar to vimentin [32]. However, PMNs may express N-cadherin upon stimulation by tumor cells. Further study is required to clarify the biological significance of N-cadherin expression in PMNs and identify other PB cell types expressing N-cadherin.

There were some limitations to this study. First, a larger cohort is needed to validate the clinical utility of sequential monitoring of N-cadherin expression in PB. Next, monitoring N-cadherin mRNA levels in PB may not be suitable in the clinic because of the complication of the measurement system. Measurement of N-cadherin protein levels may be more practical because proteins in blood are easily measured by ELISA, as shown in a previous study that evaluated N-cadherin levels in the blood of patients with malignant melanoma [33]. Finally, we focused only on PLS3, vimentin, and N-cadherin, among the various EMT markers, as potential biomarkers of NM. Based on this pilot study, we plan to perform comprehensive analyses in PB using RNA sequencing or mass spectrometry.

4. Materials and Methods

4.1. Study Patients

The data from recurrent breast cancer patients treated with eribulin (56 cases) and/or S-1 (19 cases) at Ueo Breast Surgical Hospital (Oita, Japan) from January 2015 to December 2017 were extracted from the medical records. We chose the data from patients treated with S-1 (oral 5-fluorouracil derivative) as a reference control to compare with those treated with eribulin. This is because both S-1 and eribulin are often used as third-line chemotherapy agents, after anthracyclines and taxanes [1], for patients with recurrent breast cancer in Japan, based on the findings of a randomized controlled trial (SELECT BC) showing equivalent overall survival between S-1-treated and taxane-treated patients with metastatic or recurrent breast cancer [34]. Furthermore, in contrast to eribulin, 5-fluorouracil was reported to induce the conversion of MET to EMT [11,13]. Clinical data, including the progression of metastases during either eribulin or S-1 treatment, from breast cancer patients were also investigated. In this study, disease progression was defined as NM and/or growth of PEM based on systemic computed tomography findings.

4.2. Patient Cohorts in Gene Expression Analysis

We used 3 patient cohorts (Ueo, Beppu, and Kyushu) and 2 public datasets (The Cancer Genome Atlas (TCGA) and The Kaplan–Meier Plotter) for expression analysis. The details are described below. All patients provided written informed consent to participate in this study, which was approved by each institutional review board and the Ethics and Indications Committee of Kyushu University (#577-00, #29-597).

4.3. Clinical Samples

To observe genes expression levels over time in blood, PB samples were processed from blood, collected during routine examinations conducted over a 1-year period (2015–2016), from 20 patients with HER2-negative breast cancer with relapse undergoing chemotherapy (eribulin or S-1) at Ueo Breast Surgical Hospital (affiliated with Kyushu University Beppu Hospital), after obtaining written informed consent (Ueo cohort). Inclusion criteria were Eastern Cooperative Oncology Group grade ≤ 2 and adequate organ and hematological function. Eribulin was administered intravenously (1.4 mg/m^2) on days 1 and 8 of each 3-week cycle. S-1 was administered orally (40–60 mg) twice daily for 28 consecutive days, followed by a 14-day rest period.

Observation was continued until the patient experienced disease progression, according to systemic computed tomography findings, or after 1 year of chemotherapy with eribulin or S-1 if no disease progression was detected. The detailed clinical characteristics and sampling protocol are provided in Figure S2 and Table S1. Immediately after collection, each 1 mL sample of blood was mixed with 4 mL ISOGEN-LS (Nippon Gene, Toyama, Japan) and stored at −80 °C until RNA extraction. Sixteen of the blood samples from 16 cases were sent to Kyushu University Beppu Hospital for analysis without knowledge of the histopathological or clinical results. Four cases dropped out in the middle of this study. Samples from 12 cases total were processed for gene expression analysis.

Furthermore, gene expression was evaluated in three different fractions of blood cells: polymorphonuclear leukocytes (PMNs), mononuclear cells (MCs), and red blood cells (RBCs) (Beppu cohort). Blood samples were obtained from 24 breast cancer patients with invasive carcinoma who underwent primary tumor resection at Kyushu University Beppu Hospital in 2017. Control blood samples were obtained from 10 healthy volunteers (HVs) at Kyushu University Beppu Hospital. The blood samples were immediately processed for isolation of the three cell types.

To determine the clinical significance of preoperative gene expression in the PB of patients with primary breast cancer undergoing curative surgery, we used clinical data from breast cancer patients described previously (Kyushu cohort) [35]. Briefly, a total of 594 patients with breast cancer underwent primary tumor resection at the Department of Breast Oncology, National Kyushu Cancer

Center (Fukuoka, Japan), from 2000 to 2008. Of these, 356 female patients with breast cancer without distant metastases, preoperative therapy, or previous treatment for other cancers were included in this study. Among these patients, 326 with invasive carcinoma were included in a survival analysis. The observation period ranged from 0.3 to 6.9 years (median 3.8 years). Postoperative adjuvant therapy was performed according to the guidelines set by the St. Gallen Consensus Conference [36]. The patients underwent clinical examinations at least every 3 months and mammography annually and were further evaluated only if they exhibited symptoms.

The stages and grades of the tumors were classified according to the AJCC/UICC TNM classification and stage groupings. All data including age, pathological tumor size, nuclear grade, venous involvement, lymphatic involvement, lymph node metastasis, and the statuses of estrogen receptor (ER), progesterone receptor (PgR), and human epidermal growth factor receptor (HER2) expression were obtained from medical records. Recurrence-free survival (RFS) was defined as the period from surgical treatment for cancer to detection of any sign of recurrence.

4.4. Cell Lines

Twelve human cell lines were used in this study: CRL1500, MCF-7, MDA-MB231, Mrknu1, SKBR3, YMB1 (all breast intraductal carcinoma cell lines), HMEC (mammary epithelial cell line), Raji (B lymphocytes), Jurkat (T lymphocytes), HT1080 (fibrosarcoma), THP-1 (monocytes), and KMST-6 (fibroblasts). The latter five are all non-epithelial tumor cell lines. Among the breast cancer cell lines, CRL1500 and MCF7 cells are ER positive, MDA-MB231 cells are ER, PgR and HER2 negative (triple negative), Mrknu1 cells are ER negative, SKBR3 cells are ER negative and HER2 positive, and the ER/PgR/HER2 status of YMB1 cells is unknown. The cell lines were obtained from the Cell Resource Center for Biomedical Research Institute of Development, Aging and Cancer, Tohoku University, and were maintained in RPMI-1640 supplemented with 10% fetal bovine serum at 37 °C in a 5% humidified CO_2 atmosphere. Upon reaching a subconfluent state, the cell cultures were homogenized and the lysates stored at −80 °C.

4.5. Separation of Three Blood Cell Fractions

We used Polymorphprep™ (Alere Technologies AS, Oslo, Norway) to isolate three blood cell fractions (PMNs, MCs, and RBCs) according to the manufacturer's instructions, as described previously [35]. Each fraction was mixed with ISOGEN II (Nippon Gene) and stored at −80 °C until RNA extraction.

4.6. Total RNA Extraction

Total RNA was extracted from cell lines or blood samples using ISOGEN II or ISOGEN-LS, respectively, according to the manufacturer's instructions (Nippon Gene, Toyama, Japan).

4.7. Mesenchymal and Epithelial Markers

Plastin-3 (PLS3), vimentin, and N-cadherin were selected for evaluation as mesenchymal markers [37]. PLS3 is a marker of circulating tumor cells, including not only epithelial cells but also tumor cells undergoing EMT [18,38]. CK18 and CK19 were selected for evaluation as epithelial markers, as the controls [23,37].

4.8. RT-qPCR

RT-qPCR of CK18, CK19, PLS3, vimentin, N-cadherin, RNA18S5, and GAPDH mRNA levels was performed as described previously [39]. In brief, reverse transcription was performed using random hexamers and M-MLV reverse transcriptase (Invitrogen, Carlsbad, CA, USA). qPCR was performed using LightCycler® FastStart DNA Master SYBR Green I (Roche Diagnostics, Basel, Switzerland). The raw data are presented as the relative cDNA level from Human Universal Reference Total RNA (Clontech

Laboratories, Palo Alto, CA, USA) and normalized to the level of the internal control gene (RNA18S5 or GAPDH). Relative quantification of gene expression was calculated using the $2^{-\Delta\Delta Ct}$ method. The primer sequences used for RT-PCR were as follows: CK18, forward 5′-ATCTTGGTGATGCCTTGGAC-3′ and reverse 5′-CCTGCTTCTGCTGGCTTAAT-3′; CK19, forward 5′-CATGAAAGCTCCCTTGGAAGA-3′ and reverse 5′-TGATTCTGCCGCTCACTATCAG-3′; PLS3, forward 5′-CCTTCCGTAACTGGATGAACTC-3′ and reverse 5′-GGATGCTTCCCTAATTCAACAG-3′; vimentin, forward 5′-TACAGGAAGCTGCTGGA AGG-3′ and reverse 5′-ACCAGAGGGAGTGAATCCAG-3′; N-cadherin, forward 5′-ATTGGACCATCA CTCGGCTTA-3′ and reverse 5′-CACACTGGCAAACCTTCACG-3′; RNA18S5, forward 5′-AGTCCCT GCCCTTTGTACACA-3′ and reverse 5′-CGATCCGAGGGCCTCACTA-3′; and GAPDH, forward 5′-TTGGTATCGTGGAAGGACTC-3′ and reverse 5′-AGTAGAGGCAGGGATGATGT-3′.

4.9. TCGA Analysis

We used TCGA data to analyze the expression of CK18, CK19, PLS3, vimentin, and N-cadherin in breast cancer tissues. The expression data of these genes in tumor tissues from 1093 breast cancer cases and in normal tissues from 112 cases with breast cancer available in TCGA were obtained from the Broad Institute's Firehose pipeline (http://gdac.broadinstitute.org/runs/stddata__2016_01_28/data/ BRCA/20160128/). The sequencing data were normalized by quantile normalization, as described previously [40].

4.10. Kaplan–Meier Plotter Survival Analysis

The Kaplan–Meier Plotter (www.kmplot.com), an online database that includes gene expression and clinical datasets, was used to generate Kaplan–Meier plots for RFS as described previously [41].

4.11. Statistical Analysis

For the clinical analysis, the cases were divided into two groups according to N-cadherin expression using the minimum p value approach, which is a comprehensive method to determine the optimal cutoff point for survival risk classification among continuous gene expression measurements from multiple datasets [42]. The variables were compared using the Mann–Whitney U test or Fisher's exact test. Survival curves were generated using the Kaplan–Meier method and compared by log-rank test. Cox proportional hazards regression was used for univariate and multivariate analyses to calculate hazard ratios for the factors associated with survival. A two-sided $p < 0.05$ was deemed statistically significant. Statistical analyses were performed using JMP Pro 13 software (SAS Institute, Cary, NC, USA).

5. Conclusions

We demonstrated that N-cadherin mRNA expression in blood serves as a novel prognostic biomarker for predicting NM and cancer recurrence in patients with breast cancer. Furthermore, we provide clinical evidence that eribulin inhibits NM possibly via suppression of EMT. These findings highlight the role of N-cadherin in NM, which serves as a guide for the treatment of breast cancer, and provide a better understanding of the molecular mechanism of breast cancer recurrence.

Supplementary Materials: The supplementary materials are available online at http://www.mdpi.com/1422-0067/ 21/2/511/s1.

Author Contributions: conceptualization, T.M.; data curation, Y.K.; formal analysis, T.M., Y.K., M.N., Q.H., K.S., A.F., N.H., Y.T., H.O., Y.K., and H.E.; funding acquisition, T.M. and K.M.; investigation, H.U. (Hiroki Ueo); resources, S.O. and H.U.(Hiroki Ueo); supervision, K.M. and H.U. (Hiroki Ueo); writing—original draft, T.M.; writing—review and editing, K.M. All authors have read and agreed to the published version of the manuscript.

Funding: This work was supported in part by the following grants and foundations: Japan Society for the Promotion of Science (JSPS) Grant-in-Aid for Science Research (grant numbers: JP16K07177, JP16K10543, JP16K19197, JP17K16454, JP17K16521, JP17K10593, and JP17K19608), OITA Cancer Research Foundation, Daiwa Securities Health Foundation, Grant-in-Aid for Scientific Research on Innovative Areas (15H0912), Priority Issue on Post-K computer (hp170227 and hp160219), JSPS KAKENHI (15H05707), Eli Lilly Japan K.K. Grant, and Japanese Foundation for Multidisciplinary Treatment of Cancer.

Acknowledgments: We thank Tyler Lahusen for the helpful comments and English proofreading. We also thank Kazumi. Oda, Michiko. Kasagi, Sachiko. Sakuma, Noriko. Mishima, and Tomoko. Kawano for their excellent technical assistance.

Conflicts of Interest: The authors declare that they have no competing interests.

References

1. Pivot, X.; Marme, F.; Koenigsberg, R.; Guo, M.; Berrak, E.; Wolfer, A. Pooled analyses of eribulin in metastatic breast cancer patients with at least one prior chemotherapy. *Ann. Oncol.* **2016**, *27*, 1525–1531. [CrossRef]

2. Litiere, S.; de Vries, E.G.; Seymour, L.; Sargent, D.; Shankar, L.; Bogaerts, J.; Committee, R. The components of progression as explanatory variables for overall survival in the Response Evaluation Criteria in Solid Tumours 1.1 database. *Eur. J. Cancer* **2014**, *50*, 1847–1853. [CrossRef]

3. Twelves, C.; Cortes, J.; Kaufman, P.A.; Yelle, L.; Awada, A.; Binder, T.A.; Olivo, M.; Song, J.; O'Shaughnessy, J.A.; Jove, M.; et al. "New" metastases are associated with a poorer prognosis than growth of pre-existing metastases in patients with metastatic breast cancer treated with chemotherapy. *Breast Cancer Res.* **2015**, *17*, 150. [CrossRef]

4. Demetri, G.D.; Schoffski, P.; Grignani, G.; Blay, J.Y.; Maki, R.G.; Van Tine, B.A.; Alcindor, T.; Jones, R.L.; D'Adamo, D.R.; Guo, M.; et al. Activity of Eribulin in Patients With Advanced Liposarcoma Demonstrated in a Subgroup Analysis From a Randomized Phase III Study of Eribulin Versus Dacarbazine. *J. Clin. Oncol* **2017**, *35*, 3433–3439. [CrossRef]

5. Kashiwagi, S.; Tsujio, G.; Asano, Y.; Goto, W.; Takada, K.; Takahashi, K.; Morisaki, T.; Fujita, H.; Takashima, T.; Tomita, S.; et al. Study on the progression types of cancer in patients with breast cancer undergoing eribulin chemotherapy and tumor microenvironment. *J. Transl. Med.* **2018**, *16*, 54. [CrossRef]

6. Mori, R.; Futamura, M.; Morimitsu, K.; Asano, Y.; Tokumaru, Y.; Kitazawa, M.; Yoshida, K. The mode of progressive disease affects the prognosis of patients with metastatic breast cancer. *World J. Surg. Oncol* **2018**, *16*, 169. [CrossRef]

7. Eisenhauer, E.A.; Therasse, P.; Bogaerts, J.; Schwartz, L.H.; Sargent, D.; Ford, R.; Dancey, J.; Arbuck, S.; Gwyther, S.; Mooney, M.; et al. New response evaluation criteria in solid tumours: Revised RECIST guideline (version 1.1). *Eur. J. Cancer* **2009**, *45*, 228–247. [CrossRef]

8. Cortes, J.; O'Shaughnessy, J.; Loesch, D.; Blum, J.L.; Vahdat, L.T.; Petrakova, K.; Chollet, P.; Manikas, A.; Dieras, V.; Delozier, T.; et al. Eribulin monotherapy versus treatment of physician's choice in patients with metastatic breast cancer (EMBRACE): A phase 3 open-label randomised study. *Lancet* **2011**, *377*, 914–923. [CrossRef]

9. Kaufman, P.A.; Awada, A.; Twelves, C.; Yelle, L.; Perez, E.A.; Velikova, G.; Olivo, M.S.; He, Y.; Dutcus, C.E.; Cortes, J. Phase III open-label randomized study of eribulin mesylate versus capecitabine in patients with locally advanced or metastatic breast cancer previously treated with an anthracycline and a taxane. *J. Clin. Oncol.* **2015**, *33*, 594–601. [CrossRef]

10. Cortes, J.; Schoffski, P.; Littlefield, B.A. Multiple modes of action of eribulin mesylate: Emerging data and clinical implications. *Cancer Treat Rev.* **2018**, *70*, 190–198. [CrossRef]

11. Yoshida, T.; Ozawa, Y.; Kimura, T.; Sato, Y.; Kuznetsov, G.; Xu, S.; Uesugi, M.; Agoulnik, S.; Taylor, N.; Funahashi, Y.; et al. Eribulin mesilate suppresses experimental metastasis of breast cancer cells by reversing phenotype from epithelial-mesenchymal transition (EMT) to mesenchymal-epithelial transition (MET) states. *Br. J. Cancer* **2014**, *110*, 1497–1505. [CrossRef]

12. Oba, T.; Ito, K.I. Combination of two anti-tubulin agents, eribulin and paclitaxel, enhances anti-tumor effects on triple-negative breast cancer through mesenchymal-epithelial transition. *Oncotarget* **2018**, *9*, 22986–23002. [CrossRef]

13. Terashima, M.; Sakai, K.; Togashi, Y.; Hayashi, H.; De Velasco, M.A.; Tsurutani, J.; Nishio, K. Synergistic antitumor effects of S-1 with eribulin in vitro and in vivo for triple-negative breast cancer cell lines. *Springerplus* **2014**, *3*, 417. [CrossRef]

14. Floor, S.; van Staveren, W.C.; Larsimont, D.; Dumont, J.E.; Maenhaut, C. Cancer cells in epithelial-to-mesenchymal transition and tumor-propagating-cancer stem cells: Distinct, overlapping or same populations. *Oncogene* **2011**, *30*, 4609–4621. [CrossRef]

15. Kolbl, A.C.; Jeschke, U.; Andergassen, U. The Significance of Epithelial-to-Mesenchymal Transition for Circulating Tumor Cells. *Int. J. Mol. Sci.* **2016**, *17*, 1308. [CrossRef]

16. Di Meo, A.; Bartlett, J.; Cheng, Y.; Pasic, M.D.; Yousef, G.M. Liquid biopsy: A step forward towards precision medicine in urologic malignancies. *Mol. Cancer* **2017**, *16*, 80. [CrossRef]

17. Marrugo-Ramirez, J.; Mir, M.; Samitier, J. Blood-Based Cancer Biomarkers in Liquid Biopsy: A Promising Non-Invasive Alternative to Tissue Biopsy. *Int. J. Mol. Sci.* **2018**, *19*, 2877. [CrossRef]

18. Yokobori, T.; Iinuma, H.; Shimamura, T.; Imoto, S.; Sugimachi, K.; Ishii, H.; Iwatsuki, M.; Ota, D.; Ohkuma, M.; Iwaya, T.; et al. Plastin3 is a novel marker for circulating tumor cells undergoing the epithelial-mesenchymal transition and is associated with colorectal cancer prognosis. *Cancer Res.* **2013**, *73*, 2059–2069. [CrossRef]

19. Parl, F.F.; Schmidt, B.P.; Dupont, W.D.; Wagner, R.K. Prognostic significance of estrogen receptor status in breast cancer in relation to tumor stage, axillary node metastasis, and histopathologic grading. *Cancer* **1984**, *54*, 2237–2242. [CrossRef]

20. Pichon, M.F.; Broet, P.; Magdelenat, H.; Delarue, J.C.; Spyratos, F.; Basuyau, J.P.; Saez, S.; Rallet, A.; Courriere, P.; Millon, R.; et al. Prognostic value of steroid receptors after long-term follow-up of 2257 operable breast cancers. *Br. J. Cancer* **1996**, *73*, 1545–1551. [CrossRef]

21. Kinne, D.W.; Butler, J.A.; Kimmel, M.; Flehinger, B.J.; Menendez-Botet, C.; Schwartz, M. Estrogen receptor protein of breast cancer in patients with positive nodes. High recurrence rates in the postmenopausal estrogen receptor-negative group. *Arch. Surg.* **1987**, *122*, 1303–1306. [CrossRef] [PubMed]

22. Maynard, P.V.; Davies, C.J.; Blamey, R.W.; Elston, C.W.; Johnson, J.; Griffiths, K. Relationship between oestrogen-receptor content and histological grade in human primary breast tumours. *Br. J. Cancer* **1978**, *38*, 745–748. [CrossRef] [PubMed]

23. Masuda, T.; Hayashi, N.; Iguchi, T.; Ito, S.; Eguchi, H.; Mimori, K. Clinical and biological significance of circulating tumor cells in cancer. *Mol. Oncol.* **2016**, *10*, 408–417. [CrossRef]

24. Mrozik, K.M.; Blaschuk, O.W.; Cheong, C.M.; Zannettino, A.C.W.; Vandyke, K. N-cadherin in cancer metastasis, its emerging role in haematological malignancies and potential as a therapeutic target in cancer. *BMC Cancer* **2018**, *18*, 939. [CrossRef]

25. Cao, Z.Q.; Wang, Z.; Leng, P. Aberrant N-cadherin expression in cancer. *Biomed Pharm.* **2019**, *118*, 109320. [CrossRef]

26. Takeichi, M. Cadherins: A molecular family important in selective cell-cell adhesion. *Annu. Rev. Biochem* **1990**, *59*, 237–252. [CrossRef]

27. Rieger-Christ, K.M.; Lee, P.; Zagha, R.; Kosakowski, M.; Moinzadeh, A.; Stoffel, J.; Ben-Ze'ev, A.; Libertino, J.A.; Summerhayes, I.C. Novel expression of N-cadherin elicits in vitro bladder cell invasion via the Akt signaling pathway. *Oncogene* **2004**, *23*, 4745–4753. [CrossRef]

28. Nguyen, T.; Mege, R.M. N-Cadherin and Fibroblast Growth Factor Receptors crosstalk in the control of developmental and cancer cell migrations. *Eur. J. Cell Biol.* **2016**, *95*, 415–426. [CrossRef]

29. Ashaie, M.A.; Chowdhury, E.H. Cadherins: The Superfamily Critically Involved in Breast Cancer. *Curr. Pharm. Des.* **2016**, *22*, 616–638. [CrossRef]

30. Yoshinaga, K.; Inoue, H.; Utsunomiya, T.; Sonoda, H.; Masuda, T.; Mimori, K.; Tanaka, Y.; Mori, M. N-cadherin is regulated by activin A and associated with tumor aggressiveness in esophageal carcinoma. *Clin Cancer Res.* **2004**, *10*, 5702–5707. [CrossRef]

31. Luo, Y.; Yu, T.; Zhang, Q.; Fu, Q.; Hu, Y.; Xiang, M.; Peng, H.; Zheng, T.; Lu, L.; Shi, H. Upregulated N-cadherin expression is associated with poor prognosis in epithelial-derived solid tumours: A meta-analysis. *Eur. J. Clin. Invest.* **2018**, *48*, e12903. [CrossRef]

32. Horimoto, Y.; Tokuda, E.; Murakami, F.; Uomori, T.; Himuro, T.; Nakai, K.; Orihata, G.; Iijima, K.; Togo, S.; Shimizu, H.; et al. Analysis of circulating tumour cell and the epithelial mesenchymal transition (EMT) status during eribulin-based treatment in 22 patients with metastatic breast cancer: A pilot study. *J. Transl. Med.* **2018**, *16*, 287. [CrossRef]

33. Vandyke, K.; Chow, A.W.; Williams, S.A.; To, L.B.; Zannettino, A.C. Circulating N-cadherin levels are a negative prognostic indicator in patients with multiple myeloma. *Br. J. Haematol* **2013**, *161*, 499–507. [CrossRef]

34. Takashima, T.; Mukai, H.; Hara, F.; Matsubara, N.; Saito, T.; Takano, T.; Park, Y.; Toyama, T.; Hozumi, Y.; Tsurutani, J.; et al. Taxanes versus S-1 as the first-line chemotherapy for metastatic breast cancer (SELECT BC): An open-label, non-inferiority, randomised phase 3 trial. *Lancet Oncol* **2016**, *17*, 90–98. [CrossRef]

35. Masuda, T.; Shinden, Y.; Noda, M.; Ueo, H.; Hu, Q.; Yoshikawa, Y.; Tsuruda, Y.; Kuroda, Y.; Ito, S.; Eguchi, H.; et al. Circulating Pre-microRNA-488 in Peripheral Blood Is a Potential Biomarker for Predicting Recurrence in Breast Cancer. *Anticancer Res.* **2018**, *38*, 4515–4523. [CrossRef]

36. Coates, A.S.; Winer, E.P.; Goldhirsch, A.; Gelber, R.D.; Gnant, M.; Piccart-Gebhart, M.; Thurlimann, B.; Senn, H.J.; Panel, M. Tailoring therapies—Improving the management of early breast cancer: St Gallen International Expert Consensus on the Primary Therapy of Early Breast Cancer 2015. *Ann. Oncol.* **2015**, *26*, 1533–1546. [CrossRef]

37. Francart, M.E.; Lambert, J.; Vanwynsberghe, A.M.; Thompson, E.W.; Bourcy, M.; Polette, M.; Gilles, C. Epithelial-mesenchymal plasticity and circulating tumor cells: Travel companions to metastases. *Dev. Dyn.* **2018**, *247*, 432–450. [CrossRef]

38. Ueo, H.; Sugimachi, K.; Gorges, T.M.; Bartkowiak, K.; Yokobori, T.; Muller, V.; Shinden, Y.; Ueda, M.; Ueo, H.; Mori, M.; et al. Circulating tumour cell-derived plastin3 is a novel marker for predicting long-term prognosis in patients with breast cancer. *Br. J. Cancer* **2015**, *112*, 1519–1526. [CrossRef]

39. Takano, Y.; Masuda, T.; Iinuma, H.; Yamaguchi, R.; Sato, K.; Tobo, T.; Hirata, H.; Kuroda, Y.; Nambara, S.; Hayashi, N.; et al. Circulating exosomal microRNA-203 is associated with metastasis possibly via inducing tumor-associated macrophages in colorectal cancer. *Oncotarget* **2017**, *8*, 78598–78613. [CrossRef]

40. Wakiyama, H.; Masuda, T.; Motomura, Y.; Hu, Q.; Tobo, T.; Eguchi, H.; Sakamoto, K.; Hirakawa, M.; Honda, H.; Mimori, K. Cytolytic Activity (CYT) Score Is a Prognostic Biomarker Reflecting Host Immune Status in Hepatocellular Carcinoma (HCC). *Anticancer Res.* **2018**, *38*, 6631–6638. [CrossRef]

41. Nambara, S.; Masuda, T.; Nishio, M.; Kuramitsu, S.; Tobo, T.; Ogawa, Y.; Hu, Q.; Iguchi, T.; Kuroda, Y.; Ito, S.; et al. Antitumor effects of the antiparasitic agent ivermectin via inhibition of Yes-associated protein 1 expression in gastric cancer. *Oncotarget* **2017**, *8*, 107666–107677. [CrossRef]

42. Mizuno, H.; Kitada, K.; Nakai, K.; Sarai, A. PrognoScan: A new database for meta-analysis of the prognostic value of genes. *BMC Med. Genom.* **2009**, *2*, 18. [CrossRef]

International Journal of
Molecular Sciences

Review

Beyond the Genomic Mutation: Rethinking the Molecular Biomarkers of K-RAS Dependency in Pancreatic Cancers

Carla Mottini [1] and Luca Cardone [1,2,*]

[1]   Department of Research, Advanced Diagnostics and Technological Innovations, IRCCS Regina Elena
     National Cancer Institute, 00144 Rome, Italy; carla.mottini@ifo.gov.it
[2]   Institute of Biochemistry and Cellular Biology, CNR National Research Council, 00015 Rome, Italy
*   Correspondence: luca.cardone@ifo.gov.it; Tel.: +39-0652-662-939

Received: 15 June 2020; Accepted: 14 July 2020; Published: 16 July 2020

Abstract: Oncogenic v-Ki-ras2 Kirsten rat sarcoma viral oncogene homolog (K-RAS) plays a key role in the development and maintenance of pancreatic ductal adenocarcinoma (PDAC). The targeting of K-RAS would be beneficial to treat tumors whose growth depends on active K-RAS. The analysis of *K-RAS* genomic mutations is a clinical routine; however, an emerging question is whether the mutational status is able to identify tumors effectively dependent on K-RAS for tailoring targeted therapies. With the emergence of novel K-RAS inhibitors in clinical settings, this question is relevant. Several studies support the notion that the K-RAS mutation is not a sufficient biomarker deciphering the effective dependency of the tumor. Transcriptomic and metabolomic profiles of tumors, while revealing K-RAS signaling complexity and K-RAS-driven molecular pathways crucial for PDAC growth, are opening the opportunity to specifically identify K-RAS-dependent- or K-RAS-independent tumor subtypes by using novel molecular biomarkers. This would help tumor selection aimed at tailoring therapies against K-RAS. In this review, we will present studies about how the K-RAS mutation can also be interpreted in a state of K-RAS dependency, for which it is possible to identify specific K-RAS-driven molecular biomarkers in certain PDAC subtypes, beyond the genomic K-RAS mutational status.

Keywords: pancreatic cancer; *K-RAS* oncogene; oncogene dependency; targeted therapies; biomarkers; genomic mutations; transcriptomics; metabolomics

---

1. Oncogenic K-RAS: A Critical Driver for Pancreatic Cancer

Pancreatic ductal adenocarcinoma (PDAC) is a major cause of cancer-related death with an overall five-year survival rate of only 8% [1,2]. PDAC is diagnosed at an advanced, inoperable stage in the vast majority of cases and most of the patients diagnosed with surgically resectable disease recur within the first 2–3 years after the operation [3]. Current systemic first-line treatment for advanced inoperable PDAC includes polychemotherapy regimens such as folinic acid/ 5-fluorouracil/irinotecan/oxaliplatin,(FOLFIRINOX),cisplatin/nab-paclitaxel/capecitabine/gemcitabine (PAXG), gemcitabine/nab-paclitaxel, and gemcitabine monotherapy in a small sub-group of elderly, frail, or unfit patients. Primary chemoresist/ance or recurrence rates in PDAC remain high, and overall survival from the start of first-line ranges approximately from 8 to 12 months [4–6]. Currently, no validated prognostic or predictive biomarkers exist for PDAC, except for general clinical criteria (performance status, disease burden, CA19.9 levels), and no targeted or immune-based therapies have proven to be effective so far, although a large number of clinical trials are ongoing and efficacy data for novel treatments are awaited [7–9].

The RAS pathway is one of the most frequently altered pathways in cancer, found in approximately 19% of all human cancer harboring *RAS* gene mutations [10]. Among the three major isoforms of oncogenic *RAS*, *K-RAS* is the most frequently mutated [11–13]. Mutation of *K-RAS* is the initiating genetic event of pancreatic intraepithelial neoplasias (PanINs) and is required to drive PDAC development and tumor maintenance [14–18] Oncogenic mutant *K-RAS* is found in about 88% of PDAC [10]. Oncogenic mutation in K-RAS protein leads to aberrant or constitutive signaling even in the absence of growth factors, leading to increased proliferation, invasion, and metastasis [19]. Inactivating mutations in crucially tumor suppressor genes, particularly *CDKN2A/p16*, *TP53*, and *SMAD4*, cooperate with oncogenic *K-RAS* to promote aggressive PDAC tumor growth and metastasis [19–26].

K-RAS is a member of the RAS family of Guanosine Tri-Phosphate(GTP)-ases that regulates several cellular processes including survival, proliferation, differentiation, migration, and apoptosis [27]. RAS proteins function as molecular switches promoting conversion from an inactive to an active GTP-bound state. Though tightly controlled in normal cells, the mutation in K-RAS gene leads to constitutive GTP-bound K-RAS, rendering constitutively activated RAS protein and determining the persistent activation of downstream signaling pathways resulting in uncontrolled activation of proliferation and survival pathways [28–31]. The mutations in K-RAS consist of single amino acid substitutions and are predominant at residues G12, G13, and Q61. Oncogenic mutations of G12 or G13 create a steric block that prevents the hydrolysis of GTP, whereas substitutions of Q61 interfere with the coordination of a water molecule required for GTP hydrolysis; these point mutations lead to a prevalence of the GTP-bound state and to the constitutive activation of K-RAS [19].

Once in its active form, K-RAS engages complex and dynamic downstream effectors such as the RAF/MEK and the phosphatidylinositol 3-kinase (PI3K)/AKT pathway. The Mitogen-Activated Protein Kinase (MAPK) pathway is a key mediator of oncogenic K-RAS signaling and BRAF is the principal mediator of MAPK signaling in K-RAS dependent cancer growth. The BRAF V600E mutations are mutually exclusive with K-RAS mutations [32]. However, genetic studies in mice models revealed that *BRAF* ^{V600E} mutation is sufficient to induce PanIN formation in the pancreas of *K-RAS* wild-type (WT) mice, and to develop lethal PDAC when combined with a *TP53* mutation [33]. The PI3K-dependent pathway drives tumor growth and cooperates with oncogenic K-RAS to develop PDAC [34,35]. The major driver mutations in this pathway that promote pancreatic tumor development include mutations in the catalytic and regulatory PI3K subunit, amplification of the PI3K downstream effector AKT2, and deletion/loss of tumor suppressor Phosphatase/TENsin homolog deleted on chromosome 10 (PTEN), a negative regulator of PI3K/AKT signaling [36–38].

It is important to mention that a relatively large proportion of patients with PDAC display germline mutations of some DNA damage repair (DDR) genes. Specifically, 18% of PDAC harbor mutations in homologous recombination (HR) DDR pathways such as *BRCA1* and *BRCA2* [39], and the BRCA2 inactivation in combination with p53 deficiency promotes K-RAS driven PDAC development [40,41].

It goes without saying that the genomic landscape of PDAC shows multiple genetic events, most of them contributing to tumor maintenance in cooperation with the K-RAS activation, most likely with a different degree of dependency according to the history of the tumor development, staging, or treatments. Deciphering the effective dependency of the tumor on K-RAS or on alternative oncogenes is key to promote targeted therapies in PDAC.

2. Defining the K-RAS Dependency in PDAC

K-RAS mutation represents a common genetic event in PDAC, being mutated in almost 88% of cases [10,42]. However, contrary to preclinical studies, clinical approaches have demonstrated poor efficacy of treatments targeting the K-RAS pathway in PDAC tumors carrying a K-RAS mutation. One of the potential explanations is the possibility that the genomic *K-RAS* mutation is not an efficacious molecular determinant for tumor dependency on K-RAS activation. Indeed, the absence of K-RAS gene mutations does not always correlate with K-RAS pathway inactivity due to the activation of the other components of the network [43,44], and conversely, the presence of *RAS* mutations does not necessarily predict for dependency. This can depend on the activation of additional active molecular pathways that can complement or subside for K-RAS activation. Thus, determining the genomic mutational status of specific genes is not always beneficial for predicting pathway activation and the drug response with targeted compounds [45].

Assessing K-RAS pathway activation status by more comprehensive methods will help better predict the K-RAS dependency of tumors. Years have passed since the concept of oncogene addiction was first proposed, linking single dominant oncogene to tumor growth and survival [46]. Omics studies, such as genomics, transcriptomics, and metabolomics can lead to extensive molecular profiles, which act as tools to reevaluate the traditional definition of addiction and oncogene dependency as a functional definition based on the oncogene-driven phenotype, regardless of the presence or not of a specific oncogenic gene mutation. A large number of observations in animal models and pancreatic cancer cell lines revealed that the *K-RAS* gene, although mutated or overexpressed, is dispensable in a subset of human and mouse K-RAS mutant PDAC cell lines. By using RNA interference, inducible transgenic models or Clustered Regularly Interspaced Short Palindromic Repeats (CRISPR)-Cas9 technology, it has been possible to classify two subtypes of PDACs harboring the K-RAS mutation: tumors in which a K-RAS depletion led to apoptosis and thus they are considered as "K-RAS-dependent" and others that are resistant to K-RAS depletion, without a sign of apoptosis, and considered as "K-RAS-independent" [47–53]. The extensive molecular characterization of such models shed a light on additional features that would be missed based on simple genomic classification of the tumor, with the potential of a profound implication from a therapeutic and prognostic point of view.

The goal of this review is to provide an overview of emerging molecular markers of K-RAS oncogene dependency, regardless of the genomic mutation status. Gene expression profile studies, in particular, allow to understand if the K-RAS pathway could be activated by mutations of the *K-RAS* gene or by many other mechanisms, and they help to deconstruct the K-RAS network contribution in tumor progression [45,48,54]. In addition, metabolomics studies identified pathways and metabolites that are specifically enriched in K-RAS-dependent PDAC to mediate a metabolic reprogramming relevant to tumor growth. Thus, multiple and specific molecular biomarkers underlining the oncogenic phenotype associated with a real dependency on K-RAS oncogene in PDAC are emerging. The translational value of such information is manifold since i) it helps to find novel diagnostic biomarkers that could overcome the limitation of a genomic-based approach for an effective determination of K-RAS dependency and ii) it provides the ground for novel therapeutic strategies to define effective targeted therapy against a subclass of PDAC patients, whose tumors have K-RAS dependency and actionable vulnerabilities.

In the next paragraphs, we will discuss molecular profiling based on transcriptomic and metabolomics studies that provided novel markers for K-RAS dependency in PDAC.

3. Scores of K-RAS Dependency Based on Gene Expression Signatures

Specific gene expression signatures can be associated with oncogenic mutations and deregulated signaling pathways in tumors [45,55]. Several studies demonstrated the potential of using gene expression profiles of cancer cells to analyze oncogenic pathways [45,56]. Gene expression signature is also a powerful tool for predicting drug response in vitro linked to specific molecular pathways [45,57,58]. In tumor samples, gene expression signature can reveal molecular pathways that are activated independently from mutation status, confirming that mutations were no obvious predictors for pathway activation [44]. As an example, in bladder urothelial carcinoma, Epidermal Growth Factor Receptor (EGFR)-, K-RAS-, and RAF-dependent pathways were activated in 42%, 22%, and 38% of cases, but only three patients carried the *EGFR* mutation and no mutation in *K-RAS* or *RAF* was found. Furthermore, reverse phase protein array (RPPA) demonstrated significant MAPK pathway activity in both *K-RAS*-mutant lung adenocarcinoma and *K-RAS* wild-type samples [44].

Oncogenic K-RAS-specific signatures have been derived from cancer models [48,51,59]. Gene expression profiling studies in some cell lines and human tumors have allowed identifying a more comprehensive K-RAS-dependent network able to understand the pathway activation's status and to better describe the K-RAS-dependent molecular network [47,48]. In their studies, Singh et al. [47] stratified K-RAS mutant pancreatic and lung cancer cell lines into K-RAS-dependent or K-RAS-independent subtypes, according to the ability of K-RAS to support cell viability. Specifically, the ablation of mutant *K-RAS* by RNAi affected cell viability and induced apoptosis with differential sensitivity between cell lines. A K-RAS dependency index, in a subset of pancreatic and lung cancer cell lines harboring oncogenic *K-RAS*, was defined based on caspase-3 cleavage values. To better delineate the molecular mechanisms that distinguish K-RAS-dependent and independent cell lines, a gene expression signature identifying genes differentially expressed in these two groups was derived. The authors identified a 250 gene-based gene signature to define dependency in PDAC (Table 1).

With a different approach and models, Loboda et al. identified a 147 gene-based *K-RAS* gene signature [48] that was associated with tumors with a *K-RAS* dependency (Table 1). By using a *K-RAS* knockdown-based strategy, the authors demonstrated that in a panel of lung and breast tumor cell lines, not all K-RAS mutated cells were dependent on K-RAS signaling, and some cells carrying wild-type *K-RAS* exhibited a *K-RAS* dependency. Furthermore, the K-RAS dependency gene signature was highly correlated with MEK and ERK phosphorylation and with the cellular response after a MEK inhibitor treatment, suggesting the clinical relevance of such gene signature to predict the response to K-RAS pathway inhibitors [48].

The potential translational value of these scores was validated by Mottini et al. [54] who showed that the gene expression signatures identified by Loboda and Singh showed a high correlation among them to classify K-RAS-dependent or K-RAS-independent PDAC cell lines according to signature similarity scores. In addition, the authors demonstrated, for the first time, that both genetic signatures derived by Loboda and Singh were able to identify a dependency on K-RAS in patient-derived xenograft (PDX) models of PDAC, which is a more reliable, patient-like experimental system, demonstrating the predictive capability of these two gene signatures identified by in-vitro cancer models. The authors also reported an in-vivo discrepancy between the two K-RAS-dependent gene signatures; they argued that the discrepancy could reflect the different approaches used to derive such signatures and potentially be associated with differences in how the microenvironment can influence gene signatures, in particular, the one identified by Singh et al. [47]. Indeed, classification of PDAC-PDXs using the ones identified by Singh better match the molecular subtypes recently identified in PDAC patients by applying tumor-specific factors [60,61]. On the other hand, the signature identified by Loboda might be associated with a higher degree of cell-autonomous signature, which is potentially less influenced by microenvironmental components. The authors concluded that, based on the evidence of PDX data and The Cancer Genome Atlas (TCGA) data, using combined molecular signatures might be a robust predictive tool to infer oncogene dependency from tumor biopsy or PDX models.

Table 1. Listing of upregulated and downregulated genes related to K-RAS pathway activation [45] and to K-RAS dependency [44].

Ref	Gene Symbol		Methodologies
	Up Regulated Genes	**Down Regulated Genes**	
Loboda et al., 2010	*ADAM8, ADRB2, ANGPTL4, ARNTL2,C19orf10, C20orf42, CALM2,CALU, CAPZA1, CCL20, CD274, CDCP1, CLCF1, CSNK1D, CXCL1, CXCL2,CXCL3, CXCL5, DENND2C, DUSP1, DUSP4, DUSP5, DUSP6, EFNB1, EGR1, EHD1, ELK3, EREG, FOS, FOXQ1, G0S2, GDF15, GLTP, HBEGF, IER3, IL13RA2, IL1A, IL1B, IL8, ITGA2, ITPR3, KCNK1, KCNN4, KLF5, KLF6, LAMA3, LDLR, LHFPL2, LIF, MALL, MAP1LC3B, MAST4, MMP14, MXD1, NAV3, NDRG1, NFKBIZ, NPAL1, NT5E, OXSR1, PBEF1, PHLDA1, PHLDA2, PI3, PIK3CD, PIM1, PLAUR, PNMA2, PPP1R15A, PRNP, PTGS2, PTHLH, PTPRE, PTX3, PVR, RPRC1, S100A6, SDC1, SDC4, SEMA4B, SERPINB1, SERPINB2, SERPINB5, SESN2, SFN, SLC16A3, SLC2A14, SLC2A3, SLC9A1, SPRY4, TFPI2, TGFA, TIMP1, TMEM45B, TNFRSF10A, TNFRSF10B, TNFRSF12A, TNS4, TOR1AIP1, TSC22D1, TUBA1, UAP1, UPP1, VEGF, ZFP36*	*ABCC5, ARMC8, ATPAF1, AUTS2, C1orf96, C6orf182, CELSR2, CENTB2, COQ7, DRD4, ENAH, HNRPU, HTATSF1, ID4, ITSN1, JMJD2C, KIAA1772, MIB1, MRPS14, MSI1, MSI2, NUP133, OGN, PARP1, PIAS1, RASL10B, RFPL3S, RTN3, SEC63, SF4, SH3GL2, SMAD9, STARD7, TBC1D24,TMEFF1, TTC28, TXNDC4, ZNF292, ZNF441, ZNF493, ZNF669, ZNF672*	K-RAS pathway signature derived from a superset of lung cancer, breast cancer, and colon cancer gene expression data

Table 1. *Cont.*

Ref	Gene Symbol		Methodologies
	Up Regulated Genes	**Down Regulated Genes**	
Singh et al., 2009	*SYK, ST14, TMEM30B, SPINT1, RAB25, C1orf172, GRHL2, GALNT3, SCNN1A, EVA1, ITGB6, C1orf74, PCDH1, C6orf141, HS3ST1, CDS1, DNAJA4, CLDN7, SCEL, SCIN, ANKRD22, MAL2, EHF, RAB17, C1orf106, TTC9, DENND1C, CEACAM6, MAPK13, LOC196264, BSPRY, C1orf116, VSIG1, KIAA0703, TMPRSS4, TGFA, EPN3, ERBB3, C1orf210, TMEM45B, RALGPS2, CDA, CDH1, SYTL5, FRK, OVOL2, RDHE2, LOC653857, B3GNT3, DPP4, PRSS22, EPS8L1, RBM35B, EFNB2, CGNL1, LAMA3, PGM2L1, ELF3, PLEKHA7, TIAF1, C11orf52, EPB41L5, KRTCAP3, RAB11FIP4, PPL, DSC2, TACSTD1, FER1L4, IRF6, TSPAN1, MAOA, CLDN4 TMEM154 MYO1D, GPR115, PPP1R14C, PKIB, TSPAN15, SH2D3A, AMPD3, UBD, MTAC2D1, TMC5, AIM1, ACP6, AREG, FAM102A, ZNF608, TMEM65, KIAA1522, C5orf4, NFATC3, KLF7, ELL2, OTUB2, PLEKHG1, FUT2, SORL1, MST1R, IKZF2, KRT7, C4orf34, JAG1, HOOK1, DLG3, KCNMB4, C12orf46, FLJ20273, RAC2, Gcom1, KIAA1107, STAP2, TACSTD2, SCARB2, CGN, PRSS8, DHRS3, C1orf34, FBP1, ZNF468, GDPD3, EGLN3, SEMA4B, ARHGEF3, LOC146795, RIPK4, RASEF, PRKCH, SLC37A1, EPPK1, PROM2, STON2, JUP, EPHB3, RPS6KA2, ALDH1A3, ROD1, PAK6, WFDC2, TMEM87B, SP110, C19orf21, TNFSF13, HPGD, ERO1L, ADAM8, ARSD, CYB561, FAM84B, FA2H, F11R, ALAD, EMG1, IL13RA1, TNFRSF21, PON3, FAM83H, GNA15, VEGF, YWHAZ, ARHGEF10L, SLC41A2, ACOT11, NR3C2, KIAA1217, GCHFR, KALRN, INPP4B, ST3GAL5, SAMD9, LMCD1, CD24, WFDC3, TMEM49, DOC1, AMDD, CTNND1, TGOLN2, MCTP2, CST6, CSPG2, CHCHD7, TMC6, TMEM125, PRRG4, GSN, DKFZP779L1068 CEACAM1, CAB39, MXD1, SHROOM3, LYPD3, LAMC2, ENTPD3, PADI1, ADAM28, TMC4, DAAM1, IL23A, SNN, SOX4, TXNIP, LLGL2, PRSS16, IDS, PTK6, CDH3, CAPN8, MTUS1, STOM, CEACAM19, S100A16, HOOK2, CDKN2A, APRIN, KLF5, DAPP1, ABLIM3, PDE5A, REPS2, LRRC1, JUNB, SLC40A1, ZNRF1, PSD4, KIAA1815, PAK1, KIF21B, SLC44A3, ELF1, F5, SPINT2, FGFBP1, TRIOBP, ROR1, ATP8B1, KRAS, IFIH1, TSGA10, FUT3, EDG4, ZBTB25, TJP2, MALAT, 1 B3GNT5, FUCA1, FOXP1, MET, GBP2, RPL41, NRP2, SHROOM2, SERPINA1, TMTC2, GRK5, UCA1, LOC58489, CEACAM5, RASD1, TSC22D3, CBR3, ARHGDIB, FRMD4B, S100A6, ZNF626, F3, EPHA1, PLS1, TAF9, RPH3A, SLC44A2, FAM83A, CNKSR1, KIAA0251, GPR110, DENND2D, BIK, KIAA0284, CAMP2, AZGP1, BMF, CHMP4C.*	*HNRPU, SLC39A14, PARVB, SH2B3, FLJ45482, NEDD4, IPO7, SGPP1, USP47 HIST1H1D,, FGFR1, MRC2, MSX1, FGF2, TEAD4, AGPAT5,, WDHD1, B4GALT6, TTC28, NFIC, RAPGEF1, ZIC2, RAB6IP1, RECK, LHFP, ST3GAL3 MSRB3, SLC26A2, PMP22, MAGEH1, BMP6, ROBO3, GJA7, TMEM20, MCOLN2, SEC61A2, IL11RA, COPZ2, NIN, ANTXR1, RSAD1, EEF2K, ITPR2, C14orf135, CWF19L1, ANKRD28, PPP4R2 TMEM118, TSPAN4, RAGE, DYRK4, FLJ36166, ALPK2, BCAP29, C14orf139, CSPG5, TTC7B SATB2, TCF8, SLC35B4, OSTM1, IKIP, SFXN1, TRIM7, KIAA1212, MGC39900, NFIX, PDLIM3, MIB1, MLSTD2, LOC401068, ALS2CR4, PRG1, APLN, FAM101B, LOC541471, HNRPA2B1, RHOT1, LOC153346, DYRK3, EML1, RYK KCTD15, PAX6, PLCB4, WDR35, CHRNA7, LIX1L, ACTA2, HTRA1, ABP1, ANXA6, HSPA12A, MAGEE1, SYDE1, TUB, SMARCD3, NUDT11, SYNGR1, MPHOSPH9, ADRA2C, TXNRD1, EPB41L5, MPPE1, SLC1A3, LOC439949, FLJ10847*	K-RAS dependency signature derived from a subset of K-RAS dependent primary lung tumors of squamous carcinoma and adenocarcinoma subtypes

In an attempt to identify the mechanism for K-RAS independency in PDAC, Kapoor and colleagues [50] discovered that, upon K-RAS suppression, about half of the tumors relapsed thanks to activation of a transcriptional program controlled by the cooperation of the Yes-Associated Protein 1/TEA Domain Transcription Factor 2 (YAP1/TEAD2) transcription factors complex to promote cell cycle, DNA replication, and tumor maintenance in the absence of oncogenicK-RAS. Transcriptomic and network analysis profiles in the K-RAS-independent YAP1-driven tumors provided substantial evidence on how YAP activation bypasses K-RAS mutations by supporting the transcriptional pathways that are key K-RAS targets. Thus, YAP1 supports pathways activated by mutated K-RAS, and active YAP1 pathway emerged as a putative marker and therapeutic target against pancreatic cancers showing K-RAS independency [50].

It is worth noting that the above-cited studies indicate that, in certain cell lines, the definition of K-RAS dependency might suffer a discrepancy, due to the type of in vitro biological assay. This likely depends on cell-specific genetic mutations and molecular pathways activations, within the K-RAS network, that are engaged under the relevant biological assay. For example, growing cells into two-dimensional (2D) or 3D cultures, or instead, a clonogenic assay, might differentially engage K-RAS activation, generating differences and discrepancy for the K-RAS dependency outcome in the same cell line. Such discrepancy indicates the need for an accurate and robust validation for biomarkers of dependency, and it suggests that, besides in vitro-based studies, further validation of dependency biomarkers in a more physiological context such as in vivo tumor models is required. Moreover, the implementation of multiple transcriptional molecular scores for dependency would be of advantage for an accurate definition of *K-RAS* dependency, in particular when this information aims to predict drug response [54].

Transcriptomic approaches have also been instrumental to different molecular classifications of pancreatic ductal adenocarcinoma subtypes. By examining expression data from human and mouse cell lines, Collisson et al. classified PDAC into three subtypes termed "classical", "quasi mesenchymal", and exocrine-like [49]. Notably, the classical subtype was defined by high expression of adhesion specific and epithelial genes and was reported to confer the best chance of survival. On the other hand, the quasi-mesenchymal subtype showed a higher expression of mesenchymal associated genes with a poorly differentiated phenotype and related to a poor prognosis. Bailey et al. [62] analyzed transcriptomic data from tumor tissues containing the tumor microenvironment and identified a new "immunogenic" subgroup associated with immune stroma cell populations. Expression analysis defined four subtypes as "squamous", "pancreatic progenitor", "immunogenic", and "aberrantly differentiated endocrine exocrine". The squamous subtype showed the worst overall survival and overlaps with the quasi-mesenchymal tumor subtype defined by Collison et al. [49]. Tumors of the squamous subtype were reported to be associated by the presence of gene expression programs and regulatory networks involved in the inflammatory response, hypoxia, TGFβ signaling, metabolic reprogramming, and MYC activation, while immunogenic tumors were associated with a significant immune infiltrate and upregulation of immune regulatory networks involved in acquired immune suppression. In addition, Moffitt et al. [63] performed a microarray analysis of primary and metastatic tumors, identified two tumor subtypes ("classical" subtype and a "basal-like" subtype), and two stromal subtypes ("normal" subtype and "activated" subtype). The classical subtype was associated with poor prognosis and most of the identified genes overlap with the "classical" group gene by Collison et al. [49].

The above-described genetic classifications reflect the intrinsic molecular characteristics of tumors, allowing a better definition of the clinical heterogeneity of cancers. A relationship between K-RAS dependency and some of the above-classified subtypes have been investigated [49]. Specifically, by using an RNAi-based assay to deplete *K-RAS*, it was observed that classical subtype tumors cells were more dependent on *K-RAS* than quasi-mesenchymal PDAC cell lines. Interestingly, the gene expression signature associated with K-RAS addiction validated by Singh et al. [47] suggests that genes associated with an epithelial phenotype might represent potential biomarkers of K-RAS dependency in PDAC. Kapoor et al. [50] demonstrated a significant association between K-RAS-independent tumors and cells with the quasi-mesenchymal subtype; yet the ability of these genetic classifications to act as predictive biomarkers of K-RAS dependency for translational purposes needs to be validated.

4. Scores of K-RAS Dependency Based on Metabolic Phenotypes Analyses

K-RAS oncogenic activation orchestrates metabolic reprogramming crucial for tumor growth, proliferation, and survival. Metabolic phenotypes associated with K-RAS might represent a fingerprint of K-RAS activation status as well as vulnerability at the metabolic level. The tumor microenvironment of the PDAC is characterized by dense desmoplastic regions that make it poorly vascularized causing a reduction in oxygen supply and reduced nutrient delivery to cells [64]. In this context, K-RAS promotes glucose uptake and enhances glycolysis by inducing the expression of GLUT1, the glucose transporter, and other key glycolysis enzymes including Hk1, Hk2, Pfk1, and LdhA [18]. In addition, in order to promote the increase of glycolysis and support tumor cell viability, K-RAS increases the hexosamine biosynthesis pathway (HBP) and the non-oxidative arm of the pentose phosphate pathway (PPP);

the latter generates ribose-5-phosphate for de novo nucleotide biosynthesis [18,65]. Mitochondrial reactive oxygen species (ROS) levels are also essential for K-RAS mediated transformation and growth of PDAC cells [66,67]. Interestingly, in PDAC, low intracellular ROS levels were associated with tumorigenesis compared to other tumors in which increased ROS levels were associated with tumor progression [68,69]. A non-canonical glutamine metabolism pathway to maintain redox homeostasis and instrumental to PDAC growth has also been described [66]. Beyond glycolysis and glutamine metabolism, oncogenic K-RAS also promotes autophagy to recycle the metabolites and to maintain cell viability and survival. A study from K-RAS-driven tumor cell lines and from PDAC patients highlighted an increased expression of gene encoding protein related to autophagosome formation, and the expression of these genes correlated with a worse clinical outcome in cancer patients [70]. Indeed, the pharmacological or genetic inhibition of core components of the autophagy process impaired the growth of PDAC cell lines and PDAC development in a K-RAS-driven mouse model as well as in PDAC-PDX models [71]. Furthermore, oncogenic K-RAS promoted macropinocytosis to transport extracellular proteins as an amino acid source for the tricarboxylic acid (TCA) cycle to sustain tumor growth [72]. Although the above-mentioned studies underline metabolic phenotypes that have been associated with the activation of K-RAS, whether these pathways might represent biomarkers or vulnerability of K-RAS-dependent tumors requires further investigation.

On the other hand, other studies performed a metabolomics approach in selected K-RAS-dependent and K-RAS-independent PDAC cell lines after K-RAS depletion, in order to more specifically decipher the key metabolic signatures associated with K-RAS dependency or independency. A widely integrated transcriptomic and metabolomic analysis showed a significant gene expression profile of metabolic features of K-RAS-independent tumors [73]. In this study, a subpopulation of cells that relapsed after genetic and pharmacologic ablation of the K-RAS pathway was isolated. Further characterization of these K-RAS resistant cells by transcriptomic analysis revealed the expression of key regulators of mitochondrial functions, autophagy, and lysosome activity, while metabolomic assays revealed a decrease of metabolic intermediates involved in tricarboxylic acid, impaired glycolysis, and dependency on oxidative phosphorylation (OXHPOS) as energy source and for survival [73].

K-RAS-dependent PDAC cells showed a clear upregulation of the pyrimidine biosynthetic pathways, as demonstrated by several reports. Codina et al. showed that after treatment with MEK inhibitors, K-RAS-dependent PDAC cells exhibited MYC protein downregulation, which compromised nucleotide biosynthesis, essential to support growth and the survival of K-RAS-dependent cell lines [52]. Metabolomics analysis conducted in K-RAS-dependent and independent PDAC cell lines further validated the key role of pyrimidine biosynthesis and glutamine metabolism to characterize PDAC cells with a K-RAS addiction [52]. In a different study, a 3D clonogenic synthetic lethal screening exploiting a wide range of compounds identified dihydroorotate dehydrogenase (DHODH) inhibitors as major pharmacological suppressors for the growth of K-RAS- dependent cell line. DHODH enzyme regulates de novo pyrimidine biosynthesis and its inhibition reduced, beyond pyrimidine biosynthesis, cellular levels of glutamine and glutamate, showing an antitumor effect in mice models [74]. Mottini et al. [54] showed that the cytosine analog and U.S. Food and Drug Administration (FDA)-approved drug decitabine induced the impairment of nucleotide biosynthesis and deoxyriboNucleotide TriPhosphate (dNTP) pool homeostasis selectively in K-RAS-dependent PDAC but not in K-RAS-independent PDAC cells, showing a selective cytotoxic effect. This depended on an intimate connection between K-RAS-dependent tumor cells and the pyrimidine metabolism. Overall, these studies indicated that the upregulation of key metabolites or enzymes belonging to the pyrimidine biosynthesis can act as a metabolic biomarker to score a phenotypic dependency of PDAC cells on K-RAS (Table 2). In this context, further investigation into the expression or activity of key enzymes of the pyrimidine biosynthetic pathways as prognostic markers in PDAC is needed.

Table 2. Comparison of previously published metabolic profiles of K-RAS-dependent and independent pancreatic cancer cell lines and tumors.

Ref	Methodologies	Pathways/Metabolites Analyzed
Santana Codina N et al., 2018	LC-MS/MS analyisis in K-RAS sensitive and resistant cells	pentose phosphate pathway (PPP) and nucleotide biosynthesis and glycolysis
Mottini C et al., 2019	LC/MS analysis from both dependent and independent PDAC cell lines	nucleotide metabolism and pyrimidine biosynthesis
Koundinya M et al., 2018	Mass spectrometric analysis for K-RAS dependent and independent cells and tumor tissues	de novo pyrimidine biosynthetic pathway
Viale A et al., 2014	metabolomic analysis using a LC-MS/MS in a subpopulation of dormant tumor cells surviving K-RAS ablation	Tricarboxylic acid cycle (TCA) intermediates, nucleotide triphosphates, deoxynucleotide triphosphates, glutathione (GSH) and glutathione disulphide levels

5. Scores of KRAS Dependency Based on Tumor Microenvironment and Immunogenicity

In the last few years, there has been growing interest in the role of the tumor microenvironment in PDAC development due to its role in tumor progression, invasiveness, and promoting therapies resistance. The pancreatic cancer microenvironment consists of cancer cells and non-neoplastic cells including pancreatic stellate cells (PSCs), regulatory T cells (Tregs), tumor-associated macrophages (TAMs), myeloid-derived suppressor cells (MDSCs), fibroblasts, and extracellular matrix components [75,76]. The crosstalk between the microenvironment and cancer cells induces the release of cytokines, chemokines, growth factors, and metalloproteases that help to recruit active TAMs and cancer-associated fibroblasts (CAFs), which support tumor progression and metastasis [77]. Notably, these dynamic and reciprocal interactions promote desmoplasia, with a low tumor perfusion microenvironment characterized by a dense desmoplastic stroma, which hinders drug delivery and suppresses antitumor immune response [78,79]. In addition, PDACs are often immunologically cold, as defined by the lack of effector T cells infiltrating the tumor. Traditionally, this phenomenon is partially attributed to the low mutational load of PDAC. In this scenario, it is worth asking to what extent mechanisms linked with K-RAS dependency or independency might play a role in the immunogenicity and the immune response of PDAC, and thus, whether these immunogenic phenotypes might act as biomarkers of tumor dependency. Various high-throughput genomic and transcriptomic analyses have revealed a key role of K-RAS signaling in promoting mechanisms of escape from immune surveillance [80]. K-RAS-mutated cancer cells secrete tumor-derived granulocyte-macrophage colony-stimulating factor (GM-CSF), promoting the recruitment and expansion of MDSCs in the microenvironment, which are able to suppress the antitumor activity of CD8 cytotoxic T cells [81]. Oncogenic K-RAS recruits myeloid cells through the secretion of cytokines including interleukin 6 (IL-6), IL-13, CCL2, Granulocyte Colony-Stimulating Factor (G-CSF), Macrophage Colony-Stimulating Factor (M-CSF), and GM-CSF [75]. Furthermore, evidence in mouse models showed that K-RAS-dependent signaling activated CAFs through the induction of Hedgehog ligands, regulated extracellular matrix (ECM) remodeling, and promoted collagen degradation by matrix metalloproteinases (MMPs), facilitating angiogenesis, tumor cells invasion, and metastasis [17,18,80,82,83]. Integrated genomic, transcriptomic, and immunological analysis on some PDAC subtypes revealed an immunosuppressive microenvironment in the quasi-mesenchymal subtype described by Collison [49] that correlated with high numbers of Tregs and M2 macrophages, and a low number of effector T cells. On the contrary, the classical subtype consisted of abundant M1 macrophages, CD4 and CD8 T cells, and a low number of Tregs and M2 macrophages, suggesting a immune responsive microenvironment [62,63]. Since classical and quasi-mesenchymal subtypes have a strong correlation with *K-RAS* dependency and independency, respectively, it would be important to further investigate if the immune-phenotypes correlate directly with *K-RAS* dependency and how K-RAS modulation affects the immunogenic phenotypes of PDAC.

6. Therapeutic Opportunities Against K-RAS-Dependent PDAC

Strategies developed to target K-RAS and its downstream effectors are likely to elicit a stronger therapeutic response against K-RAS-dependent tumors. Far from exhaustive, this section will provide some examples of these strategies, including direct K-RAS inhibitors, inhibitors of plasma membrane association, inhibitors of downstream signaling, and of metabolic phenotypes. The first compounds identified as capable of directly inhibiting mutant K-RAS proteins were small molecules able to interfere with the K-RAS-Guanosine Diphosphate (GDP) complex and inhibit Son of Sevenless homolog (SOS)-mediated nucleotide exchange [84–87]; other compounds instead efficiently were able to bind to RAS-GTP, thus inhibiting signaling cascades downstream of K-RAS. However, these compounds have not yet been investigated in clinical settings [88,89]. The targeting of enzymes involved in the post-translational modifications of K-RAS, necessary for protein activation, has been also investigated. Farnesylation is a post-translational modification crucial for the proper plasma membrane localization of K-RAS and downstream pathways activation. In this context, a farnesyltransferase inhibitor (FTI) termed tipifarnib was developed as a potential inhibitor of K-RAS [90]. Moreover, deltarasin, a small molecule that binds the prenyl-binding protein PDEδ, that is crucial for plasma membrane localization of farnesylated K-RAS, has also been developed [91,92]. However, clinical trials did not show a significant anti-tumor effect and any survival benefit for patients [93,94].

Current efforts to block activated K-RAS are also focused on downstream K-RAS-dependent pathways. One of the commonly studied pathways is the RAF-MEK-ERK pathway, and several MEK inhibitors have been developed including trametinib and selumetinib [95,96]. Clinical trials' results related to these inhibitors failed to show clinical benefit and effect on survival in patients [97,98]. However, a few phase I/II studies are underway to test the efficacy of other MEK inhibitors including pimasertib and refametinib in combination with gemcitabine [96,99,100]. Several small molecules have been developed to target PI3K-, AKT-, and/or the Mammalian Target Of Rapamycin (mTOR)-dependent pathway, but monotherapies with PI3K-dependent pathway inhibitors alone failed to show efficacy in K-RAS-mutant cancers [101]. However, the combination of PI3K with RAF-MEK-ERK inhibitors exhibited potent tumor growth inhibitory activity [33,87,102], but clinical results do not match with those seen in preclinical models [103]. Importantly, in most of the clinical studies cited above, the assessment for K-RAS dependency has not been performed before treatments, thus therapies were not tailored for the patients' population which were highly likely to respond.

Recently, preclinical evidence revealed a specific covalent inhibitor with high selectivity for K-RAS^{G12C} able to trap the inactive K-RAS-GDP complex, thus blocking nucleotide exchange and RAS downstream signaling [104]. Currently, the agent is being evaluated in a phase I/II clinical trial (NCT03600883) for patients with advanced solid tumors harboring a *K-RAS*^{G12C} mutation. Nonetheless, G12C mutations are rarely observed in PDAC (1%), and similar approaches targeting K-RAS^{G12D} and K-RAS^{G12V} mutations, which constitute the prevalent K-RAS mutations in PDAC are needed [19].

Autophagy and macropinocytosis are both biological mechanisms that contribute to the growth and survival of K-RAS mutant pancreatic cancer cells [71,72], and clinical studies are evaluating hydroxychloroquine as autophagy inhibitors in combination with other chemotherapeutic drugs [105].

Finally, the dependency on pyrimidine metabolism in K-RAS-dependent PDAC has been exploited in preclinical models by Mottini et al. [54]. Thanks to a computational drug repositioning approach using K-RAS-driven signatures, authors repurposed 5-aza-2′-deoxycytidine (decitabine), an FDA-approved drug, to inhibit K-RAS-dependent PDAC tumor growth. K-RAS-dependent PDACs were highly sensitive to decitabine treatment, showing reduced cell viability and impaired tumor growth. On the contrary, decitabine treatment in K-RAS-independent cell lines and tumors did show minimal or no effect.

In conclusion, several therapeutics have been developed especially for treating K-RAS-driven PDAC and tested in preclinical or clinical settings. However, in most cases, K-RAS dependency has not been assessed on the treated population, and the response rate upon treatments has not been evaluated on the basis of the effective K-RAS dependency of tumors. Based on the emergency of biomarkers for K-RAS dependency, as described in this review, the results of clinical trials and drug effectiveness should be reevaluated for a complete assessment of drug efficacy in PDAC.

7. Conclusive Remarks

Mutated K-RAS is one of the most important and validated molecular antitumor targets in PDAC and the development of therapeutics against K-RAS is under active preclinical and clinical investigation. Treating tumors with an effective phenotypic dependency on K-RAS will result in an increased likelihood to observe a therapeutic response. Scoring K-RAS dependency by means of transcriptomic or metabolomics profiling has the potential to move forward to a new generation of molecular stratification of tumors for diagnostic purposes. To accomplish this, key questions need to be addressed: What are the tumor phenotypes under the active control of oncogenic K-RAS? Could such phenotypes be readily probed into the clinical routine? Another important open question is whether tumor heterogeneity, a key aspect limiting the effectiveness of targeted therapies, is relevant in the case of K-RAS dependency, and therefore if a small percentage of K-RAS-independent cells might co-exist in the bulk of dependent tumor cells, a circumstance that could potentially limit the efficacy of any targeted therapies. Single cell-based resolution methodologies will be necessary to solve such a question. Finally, it is important to understand how dependency would be also layered by stroma and other tumor microenvironment components, thus increasing the arsenal of specific K-RAS-driven molecular markers for certain PDAC subtypes.

Funding: This work was supported by the Italian Ministry of Health funds "Ricerca Corrente" to IRCCS Istituto Nazionale Tumori "Regina Elena, the "Nastro Viola" Association, and the LILT IG grant, Bando 5 × 1000.

Conflicts of Interest: The authors declare no conflict of interest.

References

1. Rahib, L.; Smith, B.D.; Aizenberg, R.; Rosenzweig, A.B.; Fleshman, J.M.; Matrisian, L. Projecting Cancer Incidence and Deaths to 2030: The Unexpected Burden of Thyroid, Liver, and Pancreas Cancers in the United States. *Cancer Res.* **2014**, *74*, 2913–2921. [CrossRef]
2. Siegel, R.L.; Miller, K.D.; Jemal, A. Cancer statistics, 2019. *CA Cancer J. Clin.* **2019**, *69*, 7–34. [CrossRef]
3. Kamarajah, S.K.; Sutandi, N.; Robinson, S.R.; French, J.J.; White, S.A. Robotic versus conventional laparoscopic distal pancreatic resection: A systematic review and meta-analysis. *HPB* **2019**, *21*, 1107–1118. [CrossRef]
4. Conroy, T.; Paillot, B.; François, E.; Bugat, R.; Jacob, J.-H.; Stein, U.; Nasca, S.; Metges, J.-P.; Rixe, O.; Michel, P.; et al. Irinotecan Plus Oxaliplatin and Leucovorin-Modulated Fluorouracil in Advanced Pancreatic Cancer—A Groupe Tumeurs Digestives of the Fédération Nationale des Centres de Lutte Contre le Cancer Study. *J. Clin. Oncol.* **2005**, *23*, 1228–1236. [CrossRef]
5. Goldstein, D.; El-Maraghi, R.H.; Hammel, P.; Heinemann, V.; Kunzmann, V.; Sastre, J.; Scheithauer, W.; Siena, S.; Tabernero, J.; Teixeira, L.; et al. nab-Paclitaxel plus gemcitabine for metastatic pancreatic cancer: Long-term survival from a phase III trial. *J. Natl. Cancer Inst.* **2015**, *107*, djv204. [CrossRef]
6. Reni, M.; Zanon, S.; Peretti, U.; Chiaravalli, M.; Barone, D.; Pircher, C.; Balzano, G.; Macchini, M.; Romi, S.; Gritti, E.; et al. Nab-paclitaxel plus gemcitabine with or without capecitabine and cisplatin in metastatic pancreatic adenocarcinoma (PACT-19): A randomised phase 2 trial. *Lancet Gastroenterol. Hepatol.* **2018**, *3*, 691–697. [CrossRef]
7. Sarantis, P.; Koustas, E.; Papadimitropoulou, A.; Papavassiliou, A.G.; Karamouzis, M.V. Pancreatic ductal adenocarcinoma: Treatment hurdles, tumor microenvironment and immunotherapy. *World J. Gastrointest. Oncol.* **2020**, *12*, 173–181. [CrossRef] [PubMed]
8. Fan, J.-Q.; Wang, M.-F.; Chen, H.-L.; Shang, D.; Das, J.K.; Song, J. Current advances and outlooks in immunotherapy for pancreatic ductal adenocarcinoma. *Mol. Cancer* **2020**, *19*, 1–22. [CrossRef] [PubMed]
9. Li, K.-Y.; Yuan, J.-L.; Trafton, D.; Wang, J.-X.; Niu, N.; Yuan, C.-H.; Liu, X.-B.; Zheng, L. Pancreatic ductal adenocarcinoma immune microenvironment and immunotherapy prospects. *Chronic Dis. Transl. Med.* **2020**, *6*, 6–17. [CrossRef] [PubMed]
10. Prior, I.; Hood, F.E.; Hartley, J.L. The Frequency of Ras Mutations in Cancer. *Cancer Res.* **2020**. [CrossRef]

11. Jemal, A.; Siegel, R.; Ward, E.; Hao, Y.; Xu, J.; Murray, T.; Thun, M.J. Cancer statistics, 2008. *CA Cancer J. Clin.* **2008**, *58*, 71–96. [CrossRef]

12. Heinemann, V.; Stintzing, S.; Kirchner, T.; Boeck, S.; Jung, A. Clinical relevance of EGFR- and KRAS-status in colorectal cancer patients treated with monoclonal antibodies directed against the EGFR. *Cancer Treat. Rev.* **2009**, *35*, 262–271. [CrossRef]

13. Cox, A.D.; Fesik, S.W.; Kimmelman, A.C.; Luo, J.; Der, C.J. Drugging the undruggable RAS: Mission Possible? *Nat. Rev. Drug Discov.* **2014**, *13*, 828–851. [CrossRef]

14. Hingorani, S.; Petricoin, E.F.; Maitra, A.; Rajapakse, V.; King, C.; Jacobetz, M.A.; Ross, S.; Conrads, T.P.; Veenstra, T.D.; Hitt, B.A.; et al. Preinvasive and invasive ductal pancreatic cancer and its early detection in the mouse. *Cancer Cell* **2003**, *4*, 437–450. [CrossRef]

15. Guerra, C.; Schuhmacher, A.J.; Cañamero, M.; Grippo, P.J.; Verdaguer, L.; Pérez-Gallego, L.; Dubus, P.; Sandgren, E.P.; Barbacid, M. Chronic Pancreatitis Is Essential for Induction of Pancreatic Ductal Adenocarcinoma by K-Ras Oncogenes in Adult Mice. *Cancer Cell* **2007**, *11*, 291–302. [CrossRef]

16. Morris, J.P.; Wang, S.C.; Hebrok, M. KRAS, Hedgehog, Wnt and the twisted developmental biology of pancreatic ductal adenocarcinoma. *Nat. Rev. Cancer* **2010**, *10*, 683–695. [CrossRef]

17. Collins, M.A.; Bednar, F.; Zhang, Y.; Brisset, J.-C.; Galbán, S.; Galbán, C.J.; Rakshit, S.; Flannagan, K.S.; Adsay, N.V.; Di Magliano, M.P. Oncogenic Kras is required for both the initiation and maintenance of pancreatic cancer in mice. *J. Clin. Investig.* **2012**, *122*, 639–653. [CrossRef]

18. Ying, H.; Kimmelman, A.C.; Lyssiotis, C.A.; Hua, S.; Chu, G.C.; Fletcher-Sananikone, E.; Locasale, J.W.; Son, J.; Zhang, H.; Coloff, J.L.; et al. Oncogenic Kras Maintains Pancreatic Tumors through Regulation of Anabolic Glucose Metabolism. *Cell* **2012**, *149*, 656–670. [CrossRef]

19. Bryant, K.L.; Mancias, J.D.; Kimmelman, A.C.; Der, C.J. KRAS: Feeding pancreatic cancer proliferation. *Trends Biochem. Sci.* **2014**, *39*, 91–100. [CrossRef] [PubMed]

20. Maitra, A.; Adsay, N.V.; Argani, P.; Iacobuzio-Donahue, C.; De Marzo, A.; Cameron, J.L.; Yeo, C.J.; Hruban, R.H. Multicomponent Analysis of the Pancreatic Adenocarcinoma Progression Model Using a Pancreatic Intraepithelial Neoplasia Tissue Microarray. *Mod. Pathol.* **2003**, *16*, 902–912. [CrossRef]

21. Jones, S.; Zhang, X.; Parsons, D.W.; Lin, J.C.-H.; Leary, R.J.; Angenendt, P.; Mankoo, P.; Carter, H.; Kamiyama, H.; Jimeno, A.; et al. Core Signaling Pathways in Human Pancreatic Cancers Revealed by Global Genomic Analyses. *Science* **2008**, *321*, 1801–1806. [CrossRef] [PubMed]

22. Biankin, A.V.; Initiative, A.P.C.G.; Waddell, N.; Kassahn, K.; Gingras, M.-C.; Muthuswamy, L.B.; Johns, A.L.; Miller, D.K.; Wilson, P.J.; Patch, A.-M.; et al. Pancreatic cancer genomes reveal aberrations in axon guidance pathway genes. *Nature* **2012**, *491*, 399–405. [CrossRef]

23. Siegel, R.L.; Miller, K.D.; Jemal, A. Cancer Statistics, 2017. *CA Cancer J. Clin.* **2017**, *67*, 7–30. [CrossRef]

24. Hezel, A.F.; Kimmelman, A.C.; Stanger, B.Z.; Bardeesy, N.; DePinho, R.A. Genetics and biology of pancreatic ductal adenocarcinoma. *Genes Dev.* **2006**, *20*, 1218–1249. [CrossRef]

25. Vincent, A.; Herman, J.; Schulick, R.; Hruban, R.H.; Goggins, M. Pancreatic cancer. *Lancet* **2011**, *378*, 607–620. [CrossRef]

26. Ryan, D.P.; Hong, T.S.; Bardeesy, N. Pancreatic adenocarcinoma. *N. Engl. J. Med.* **2014**, *371*, 2140–2141. [CrossRef]

27. Castellano, E.; Santos, E. Functional Specificity of Ras Isoforms. *Genes Cancer* **2011**, *2*, 216–231. [CrossRef] [PubMed]

28. Colicelli, J. Human RAS Superfamily Proteins and Related GTPases. *Sci. Signal.* **2004**, *2004*, re13. [CrossRef]

29. Rajalingam, K.; Schreck, R.; Rapp, U.R.; Štefan, A. Ras oncogenes and their downstream targets. *Biochim. Biophys. Acta BBA Bioenerg.* **2007**, *1773*, 1177–1195. [CrossRef]

30. Karnoub, A.E.; Weinberg, R.A. Ras oncogenes: Split personalities. *Nat. Rev. Mol. Cell Biol.* **2008**, *9*, 517–531. [CrossRef]

31. Vigil, D.; Cherfils, J.; Rossman, K.L.; Der, C.J. Ras superfamily GEFs and GAPs: Validated and tractable targets for cancer therapy? *Nat. Rev. Cancer* **2010**, *10*, 842–857. [CrossRef] [PubMed]

32. Witkiewicz, A.K.; McMillan, E.A.; Balaji, U.; Baek, G.; Lin, W.-C.; Mansour, J.; Mollaee, M.; Wagner, K.-U.; Koduru, P.; Yopp, A.; et al. Whole-exome sequencing of pancreatic cancer defines genetic diversity and therapeutic targets. *Nat. Commun.* **2015**, *6*, 6744. [CrossRef] [PubMed]

33. Collisson, E.A.; Trejo, C.L.; Silva, J.M.; Gu, S.; Korkola, J.E.; Heiser, L.M.; Charles, R.-P.; Rabinovich, B.A.; Hann, B.; Dankort, D.; et al. A Central Role for RAF→MEK→ERK Signaling in the Genesis of Pancreatic Ductal Adenocarcinoma. *Cancer Discov.* **2012**, *2*, 685–693. [CrossRef] [PubMed]

34. Eser, S.; Reiff, N.; Messer, M.; Seidler, B.; Gottschalk, K.; Dobler, M.; Hieber, M.; Arbeiter, A.; Klein, S.; Kong, B.; et al. Selective Requirement of PI3K/PDK1 Signaling for Kras Oncogene-Driven Pancreatic Cell Plasticity and Cancer. *Cancer Cell* **2013**, *23*, 406–420. [CrossRef] [PubMed]

35. Torres, C.; Mancinelli, G.; Cordoba-Chacon, J.; Viswakarma, N.; Castellanos, K.; Grimaldo, S.; Kumar, S.; Principe, D.; Dorman, M.J.; McKinney, R.; et al. p110gamma deficiency protects against pancreatic carcinogenesis yet predisposes to diet-induced hepatotoxicity. *Proc. Natl. Acad. Sci. USA* **2019**, *116*, 14724–14733. [CrossRef] [PubMed]

36. Su, G.H.; Qiu, W.; Ciau, N.T.; Ho, D.J.; Li, X.; Allendorf, J.D.; Remotti, H.; Su, G.H. PIK3CA mutations in intraductal papillary mucinous neoplasm/carcinoma of the pancreas. *Clin. Cancer Res.* **2006**, *12*, 3851–3855. [CrossRef]

37. Jaiswal, B.S.; Janakiraman, V.; Kljavin, N.M.; Chaudhuri, S.; Stern, H.M.; Wang, W.; Kan, Z.; Dbouk, H.A.; Peters, B.A.; Waring, P.; et al. Somatic Mutations in p85α Promote Tumorigenesis through Class IA PI3K Activation. *Cancer Cell* **2009**, *16*, 463–474. [CrossRef]

38. Ying, H.; Elpek, K.G.; Vinjamoori, A.; Zimmerman, S.M.; Chu, G.C.; Yan, H.; Fletcher-Sananikone, E.; Zhang, H.; Liu, Y.; Wang, W.; et al. PTEN is a major tumor suppressor in pancreatic ductal adenocarcinoma and regulates an NF-kappaB-cytokine network. *Cancer Discov.* **2011**, *1*, 158–169. [CrossRef]

39. Ying, H.; Dey, P.; Yao, W.; Kimmelman, A.C.; Draetta, G.F.; Maitra, A.; Depinho, R.A. Genetics and biology of pancreatic ductal adenocarcinoma. *Genes Dev.* **2016**, *30*, 355–385. [CrossRef]

40. Rowley, M.; Ohashi, A.; Mondal, G.; Mills, L.; Yang, L.; Zhang, L.; Sundsbak, R.; Shapiro, V.; Muders, M.H.; Smyrk, T.; et al. Inactivation of Brca2 Promotes Trp53-Associated but Inhibits KrasG12D-Dependent Pancreatic Cancer Development in Mice. *Gastroenterology* **2011**, *140*, 1303–1313. [CrossRef] [PubMed]

41. Skoulidis, F.; Cassidy, L.D.; Pisupati, V.; Jonasson, J.G.; Bjarnason, H.; Eyfjörd, J.E.; Karreth, F.A.; Lim, M.; Barber, L.M.; Clatworthy, S.A.; et al. Germline Brca2 Heterozygosity Promotes KrasG12D-Driven Carcinogenesis in a Murine Model of Familial Pancreatic Cancer. *Cancer Cell* **2010**, *18*, 499–509. [CrossRef] [PubMed]

42. Singhi, A.; George, B.; Greenbowe, J.R.; Chung, J.; Suh, J.; Maitra, A.; Klempner, S.J.; Hendifar, A.; Milind, J.M.; Golan, T.; et al. Real-Time Targeted Genome Profile Analysis of Pancreatic Ductal Adenocarcinomas Identifies Genetic Alterations That Might Be Targeted With Existing Drugs or Used as Biomarkers. *Gastroenterology* **2019**, *156*, e4. [CrossRef] [PubMed]

43. Downward, J. Cancer biology: Signatures guide drug choice. *Nature* **2006**, *439*, 274–275. [CrossRef]

44. Shrestha, G.; MacNeil, S.M.; McQuerry, J.A.; Jenkins, D.; Sharma, S.; Bild, A. The value of genomics in dissecting the RAS-network and in guiding therapeutics for RAS-driven cancers. *Semin. Cell Dev. Biol.* **2016**, *58*, 108–117. [CrossRef]

45. Bild, A.H.; Yao, G.; Chang, J.T.; Wang, Q.; Potti, A.; Chasse, D.; Joshi, M.-B.; Harpole, D.; Lancaster, J.M.; Berchuck, A.; et al. Oncogenic pathway signatures in human cancers as a guide to targeted therapies. *Nature* **2005**, *439*, 353–357. [CrossRef]

46. Sharma, S.V.; Settleman, J. Oncogene addiction: Setting the stage for molecularly targeted cancer therapy. *Genes Dev.* **2007**, *21*, 3214–3231. [CrossRef]

47. Singh, A.; Greninger, P.; Rhodes, D.; Koopman, L.; Violette, S.; Bardeesy, N.; Settleman, J. A Gene Expression Signature Associated with "K-Ras Addiction" Reveals Regulators of EMT and Tumor Cell Survival. *Cancer Cell* **2009**, *15*, 489–500. [CrossRef]

48. Loboda, A.; Nebozhyn, M.; Klinghoffer, R.; Frazier, J.; Chastain, M.; Arthur, W.; Roberts, B.; Zhang, T.; Chenard, M.; Haines, B.B.; et al. A gene expression signature of RAS pathway dependence predicts response to PI3K and RAS pathway inhibitors and expands the population of RAS pathway activated tumors. *BMC Med. Genom.* **2010**, *3*, 26. [CrossRef]

49. Collisson, E.A.; Sadanandam, A.; Olson, P.; Gibb, W.J.; Truitt, M.; Gu, S.; Cooc, J.; Weinkle, J.; Kim, G.E.; Jakkula, L.; et al. Subtypes of pancreatic ductal adenocarcinoma and their differing responses to therapy. *Nat. Med.* **2011**, *17*, 500–503. [CrossRef]

50. Kapoor, A.; Yao, W.; Ying, H.; Hua, S.; Liewen, A.; Wang, Q.; Zhong, Y.; Wu, C.-J.; Sadanandam, A.; Hu, B.; et al. Yap1 Activation Enables Bypass of Oncogenic Kras Addiction in Pancreatic Cancer. *Cell* **2014**, *158*, 185–197. [CrossRef]

51. Tsang, Y.H.; Dogruluk, T.; Tedeschi, P.M.; Wardwell-Ozgo, J.; Lu, H.; Espitia, M.; Nair, N.; Minelli, R.; Chong, Z.; Chen, F.; et al. Functional annotation of rare gene aberration drivers of pancreatic cancer. *Nat. Commun.* **2016**, *7*, 10500. [CrossRef]

52. Santana-Codina, N.; Roeth, A.A.; Zhang, Y.; Yang, A.; Mashadova, O.; Asara, J.M.; Wang, X.; Bronson, R.T.; Lyssiotis, C.A.; Ying, H.; et al. Oncogenic KRAS supports pancreatic cancer through regulation of nucleotide synthesis. *Nat. Commun.* **2018**, *9*, 4945. [CrossRef] [PubMed]

53. Muzumdar, M.D.; Chen, P.-Y.; Dorans, K.J.; Chung, K.M.; Bhutkar, A.; Hong, E.; Noll, E.M.; Sprick, M.R.; Trumpp, A.; Jacks, T. Survival of pancreatic cancer cells lacking KRAS function. *Nat. Commun.* **2017**, *8*, 1090. [CrossRef] [PubMed]

54. Mottini, C.; Tomihara, H.; Carrella, D.; Lamolinara, A.; Iezzi, M.; Huang, J.K.; Amoreo, C.A.; Buglioni, S.; Manni, I.; Robinson, F.S.; et al. Predictive Signatures Inform the Effective Repurposing of Decitabine to Treat KRAS–Dependent Pancreatic Ductal Adenocarcinoma. *Cancer Res.* **2019**, *79*, 5612–5625. [CrossRef]

55. Furge, K.A.; Tan, M.H.; Dykema, K.; Kort, E.; Stadler, W.; Yao, X.; Zhou, M.; Teh, B.T. Identification of deregulated oncogenic pathways in renal cell carcinoma: An integrated oncogenomic approach based on gene expression profiling. *Oncogene* **2007**, *26*, 1346–1350. [CrossRef] [PubMed]

56. Nevins, J.R.; Potti, A. Mining gene expression profiles: Expression signatures as cancer phenotypes. *Nat. Rev. Genet.* **2007**, *8*, 601–609. [CrossRef] [PubMed]

57. Chang, J.T.; Carvalho, C.; Mori, S.; Bild, A.H.; Gatza, M.L.; Wang, Q.; Lucas, J.E.; Potti, A.; Febbo, P.G.; West, M.; et al. A Genomic Strategy to Elucidate Modules of Oncogenic Pathway Signaling Networks. *Mol. Cell* **2009**, *34*, 104–114. [CrossRef]

58. Connor, A.A.; Denroche, R.E.; Jang, G.H.; Lemire, M.; Zhang, A.; Chan-Seng-Yue, M.; Wilson, G.; Grant, R.C.; Merico, D.; Lungu, I.; et al. Integration of Genomic and Transcriptional Features in Pancreatic Cancer Reveals Increased Cell Cycle Progression in Metastases. *Cancer Cell* **2019**, *35*, e7. [CrossRef]

59. Qian, J.; Niu, J.; Li, M.; Chiao, P.J.; Tsao, M.-S. In vitroModeling of Human Pancreatic Duct Epithelial Cell Transformation Defines Gene Expression Changes Induced by K-rasOncogenic Activation in Pancreatic Carcinogenesis. *Cancer Res.* **2005**, *65*, 5045–5053. [CrossRef]

60. Siolas, D.; Hannon, G.J. Patient-derived tumor xenografts: Transforming clinical samples into mouse models. *Cancer Res.* **2013**, *73*, 5315–5319. [CrossRef]

61. Knudsen, E.S.; Balaji, U.; Mannakee, B.; Vail, P.; Eslinger, C.; Moxom, C.; Mansour, J.; Witkiewicz, A.K. Pancreatic cancer cell lines as patient-derived avatars: Genetic characterisation and functional utility. *Gut* **2017**, *67*, 508–520. [CrossRef] [PubMed]

62. Bailey, P.J.; Initiative, A.P.C.G.; Chang, D.K.; Nones, K.; Johns, A.L.; Patch, A.-M.; Gingras, M.-C.; Miller, D.K.; Christophi, C.; Bruxner, T.J.; et al. Genomic analyses identify molecular subtypes of pancreatic cancer. *Nature* **2016**, *531*, 47–52. [CrossRef] [PubMed]

63. Moffitt, R.A.; Marayati, R.; Flate, E.L.; Volmar, K.E.; Loeza, S.G.H.; Hoadley, K.A.; Rashid, N.U.; Williams, L.A.; Eaton, S.C.; Chung, A.H.; et al. Virtual microdissection identifies distinct tumor- and stroma-specific subtypes of pancreatic ductal adenocarcinoma. *Nat. Genet.* **2015**, *47*, 1168–1178. [CrossRef] [PubMed]

64. Sousa, C.M.; Kimmelman, A.C. The complex landscape of pancreatic cancer metabolism. *Carcinogenesis* **2014**, *35*, 1441–1450. [CrossRef]

65. Guillaumond, F.; Leca, J.; Olivares, O.; Lavaut, M.-N.; Vidal, N.; Berthezène, P.; Dusetti, N.J.; Loncle, C.; Calvo, E.; Turrini, O.; et al. Strengthened glycolysis under hypoxia supports tumor symbiosis and hexosamine biosynthesis in pancreatic adenocarcinoma. *Proc. Natl. Acad. Sci. USA* **2013**, *110*, 3919–3924. [CrossRef]

66. Son, J.; Lyssiotis, C.A.; Ying, H.; Wang, X.; Hua, S.; Ligorio, M.; Perera, R.M.; Ferrone, C.R.; Mullarky, E.; Shyh-Chang, N.; et al. Glutamine supports pancreatic cancer growth through a KRAS-regulated metabolic pathway. *Nature* **2013**, *496*, 101–105. [CrossRef]

67. Kong, B.; Qia, C.; Erkan, M.; Kleeff, J.; Michalski, C.W. Overview on how oncogenic Kras promotes pancreatic carcinogenesis by inducing low intracellular ROS levels. *Front. Physiol.* **2013**, *4*, 246. [CrossRef]

68. Weinberg, F.; Hamanaka, R.; Wheaton, W.W.; Weinberg, S.; Joseph, J.; Lopez, M.; Kalyanaraman, B.; Mutlu, G.M.; Budinger, G.R.S.; Chandel, N.S. Mitochondrial metabolism and ROS generation are essential for Kras-mediated tumorigenicity. *Proc. Natl. Acad. Sci. USA* **2010**, *107*, 8788–8793. [CrossRef]

69. Matés, J.M.; Segura, J.A.; Alonso, F.J.; Márquez, J. Oxidative stress in apoptosis and cancer: An update. *Arch. Toxicol.* **2012**, *86*, 1649–1665. [CrossRef] [PubMed]

70. Fujii, S.; Mitsunaga, S.; Yamazaki, M.; Hasebe, T.; Ishii, G.; Kojima, M.; Kinoshita, T.; Ueno, T.; Esumi, H.; Ochiai, A. Autophagy is activated in pancreatic cancer cells and correlates with poor patient outcome. *Cancer Sci.* **2008**, *99*, 1813–1819. [CrossRef] [PubMed]

71. Yang, A.; Herter-Sprie, G.; Zhang, H.; Lin, E.Y.; Biancur, D.; Wang, X.; Deng, J.; Hai, J.; Yang, S.; Wong, K.K.; et al. Autophagy Sustains Pancreatic Cancer Growth through Both Cell-Autonomous and Nonautonomous Mechanisms. *Cancer Discov.* **2018**, *8*, 276–287. [CrossRef]

72. Commisso, C.; Davidson, S.M.; Soydaner-Azeloglu, R.G.; Parker, S.J.; Kamphorst, J.J.; Hackett, S.; Grabocka, E.; Nofal, M.; Drebin, J.A.; Thompson, C.B.; et al. Macropinocytosis of protein is an amino acid supply route in Ras-transformed cells. *Nature* **2013**, *497*, 633–637. [CrossRef] [PubMed]

73. Viale, A.; Pettazzoni, P.; Lyssiotis, C.A.; Ying, H.; Sanchez, N.; Marchesini, M.; Carugo, A.; Green, T.; Seth, S.; Giuliani, V.; et al. Oncogene ablation-resistant pancreatic cancer cells depend on mitochondrial function. *Nature* **2014**, *514*, 628–632. [CrossRef]

74. Koundinya, M.; Sudhalter, J.; Courjaud, A.; Lionne, B.; Touyer, G.; Bonnet, L.; Menguy, I.; Schreiber, I.; Perrault, C.; Vougier, S.; et al. Dependence on the Pyrimidine Biosynthetic Enzyme DHODH Is a Synthetic Lethal Vulnerability in Mutant KRAS-Driven Cancers. *Cell Chem. Biol.* **2018**, *25*, e11. [CrossRef] [PubMed]

75. Dougan, S.K. The Pancreatic Cancer Microenvironment. *Cancer J.* **2017**, *23*, 321–325. [CrossRef] [PubMed]

76. Murakami, T.; Hiroshima, Y.; Matsuyama, R.; Homma, Y.; Hoffman, R.M.; Endo, I. Role of the tumor microenvironment in pancreatic cancer. *Ann. Gastroenterol. Surg.* **2019**, *3*, 130–137. [CrossRef]

77. Ungefroren, H.; Sebens, S.; Seidl, D.; Lehnert, H.; Hass, R. Interaction of tumor cells with the microenvironment. *Cell Commun. Signal.* **2011**, *9*, 18. [CrossRef]

78. Feig, C.; Gopinathan, A.; Neesse, A.; Chan, D.S.; Cook, N.; Tuveson, D.A. The pancreas cancer microenvironment. *Clin. Cancer Res.* **2012**, *18*, 4266–4276. [CrossRef]

79. Cannon, A.; Thompson, C.; Hall, B.R.; Jain, M.; Kumar, S.; Batra, S.K. Desmoplasia in pancreatic ductal adenocarcinoma: Insight into pathological function and therapeutic potential. *Genes Cancer* **2018**, *9*, 78–86. [CrossRef]

80. Carvalho, P.; Guimarães, C.F.; Cardoso, A.P.; Mendonça, S.; Costa, A.; Oliveira, M.J.; Velho, S. KRAS Oncogenic Signaling Extends beyond Cancer Cells to Orchestrate the Microenvironment. *Cancer Res.* **2017**, *78*, 7–14. [CrossRef]

81. Bayne, L.J.; Beatty, G.L.; Jhala, N.; Clark, C.E.; Rhim, A.D.; Stanger, B.Z.; Vonderheide, R.H. Tumor-Derived Granulocyte-Macrophage Colony-Stimulating Factor Regulates Myeloid Inflammation and T Cell Immunity in Pancreatic Cancer. *Cancer Cell* **2012**, *21*, 822–835. [CrossRef] [PubMed]

82. Ji, Z.; Mei, F.C.; Xie, J.; Cheng, X. Oncogenic KRAS Activates Hedgehog Signaling Pathway in Pancreatic Cancer Cells. *J. Biol. Chem.* **2007**, *282*, 14048–14055. [CrossRef] [PubMed]

83. Mills, L.D.; Zhang, Y.; Marler, R.J.; Herreros-Villanueva, M.; Zhang, L.; Almada, L.L.; Couch, F.; Wetmore, C.; Di Magliano, M.P.; Fernandez-Zapico, M.E. Loss of the Transcription Factor GLI1 Identifies a Signaling Network in the Tumor Microenvironment Mediating KRAS Oncogene-induced Transformation. *J. Biol. Chem.* **2013**, *288*, 11786–11794. [CrossRef] [PubMed]

84. Maurer, T.; Garrenton, L.S.; Oh, A.; Pitts, K.; Anderson, D.J.; Skelton, N.J.; Fauber, B.P.; Pan, B.; Malek, S.; Stokoe, D.; et al. Small-molecule ligands bind to a distinct pocket in Ras and inhibit SOS-mediated nucleotide exchange activity. *Proc. Natl. Acad. Sci. USA* **2012**, *109*, 5299–5304. [CrossRef] [PubMed]

85. Sun, Q.; Burke, J.P.; Phan, J.; Burns, M.C.; Olejniczak, E.T.; Waterson, A.G.; Lee, T.; Rossanese, O.W.; Fesik, S.W. Discovery of small molecules that bind to K-Ras and inhibit Sos-mediated activation. *Angew. Chem. Int. Ed. Engl.* **2012**, *51*, 6140–6143. [CrossRef]

86. Winter, J.J.G.; Anderson, M.; Blades, K.; Brassington, C.; Breeze, A.; Chresta, C.; Embrey, K.; Fairley, G.; Faulder, P.; Finlay, M.R.V.; et al. Small Molecule Binding Sites on the Ras: SOS Complex Can Be Exploited for Inhibition of Ras Activation. *J. Med. Chem.* **2015**, *58*, 2265–2274. [CrossRef]

87. Zeitouni, D.; Pylayeva-Gupta, Y.; Der, C.J.; Bryant, K.L. KRAS Mutant Pancreatic Cancer: No Lone Path to an Effective Treatment. *Cancers* **2016**, *8*, 45. [CrossRef]

88. Hillig, R.C.; Sautier, B.; Schroeder, J.; Moosmayer, D.; Hilpmann, A.; Stegmann, C.M.; Werbeck, N.D.; Briem, H.; Boemer, U.; Weiske, J.; et al. Discovery of potent SOS1 inhibitors that block RAS activation via disruption of the RAS–SOS1 interaction. *Proc. Natl. Acad. Sci. USA* **2019**, *116*, 2551–2560. [CrossRef]

89. Mattox, T.E.; Chen, X.; Maxuitenko, Y.Y.; Keeton, A.B.; Piazza, G.A. Exploiting RAS Nucleotide Cycling as a Strategy for Drugging RAS-Driven Cancers. *Int. J. Mol. Sci.* **2019**, *21*, 141. [CrossRef]

90. Palsuledesai, C.C.; Distefano, M.D. Protein Prenylation: Enzymes, Therapeutics, and Biotechnology Applications. *ACS Chem. Biol.* **2014**, *10*, 51–62. [CrossRef]

91. Chandra, A.; Grecco, H.E.; Pisupati, V.; Perera, D.; Cassidy, L.; Skoulidis, F.; Ismail, S.A.; Hedberg, C.; Hanzal-Bayer, M.; Venkitaraman, A.R.; et al. The GDI-like solubilizing factor PDEdelta sustains the spatial organization and signalling of Ras family proteins. *Nat. Cell Biol.* **2011**, *14*, 148–158. [CrossRef] [PubMed]

92. Zimmermann, G.; Papke, B.; Ismail, S.; Vartak, N.; Chandra, A.; Hoffmann, M.; Hahn, S.A.; Triola, G.; Wittinghofer, A.; Bastiaens, P.I.; et al. Small molecule inhibition of the KRAS-PDEdelta interaction impairs oncogenic KRAS signalling. *Nature* **2013**, *497*, 638–642. [CrossRef] [PubMed]

93. Van Cutsem, E.; Van De Velde, H.; Karasek, P.; Oettle, H.; Vervenne, W.; Szawlowski, A.; Schöffski, P.; Post, S.; Verslype, C.; Neumann, H.; et al. Phase III Trial of Gemcitabine Plus Tipifarnib Compared With Gemcitabine Plus Placebo in Advanced Pancreatic Cancer. *J. Clin. Oncol.* **2004**, *22*, 1430–1438. [CrossRef] [PubMed]

94. Macdonald, J.S.; McCoy, S.; Whitehead, R.P.; Iqbal, S.; Wade, J.L.; Giguere, J.K.; Abbruzzese, J.L.; Iii, J.L.W. A phase II study of farnesyl transferase inhibitor R115777 in pancreatic cancer: A Southwest oncology group (SWOG 9924) study. *Investig. N. Drugs* **2005**, *23*, 485–487. [CrossRef] [PubMed]

95. Rutkowski, P.; Lugowska, I.; Kosela-Paterczyk, H.; Kozak, K. Trametinib: A MEK inhibitor for management of metastatic melanoma. *OncoTargets Ther.* **2015**, *8*, 2251–2259. [CrossRef] [PubMed]

96. Asati, V.; Mahapatra, D.K.; Bharti, S.K. K-Ras and its inhibitors towards personalized cancer treatment: Pharmacological and structural perspectives. *Eur. J. Med. Chem.* **2017**, *125*, 299–314. [CrossRef]

97. Bodoky, G.; Timcheva, C.; Spigel, D.R.; La Stella, P.J.; Ciuleanu, T.E.; Pover, G.; Tebbutt, N.C. A phase II open-label randomized study to assess the efficacy and safety of selumetinib (AZD6244 [ARRY-142886]) versus capecitabine in patients with advanced or metastatic pancreatic cancer who have failed first-line gemcitabine therapy. *Investig. N. Drugs* **2011**, *30*, 1216–1223. [CrossRef]

98. Infante, J.R.; Somer, B.G.; Park, J.O.; Li, C.-P.; Scheulen, M.E.; Kasubhai, S.M.; Oh, -Y.; Liu, Y.; Redhu, S.; Steplewski, K.; et al. A randomised, double-blind, placebo-controlled trial of trametinib, an oral MEK inhibitor, in combination with gemcitabine for patients with untreated metastatic adenocarcinoma of the pancreas. *Eur. J. Cancer* **2014**, *50*, 2072–2081. [CrossRef]

99. Van Laethem, J.-L.; Riess, H.; Jassem, J.; Haas, M.; Martens, U.M.; Weekes, C.; Peeters, M.; Ross, P.; Bridgewater, J.; Melichar, B.; et al. Phase I/II Study of Refametinib (BAY 86-9766) in Combination with Gemcitabine in Advanced Pancreatic cancer. *Target. Oncol.* **2016**, *12*, 97–109. [CrossRef]

100. Van Cutsem, E.; Hidalgo, M.; Canon, J.-L.; Macarulla, T.; Bazin, I.; Poddubskaya, E.V.; Manojlović, N.; Radenković, D.; Verslype, C.; Raymond, E.; et al. Phase I/II trial of pimasertib plus gemcitabine in patients with metastatic pancreatic cancer. *Int. J. Cancer* **2018**, *143*, 2053–2064. [CrossRef]

101. Junttila, M.R.; Devasthali, V.; Cheng, J.H.; Castillo, J.; Metcalfe, C.; Clermont, A.C.; Otter, D.D.; Chan, E.; Bou-Reslan, H.; Cao, T.; et al. Modeling Targeted Inhibition of MEK and PI3 Kinase in Human Pancreatic Cancer. *Mol. Cancer Ther.* **2014**, *14*, 40–47. [CrossRef]

102. Ning, C.; Liang, M.; Liu, S.; Wang, G.; Edwards, H.; Xia, Y.; Polin, L.; Dyson, G.; Taub, J.W.; Mohammad, R.M.; et al. Targeting ERK enhances the cytotoxic effect of the novel PI3K and mTOR dual inhibitor VS-5584 in preclinical models of pancreatic cancer. *Oncotarget* **2017**, *8*, 44295–44311. [CrossRef]

103. Bournet, B.; Muscari, F.; Buscail, C.; Assenat, E.; Barthet, M.; Hammel, P.; Selves, J.; Guimbaud, R.; Cordelier, P.; Buscail, L. KRAS G12D Mutation Subtype Is A Prognostic Factor for Advanced Pancreatic Adenocarcinoma. *Clin. Transl. Gastroenterol.* **2016**, *7*, e157. [CrossRef] [PubMed]

104. Janes, M.R.; Zhang, J.; Li, L.-S.; Hansen, R.; Peters, U.; Guo, X.; Chen, Y.; Babbar, A.; Firdaus, S.J.; Darjania, L.; et al. Targeting KRAS Mutant Cancers with a Covalent G12C-Specific Inhibitor. *Cell* **2018**, *172*, e17. [CrossRef] [PubMed]

105. Amaravadi, R.K.; Lippincott-Schwartz, J.; Yin, X.-M.; Weiss, W.A.; Takebe, N.; Timmer, W.; DiPaola, R.S.; Lotze, M.T.; White, E. Principles and current strategies for targeting autophagy for cancer treatment. *Clin. Cancer Res.* **2011**, *17*, 654–666. [CrossRef] [PubMed]

International Journal of
Molecular Sciences

Article

Immunohistochemical Analysis Revealed a Correlation between Musashi-2 and Cyclin-D1 Expression in Patients with Oral Squamous Cells Carcinoma

Giuseppe Troiano [1], Vito Carlo Alberto Caponio [1], Gerardo Botti [2], Gabriella Aquino [2], Nunzia Simona Losito [2], Maria Carmela Pedicillo [1], Khrystyna Zhurakivska [1,*], Claudia Arena [1], Domenico Ciavarella [1], Filiberto Mastrangelo [1], Lucio Lo Russo [1], Lorenzo Lo Muzio [1] and Giuseppe Pannone [1]

[1] Department of Clinical and Experimental Medicine, University of Foggia, Via Rovelli 50, 71122 Foggia, Italy; giuseppe.troiano@unifg.it (G.T.); vito_caponio.541096@unifg.it (V.C.A.C.); mariacarmela.pedicillo@unifg.it (M.C.P.); claudia.arena@unifg.it (C.A.); domenico.ciavarella@unifg.it (D.C.); filiberto.mastrangelo@unifg.it (F.M.); lucio.lorusso@unifg.it (L.L.R.); lorenzo.lomuzio@unifg.it (L.L.M.); giuseppe.pannone@unifg.it (G.P.)

[2] Pathology Unit, Istituto Nazionale per lo Studio e la Cura dei Tumori, "Fondazione G. Pascale", IRCCS, 80131 Naples, Italy; g.botti@istitutotumori.na.it (G.B.); g.aquino@istitutotumori.na.it (G.A.); n.losito@istitutotumori.na.it (N.S.L.)

* Correspondence: Khrystyna.zhurakivska@unifg.it

Received: 18 November 2019; Accepted: 20 December 2019; Published: 23 December 2019

Abstract: Aim: Musashi 2 (MSI2), which is an RNA-binding protein, plays a fundamental role in the oncogenesis of several cancers. The aim of this study is to investigate the expression of MSI2 in Oral Squamous Cell Carcinoma (OSCC) and evaluate its correlation to clinic-pathological variables and prognosis. Materials and Methods: A bioinformatic analysis was performed on data downloaded from The Cancer Genome Atlas (TCGA) database. The MSI2 expression data were analysed for their correlation with clinic-pathological and prognostic features. In addition, an immmunohistochemical evaluation of MSI2 expression on 108 OSCC samples included in a tissue microarray and 13 healthy mucosae samples was performed. Results: 241 patients' data from TCGA were included in the final analysis. No DNA mutations were detected for the MSI2 gene, but a hyper methylated condition of the gene emerged. MSI2 mRNA expression correlated with Grading ($p = 0.009$) and overall survival ($p = 0.045$), but not with disease free survival ($p = 0.549$). Males presented a higher MSI2 mRNA expression than females. The immunohistochemical evaluation revealed a weak expression of MSI2 in both OSCC samples and in healthy oral mucosae. In addition, MSI2 expression directly correlated with Cyclin-D1 expression ($p = 0.022$). However, no correlation has been detected with prognostic outcomes (overall and disease free survival). Conclusions: The role of MSI2 expression in OSCC seems to be not so closely correlated with prognosis, as in other human neoplasms. The correlation with Cyclin-D1 expression suggests an indirect role that MSI2 might have in the proliferation of OSCC cells, but further studies are needed to confirm such results.

Keywords: MSI2; OSCC; oral cancer; musashi 2; prognosis

---

1. Introduction

Oral cancer (OC) belongs to the wider family of Head and Neck Cancers (HNCs). OC is a highly relevant problem for global public health, with a clinical impact in terms of incidence, prevalence, and mortality rates that do not tend to improve. It is reported to be the 11th most common malignancy worldwide [1]. Around 90% of OCs are histologically classified as squamous cell carcinoma (OSCC), involving the mucosal surface of the oral cavity and tongue [2]. Oral carcinogenesis encompasses multistep processes that drive the progression from normal mucosa to OSCC [3]. Changes in the DNA sequence, accumulation of somatic mutations and epigenetic events are the main mechanisms that are involved in tumor progression. In particular, epigenetic and post-transcriptional events, gained an important role in cancer [4]. Key-regulators of these mechanisms are the RNA-Binding Proteins (RBPs) that cause variations in protein expression, due to their involvement in splicing, mRNA-polyadenylation, editing, and r-tRNA stabilization [5]. Musashi-2 (MSI2) is one of the most studied RBPs. In particular, different studies evaluated its role in cancer. For example, MSI2 overexpression was linked to an increase of invasion and metastasis in non-small cell lung carcinoma, whereas its depletion showed a decrease of epithelia-mesenchymal transition [6]. In bladder cancer, the differentiation antagonizing non-protein noding RNA (DANCR) long non-coding RNA (lncRNA) acts by sponging miR-149 increasing the expression of MSI2, getting worse a malignant phenotype [7]. In addition, MSI2 seems to be involved in patients' prognosis, resulting as prognostic factor in gastric [8] and cervical cancer [9]; in lung cancer, MSI2 emerged as a novel therapeutic target [10]. Several other factors that are responsible of the regulation of cell proliferation and cell cycle control have been proposed as diagnostic, prognostic, and therapeutic markers for certain malignancies. Among these, the cyclin D1 has been deeply investigated and were shown to be essential for the tumorigenesis of melanoma, breast cancer, and colon and oral squamous cell carcinoma (OSCC) [11]. Cyclin D1 belongs to the family of Cyclins and it is essential in the regulation of cell proliferation, DNA repair, and cell migration control [12]. The aim of this study was to investigate the expression of MSI2 in OSCC samples, through a histologic and bioinformatics analysis in order to evaluate its correlation to clinic-pathological variables and prognosis. Furthermore, a staining for Cyclin-D1 has been performed on OSCC tissue microarray (TMA) and the correlation of Cyclin D1 expression with MSI2 was investigated.

2. Results

2.1. Analysis of MSI2 Mutations, Gene Methylation and mRNA Expression in TCGA Database

A total 241 patients' records were included in this analysis after extracting and matching clinic-pathological data from the TCGA database. Table 1 summarizes the main clinical-pathological characteristics of the included patients. DNA mutations and copy number alterations were not detected for the MSI2 gene in patients with OSCC included in the TCGA database (0/241, 0%). The expression of MSI2 mRNA (log2 (fpkm+1)) was relatively low ranging from 0.2785 to 2.7117 with a mean of 1.270734 (S.E. 0.030) and a median of 1,40673. According to the median value, patients were divided in low (≤1.40673) and high (>1.40673) MSI2 mRNA expression. Methylation status, measured in beta unit, showed a hyper methylated condition of the gene in all of the patients analyzed, the values of gene methylation ranged from 0.6802 to 0.9910 with a mean of 0.974483 (S.E. 0.0018993). The Spearman rank correlation test did not show a significant correlation between mRNA expression and methylation status of the gene ($\rho = -0.44$; $p = 0.498$); however, a higher methylation status was detected for the low-expression group with results that were close to the statistical significance (Mann–Whitney $p = 0.095$). MSI2 mRNA expression correlated with Grading ($\rho = 0.169$; $p = 0.009$) and showed a differential expression according to the gender (Mann–Whitney $p = 0.001$) with males' samples showing a higher expression, while MSI2 methylation profile correlated to the age of patients ($\rho = 0.140$; $p = 0.03$) (Table 2). Univariate and multivariate analyses were performed, aiming to investigate whether MSI2 mRNA expression in the TCGA database was able to predict prognosis. The results of the univariate analysis were promising, showing a significant association between MSI2 mRNA expression (High vs low) and overall survival (Hazard Ratio, HR = 1.488; 95% C.I. 1.013–2.185; $p = 0.045$); furthermore, the results of the multivariate analysis (HR = 1.437; 95% C.I. 0.952–1.970; $p = 0.084$) were close to the threshold of statistical significance. Conversely MSI2 mRNA expression did not correlate with disease free survival (HR = 0.827; 95% C.I. 0.443–1.542; $p = 0.549$) in OSCC patients.

Table 1. Clinical-pathological characteristics of patients included in The Cancer Genome Atlas (TCGA) analysis.

Clinic-Pathological Information	Groups	Number of Patients
Age	≤ 65 years old	144/241
	> 65 years old	97/241
Gender	Male	158/241
	Female	83/241
Grade	1	43/241
	2	147/241
	3	51/241
Stage	1–2	76/241
	3–4	165/241
Subsite	Tongue	103/241
	Gingivo-buccal	30/241
	Floor of the mouth	46/241
	Others	62/241

Table 2. Spearman rank correlation for the 241 Oral Squamous Cell Carcinoma (OSCC) patients included in the TCGA database. $* p < 0.05$; $** p < 0.001$.

Variable	Age	Grade	Stage	Gender	Perineural Invasion	MSI2 Methylation	MSI2 mRNA Expression	Ki-67 mRNA Expression	Cyclin-D mRNA Expression
Age	$\rho = 1$ p-value = 1	0.088 0.175	−0.084 0.197	**0.243** **0.001 ****	0.060 0.414	**0.140** **0.03 ***	−0.121 0.06	−0.069 0.287	0.033 0.607
Grade		$\rho = 1$ p-value = 1	−0.003 0.959	−0.066 0.307	0.100 0.176	−0.094 0.146	0.169 0.009 **	−0.105 0.105	0.053 0.418
Stage			$\rho = 1$ p-value = 1	−0.026 0.686	0.199 0.006 **	−0.031 0.632	−0.035 0.587	0.038 0.561	−0.062 0.340
Gender				$\rho = 1$ p-value = 1	−0.090 0.907	−0.045 0.485	−0.220 0.001 **	0.042 0.520	0.043 0.505
Perineural Invasion					$\rho = 1$ p-value = 1	−0.178 0.014 *	0.033 0.650	−0.132 0.071	−0.038 0.606
MSI2 Methylation						$\rho = 1$ p-value = 1	−0.044 0.498	0.094 0.144	0.027 0.673
MSI2 mRNA expression							$\rho = 1$ p-value = 1	−0.093 0.150	0.003 0.963
Ki-67 mRNA expression								$\rho = 1$ p-value = 1	0.253 0.000 **
Cyclin-D mRNA expression									$\rho = 1$ p-value = 1

2.2. Immunohistochemical Analysis of MSI2 Expression on TMA

The IHC analysis of MSI2 protein expression was performed on a total of 108 patient' samples included in the TMA; such patients had been treated at the National Cancer Institute "Giovanni Pascale" between 1997 and 2012. Table 3 reports the clinical pathological information of patients included in the cohort. An analysis of protein expression in the TMA samples revealed that MSI2 is not frequently expressed in OSCC, in fact 58.3% (63/108) cases analyzed resulted in being negative for MSI2 expression. Of the remaining 41.7% (45/108) samples, only 5.6% (6/108) showed higher level of MSI2 expression. The presence of MSI2 expression directly correlated with Cyclin-D1 expression ($\rho = 0.279$; Chi-Squared p-value = 0.022) (Table 4), this last one resulted to be higher expressed in males than in females (Mann–Whitney p-value = 0.024). The presence of MSI2 expression in the TMA cohort did not correlate with overall survival (HR = 0.575; 95% C.I. 0.278–1.190; $p = 0.136$) (Table 5). In the 13 oral healthy mucosae analyzed the expression of MSI2 was faint and mainly confined to the basal layer with a percentage of expression lower than 5% of the whole number of epithelial cells (Figure 1).

Table 3. Clinical-pathological characteristics of patients included in the immunoistochemical analysis.

Clinic-Pathological Information	Groups	Number of Patients
Age	≤ 65 years old	45/108
	> 65 years old	63/108
Gender	Male	79/108
	Female	29/108
Grade	1	22/108
	2	50/108
	3	36/108
Stage	1–2	35/108
	3–4	73/108
Subsite	Tongue	67/108
	Gingivo-buccal	23/108
	Floor of the mouth	13/108
	Others	5/108

Table 4. Spearman rank correlation for patients included in the TMA. Age. Cyclin-D1 and Ki-67 were included as continuous variables. while Grade. Stage. Gender. and MSI2 as categorical variables. $^* p < 0.05;$ $^{**} p < 0.001$.

Variable	Age	Grade	Stage	Gender	MSI2 Expression (Neg/Pos)	Cyclin-D1 Expression	Ki-67 Expression
Age	ρ = 1 *p*-value = 1	0.049 0.617	−0.151 0.122	−0.062 0.528	−0.006 0.955	0.021 0.865	0.148 0.226
Grade		ρ = 1 *p*-value = 1	0.098 0.313	0.034 0.726	0.223 −0.066	0.060 −0.044	0.398 0.001 *
Stage			ρ = 1 *p*-value = 1	−0.008 0.993	0.497 0.676	0.721 0.468	−0.004 0.977
Gender				ρ = 1 *p*-value = 1	−0.088 0.364	−0.277 0.023 *	−0.285 0.031 *
MSI2 expression (Neg/Pos)					ρ = 1 *p*-value = 1	0.279 0.022 *	0.122 0.315
Cyclin-D1 expression						ρ = 1 *p*-value = 1	0.485 0.000 **
Ki-67 expression							ρ = 1 *p*-value = 1

Table 5. Multivariate Cox regression analysis for MSI2 adjusted for other clinic-pathological parameters. $^* p < 0.05$.

	Overall Survival		
Variables	Hazard Ratio	95.0% C.I.	*p*-Value
Gender	1.904	0.964–3.759	0.064
Grade	1.162	0.706–1.911	0.555
Age	1.018	0.986–1.052	0.274
Stage	2.968	1.844–4.779	0.000 *
MSI2 (Pos/Neg)	0.621	0.299–1.288	0.201

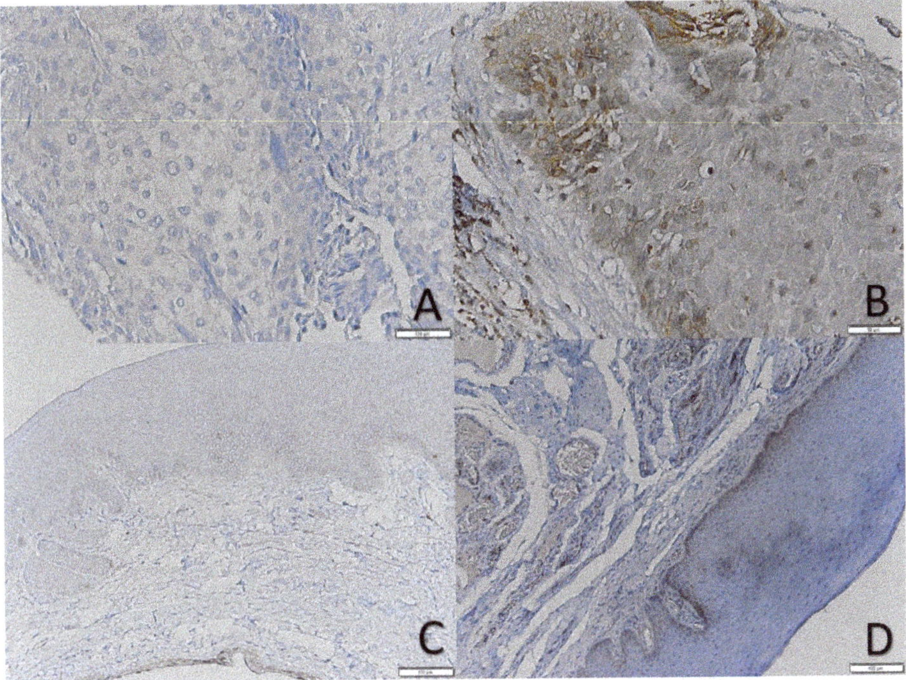

Figure 1. (**A**) Negative OSCC for MSI2; (**B**) Basal MSI2 positivity intensity +2; (**C**) 40% Normal mucosa negative for MSI2 with mild positivity in a ductal epithelium; and, (**D**) normal mucosa mild positive for MSI2 in the basal layer.

3. Discussion

OSCC represents more than 90% of oral cancers and it is one of the most aggressive cancers, being characterized by a mortality rate reaching 50% of patients on average [13]. Continuous efforts are made to better understand the processes that lead to its onset and progression, as well as to the discovering of potential targets for its therapy. Similarly to other solid tumors, the onset of OSCC results from the accumulation of a certain number of genetic or epigenetic alterations into the cells, which cause cell cycle dysregulation and uncontrolled cell proliferation [14] Recently, the new "omic" sciences provided a great amount of data that characterize the tumors at the molecular level and that can lead to discovering specific biomarkers that could make the tumor treatment more efficient, precise, and predictable [15]. A great contribution has been obtained from bioinformatics that allowed for analyzing this enormous amount of data, giving them a clinical significance [16].

Musashi RNA-binding protein 2 (MSI-2) has been demonstrated to be involved in several solid and blood cancers, where its expression emerged to be higher than in normal tissues and correlate with the prognosis [17]. Its role seems to be explicated in different processes, among which: epithelial-mesenchymal transition, migration, invasion, cell proliferation, and drug resistance [17]. While for many tumors, such as those arising from breast [18], cervical [9], colon [19], lung [6], etc. the role of MSI-2 proteins has been extensively studied and some target therapies proposed, no results regarding the role of MSI-2 in OSCC are reported in the literature. In this work, we combined both a bioinformatics analysis of data that were extracted from electronic TCGA database and an immunohistochemical evaluation of our samples to better understand the role of MSI-2 in oral cancer. The first interesting result emerged from the analysis of genomic data revealed that no DNA mutations and copy number alterations are detected for the MSI2 gene in patients with OSCC. A hyper methylated condition of the gene emerged in all the patients from the investigation of epigenetic modifications. The analysis of transcriptomic data showed a relatively low expression of MSI2 mRNA in the OSCC samples. However, the mRNA expression did not result to be significantly correlated to the methylation status of the gene.

The associations between its mRNA expression and some clinicopathological characteristics of patients have been analyzed to explore the clinical value of MSI2. The analysis revealed that MSI2 mRNA expression correlates with tumor grading and males show a higher level of MSI2 mRNA expression when compared to female patients. No other clinical features seemed to significantly correlate with MSI2 mRNA expression.

As to our knowledge, no difference between sexes has been highlighted in the literature until now regarding the expression of these regulating molecules. Nevertheless, in our study Males showed significantly higher expression levels of both MSI2 and Cyclin-D1. Such results emerged from the analysis of TCGA database and they have been confirmed by our samples. This could be the decisive contribution that big data can give us in the path towards increasingly personalized medicine [16].

Another question of this work aimed to investigate whether the expression of MSI2 in OSCC patients could predict the prognosis. The results emerged from correlation analysis lay for a significant association between high MSI2 mRNA expression and poor overall survival rate; meanwhile, the disease free survival seems not to be correlated with the MSI2 mRNA expression level.

We performed the immunohistochemical evaluation of healthy mucosae and OSCC samples to determine the expression levels of MSI2, in parallel with this bioinformatic analysis. It was of great interest to discover that the protein expression of MSI2 was low or even absent both in healthy samples and OSCC TMA and it did not correlate with any prognostic behavior. These data, combined with those deriving from TCGA analysis, lead us to affirm that, unlike many other cancers [8,20,21], for OSCC the expression of MSI2 appears to be a poor prognostic biomarker.

In addition to these results, some information regarding the potential biological functions of MSI2 in oral cancer emerged from the evaluation of the TMAs. In particular, the presence of MSI2 expression directly correlated with Cyclin-D1 expression. Cyclin-D1 is a protein that plays a crucial role in cell cycle regulation, including cell proliferation and growth, as well as DNA repair and cell migration control [22]. Its key role in tumorigenesis of several tumors, among which oral cancer, has been proposed [12], and a poor prognosis correlated to its overexpression [23]. Several mechanisms of cyclin D1 overexpression in OSCC have been identified. They range from amplification to polymorphisms and mutational events involving the oncogene CCND1, but a fundamental role of some signaling pathway intermediaries has been also suggested [24]. Zhang et al. [25] demonstrated in their study how MSI2 silencing inhibited leukemic cell growth and caused a decreasing of Cyclin D1 expression. Han et al. [26] drew the same conclusions, showing how the MSI2 silencing induced cell cycle arrest in G0/G1 phase, with decreased Cyclin D1 and increased p21 expression. In the same way, a study investigating the role of MSI2 in Hematopoietic stem cell activity discovered a close correlation between the expression profile of MSI2 and that of Cyclin D1 [27]. Given the results of our study, in a similar manner, MSI2 could affect the Cyclin D1 expression in the cells of OSCC, but further studies are needed to affirm this.

4. Methods and Materials

4.1. Analysis of MSI2 Expression and Methylation in The Cancer Genome Atlas (TCGA)

The gene expression RNAseq data HTSq-Fragments Per Kilobase Million (FPKM) were downloaded from UCSC Xena Browser (https://xena.ucsc.edu/) [28]. Data were downloaded for MSI2, MKI67, and CCND1 mRNA expression. This platform was also used to access and download the methylation profile quantification for MSI2 (https://gdc.xenahubs.net/download/TCGA-HNSC/Xena_Matrices/TCGA-HNSC.methylation450.tsv.gz; Full metadata—Illumina Human Methylation 450 expressed as beta unit). Data were organized in Microsoft Excel sheet and then pasted in cBioPortal for Cancer genomics in order to display visually the patients' profile (http://www.cbioportal.org) [29]. Clinic-pathological and follow-up information were downloaded from Genomic Data Commons (GDC) Data Portal (https://portal.gdc.cancer.gov/) [30].

4.2. Immunohistochemistry of MSI2 Expression in OSCC Tissue Microarray

The reporting recommendations for tumor marker prognostic studies (REMARK guidelines) [31] were taken as a reference for carrying out this study. All of the patients filled written informed consent for the use of their samples, according to the institutional regulations and the ethics committee of the National Cancer Institute "Giovanni Pascale", as "Bio-Banca Istituzionale BBI" Deliberation NO. 15 del 20 Jan. 2016, approved and registered the study. We decided to exclude patients with HPV-positive tumors and those arising from the base of the tongue, tonsils, oropharynx, and lips. Patients with a follow-up lower than eight months were also excluded. A total of 122 patients were included in this study. Of them, 103 reported follow-up information (from 8 to 150 months—mean of 47.34 S.D. 34.609), meanwhile the microarray tissue was not evaluable for 14 patients. None of the patients had undergone treatments prior to tissue collection. The patients were diagnosed of OSCC and 7th American Joint Committee on Cancer (AJCC) staging system was applied. An evaluation of 13 mucosae samples from healthy subjects has been also performed. The paraffin blocks were cored in a 0.6 mm support (area of 0.28 mm^2) and then transferred to the recipient master block while using Galileo TMA CK 3500® Tissue Microarrayer. As the control, we used an H&E staining of a 4-μm TMA section. Immuno-histochemical staining was performed by using a mouse monoclonal antibody (Ab), which was supplied by abcam (Mouse monoclonal Anti-MSI2 antibody [OTI2F10]—ab156770), in addition staining for Cyclin-D1 (Ventana-Roche, SP4-R) and Ki-67 (Ventana-Roche) was also performed. We used an automated staining device (Ventana-Roche), with a streptavidin-biotin horseradish peroxidase technique (LSABHRP), in order to uncover the primary Abs. An optical microscope (OLYMPUS BX53, at ×200) detected immune-stained spots in four high power fields (HPFs) and they were analyzed by ISE TMA Software (Integrated System Engineering, Milan, Italy). Two of the authors (GP and GT) performed the observational quantification analysis in a joint session. Detre S. et al. method was applied to assess the scoring of immunostaining [32]. The intensity (I) of expression was scored from 0 to 3 (0 = no staining; 1 = yellow; 2 = light brown; and, 3 = black brown/black). The relative number of the positive stained cells (%) was scored from 0 to 4 (0 = 0%; 1 < 10%; 2 = 10-50%; 3 = 51-80%; 4 > 80%). All of the samples resulted in only being stained in the cytoplasm. We decided to categorize patients in negative/positive tumors because of the relative low expression of MSI2 in OSCC.

4.3. Statistical Analysis

SPSS statistical software 21.0 was used to perform all of the statistical analyses. Spearman rank correlation analysis was performed to investigate the correlation of MSI2 expression to clinic-pathological variables. We decided to use non-parametric tests because of the non-normal distribution of the variable checked by means of the Shapiro–Wilk normality test. For such reason, the Mann–Whitney test was applied to explore the difference in expression between groups. Univariate survival analysis was performed with the Kaplan–Meier method to estimate both overall survival rates and disease-free survival, while comparing results between groups with the log-Rank test. A multivariate Cox Proportional Hazard Model was built in order to assess the prognostic significance of MSI2 expression after adjusting for covariates, including the clinic-pathological variables: age, grading, staging (8th AJCC edition), and gender, as covariates.

5. Conclusions

The role of MSI2 expression in OSCC seems to be not so closely correlated with prognosis, as in other solid and blood tumors. The MSI2 mRNA expression in oral cancer is higher in males and it is correlated with tumor grade. The protein expression of MSI2 in OSCC samples is relatively low, but significantly correlated with Cyclin D1 expression. Further studies are necessary to investigate its role in OSCC genesis, progression, and prognosis.

Author Contributions: G.T.: V.C.A.C., L.L.R. performed bioinformatic analysis, G.B., G.A., N.S.L., M.C.P., and G.P. collected, prepared and analyzed semples, K.Z., C.A., D.C., and F.M. contributed to write the manuscript, G.T., L.L.M., and G.P. conceived and supervised the project. All authors have read and agreed to the published version of the manuscript.

Funding: The research has been funded by authors' own institutions.

Conflicts of Interest: The authors declare no conflict of interest.

References

1. Ghantous, Y.; Abu Elnaaj, I. Global Incidence and Risk Factors of Oral Cancer. *Harefuah* **2017**, *156*, 645–649.
2. Attar, E.; Dey, S.; Hablas, A.; Seifeldin, I.A.; Ramadan, M.; Rozek, L.S.; Soliman, A.S. Head and neck cancer in a developing country: A population-based perspective across 8 years. *Oral Oncol.* **2010**, *46*, 591–596. [CrossRef]
3. Lingen, M.W.; Pinto, A.; Mendes, R.A.; Franchini, R.; Czerninski, R.; Tilakaratne, W.M.; Partridge, M.; Peterson, D.E.; Woo, S.B. Genetics/epigenetics of oral premalignancy: Current status and future research. *Oral Dis.* **2011**, *17* (Suppl. 1), 7–22. [CrossRef]
4. Fritzell, K.; Xu, L.D.; Lagergren, J.; Ohman, M. ADARs and editing: The role of A-to-I RNA modification in cancer progression. *Semin Cell Dev. Biol.* **2018**, *79*, 123–130. [CrossRef]
5. Soni, S.; Anand, P.; Padwad, Y.S. MAPKAPK2: The master regulator of RNA-binding proteins modulates transcript stability and tumor progression. *J. Exp. Clin. Cancer Res.* **2019**, *38*, 121. [CrossRef]
6. Kudinov, A.E.; Deneka, A.; Nikonova, A.S.; Beck, T.N.; Ahn, Y.H.; Liu, X.; Martinez, C.F.; Schultz, F.A.; Reynolds, S.; Yang, D.H.; et al. Musashi-2 (MSI2) supports TGF-beta signaling and inhibits claudins to promote non-small cell lung cancer (NSCLC) metastasis. *Proc. Natl. Acad. Sci. USA* **2016**, *113*, 6955–6960. [CrossRef]
7. Zhan, Y.; Chen, Z.; Li, Y.; He, A.; He, S.; Gong, Y.; Li, X.; Zhou, L. Long non-coding RNA DANCR promotes malignant phenotypes of bladder cancer cells by modulating the miR-149/MSI2 axis as a ceRNA. *J. Exp. Clin. Cancer Res.* **2018**, *37*, 273. [CrossRef]
8. Yang, Z.; Li, J.; Shi, Y.; Li, L.; Guo, X. Increased musashi 2 expression indicates a poor prognosis and promotes malignant phenotypes in gastric cancer. *Oncol. Lett.* **2019**, *17*, 2599–2606. [CrossRef]
9. Liu, Y.; Fan, Y.; Wang, X.; Huang, Z.; Shi, K.; Zhou, B. Musashi-2 is a prognostic marker for the survival of patients with cervical cancer. *Oncol. Lett.* **2018**, *15*, 5425–5432. [CrossRef]
10. Zhang, M.R.; Xi, S.; Shukla, V.; Hong, J.A.; Chen, H.; Xiong, Y.; Ripley, R.T.; Hoang, C.D.; Schrump, D.S. The Pluripotency Factor Musashi-2 Is a Novel Target for Lung Cancer Therapy. *Ann. Am. Thorac. Soc.* **2018**, *15*, S124. [CrossRef]
11. Santarius, T.; Shipley, J.; Brewer, D.; Stratton, M.R.; Cooper, C.S. A census of amplified and overexpressed human cancer genes. *Nat. Rev. Cancer* **2010**, *10*, 59–64. [CrossRef]
12. Ramos-Garcia, P.; Gil-Montoya, J.A.; Scully, C.; Ayen, A.; Gonzalez-Ruiz, L.; Navarro-Trivino, F.J.; Gonzalez-Moles, M.A. An update on the implications of cyclin D1 in oral carcinogenesis. *Oral Dis.* **2017**, *23*, 897–912. [CrossRef]
13. Ferlay, J.; Soerjomataram, I.; Dikshit, R.; Eser, S.; Mathers, C.; Rebelo, M.; Parkin, D.M.; Forman, D.; Bray, F. Cancer incidence and mortality worldwide: Sources, methods and major patterns in GLOBOCAN 2012. *Int. J. Cancer* **2015**, *136*, E359–E386. [CrossRef]
14. Feller, L.L.; Khammissa, R.R.; Kramer, B.B.; Lemmer, J.J. Oral squamous cell carcinoma in relation to field precancerisation: Pathobiology. *Cancer Cell Int.* **2013**, *13*, 31. [CrossRef]
15. Tonella, L.; Giannoccaro, M.; Alfieri, S.; Canevari, S.; De Cecco, L. Gene Expression Signatures for Head and Neck Cancer Patient Stratification: Are Results Ready for Clinical Application? *Curr. Treat. options in Oncol.* **2017**, *18*, 32. [CrossRef]
16. Chin, L.; Andersen, J.N.; Futreal, P.A. Cancer genomics: From discovery science to personalized medicine. *Nat. Med.* **2011**, *17*, 297–303. [CrossRef]

17.	Kudinov, A.E.; Karanicolas, J.; Golemis, E.A.; Boumber, Y. Musashi RNA-Binding Proteins as Cancer Drivers and Novel Therapeutic Targets. *Clin. Cancer Res.* **2017**, *23*, 2143–2153. [CrossRef]

18.	Kang, M.H.; Jeong, K.J.; Kim, W.Y.; Lee, H.J.; Gong, G.; Suh, N.; Gyorffy, B.; Kim, S.; Jeong, S.Y.; Mills, G.B.; et al. Musashi RNA-binding protein 2 regulates estrogen receptor 1 function in breast cancer. *Oncogene* **2017**, *36*, 1745–1752. [CrossRef]

19.	Ouyang, S.W.; Liu, T.T.; Liu, X.S.; Zhu, F.X.; Zhu, F.M.; Liu, X.N.; Peng, Z.H. USP10 regulates Musashi-2 stability via deubiquitination and promotes tumour proliferation in colon cancer. *FEBS Lett.* **2019**, *593*, 406–413. [CrossRef]

20.	Zong, Z.; Zhou, T.; Rao, L.; Jiang, Z.; Li, Y.; Hou, Z.; Yang, B.; Han, F.; Chen, S. Musashi2 as a novel predictive biomarker for liver metastasis and poor prognosis in colorectal cancer. *Cancer Med.* **2016**, *5*, 623–630. [CrossRef]

21.	Dong, P.; Xiong, Y.; Hanley, S.J.B.; Yue, J.; Watari, H. Musashi-2, a novel oncoprotein promoting cervical cancer cell growth and invasion, is negatively regulated by p53-induced miR-143 and miR-107 activation. *J. Exp. Clin. Cancer Res.* **2017**, *36*, 150. [CrossRef]

22.	Li, Z.; Wang, C.; Jiao, X.; Lu, Y.; Fu, M.; Quong, A.A.; Dye, C.; Yang, J.; Dai, M.; Ju, X.; et al. Cyclin D1 regulates cellular migration through the inhibition of thrombospondin 1 and ROCK signaling. *Mol. Cell. Biol.* **2006**, *26*, 4240–4256. [CrossRef]

23.	Zhao, Y.; Yu, D.; Li, H.; Nie, P.; Zhu, Y.; Liu, S.; Zhu, M.; Fang, B. Cyclin D1 overexpression is associated with poor clinicopathological outcome and survival in oral squamous cell carcinoma in Asian populations: Insights from a meta-analysis. *PLoS ONE* **2014**, *9*, e93210. [CrossRef]

24.	Knudsen, K.E.; Diehl, J.A.; Haiman, C.A.; Knudsen, E.S. Cyclin D1: Polymorphism, aberrant splicing and cancer risk. *Oncogene* **2006**, *25*, 1620–1628. [CrossRef]

25.	Zhang, H.; Tan, S.; Wang, J.; Chen, S.; Quan, J.; Xian, J.; Zhang, S.; He, J.; Zhang, L. Musashi2 modulates K562 leukemic cell proliferation and apoptosis involving the MAPK pathway. *Exp. Cell Res.* **2014**, *320*, 119–127. [CrossRef]

26.	Han, Y.; Ye, A.; Zhang, Y.; Cai, Z.; Wang, W.; Sun, L.; Jiang, S.; Wu, J.; Yu, K.; Zhang, S. Musashi-2 Silencing Exerts Potent Activity against Acute Myeloid Leukemia and Enhances Chemosensitivity to Daunorubicin. *PLoS ONE* **2015**, *10*, e0136484. [CrossRef]

27.	Hope, K.J.; Cellot, S.; Ting, S.B.; MacRae, T.; Mayotte, N.; Iscove, N.N.; Sauvageau, G. An RNAi screen identifies Msi2 and Prox1 as having opposite roles in the regulation of hematopoietic stem cell activity. *Cell Stem Cell* **2010**, *7*, 101–113. [CrossRef]

28.	Casper, J.; Zweig, A.S.; Villarreal, C.; Tyner, C.; Speir, M.L.; Rosenbloom, K.R.; Raney, B.J.; Lee, C.M.; Lee, B.T.; Karolchik, D.; et al. The UCSC Genome Browser database: 2018 update. *Nucleic Acids Res.* **2018**, *46*, D762–D769.

29.	Gao, J.; Aksoy, B.A.; Dogrusoz, U.; Dresdner, G.; Gross, B.; Sumer, S.O.; Sun, Y.; Jacobsen, A.; Sinha, R.; Larsson, E.; et al. Integrative analysis of complex cancer genomics and clinical profiles using the cBioPortal. *Sci. Signal.* **2013**, *6*, pl1. [CrossRef]

30.	Grossman, R.L.; Heath, A.P.; Ferretti, V.; Varmus, H.E.; Lowy, D.R.; Kibbe, W.A.; Staudt, L.M. Toward a Shared Vision for Cancer Genomic Data. *N. Engl. J. Med.* **2016**, *375*, 1109–1112. [CrossRef]

31.	McShane, L.M.; Altman, D.G.; Sauerbrei, W.; Taube, S.E.; Gion, M.; Clark, G.M.; Statistics Subcommittee of the NCI-EORTC Working Group on Cancer Diagnostics. REporting recommendations for tumor MARKer prognostic studies (REMARK). *Nat. Clin. Pract. Oncol.* **2005**, *2*, 416–422. [CrossRef]

32.	Detre, S.; Saclani Jotti, G.; Dowsett, M. A "quickscore" method for immunohistochemical semiquantitation: Validation for oestrogen receptor in breast carcinomas. *J. Clin. Pathol.* **1995**, *48*, 876–878. [CrossRef]

MDPI

St. Alban-Anlage 66

4052 Basel

Switzerland

Tel. +41 61 683 77 34

Fax +41 61 302 89 18

www.mdpi.com

International Journal of Molecular Sciences Editorial Office

E-mail: ijms@mdpi.com

www.mdpi.com/journal/ijms